Sustainability Sciences in Asia and Africa

Agroecosystem Dynamics and Sustainable Practices

Sustainability as a scientific endeavor requires broad understanding of the interconnections in our global environment. It is therefore positioned at the crossroads of research in natural sciences, technology and human sciences. This book series supports the global efforts by providing timely coverage of the progress, opportunities, and challenges towards achieving sustainable development goals (SDGs) in Asia and Africa. It fosters the research in transdisciplinary academic fields spanning across agriculture and global food security, health and diseases, access to clean water and affordable energy, climate resilience and effects of global climate change on biodiversity, water and land. This series is planned to narrate the stories of scientific development and progression towards SDGs from the regions of Asia and Africa. The overall aim of the series is to provide a comprehensive publication platform and act as a knowledge engine in the growth of sustainability sciences with a special focus on developing nations. The series constitute of four sub-series each focusing on a specific theme closely linked to a set of SDGs. Each sub-series is leaders and experts in agricultural sciences, environmental sciences, biomedicine and medical sciences respectively. Books in this series are useful for environmentalists, biotechnologists, plant scientists, medical practitioners, conservation biologists, policy makers, NGOs and scientific bodies working in the areas of sustainability sciences in the developing regions.

This book series publishes research associated with the prevailing opportunities and challenges faced by agroecosystems in Asia and Africa. It also explores the future scope of agroecological techniques and offers guidance on sustainable management of these systems. The series discusses important areas such as sustainable agricultural production, soil health, water resource management, biodiversity conservation, animal health and productivity, insect management strategies, genome editing and climate change mitigation. Further, it includes the challenges faced by African agriculture in arid and semi-arid regions, particularly in response to climate change. It highlights issues such as water shortages, soil degradation, desertification, and the need for sustainable farming techniques.

Sustainable agroecosystem maintenance entails strategies that improve overall environmental health and economic viability while limiting negative environmental impacts. The series takes a multidisciplinary approach to investigate creative ways for boosting sustainable agriculture, increasing food security, and improving lives for people in global south. This book series acts as a global publishing platform to collate and disseminate knowledge on fostering resilient and productive agroecosystems, thus enhancing accessibility, and fostering informed decision-making.

Adornis D. Nciizah
Hupenyu Allan Mupambwa
Patrick Nyambo
Editors

Biofertilizers for Sustainable Agriculture in Africa

Springer

Editors
Adornis D. Nciizah
Agricultural Research Council - Natural Resources & Engineering
Pretoria, South Africa

Patrick Nyambo
Vegetables, Industrial and Medicinal Plants
Agricultural Research Council
Pretoria, South Africa

Hupenyu Allan Mupambwa
Sam Nujoma Marine and Coastal Resources Research Centre
Sam Nujoma Campus University of Namibia
Henties Bay, Namibia

ISSN 2730-6771 ISSN 2730-678X (electronic)
Sustainability Sciences in Asia and Africa
ISSN 3059-247X ISSN 3059-2488 (electronic)
Agroecosystem Dynamics and Sustainable Practices
ISBN 978-981-95-6008-0 ISBN 978-981-95-6009-7 (eBook)
https://doi.org/10.1007/978-981-95-6009-7

This Springer imprint is published by the registered company Springer Nature Singapore Pte Ltd.
The registered company address is: 152 Beach Road, #21-01/04 Gateway East, Singapore 189721, Singapore

Preface

In recent years, the challenges of soil degradation, climate change, and food insecurity have underscored the urgency of transitioning toward more sustainable agricultural practices across the African continent. Among the promising solutions, biofertilizers have emerged as a critical component for improving soil health, enhancing crop productivity, and reducing dependency on chemical inputs. However, although their potential is widely acknowledged, research, innovation, and adoption remain uneven across Africa. This edited volume, *Biofertilizers for Sustainable Agriculture in Africa*, was conceived to bridge this gap by bringing together current research, case studies, and forward-looking insights into the role of biofertilizers in transforming African agriculture.

This book is the result of a collaborative effort by researchers, practitioners, and policy thinkers across Africa and beyond, whose contributions reflect the breadth of expertise and innovation in this field. The chapters were reviewed for their scientific rigor and practical relevance. They cover a wide range of topics, from foundational overviews to soil fertility management strategies, advanced technologies, and socioeconomic considerations.

The volume is organized into five thematic sections:

- *Section 1* introduces the fundamentals and offers a broad overview of biofertilizer types and their potential.
- *Section 2* explores biofertilizer applications for soil fertility management.
- *Section 3* discusses their roles in crop production and building resilience to biotic and abiotic stressors.
- *Section 4* presents cutting-edge innovations, including nanotechnology and omics.
- *Section 5* addresses broader socioeconomic, policy, and environmental implications.

We hope this book will serve as a resource for students, researchers, extension workers, and policymakers seeking to harness the power of biofertilizers to foster a more resilient and ecologically sound agricultural future in Africa.

We are deeply grateful to all the contributing authors for their high-quality work and commitment and to the peer reviewers for their valuable feedback.

We believe that this volume will inspire further research, dialogue, and innovation in the pursuit of sustainable agricultural systems on the continent.

Pretoria, South Africa Adornis D. Nciizah
Henties Bay, Namibia Hupenyu Allan Mupambwa
Pretoria, South Africa Patrick Nyambo
July 2025

Acknowledgments

We extend our heartfelt gratitude to all the contributors whose expertise and dedication made this volume possible. Each chapter reflects a significant body of work that brings valuable insight into the evolving role of biofertilizers in sustainable agricultural development across the African continent.

We are especially grateful to the anonymous peer reviewers who provided constructive feedback that helped improve the quality and coherence of the chapters. Their input was vital in ensuring the scientific rigor and practical relevance of this book.

We would like to recognize the role of Springer Nature's editorial and production teams, particularly our publishing editor, for their professional guidance, patience, and commitment to excellence throughout the publication process.

Finally, we acknowledge the many smallholder farmers, extension workers, and development practitioners whose lived experiences and insights continue to inspire our research and commitment to transforming agriculture through sustainable practices.

We hope this book will serve as a useful resource and a platform for future collaboration, innovation, and knowledge exchange in the field of biofertilizers and beyond.

July 2025

Adornis D. Nciizah
Hupenyu Allan Mupambwa
Patrick Nyambo

Contents

Editors and Contributors

About the Editors

Adornis D. Nciizah is a Specialist Researcher in Soil Science at the Agricultural Research Council—Natural Resources and Engineering (ARC–NRE), South Africa. His expertise spans soil health, agroecology, climate-resilient agriculture, and sustainable land management. He has led numerous multidisciplinary projects addressing climate-smart agriculture, erosion control, and soil carbon sequestration, positioning him at the forefront of innovation in sustainable soil management. Through these initiatives, he continues to advance practical approaches to soil fertility and ecosystem resilience. His editorial leadership brings both scientific depth and practical relevance to this volume, making it a valuable resource for researchers, practitioners, and policymakers committed to sustainable agriculture.

Hupenyu Allan Mupambwa holds a PhD in Soil Science, where his research focused on vermicomposting as a waste beneficiation technology important in driving organic soil fertility. Currently, he leads the Desert and Coastal Agriculture research program at the University of Namibia. He has more than 17 years' experience in agriculture research and training as well as university undergraduate and postgraduate teaching and research. He has published more than 35 peer-reviewed international publications and 27 book chapters and edited 3 books. His research focuses on waste beneficiation, vermi-technology, and organic soil fertility management with a drive toward improving the soil quality among resource poor farmers.

Patrick Nyambo is a seasoned researcher and systems agronomist with extensive expertise in agronomy, water-energy-food nexus, and climate-smart agriculture. He holds a PhD in Agricultural Science (Crop Science) and has over a decade of experience leading multidisciplinary research on soil fertility, crop productivity, biochar applications, and sustainable land management. He has made significant contributions to the scientific community through peer-reviewed publications, book chapters, and collaborative research projects. He is affiliated with the Agricultural

Research Council—Vegetable Industrial and Medicinal Plants Institute (ARC-VIMP) in South Africa. He is a dedicated academic mentor and reviewer, actively supporting capacity building and knowledge dissemination across African agricultural research networks. His editorial interests span agronomic innovations, soil-plant interactions, and climate resilience.

Contributors

Akanksha Department of Agronomy, Faculty of Agricultural Sciences, SGT University Gurugram, Gurugram, India

Dhitri Borah Department of Zoology, Biswanath College, Gauhati University, Biswanath, Assam, India

Ankita Chatterjee Department of Agronomy, Faculty of Agricultural Sciences, SGT University Gurugram, Gurugram, India

Tavengwa Chitata Department of Water Governance, IHE Delft, Institute for Water Education, Delft, The Netherlands

Sucheta Dahiya Department of Agronomy, Faculty of Agricultural Sciences, SGT University Gurugram, Gurugram, India

Suman Samilita Dash Department of Entomology, OUAT, Bhubaneswar, Odisha, India

Crop Protection Division, ICAR-CRRI, Cuttack, Odisha, India

Deekshith H. N. Chaudhary Sarwan Kumar Himachal Pradesh Krishi Vishvavidyalaya, Palampur, Himachal Pradesh, India

D. Dlamini Department of Agronomy & Horticulture, Faculty of Agriculture, Environment & Natural Resources Management, Midlands State University, Senga, Gweru, Zimbabwe

Aastha Dubey School of Agriculture, Galgotias University, Greater Noida, UP, India

Admire R. Dzvene Centre for Global Change, Faculty of Science and Agriculture, University of Fort Hare, Alice, South Africa

Department of Agronomy, Faculty of Science and Agriculture, University of Fort Hare, Alice, South Africa

C. T. Gadzirayi CIRAD Zimbabwe, Belgravia, Harare, Zimbabwe

Priti Giri Department of Botany, University of Delhi, Delhi, India

Ravi Kumar Goswami Department of Zoology, Hindu College, University of Delhi, Delhi, India

Tatenda Justice Gunda Golden Helix Life Sciences Ltd, UNN Hse, Community 25 Peace Be Tema, Greater Accra, Ghana

T. T. Haripo Department of Agronomy & Horticulture, Faculty of Agriculture, Environment & Natural Resources Management, Midlands State University, Senga, Gweru, Zimbabwe

Jyoti Ranjan Jena Department of Agricultural Economics, OUAT, Bhubaneswar, Odisha, India

Rameswar Jena Department of Agronomy, OUAT, Bhubaneswar, Odisha, India

Amani Gabriel Kaningini Department of Agriculture and Animal Health, College of Agriculture and Environmental Sciences, University of South Africa, Florida, South Africa

Keletso C. Mohale Department of Agriculture and Animal Health, Science Campus, University of South Africa (UNISA), Johannesburg, Gauteng, South Africa

Sachin Kumar Department of Agronomy, Faculty of Agricultural Sciences, SGT University Gurugram, Gurugram, India

Gemina Abigail Boitumelo Letsoalo Microbiology and Environmental Biotechnology Research Group, Soil Science Division, Agricultural Research Council-Natural Resources and Engineering, Pretoria, South Africa

M. M. Mabitsela Department of Agriculture, University of Zululand, KwaDlangezwa, South Africa

Thembeka Mabiya Life Sciences Building, Department of Biotechnology, University of the Western Cape, Bellville, South Africa

Takudzwa C. Madzivanzira Freshwater Research Group, Department of Biological Sciences and Ecology, University of Zimbabwe, Harare, Zimbabwe

South African Institute for Aquatic Biodiversity (SAIAB), Makhanda, Eastern Cape, South Africa

Never Mafuse Department of Agricultural Economics, Education and Extension, Bindura University of Science Education, Bindura, Zimbabwe

B. T. Makaure Department of Agronomy & Horticulture, Faculty of Agriculture, Environment & Natural Resources Management, Midlands State University, Senga, Gweru, Zimbabwe

V. Makuvaro Department of Agronomy & Horticulture, Faculty of Agriculture, Environment & Natural Resources Management, Midlands State University, Senga, Gweru, Zimbabwe

Mashapa Elvis Malobane Department of Agriculture and Animal Health, College of Agriculture and Environmental Sciences, University of South Africa, Florida, South Africa

Beverly Mampholo Agricultural Research Council – Roodeplaat, Vegetable, Industrial and Medicinal Plants, Pretoria, South Africa

Sudeshna Mandal Department of Medicine, Indiana University School of Medicine, Indianapolis, IN, USA

Ronald Mandumbu Department of Crop Production, Faculty of Agriculture, University of Eswatini, Luyengo, Eswatini

Tirivashe Phillip Masere Department of Land and Water Resources Management, Faculty of Agriculture, Environment and Natural Resources Management, Midlands State University, Gweru, Zimbabwe

C. V. Mashamaite Agricultural Research Council—Vegetables, Industrial and Medicinal Plants, Pretoria, South Africa

Edith Matema Department of Agribusiness, Women's University in Africa, Manresa, Harare, Zimbabwe

Kahsay Tadesse Mawcha Biopesticides Group, International Centre for Genetic Engineering and Biotechnology, Cape Town, South Africa

Department of Plant and Soil Sciences, University of Pretoria, Pretoria, South Africa

Loice Mhengera Department of Land and Water Resources Management, Faculty of Agriculture, Environment and Natural Resources Management, Midlands State University, Gweru, Zimbabwe

Rangarirai Lucia Mhindu Department of Land and Water Resources Management, Faculty of Agriculture, Environment and Natural Resources Management, Midlands State University, Gweru, Zimbabwe

N. Mkhize Agricultural Research Council—Vegetables, Industrial and Medicinal Plants, Pretoria, South Africa

Thobo Motlhalamme Department of Agriculture and Animal Health, College of Agriculture and Environmental Sciences, University of South Africa, Florida, South Africa

H. Motsi Department of Soil Science, Faculty of AgriSciences, Stellenbosch University, Matieland, South Africa

B. Muchara University of South Africa (UNISA), Graduate School of Business Leadership, Pretoria, South Africa

N. Munyuki Department of Agronomy & Horticulture, Faculty of Agriculture, Environment & Natural Resources Management, Midlands State University, Senga, Gweru, Zimbabwe

Hupenyu Allan Mupambwa Sam Nujoma Marine and Coastal Resources Research Centre, Sam Nujoma Campus, University of Namibia, Henties Bay, Namibia

Reward Muzerengwa Institute of Medicine, Department of Microbiology and Virology, Peoples' Friendship of Russia University, Moscow, Russia

African Genome Centre (AGC), University Mohammed VI Polytechnic (UM6P), Ben Guerir, Morocco

Norman Muzhinji Department of Plant Sciences, Division of Plant Pathology, University of the Free State, Bloemfontein, South Africa

Anweshita Nayak Crop Production Division, ICAR-CRRI, Cuttack, Odisha, India

Adornis D. Nciizah Department of Agriculture and Animal Health, University of South Africa, Florida Park, Roodepoort, South Africa

Agricultural Research Council of South Africa – Natural Resources & Engineering (ARC–NRE), Pretoria, South Africa

Blessing Mirika Ndau Department of Land and Water Resources Management, Faculty of Agriculture, Environment and Natural Resources Management, Midlands State University, Gweru, Zimbabwe

Noluyolo Nogemane Department of Agriculture and Animal Health, Science Campus, University of South Africa (UNISA), Johannesburg, Gauteng, South Africa

Patrick Nyambo Agricultural Research Council—Vegetables, Industrial and Medicinal Plants Institute, Pretoria, South Africa

Makomborero Nyoni Innov Bio_M Ltd, 131 Rhodesville, Greendale, Harare, Zimbabwe

Subhra Parija Department of Soil Science and Agricultural Chemistry, OUAT, Bhubaneswar, Odisha, India

Crop Protection Division, ICAR-CRRI, Cuttack, Odisha, India

Cosmas Parwada Department of Agronomy, Midlands State University, Gweru, Zimbabwe

Tessia Rakgotho Life Sciences Building, Department of Biotechnology, University of the Western Cape, Bellville, South Africa

Agricultural Research Council – Roodeplaat, Vegetable, Industrial and Medicinal Plants, Pretoria, South Africa

Haripriya Rama Microbiology and Environmental Biotechnology Research Group, Soil Science Division, Agricultural Research Council-Natural Resources and Engineering, Pretoria, South Africa

Bharat Bhushan Rana Chaudhary Sarwan Kumar Himachal Pradesh Krishi Vishvavidyalaya, Palampur, Himachal Pradesh, India

Varunendra Singh Rawat Department of Zoology, Hindu College, University of Delhi, Delhi, India

Ashira Roopnarain Microbiology and Environmental Biotechnology Research Group, Soil Science Division, Agricultural Research Council-Natural Resources and Engineering, Pretoria, South Africa

Department of Environmental Sciences, College of Agriculture and Environmental Sciences, University of South Africa-Florida Campus, Johannesburg, South Africa

Chinmaya Sahoo Dr. Rajendra Prasad Central Agricultural University, Pusa, Bihar, India

Ankit Saini Dr. KSG Akal College of Agriculture, Eternal University, Sirmaur, Himachal Pradesh, India

Chaudhary Sarwan Kumar Himachal Pradesh Krishi Vishvavidyalaya, Palampur, Himachal Pradesh, India

Sonal Saluja Department of Zoology, Dyal Singh College, Karnal, Kurukshetra University, Kurukshetra, Haryana, India

Satyabrata Sarangi Department of Entomology, OUAT, Bhubaneswar, Odisha, India

Crop Protection Division, ICAR-CRRI, Cuttack, Odisha, India

Vinod Sharma Chaudhary Sarwan Kumar Himachal Pradesh Krishi Vishvavidyalaya, Palampur, Himachal Pradesh, India

Prasun Kumar Singh School of Agriculture, Galgotias University, Greater Noida, UP, India

Shalini Singh School of Agriculture, Galgotias University, Greater Noida, UP, India

Nguza Siyambango-Mulisa University of Namibia, Multidisciplinary Research, Centre for Research Services, University of Namibia, Main Campus, Windhoek, Namibia

Tetelo Tolo Microbiology and Environmental Biotechnology Research Group, Soil Science Division, Agricultural Research Council-Natural Resources and Engineering, Pretoria, South Africa

Ebere Lovelyn Udeh Department of Agriculture and Animal Health, University of South Africa, Florida, South Africa

Simin Emamzadeh Yazdi Department of Agriculture and Animal Health, Science Campus, University of South Africa (UNISA), Johannesburg, Gauteng, South Africa

Leocadia Zhou Centre for Global Change, Faculty of Science and Agriculture, University of Fort Hare, Alice, South Africa

Farai Solomon Zirebwa Department of Land and Water Resources Management, Faculty of Agriculture, Environment and Natural Resources Management, Midlands State University, Gweru, Zimbabwe

R. M. Zishiri Department of Agronomy & Horticulture, Faculty of Agriculture, Environment & Natural Resources Management, Midlands State University, Senga, Gweru, Zimbabwe

Munetsi Zvavahera Centre for Global Change, Walter Sisulu University, Mthatha, Eastern Cape, South Africa

Chapter 1
Overview of Biofertilizers and Their Potential Role in Sustainable Agriculture

Satyabrata Sarangi, Subhra Parija, Anweshita Nayak, Suman Samilita Dash, Rameswar Jena, and Jyoti Ranjan Jena

Abstract Modern agriculture needs latest mechanization, improved technology, and advanced pest control strategies to cope with the increasing demand for food. The world population has almost crossed the mark of 8 billion in the last decade. The higher and indefinite use of chemical pesticides and synthetic fertilizers has a direct and challenging impact on the environment, including environmental pollution, shrinkage of the micro-fauna, crop productivity, loss of biodiversity, economic backlash, and several other health hazards, due to their incomplete degradation. Due to factors like high risk of pollution, severe heat generated by greenhouse gases, and larger food demand, the need for biofertilizers is expected to grow in the future. Biofertilizer is a substance that includes living organisms, helping in higher crop productivity, creating a healthy biosphere with a sustainable environment. Several essential mineral nutrients like zinc, iron, nickel, manganese, magnesium, etc. are available in the soil in insoluble and complex forms, which are made available to the plant roots by the microbes in soluble form, which reduce the load of conventional inorganic

S. Sarangi (✉)
Department of Entomology, OUAT, Bhubaneswar, Odisha, India

Crop Protection Division, ICAR-CRRI, Cuttack, Odisha, India

S. Parija
Department of Soil Science and Agricultural Chemistry, OUAT, Bhubaneswar, Odisha, India

Crop Production Division, ICAR-CRRI, Cuttack, Odisha, India

A. Nayak
Crop Production Division, ICAR-CRRI, Cuttack, Odisha, India

S. S. Dash
Department of Entomology, OUAT, Bhubaneswar, Odisha, India

Crop Protection Division, ICAR-IIVR, Varanasi, Uttar Pradesh, India

R. Jena
Department of Agronomy, OUAT, Bhubaneswar, Odisha, India

J. R. Jena
Department of Agricultural Economics, OUAT, Bhubaneshwar, Odisha, India

A. D. Nciizah et al. (eds.), *Biofertilizers for Sustainable Agriculture in Africa*,
Sustainability Sciences in Asia and Africa, https://doi.org/10.1007/978-981-95-6009-7_1

fertilizers and help in nitrogen fixation, phosphate solubilization, phytohormone and siderophore production, lytic enzyme production, induced systemic resistance, and bioremediation, etc. with increased crop production by 20–30% and reducing the use of inorganic synthetic chemical fertilizers by 25–28%. Biofertilizers include bacteria, fungi, cyanobacteria, plant growth promoting rhizobacteria (PGPRs), and the most recent nano-biofertilizer, which helps directly as well as indirectly in restoring soil fertility, indicating improved soil physical health. These are more desirable than synthetic fertilizers due to their low cost, eco-friendly impact, qualitative produce, ecosystem restoration, and minimal health hazards. In majority of the agricultural sectors, the use of biofertilizers has been promoted over conventional inorganic fertilizers to overcome the inconsistent performance in the crop field, which has been evident from time to time. All these have been briefly explained in the following chapter.

Keywords Biofertilizers · PGPR · Nano-biofertilizers · Africa · Chemical fertilizers · Crop production

1.1 Introduction

1.1.1 Need for Intensive Agriculture due to Population Growth

Mother Earth is the base for agriculture, and soil is the greatest medium. As the world's population increased from approximately 1.6 billion in 1900 to approximately 7.6 billion in 2017, the demand for food increased significantly, placing a great deal of pressure on the agricultural sector to guarantee a sufficient supply of food (Walling and Vaneeckhaute 2020). The problem of meeting the world's food needs has become increasingly pressing as estimates place the population at approximately 9.8 billion by 2050 (Bazargani and Deemyad 2024; Kumar and Kalita 2017). According to the United Nations Food and Agriculture Organization (FAO), to adequately feed the world's population, which is expected to reach between 9.4 and 10.2 billion people by 2050, food production must increase by 70% from 2010 levels (Chandra and Collis 2021). When we talk about African subcontinents, an increasing trend in population is observed every year. According to statistical reports (Statistics South Africa 2017), the inhabitants of Africa, especially in rural areas who are living their livelihoods under the upper boundary of the poverty line, need a substantial food supply. To meet this food demand, intensive agriculture has come into play. By practicing intensive methods of agriculture, the land becomes infertile over time because of the complete exhaustion of minerals. To meet food security, modern agriculture has focused on using HYV (High-Yielding Varieties) and hybrid seeds, which respond well to high doses of chemical fertilizer. This non-judicious application of inorganic fertilizer has led to pollution and contamination of the soil and water bodies (Savci 2012).

1.1.2 Negative Impact of Inorganic Chemical Fertilizers on the Environment

By suppressing biological nitrogen fixation processes and impeding the function of beneficial nitrogen-fixing bacteria necessary for natural nutrient cycling (Tripathi et al. 2020), the overuse of chemical fertilizers in modern agriculture upsets the delicate balance of soil biodiversity, with a negative impact on mycorrhizal fungi (the symbiotic partners that enhance plant nutrient absorption and overall soil health) (Igiehon and Babalola 2018; Nagy et al. 2023). According to Chen et al. (2018), an ongoing dependence on chemical fertilizers causes soil acidification, reducing the availability of vital micronutrients like iron, zinc, and manganese, that are essential for plant physiological functions. As a result, the soil is now lacking in organic matter (Bot and Benites 2005) and vital plant nutrients. It was also mentioned by Savci (2012) that environmental deterioration, nutrient leaching, eutrophication, soil microbial flora modification, and soil pH distortion have been caused by the overuse of inorganic fertilizers. Furthermore, plants may develop nutrient imbalances as a result of the irregular nutrient ratios present in many chemical fertilizers, making them more susceptible to insect and disease attacks (Rajawat et al. 2019). When chemical fertilizers' high levels of nitrogen and phosphorus are not completely absorbed by crops, they leak into surface and groundwater, causing eutrophication, a condition marked by excessive algal blooms that lower oxygen levels in the water, which eventually kills aquatic organisms and affects fish populations and other aquatic life, lowering biodiversity and upsetting the ecosystem (Durán-Lara et al. 2020). Additionally, nitrate pollution in drinking water, which is often caused by fertilizer runoff, is a serious health problems, particularly for young children, as it can result in methemoglobinemia, often known as "blue baby syndrome" (de Andrade et al. 2023). According to research, one of the main causes of the increase in the number of coastal areas experiencing severe eutrophication is agricultural runoff enhanced with chemical fertilizers. More than 450 areas worldwide are affected, and the excess nitrogen has accelerated primary production beyond the capacity of higher-level consumers to consume, making the eutrophication problem worse (García et al. 2018).

1.1.3 Why Sustainable Agriculture?

According to Nagy et al. (2023), African soils have poor fertility, and abrupt climate, along with other problems like land acquisition, tend to be a struggle and a global concern for ecological balance. Secondly, these synthetic fertilizers are costlier, which is expensive especially for small and marginal farmers (Camara and Heineman 2006). All things rise to one thing that needs a replenishment strategy for sustainable crop productivity (Lahiff and Cousins 2005). Sustainable agricultural practice is a combination of biological, chemical, physical, and economic ways with the main objective of optimum and efficient utilization of resources and protecting ecological biodiversity (Lichtfouse et al. 2009; Patel et al. 2014). The goal of sustainable agriculture is to meet current demands for food and textiles while protecting ecosystems and natural resources for coming generations. Crop rotation, organic farming, and reduced chemical use are some of the techniques to balance ecological health and productivity. Microbial agents like biofertilizers that naturally increase soil fertility have emerged as crucial tools for reducing reliance on synthetic fertilizers and promoting long-term agricultural resilience in this regard. In the quest for sustainable agriculture, biofertilizers are becoming more and more well-liked as a practical substitute for harmful chemical fertilizers (Nosheen et al. 2021).

1.1.4 Need for Switch from Chemical to Biofertilizers for Sustainable Crop Growth

It has been proposed that using biological fertilizers is a helpful tool for sustainable agriculture and nutrient management (Kawalekar 2013; Malusà and Ciesielska 2014). Biofertilizers provide better soil conditions that include improved soil fertility, nitrogen fixation, and phosphate solubilization; enhanced efficacy of applied N and P; enhanced activities of other microbes; as well as the release of growth stimulants. Biofertilizers fix nitrogen in the soil, produce organic acids and growth chemicals, and provide complex phosphorus to plants, potentially reducing the need for inorganic fertilizers (Patel et al. 2014). The efficacy of inorganic fertilizer was pronounced when they were combined with biofertilizers. Balanced C: N ratio, increased nutrient availability, lesser nutrient retention, and more translocation to the aerial part for protoplasmic protein synthesis cause more vegetative growth in plants (Dhawale et al. 2011).

1.1.5 Role of Plant Growth Promoting Rhizobacteria (PGPRs)

The diverse group of helpful bacteria known as plant growth-promoting rhizobacteria (PGPR) live in the rhizosphere, the region of soil that surrounds plant roots, and support the growth and health of plants in many ways, both directly and indirectly. They impart a direct influence by producing plant hormones like auxins, cytokinins, and gibberellins, which promote root and shoot growth, solubilize phosphorus and other essential minerals, and transform atmospheric nitrogen into a form that can be used (Barnawal et al. 2017). By generating antibiotics, siderophores (which seize iron and prevent the growth of pathogens), and inducing systemic resistance in plants, they indirectly strengthen the plant defences against infections (Vurukonda et al. 2016). Plant growth-promoting rhizobacteria simulate plant growth substances, i.e., phytohormones such as auxin, IAA, GA, and ethylene (Kumari et al.

2016; Tahir et al. 2017) by invading the plant roots (Beneduzi et al. 2012). PGPR inoculant provides tolerance against salinity (Bharti et al. 2013), drought (Niu et al. 2018; Ilyas et al. 2020), against biotic stress (Timmusk et al. 2014), and improves seed germination (Almaghrabi et al. 2014). Biocontrol ability of biofertilizers can inhibit pathogens from causing disease and pest attacks as well (García-Fraile et al. 2015). For example, the symbiotic association between leguminous crops and rhizobium synthesizes a cyanogenic substance that provides resistance to the crop from herbivore attack (García-Fraile et al. 2015). Some strains like *Cyanobacteria*, *Azospirillum*, *Burkholderia*, *Bacillus*, *Enterobacter*, *Aspergillus*, and *Penicillium* clean up pollutants like heavy metals from the soil and water and act as bioremediates (Choudhary 2010; Jain and Khichi 2014).

1.2 Role of Biofertilizers in Crop Production and Development

As the global population rises, securing food security and ensuring food demand especially in developing countries like Africa, becomes challenging. Chemical fertilizers help plants grow faster and more efficiently to meet this demand. However, reliance on such methods raises concerns about sustainability and long-term impacts on agriculture and the environment. Excessive use of chemical fertilizers leads to environmental pollution, soil degradation, altered soil ecology, and health risks. It also contributes to declining agricultural productivity over time. Additionally, climate factors increase abiotic stress on crops, further reducing yields. These combined effects pose significant challenges to sustainable agriculture and food security. Biofertilizers are becoming more popular as a result of the drawbacks of chemical fertilizers. By fixing nitrogen, solubilizing zinc, potassium, and phosphorus, producing hormones, and releasing siderophores and enzymes, they naturally improve plant nutrition. They also support sustainable agriculture by shielding plants from stress and diseases. Biofertilizers help meet the growing demand for food worldwide by providing enough nutrients for crops to grow healthily. They are sustainable, economical, and environmentally benign substitutes for chemical fertilizers, which makes them appropriate for long-term agricultural use while safeguarding the environment and guaranteeing food security for the expanding population (Chaudhary et al. 2022). Biofertilizer is a substance that contains cultured living microorganisms that cycle nutrients by mobilizing and solubilizing the nutrient complexes, and also plays a major role in controlling disease and pests (Kawalekar 2013; Rose et al. 2014). Applying biofertilizers that include live microorganisms to plant surfaces, seeds, roots, or soil promotes plant development by making more plant nutrients and growth substances available to the host crops (Vessey 2003; Figueiredo et al. 2011). Biofertilizer is not same as organic fertilizers like compost, animal manure, and plant extracts mentioned by Mazid and Khan (2015). Biofertilizers are products that may be available in the form of liquid or carrier materials that include either dormant or living organisms, either alone or in combination. The organism in the biofertilizer help in fixing atmospheric nitrogen (*Anabaena*, *Azotobacter*, *Rhizobium*) or solubilizing phosphorous (*Pseudomonas* sp.) or sulphur (Arbuscular mycorrhiza) and produce growth promoting compounds or phytohormones and enzymes (Sharma et al. 2022) to increase crop growth and yield (Dinesh Kumar et al. 2018).

1.2.1 Role of Biofertilizer in Organic Farming

In organic farming, biofertilizers have been crucial for increasing crop yield in a sustainable way. The soil's fertility is significantly improved by applying a certain kind of biofertilizer, which raises crop yield metrics and consequently, the agricultural output. By increasing the fixation of biological nitrogen, increasing the solubilization and adsorption of nutrients for their availability and uptake, and by promoting antibiosis and accumulation of soil organic matter, biofertilizers in the soil greatly increase crop yield (Khosro and Yousef 2012). Soil-living microorganisms near the rhizosphere are the source of biofertilizers (Choudhary et al. 2017). These biofertilizers preserve the plant-water relationship, shield crops from water stress, and lower the prevalence of pests and soil-borne diseases (Mącik et al. 2020).

1.2.2 Role of Biofertilizer in Soil Health

Soil health and productivity of soil as well as plants are the effect of interaction between the plant, soil, and the beneficial microorganism present in the rhizosphere, which helps in maintaining an ecological balance (Harman et al. 2020; Kumar et al. 2021). The inherent capacity of soil to provide fertilizer to plants is inseparably connected with the balance of microorganisms and plants (Vishwakarma et al. 2020). To maintain such ecology the application of biofertilizer that has the properties of the cultured microorganism, can provide a favourable environment for the nutrient accessibility and organic matter decomposition too. Also, the effect of biofertilizer depends on the population in the rhizosphere (Hassan et al. 2019) and factors like chemotaxis, adhesion, and biofilm decides the rhizospheric colonization (Trivedi et al. 2020). By altering the bacterial composition and reducing the quantity of harmful bacteria, the use of biofertilizers containing *B. amyloliquefaciens* can successfully reduce banana fusarium wilt by increasing the abundance of bacteria and decreasing that of fungi (Fu et al. 2017). Also, it is observed that applying biofertilizers to soil can enhance its physical characteristics, particularly its ability to retain moisture because of its enhanced micro porosity (Alencar et al. 2015). With the integration of biofertilizer with the synthetic commercial fertilizer, the soil productivity can be preserved and the rate of recommended synthetic fertilizer can be reduced, as was observed in rice productivity (Ghimire et al. 2021). Biofertilizer is the material when added to soil, roots, plant surface, leaves, and seed, increases growth by a strategy that supports the host plant's accessibility to primary nutrients, which are inhabited by living microorganisms that invade the rhizosphere (Vessey 2003).

1.2.3 Benefits of Using Biofertilizers

Use of biofertilizers improves production rapidly, uses minimal energy, lowers soil and water pollution, increases soil fertility, increases biological control and the malignancy of phytopathogenic organisms. Biofertilizer can be added to soil as inoculums that divide by participating in the process of cycling nutrients, which directly increases the crop productivity. Chemical fertilizer, on the other hand, is said to get lost by 60–90% of its total applied amount outside the soil surface (Srivastava 2021). These chemical fertilizers are not affordable for farmers. Farmers may easily afford biofertilizer over chemical fertilizer since it is readily available, environmentally friendly, efficient in application, and preserves soil fertility and productivity (Kumar and Kumar 2019). Beyranvand et al. (2013) claimed that higher grain and biomass production from the use of biofertilizer is a significant advantage, leading to a decrease in crop production input as a result of increased yield and reduced input cost. The percentage of the attainable value of benefits compared to the actual cost of biofertilizer is used to determine the benefit-cost analysis of a biofertilizer. Fertilizer, which is of lower cost, reduces the cost of input and increases the yield, which ultimately increases the profitability of small and marginal farmers (Suyal et al. 2016). The benefits of biofertilizer are presented in Fig. 1.1.

1.3 Types of Biofertilizers

A biofertilizer is mainly comprised of microbial strains or inoculum and a carrier material, along with some preservatives that enhance the product's shelf life. Based on the type of microorganism used, biofertilizers can be either bacterial, fungal, or algal. Depending on the state of the carrier material, biofertilizers may be in solid or powder form or of liquid formulations (Table 1.1). Biofertilizers can also be categorized on their functional basis as nitrogen-fixing biofertilizers, phosphorus and potassium solubilizing and mobilizing biofertilizers, micronutrient biofertilizers, and plant growth-promoting rhizobacterial biofertilizers. Modern-day necessities have also led to the development of nano-biofertilizers for agricultural use.

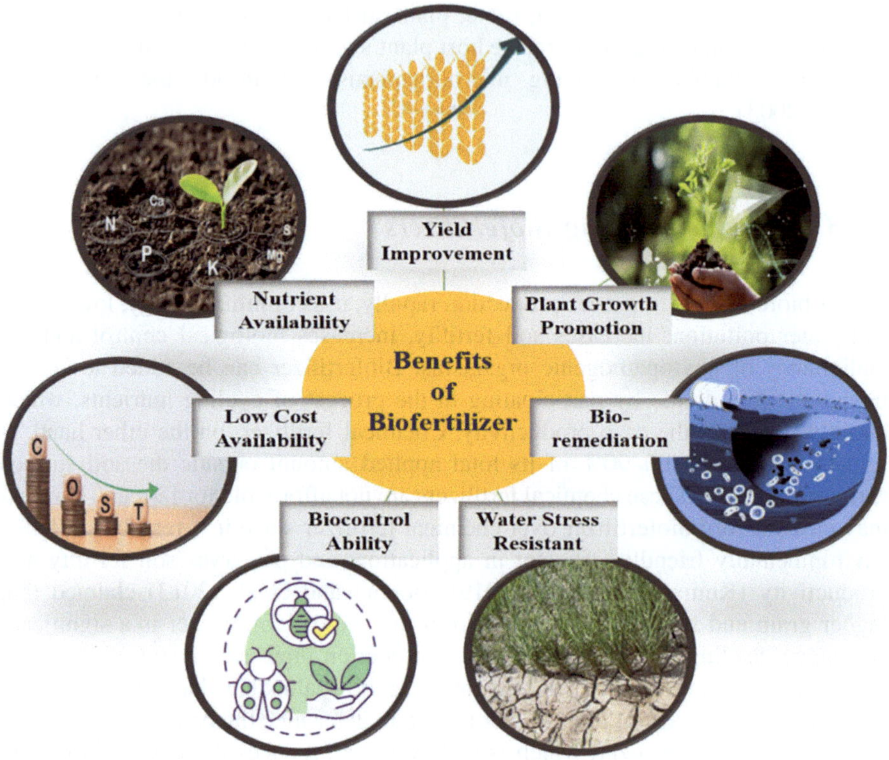

Fig. 1.1 Benefits of biofertilizer applications in agriculture for qualitative and quantitative crop health

1.3.1 Types of Biofertilizers Based on the Availability of Essential Nutrients

1.3.1.1 Nitrogen-Fixing Biofertilizers

Nitrogen is an essential macronutrient that promotes vegetative growth and forms the key constituent of chlorophyll, which is essential for photosynthesis. In soils deficient in nitrogen (N), diazotrophs fix atmospheric dinitrogen gas (N_2) in the process of biological nitrogen fixation (BNF). Diazotrophs generally develop mutual relationships with the host plants that can be symbiotic, non-symbiotic, or associative. Symbiosis is commonly observed in the bacterial family Rhizobiaceae, which are found in the root nodules of several legumes like cowpea, groundnut, and soybean. Nitrogen fixers that exhibit non-symbiotic and free-living traits include *Beijerinckia*, *Azotobacter*, and *Clostridium*, while *Azospirillum* and *Enterobacter* species are grouped under associative N-fixers (Shridhar 2012). *Azotobacter* and *Bacillus* spp. play a major role in nitrogen fixation and improving the growth of maize and forest crops (Etesami et al. 2014; Azeem et al. 2022). In Africa, the use

Table 1.1 Common characteristics of solid and liquid biofertilizers

Characteristics	Solid biofertilizer	Liquid biofertilizer
Physical state	Solid	Liquid
Carrier material	Peat, lignite, compost	Water, liquid medium
Application method	Direct soil application, seed treatment	Foliar spray, seedling treatment
Microbial concentration	Lower	Higher
Performance	Slow colonization	Faster colonization
Shelf life	Longer	Shorter
Handling and storage	Easier	More complex
Advantage	Ease of application and longer shelf life	Higher microbial count and faster action
Disadvantage	Lower microbial load and slower action	Higher cost, sensitive to temperature, higher risk of contamination
Examples	Rhizobium-based biofertilizers for legumes; *Azotobacter*-based biofertilizers for cereals	*Azospirillum*-based biofertilizer for N-fixation; *Pseudomonas*-based biofertilizers for plant growth promotion

of nitrogen-fixing biofertilizers is gaining popularity, especially for leguminous crops and the crops followed under crop rotation (Masso et al. 2017), and is critical for improving soil fertility (Aloo et al. 2021b). The bacteria *Pseudomonas protegens* is reported to promote plant growth in the nitrogen-deficient condition (Jing et al. 2020).

Rhizobium species grow symbiotically with legumes and can fix about 100 kg N/ha annually (Majumdar 2015). By colonizing the roots of leguminous crops, the bacteria help in the development of root nodules where nitrogen fixation takes place. Biofertilizer containing *Rhizobium* is reported to minimize the need for chemical fertilizers by 50% in arid and semiarid regions of Zimbabwe, Tanzania, and Kenya that have N-deficiency (Chianu et al. 2011). Mishra and Dash (2014) found that the application of *Rhizobium* and *Bradyrhizobium* replenishes soil nitrogen by fixing up to 50–200 kg N/Ha. *Bradyrhizobium japonicum* inoculation enhanced N-fixation, nodulation, and plant biomass in soybean plants (Htwe et al. 2019). Also, *Bradyrhizobium* spp. is reported to improve nitrogen fixation and IAA production in moong (Alkurtany et al. 2018).

Azotobacter are non-symbiotic, heterotrophic, free-living, and aerobic bacteria found under neutral to alkaline soil conditions and are known to fix about 20 kg N/ha annually. The most predominant species is *Azotobacter chroococcum*. However, species like *Azotobacter beijerinckii*, *Azotobacter insignis*, *Azotobacter macrocytogenes*, and *Azotobacter vinelandii* are also of much importance (Abbey et al. 2019). The bacteria *Azospirillum* from the Spirillaceae family are heterotrophic, free-living, motile, aerobic, and associative. They are known to withstand flooded conditions and fix about 40 kg N/ha annually (Mahdi et al. 2010). In this genus, *Azospirillum brasilense* and *Azospirillum lipoferum* are found all around the world and are beneficial to many non-leguminous crops and C_4 plants. According to the reports, *Azospirillum brasilense* boosts plant biomass and wheat grain production, enhances plant nutrition by decreasing N-fertilization (Galindo et al. 2022).

Blue-green algae (BGA) or cyanobacteria are phototrophic and are abundantly found in submerged rice fields. They fix about 30 kg N/ha per year (Mishra et al. 2013). Commonly used cyanobacterial biofertilizers include *Anabaena*, *Aulosira*, *Nostoc*, *Plectonema*, *Scytonema*, and *Tolypothrix*. Cyanobacterial biofertilizer consists of *Anabaena* or *Nostoc* in association with *Azolla*, which is commonly used in rice cultivation. *Azolla*, a heterosporous water fern in association with *Anabaena*, can fix about 100 kg N/ha in submerged rice cultivation (Wagner 2012). It was also found that in rice, there was a 50% reduction of nitrogen fertilizer application after using biofertilizer (Rose et al. 2014). Recent studies suggest that in African countries like South Africa, Nigeria, Togo, Ivory Coast, Kenya, and Senegal, *Azolla* is used to increase available nitrogen in soil and the yield of rice crops (Diniz et al. 2015). Other groups of bacteria that can fix atmospheric nitrogen are *Klebsiella pneumoniae*, *Azoarcus*, and *Pontoea agglomerans* (Raimi et al. 2017).

1.3.1.2 Phosphorus-Solubilizing Biofertilizers (PSB)

Phosphorus (P), a major constituent of plant cells and nucleic acids, is also responsible for the synthesis of ATP. Generally, P content in soil is enormous but the plant-available form is usually fixed as insoluble complexes resulting in its unavailability. The application of phosphorus solubilizing biofertilizer (PSB) enhances the availability and bio-accessibility of phosphorus for crop growth and development. PSB comprises phospho-bacterin that produces organic acids like gluconic acids and citric acids and increases soil acidity, thus leading to the dissolution of insoluble calcium complexes like hydroxyapatites, phosphates, and rock phosphates, improving their bio-availability (Mahanty et al. 2017). PSB can solubilize phosphorus, which has been chemically fixed as calcium phosphate and magnesium phosphate in alkaline soil and as aluminium or iron phosphate in acidic soil and keep them in a readily available form for plant absorption (Mishra et al. 2013). These microbes also produce phosphatases and phytases that release phosphate from organic compounds. PSB mainly comprises of bacteria and fungi that solubilize and mobilize P in soil. These include *Azotobacter*, *Agrobacterium*, *Aerobacter*, *Achromobacter*, *Beijerinckia*, *Bacillus*, *Burkholderia*, *Enterobacter*, *Flavobacterium*, *Microbacterium*, *Micrococcus*, *Pseudomonas*, *Rhizobium*, *Serratia*, *Aspergillus*, *Erwinia*, and *Penicillium*. PSB also promotes biological N-fixation and enhances N availability to plants. *Azotobacter chroococcum*, *Pseudomonas putida*, *Pseudomonas chlororaphis*, *Bacillus circulans*, and *Rhizobium leguminossarium* have been used as phosphate solubilizing biofertilizers (PSB) on several crops like pulses, wheat, tomato, potato, etc. (Malusà and Ciesielska 2014).

To survive in P-deficient conditions, plants develop extensive root growth with the help of mycorrhizal association where fungal hyphae help in mobilizing and making phosphorus available to the crops (Mujawar 2014). According to Adeleke et al. (2010), ectomycorrhizal fungi like *Phialocephala fortini*, *Paxillus involutus*, *Pisolithus tinctorius*, and *Suillus tomentosus* are potential P and K mobilizers under different soil conditions. The arbuscular mycorrhizal fungi (AMF) also enhance nutrient availability in the soil through specialized structures like vesicles and arbuscules (Raimi et al. 2017). Using arbuscular fungi can increase a plant's availability of P and shield it from stressful situations (Nacoon et al. 2020). The species like *Acaulospora* sp., *Glomus* sp., *Entrophospora* sp., and *Paraglomus* sp. are Arbuscular mycorrhiza that can transport both organic and inorganic forms of phosphorus to the plant (Haro and Benito 2019).

1.3.1.3 Potassium Solubilizing Biofertilizers (KSB)

Potassium regulates stomatal closing and opening and also facilitates nutrient translocation. It provides resistance to stressful situations like drought (Santosh et al. 2022). It also aids in metabolic activities like protein synthesis, enzyme activation, sugar breakdown, and photosynthesis. In soil systems, potassium (K) exists dynamically in three forms, that is, unavailable, slowly available, and readily available K (Kaiser et al. 2016). KSB application helps to solubilize the unavailable K, making it more bio-accessible. By producing organic acids and chelators, these biofertilizers solubilize K from complexes like muscovite, mica, orthoclase, illite, and biotite by secreting enzymes, biofilms, hydroxyl anions, and organic ligands (Bahadur et al. 2014). They also release acidolytic enzymes that liberates potassium in plant available form. Potassium solubilizing microbes that mobilize and solubilize K efficiently include *Clostridium*, *Bacillus*, *Aspergillus*, *Burkholderia*, *Pseudomonas*, *Paenibacillus*, and *Acidothiobacillus*. The bacteria *Frateuria aurantia* is also reported to mobilize and solubilize the unavailable forms of K for plant growth (Mazid and Khan 2015). *Bacillus mucilaginous*, *B. edaphicus*, *B. circulans*, *Acidithiobacillus ferrooxidans*, *Frateuria aurantia*, *Herbaspirillum* spp., and *Clostridium* spp., and fungi like *Aspergillus* spp. and some arbuscular mycorrhiza fungi are some of the potassium solubilizing species. The application of KSB is of much significance in African countries that experience high rates of potassium losses like Burundi, Guinea Bissau, Nigeria, Malawi, Rwanda, Swaziland, Kenya, and Uganda (Bationo et al. 2012).

1.3.1.4 Micronutrient Biofertilizer

Microbial genera *Bacillus*, *Bradyrhizobium*, *Rhizobium*, *Pseudomonas*, *Saccharomyces*, *Thiobacillus*, and *Penicillium* are reported to enhance the uptake and availability of soil micronutrients (Ahsan et al. 2012). Hafeez et al. (2013) reported that the application of *Bacillus subtilis*, *Saccharomyces*, and *Thiobacillus thioxidans* as biofertilizers resulted in mobilizing and solubilizing Zn into Zn^{2+}cation, which is predominantly taken up by the crops. These bacteria help in solubilizing insoluble Zn compounds like zinc oxides, zinc carbonates, and zinc phosphates that occur as unavailable complexes in soil (Hussain et al. 2015). Bacteria are known to obtain iron by secreting iron chelators known as siderophores (Rajkumar et al. 2010). In soils with immobilized Fe, siderophores are produced by rhizobacteria. They solubilize iron into complex forms that are easily absorbed by the roots (Ghavami et al. 2017).

A fungal species, *Trichoderma harzianum* is reported to solubilize MnO_2 and metallic Zn by chelation and reduction mechanism. Vesicular Arbuscular Mycorrhiza (VAM) is also capable of solubilizing Fe, Zn, Cu, and Mn, in soil (Raimi et al. 2017).

1.3.1.5 Plant Growth-Promoting Rhizobacterial Biofertilizers (PGPRs)

Several beneficial microbes grow in the rhizospheric soil that participate in the cycling of essential nutrients and the production of plant growth-promoting substances. These microbes are therefore called plant growth-promoting rhizobacteria (PGPR). Eichmann et al. (2021) reported that the production of hormones is an important function of beneficial microbes. Owing to their fertilization potential, some of these PGPRs are considered biofertilizers. PGPR comprises the bacterial genera *Azospirillum*, *Azotobacter*, *Burkholderia*, *Bacillus*, *Erwinia*, *Flavobacterium*, *Pseudomonas*, *Rhizobium*, and *Serratia.* This group of biofertilizers is a collection of heterogeneous groups of beneficial microbes that inhabit the rhizosphere and stimulate plant growth through different mechanisms. They function through N-fixation, nutrient solubilization, and phyto-stimulation (Hassen et al. 2016). These microbes secrete hormones and anti-metabolites that promote the decomposition of organic matter and root growth, thus enhancing nutrient availability and crop yield. They also suppress diseases through antimicrobial metabolites like tensin, phenazines, and hydrogen cyanide. The interaction of PGPR with roots induces systemic resistance in plants against all types of pathogens (Ahemad and Kibret 2014). Certain groups of biofertilizer-PGPRs secrete phytohormones like auxin, gibberellins, and cytokinin resulting in an increased plant foliage, root elongation, fruit yield, and plant-microbial symbiotic association (Hassen et al. 2016). The different phytohormones having distinguished characteristics are auxin, which helps in cell division and differentiation, cytokinin restricts senescence, ABA (Abscisic acid) marks stress conditions, and Gibberellin helps in pollen germination and shoot elongation. Salomon et al. (2014) mentioned that *Bacillus licheniformis* is a species that produces IAA, ABA, and GA and improves the growth of the grapevine. Bacteria such as *Paenibacillus* and *Bacillus* are reported to promote plant-mycorrhizal colonization (Adeleke and Dames 2014). The cultivation of rice and maize with gibberellic acid-producing PGPRs reported a significant increase in the growth and yield of the crops (Vacheron et al. 2013). There was a reported increase in the surface area of the roots and total root biomass in tomato crops inoculated with *Azospirillum* due to the production of IAA-mediated ethylene and an increment in maize growth under iron-poor soil when fertilized with siderophore-producing PGPRs (Raimi et al. 2017). PGPRs used as biofertilizer live in the rhizosphere and on the root surface and in the intercellular space as endophytes, facilitate the growth of the plant, and provide tolerance in stress conditions (Ghosh et al. 2019) (Fig. 1.2).

1.3.2 *Types of Biofertilizers Suitable for Different Cropping Systems*

Biofertilizers are widely used in Africa to promote sustainable agriculture (Bhattacharjee and Dey 2014) and obtain higher yields. Africa's diverse cropping system ranging from cereal-based systems to legume-based systems, adapted to its diverse climates, soils, and socio-economic conditions offer a great and unique opportunity for the use of biofertilizers. All of the major cereals like corn, wheat, and rice are chiefly produced in Africa, in that order of significance. Nowadays, crops only absorb 30–50% of nitrogenous fertilizers and around 10–45% of phosphorus fertilizers applied (Singh et al. 2016). Therefore, it is necessary to look for more economically and environmentally beneficial sources. In legume symbiosis, bioinoculants, an alternate source of chemical fertilizers, are a promising technology (Youseif et al. 2017). The mycorrhizal fungi that improve P uptake and the nitrogen-fixing rhizobia are well-known (Miransari 2011). The use of specific biofertilizer strains in different cropping systems in Africa has resulted in significant rise in crop yield and quality along with improvement in nutrient use efficiencies (Table 1.2). The use of bacterial genus *Acinetobacter* as a biofertilizer helps in plant growth promotion through antibiosis, auxin production, solubilization of calcium phosphate and siderophore production (Kumakech and Turyagyenda 2024). Velázquez et al. (2008) reported the association of *Acinetobacter* as rhizospheric and endophytic species in sugarcane crop.

1.3.3 *Nano-Biofertilizers*

Nano biofertilizers have come up as an emerging class of agricultural inputs that combine nanotechnology with biofertilizers to enhance nutrient use efficiency, crop productivity, and soil health.

1.3.3.1 Functions of Nano-Biofertilizers

Such a product results in slow release of nutrients to the crops, enriches the soil with beneficial microbes, increases the nutrient use efficiency and reduces nutrient loss. These fertilizers integrate beneficial microorganisms with nano material carriers (1–100 nm) like nano clays, carbon nanotubes, metal nano particles of zinc, titanium, and iron), nano silica, or biodegradable polymers like chitosan nano particles.

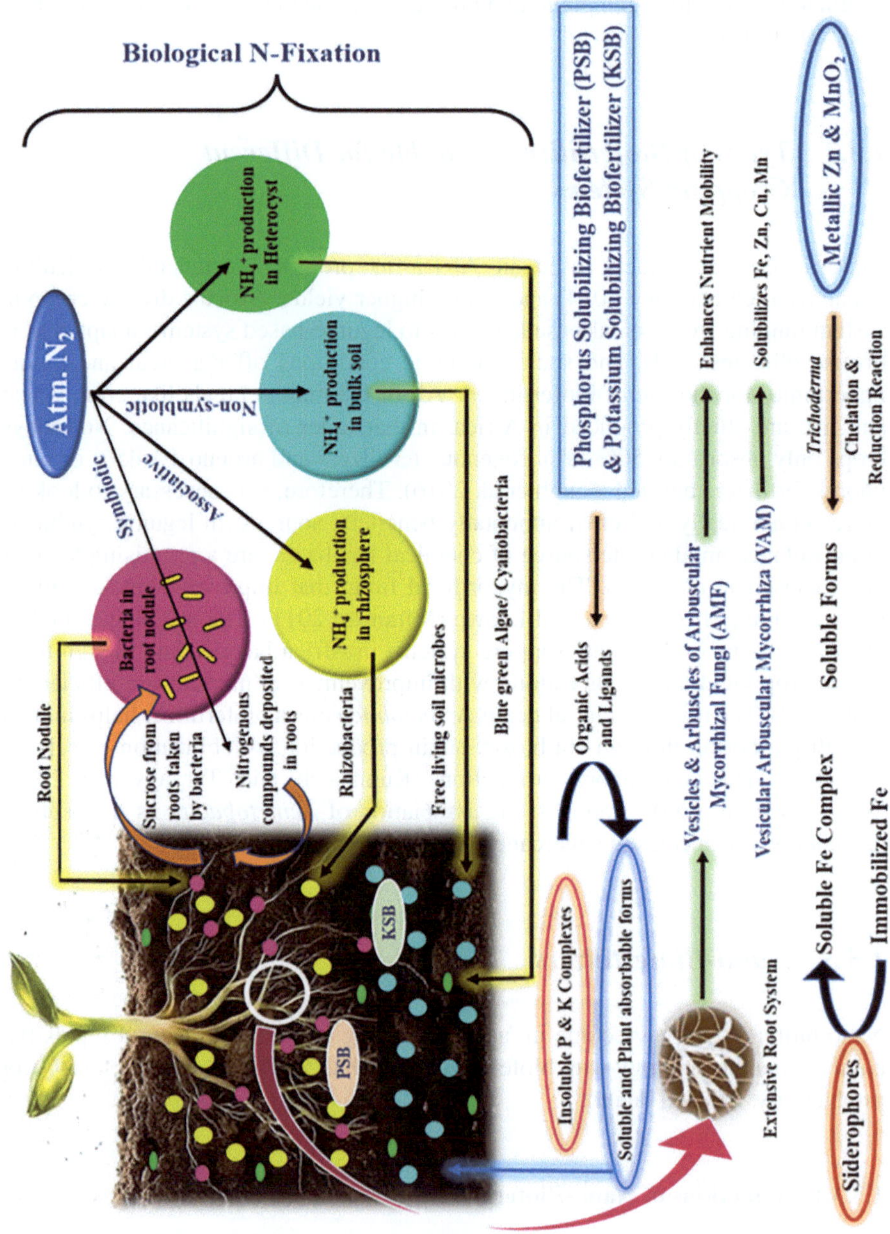

Fig. 1.2 Mode of action of different types of biofertilizers

Table 1.2 Effect of biofertilizers in different cropping systems in Africa

Cropping systems	Biofertilizer	Effect	Location	References
Cereal-based cropping systems				
Corn	*Lysinibacillus sphaericus, Paenibacillus alvei, Bacillus pumilus, Bacillus safensis* *Azospirillum, Azotobacter* *Pseudomonas, Klebsiella oxytoca, Enterobacter sakazakii*	33% increase in yield 100% reduction in chemical fertilizer input Increased agronomic characteristics	South Africa Egypt Nigeria, Kenya	Breedt et al. (2017) Naserirad et al. (2011) Babalola et al. (2007)
Rice	*Azolla pinnata, A. nilotica, A. filiculoides, A. caroliniana*	Increased yield and Nitrogen content in soil enhanced	Kenya, Nigeria, Ivory Coast, South Africa, Togo & Senegal	Aloo et al. (2021a)
Wheat	*Rhizobium radiobacter*	Good soil applicant, good quality bread production	Giza, Egypt	Hadide et al. (2019)
Legume-based Cropping Systems				
Faba bean	*Rhizobium, Agrobacterium*	Increased modulation, total Nitrogen uptake, and yield and yield attributes	Egypt	Youseif et al. (2017)
Soybean	*Rhizobium* sp., *Bradyrhizobium* sp., *Trichoderma harzianum*	Growth hormones production, biocontrol agent, effective increase in yield of soybean	Nigeria	N'Cho et al. (2015)
Green gram	*Bradyrhizobium*	Increase in nitrogen fixation and yields of green gram	Ethiopia	Raji et al. (2019)
Common beans	*Rhizobium* sp.	285% increase in iron, 67% increase in Zinc content	Tanzania	Tairo and Ndakidemi (2014)
Vegetable-based cropping system				
Tomato	*Pseudarthrobacter chlorophenolicus*	Plant growth promotion	Juja, Kiambu county, Kenya	Issifu et al. (2022)

(continued)

Table 1.2 (continued)

Cropping systems	Biofertilizer	Effect	Location	References
	Pseudomonas aeruginosa, *Bacillus subtilis*, *Citrobacter youngae*, *Klebsiella pneumonia* *P. aeruginosa*	Better growth parameters like the height of the seedling, leaf area, and number of leaves Improved growth	Nigeria Nigeria	Akintokun and Taiwo (2016) Adesemoye and Ugoji (2009)
Potato	*Serratia* sp., *Citrobacter* sp.	Increase in nutrient levels of potato tubers and rhizospheric soils	Kenya	Aloo et al. (2021a)
Okra	*Pseudomonas aeruginosa* *Azospirillum*, *Azotobacter*	Improved growth 100% reduction in chemical fertilizer input	Nigeria Egypt	Adesemoye and Ugoji (2009) Shaheen et al. (2007)
Oilseed crops-based cropping system				
Sunflower	BioNiK-Phos fertilizer (*Acinetobacter* spp.)	Soil fertility management, atmospheric nitrogen fixation, and phosphorus solubilization	Uganda	Kumakech and Turyagyenda (2024)
Peanut and sunflower	Nitrobein, Phosphorein, and Potash which contain N fixers, P solubilizers, and K mobilizers	Oil content enhanced, biomass gained, and increased seed yield	Cairo, Egypt	Ahmed and El-Araby (2012)

Green nano biofertilizers are synthesized using biogenic and microbial compounds such as plant extracts, fermented extracts, enzymes, organic polymers and other polysaccharides where metal ions are reduced to create biocompatible nanocomposites (Bahrulolum et al. 2021). A wide range of microorganisms like *Clostridium thermoaceticum*, *Desulfovibrio desulfuricans*, *Klebsiella aerogenes*, cyanobacteria like *Spirulina*, *Nostoc*, *Anabaena*, *Synechocystis*, *Phormidium*, *Lyngbya*, *Oscillatoria* and green algae like *Chlorella*, *Scenedesmus*, *Chlorococcum*, *Cosmarium*, *Chlamydomonas*, *Botryococcus* have been used to synthesize a variety of nano-biofertilizers containing metals like silver, gold, copper, zinc, titanium, palladium, and nickel in an economical and ecofriendly manner (Bairwa et al. 2023). Fungal species like *Aspergillus* and *Fusarium oxysporum* are used for the production of myco-nano biofertilizers (Sonawane et al. 2022). Nano biofertilizers are generally synthesized by physicochemical and biological methods. The synthesis involves three major steps, that is, the growth of the biofertilizer culture, its encapsulation with nanoparticles and the evaluation and assessment of its performance, purity, and shelf-life (Sharma et al. 2023).

1.3.3.2 Soil Application of Nano Biofertilizers

Soil application of nano biofertilizers can promote plant growth and restore soil fertility; however, their activity, mobility, and bioavailability in soil might be severely restricted by heterogeneous aggregation between the nano particles and soil particles. Therefore, foliar spraying and seed treatment with nano biofertilizer are encouraged over soil application. Foliar application allows the nano particles to quickly infiltrate the plant tissues and lower the reactive oxygen species drastically. The various mechanisms through which the nano particles penetrate the plant cells and move intercellularly include endocytosis through cell wall pores, plasmodesmata, aquaporins, and ion transporters (Wu and Li 2022). Nano biofertilizers can also be used in seed priming procedure that involves soaking seeds in nano biofertilizers to accelerate seed germination and seedling production. It stimulates plant growth promoting hormones and helps in upregulating stress-resistance genes (Nile et al. 2022). The synergistic function of the nano particles and biofertilizers serves as the foundation for the mechanism of nano biofertilizer mediated plant growth promotion. It promotes plant growth enhancing nutrient availability and nutrient uptake efficacy of plants, synthesis of phytohormones like auxin, gibberellin, cytokinin and abscisic acid, biocontrol of phytopathogens, alleviation of biotic and abiotic stresses such as salinity, drought, flooding, etc., enhancing photosynthetic activity and increasing photosynthate deposition in plants (Mahapatra et al. 2022).

1.3.4 Production and Quality Control of Biofertilizers

The biofertilizer production process involves the growth and mass multiplication of specific microbial strain under sterile conditions using specific growth media. The culture is then mixed with a suitable carrier material or formulated into a liquid medium. The final product is then packed in airtight containers for distribution. The step wise protocol involves:

1. Selection and isolation of efficient microbial strains
2. Growing it in specific nutrient media
3. Fermentation in bioreactors under optimal conditions
4. Packaging and labelling

Quality control ensures the effectiveness, safety, and shelf life of the product. It is the most important step before making the biofertilizer market ready. The key parameters that need to be checked to maintain qualitative aspects are: (a) Viable cell count: A minimum cell count of 10^7–10^8 CFU/g (solid) or CFU/ml (liquid) is mandatory; (b) Purity testing: There should be no contamination of unwanted microbes up to 10^{-4} dilution; (c) Moisture and pH: Must be regulated within optimal range till the expiry of the product; (d) Shelf life: Stability of the product must be ensured with specifying proper storage conditions; (e) Strain identity: the strain used must be verified using biochemical and/or molecular methods; (f) Standards:

The national/international guidelines must be followed throughout the production process. Table 1.3 enlists the detail of a number of commercial biofertilizers manufactures and used in Africa.

1.4 Differences Between Chemical Fertilizers and Biofertilizers

Current soil and agriculture management practices rely mostly on the continual application of inorganic chemical-based fertilizers which is costly and causes environmental degradation. To minimize the adverse effects of chemicals, the introduction of biofertilizers is a promising arena of nutrient management nowadays. Biofertilizers, developed as a vital aspect of the integrated nutrient management system, have considerable potential to boost agricultural yields through sustainable nutrient supply (Suhag 2016). Applying biofertilizers as seed or soil inoculants causes them to proliferate, take part in the cycling of nutrients, and increase crop productivity (Singh et al. 2011). Biofertilizers, unlike chemical and organic fertilizers, do not directly provide nutrients to crops and instead rely on specialized bacteria and fungi. A healthy plant typically has a rhizosphere rich in helpful microorganisms. Microbial inoculants can boost crop output by reducing fertilizer requirements and increasing nutrient extraction from the soil. Biofertilizer is more effective than chemical fertilizer to induce rice plant growth and yield production (Alam and Seth 2014). Glycoprotein and glomalin substances from mycorrhizal fungi protect the soil structure from erosion and improve soil stability. Also, algal biomass increases soil organic matter and nutrient contents causing enhanced microbial activity. Bacteria also play a significant role in the development of soil aggregates by the release of hydrophobic compounds such as waxes and fats, which link soil particles, resulting in better aggregate stability. Whereas, mineral fertilization deteriorates soil aggregate stability. The harmful effect of chemical fertilizers (decrease in pH) on microorganisms may have restricted their ability to build soil aggregates (Al-Zabee and Al-Maliki 2019). Biofertilizers provide better soil conditions that include improved soil fertility, nitrogen fixation, and phosphate solubilization; enhanced efficacy of applied N and P; enhanced activities of other microbes; as well as the release of growth stimulants. The efficacy of the inorganic fertilizer was pronounced when they were combined with biofertilizers. Balanced C: N ratio, increased nutrient availability, lesser nutrient retention and more translocation to the aerial part for protoplasmic protein synthesis causes more vegetative growth in plants (plant height and branches) (Dhawale et al. 2011).

Inoculation with diazotrophs in okra fixes atmospheric N, enhances phosphate availability, produces growth-stimulating and antifungal compounds, and finally increases the total yield. *Azospirillum* produces plant hormones such as IAA, GA3, and cytokinin, which promote plant development and help in nitrogen fixation. According to Ramakrishnan and Selvakumar (2012), plants treated with *Azotobacter*

Table 1.3 Commercial biofertilizers in African subcontinents

Microbial strain	Tradename	Manufacturer	Function
Arbuscular mycorrhizal fungi (AMF)	Mycoroot green, Mycoroot Super Booster, Mycoroot superGro	Mycoroot (Pty) Ltd., South Africa	P, Zn, Cu, and Fe solubilization
Azospirillum	Rhizostim	Soygro (Ltd) Ltd., South Africa	N-fixation, P solubilization, IAA, and siderophore production
Azospirillum brasilense	Mazospirflo	Soygro (Ltd) Ltd., South Africa	
Bacillus	Composter	BioControl Products SA (Ltd) Ltd.	P and K solubilization, auxin, gibberellin, and cytokinin production
Bacillus	Biostat, Firstbase, Landbac, LifeForce, Waterbac	Microbial Solution (Pty) Ltd., South Africa	
Bacillus subtilis	Bac up	BioControl Products SA (Pty) Ltd.	
Bacillus subtilis	B-RUS, Extrasol	Ag-Chem Africa (Ltd) Ltd., South Africa	
Bervibacillus laterosporus, *Lysinibacillus sphaericus*, *Paenibacillus chitinolytic*, and *Sporolactobacillus laevolacticus*	SoilFix	BioControl Products SA (Ltd) Ltd.	
Bradyrhizobium elkanii	Likuiq Semia	Microbial Solution (Pty) Ltd., South Africa	N-fixation, P solubilization, IAA, and siderophore production
Bradyrhizobia	Nodumax	IITA Business incubation platform, Nigeria	
Bradyrhizobium japonicum	N-Soy	BioControl Products SA (Ltd) Ltd.	
Bradyrhizobium japonicum	Histick	BASF SA (Pty) Ltd., South Africa	
Bradyrhizobium japonicum	Soyflo, Vault N Soy	Soygro (Ltd) Ltd., South Africa	

(continued)

Table 1.3 (continued)

Microbial strain	Tradename	Manufacturer	Function
Rhizobium phaseolus	N-Bean	BioControl Products SA (Ltd) Ltd.	N-fixation, IAA, and siderophore production
Rhizobium	SeeddQuestR	Soygro (Ltd) Ltd., South Africa	
Rhizobia	BIOFIX	MEA Fertilizer Ltd., Kenya	
Sinorhizobium meliloti	Nitrasec Alfaalfa (Leucerne)	Microbial solution (Pty) Ltd., South Africa	N-fixation
Pseudomonas florescens	NAT-P	BioControl Products SA (Ltd) Ltd.	P solubilization, IAA, and siderophore production
Azospirillum lipoferum and *Azospirillum brasilense*	Azo-N	BioControl Products SA (Pty) Ltd.	N-fixation, P solubilization, IAA, and siderophore production
Azospirillum brasilense, *Azotobacter chroococcum*, and *Azospirillum lipoferum*	Azo-N Plus	BioControl Products SA (Pty) Ltd.	N-fixation, P solubilization, IAA, kinetin, and siderophore production
Bacillus, *Enterobacter*, *Pseudomonas*, *Rhizobium*, and *Stenotophomonas*	Organico	Amka Products (Pty) Ltd. South Africa	N-fixation, P and K solubilization, auxin, gibberellin, cytokinin, and siderophore production
Azotobacter chroococcum, *Bacillus thuringiensis*, *Bacillus subtilis*, *Lactobacillus*, and *Pseudomonas florescens*	Soil Vital Q	BioControl Products SA (Pty) Ltd.	N-fixation, P and K solubilization, auxin, gibberellin, kinetin, cytokinin, and siderophore production
Glomus etunicatum, *Glomus intraradices*, and *Glomus mosseae*	Rhizatech	Dudu Tech, Naivasa, Kenya	P, Zn, Cu, Fe solubilization and increases salinity tolerance

Source: Raimi et al. (2017) and Aloo et al. (2021b)

and *Azospirillum* had the highest levels of protein and chlorophyll. Since nitrogen is the main component of proteins, it is necessary for the creation of protoplasm, which causes cells to develop and divide, ultimately leading to enhanced plant growth and fruit yield. The mismatch between the plant's N demand and fertilizer-N supply affects rice growth and yield, resulting in a 50–70% loss due to volatilization, leaching, and other environmental factors. Common applications of chemical fertilizers reduce soil fertility, change soil biology, and raise crop production costs. High usage of these chemical fertilizers also poses environmental hazards like greenhouse gas emissions and nitrate leaching, which endangers human health. So, the introduction of N_2 fixing and P solubilizing bacteria from the Indigenous source is an economical and eco-friendly approach for crop production as it escalates both macro and micronutrient uptake by nutrient mobilization, bioavailability, and decreased adsorption (Naher et al. 2016).

1.4.1 Quantitative Product Optimization by Biofertilizers over Chemical Fertilizers

The combined use of organic manure and chemical fertilizers with biofertilizers is also a promising strategy for maintaining soil microbial communities and activities, which will ultimately have synergistic effects on various soil physicochemical properties and crop production (Li et al. 2017). This is because the balance of soil microbial communities is primarily sensitive to the long-term application of chemical fertilizers (Sun et al. 2015). Adopting efficient microorganisms has the potential to inhibit soil-borne pathogens and increase soil biological fertility by encouraging the growth of beneficial bacteria. Application of azolla biofertilizer in rice cultivation has shown to increase the grain yield by 29.2%, highlighting its efficacy in enhancing crop productivity without the adverse effects associated with chemical fertilizers (Yadav and Sarkar 2019). Integrating biofertilizers with chemical fertilizers has shown to improve maize yield in wheat-maize cropping system by delaying senescence and increasing grain weight along with reducing greenhouse gas emission and water pollution (Ali et al. 2024). Studies indicate that the use of solid biofertilizers and organic ameliorants positively impacts soil biological properties and chickpea yield (Patel et al. 2024). Seed inoculation with *Azotobacter* has reported to increase the uptake of N by 45.29%, P by 45.45%, K by 57.97% and Zn by 46.80% in seeds of *Brassica napus* L. (Prajapati et al. 2024). Inoculation with biofertilizer has led to an average yield increment of 16.2% across various crops. It has enhanced the nitrogen use efficiency by 5.8 kg yield per kg N applied and phosphorus use efficiency by an average of 7.5 kg yield per kg of P applied, with legumes showing the highest improvement at 8.3 kg yield per kg of applied N and about 7.8 kg yield per kg of applied P (Schütz et al. 2018). The perspectives on chemical fertilizers and bio-fertilizers differ significantly in terms of their production, impact on the environment, sustainability, and long-term agricultural health. These differences can be examined from various standpoints, such as scientific, economic, environmental, and even cultural. The specific differences are presented in Table 1.4.

Table 1.4 Difference between chemical fertilizers and biofertilizers

Chemical fertilizers	Bio-fertilizers
Production and composition	
These are manufactured synthetically using industrial processes, typically from petroleum-based products or mined minerals (e.g., nitrogen, phosphorus, potassium)	These are natural substances that contain living microorganisms such as bacteria, fungi, or algae, which help to improve soil fertility by enhancing the natural processes of nutrient cycling
They are highly concentrated and contain specific nutrients like nitrogen (urea), phosphorus (single super phosphate), and potassium (muriate of potash) in precise ratios	Common examples include rhizobium (which fixes nitrogen for legumes), mycorrhizal fungi (which improve phosphorus uptake), and compost-based inoculants
They can be quickly absorbed by plants, providing immediate nutrient availability	Their action is more gradual and sustainable, working overtime to increase soil fertility and health
Environmental impact	
Excessive use can lead to soil degradation, water pollution, and eutrophication	Considered environmentally friendly as they work in harmony with natural processes, enhancing soil structure and microbial diversity
Chemical fertilizers reduce soil biodiversity and damage beneficial organisms, such as earthworms and microbes, disrupting the natural ecosystem	They help in reducing the need for chemical inputs, potentially lowering the risk of environmental pollution
Over-reliance can lead to soil acidification and nutrient imbalances, making the soil less productive in the long term	Carbon footprint is much lower compared to chemical fertilizers since they are derived from natural, renewable sources
Sustainability and soil health	
Tend to deplete soil health over time because they provide a narrow spectrum of nutrients, leaving out trace minerals and affecting soil microorganisms	Promote long-term soil health by enriching soil microbiota, improving organic matter content, and enhancing nutrient cycling
Can lead to dependency on continued chemical applications, as they do not regenerate soil health	They can enhance soil structure and water retention, fostering a more sustainable, resilient farming system. And over time, they can reduce the need for external inputs, as the soil becomes more fertile and self-sustaining
Economic considerations	
They can be expensive and their prices fluctuate based on market conditions, particularly as they are often derived from fossil fuels	Initial costs may be lower (especially if using locally available or homemade bio-fertilizers), but they may require more time and knowledge to implement effectively
They provide immediate results, and for short-term crop yields, they may be more cost-effective	While they may not offer immediate results like chemical fertilizers, over the long term, they can reduce the need for external inputs, potentially making farming more cost-effective and resilient

(continued)

Table 1.4 (continued)

Chemical fertilizers	Bio-fertilizers
Nutrient release and efficiency	
Nutrients in chemical fertilizers are immediately available to plants, which can lead to rapid growth and quick results	Nutrient release is slow and steady, which supports longer-term plant growth. They improve the availability of nutrients that may otherwise be locked in the soil
The nutrients are often leached away if not properly managed, leading to inefficiency and potential environmental harm	Bio-fertilizers help in nutrient cycling and increase the uptake efficiency of plants, even for nutrients that are less available in the soil
Impact on plant growth and yield	
Increase crop yields in the short term by rapidly supplying nutrients, especially when soil fertility is low or depleted	While the initial yield may not be as high as with chemical fertilizers, bio-fertilizers support overall plant health, resilience, and quality over time
They often do not support plant health as comprehensively as bio-fertilizers, potentially leading to disease susceptibility or nutrient imbalances	They can also lead to more sustainable productivity, especially in the long run, due to improved soil structure and microbiome health
Regulatory and public perception	
Widely accepted and used in conventional agriculture	Increasingly popular in organic farming and sustainable agricultural practices
The growing concerns over their health and environmental impacts have led to increased regulation in some areas, as well as rising demand for organic produce	Perceived as a more natural and eco-friendlier alternative, although they may face resistance in conventional systems where quick yields are prioritized
Cultural and agricultural practices	
Common in industrial agriculture due to their ability to rapidly boost yields and maintain productivity at scale	Often used in traditional farming systems, where practices are more focused on ecological balance, sustainable practices, and maintaining soil health
The global shift towards sustainability is challenging the dominance of chemical fertilizers, especially as farming communities begin to recognize the long-term risks	The cultural shift towards organic farming and regenerative practices is further promoting the adoption of bio-fertilizers

1.4.2 Difference in the Impact of Chemical and Biofertilizers on the Life Cycle of Crops

The impact of chemical fertilizers and bio-fertilizers on different stages of a crop's life cycle (from germination to harvest and beyond) varies significantly. Both types of fertilizers influence soil health, plant growth, nutrient uptake, and crop yield, but they do so in different ways, with distinct benefits and challenges. Table 1.5 explores their effects at each stage of the crop life cycle.

Table 1.5 Effect of Chemical and biofertilizer at each stage of the crop life cycle

Chemical fertilizers	Bio-fertilizers
Soil status and germination	
Immediate nutrient availability can help crops establish strong early growth, especially nitrogen, phosphorus, and potassium	*Gradual nutrient release* from bio-fertilizers can help seedlings establish more naturally by providing a balanced nutrient supply
Potential negative effects: High concentrations of chemical fertilizers can harm seedling roots, particularly if the soil nutrient levels are too high or not properly balanced. This can stunt early growth or cause nutrient toxicity	*Improved soil structure*: Microbial activity, such as the action of mycorrhizal fungi and nitrogen-fixing bacteria, enhances the soil's ability to retain moisture and nutrients, promoting better root development
Soil conditions: Overuse of chemical fertilizers can also lead to soil acidification or nutrient imbalances, which may negatively affect the seedling's ability to establish properly	*Beneficial soil microorganisms*: Bio-fertilizers increase soil microbial biodiversity, which can improve root health, seed germination, and early plant resilience
Vegetative growth (leaf and stem development)	
Rapid growth due to the immediate availability of nutrients (especially nitrogen) leads to lush green foliage and quick stem development	*Steady nutrient availability* helps plants develop strong, healthy stems and leaves without the risk of excessive growth
Risk of nutrient imbalances: Overuse of nitrogen can cause excessive vegetative growth at the expense of root or flower development. This might also make the plant more prone to diseases and pest infestations	*Improved soil health*: Bio-fertilizers promote the growth of beneficial soil microorganisms (e.g., rhizobium, mycorrhizal fungi), which can improve nutrient availability and overall soil fertility
Depletion of soil microbes: Continued use of chemical fertilizers can disrupt the soil microbiome, reducing microbial activity and soil health, which may lead to less resilient plants in the long run	*Enhanced drought resistance*: Bio-fertilizers improve the soil's water retention and can enhance plant tolerance to abiotic stresses like drought
Flowering and fruiting (reproductive stage)	
Rapid nutrient availability: can boost flowering and fruit sets, especially with the use of phosphorus and potassium, which are essential for reproductive development	*Balanced nutrient supply*: Bio-fertilizers help improve flowering and fruiting by enhancing nutrient cycling and the availability of micronutrients, which are often critical at this stage
Potential drawbacks: High nitrogen levels might promote excessive vegetative growth rather than fruit, which can lead to poor fruiting and quality issues (e.g., fewer or smaller fruits)	*Improved plant health*: Because bio-fertilizers support soil microbial life, they contribute to better overall plant health, reducing susceptibility to diseases that could affect flowering or fruit set
Soil fertility issues: Long-term reliance on chemical fertilizers can result in nutrient imbalances in the soil, potentially limiting the plant's ability to properly form flowers and fruits	*Slow-release nutrient support*: The slow, steady release of nutrients provided by bio-fertilizers supports continued reproductive success without the risks of over-fertilization

(continued)

Table 1.5 (continued)

Chemical fertilizers	Bio-fertilizers
Maturation and harvest	
Quick maturation: Chemical fertilizers can speed up the process of crop maturation, allowing for quicker harvesting, which can be beneficial in commercial farming where time is crucial	*Gradual nutrient release*: The steady release of nutrients promotes steady maturation and the development of crops with improved flavour, texture, and overall quality.
Potential risks: Over-fertilization, especially with nitrogen, can result in poor fruit or seed quality, delayed ripening, or lodge (falling over) of plants due to excessive vegetative growth	*Improved resilience*: Crops treated with bio-fertilizers are generally more resilient during the maturation phase, with fewer issues related to stress (e.g., drought or pest attacks) leading to better-quality harvests
Nutrient leaching: If applied excessively, chemical fertilizers can lead to nutrient leaching into the soil, reducing the final nutrient content of the harvested produce and potentially contaminating groundwater	*Long-term soil health*: Bio-fertilizers build the soil's fertility over time, leading to a more sustainable harvest and better-quality produce in the long term
Post-harvest and soil health	
Depletion of soil nutrients: Continuous use of chemical fertilizers without organic amendments can lead to soil degradation over time, reducing the ability of the soil to support healthy crops	*Sustainability and soil regeneration*: Bio-fertilizers improve soil organic matter, increase microbial diversity, and support soil structure, which helps maintain long-term fertility
Nutrient imbalances: The application of synthetic fertilizers can lead to soil nutrient imbalances, making the soil less fertile and requiring increasingly higher fertilizer inputs to maintain yields	*Reduced chemical dependency*: The use of bio-fertilizers can decrease the need for synthetic fertilizers, leading to a more sustainable cycle of nutrient cycling in the soil
Environmental pollution: Runoff from excessive fertilizer use can cause eutrophication of nearby water bodies, leading to algal blooms, hypoxia, and damage to aquatic life	*Environmental benefits*: Bio-fertilizers have a lower environmental footprint, promoting healthier ecosystems and reducing the risk of soil erosion, water contamination, and biodiversity loss

1.5 Success Stories of Biofertilizer Applications and Their Outcome in Agriculture

Biofertilizers are proving to be transformative in African agriculture, offering sustainable solutions for improved productivity and soil health. Across the subcontinent, success stories underline the potential of biofertilizers to address the challenges of conventional farming.

- In Malawi, biofertilizers have played a pivotal role in enhancing soybean production. Field trials demonstrated a 50% increase in yields and improved soil nitrogen levels, showcasing the effectiveness of biofertilizers in promoting nutrient cycling (SustInAfrica 2023). Similarly, in Kenya, smallholder farmers adopting biofertilizers reported enhanced soil health and reduced dependency on chemical inputs, leading to more sustainable farming practices (ScienceVolks 2024).

- In East Africa, Kenya and Rwanda have initiated biofertilizer programs with significant success. The Kenya Agricultural and Livestock Research Organization (KALRO) has been instrumental in testing and promoting biofertilizers, leading to improved yields and reduced chemical fertilizer dependency (Vishwakarma et al. 2020). Similarly, research in Ethiopia showed biofertilizers' ability to enhance soil health and crop productivity, while local farmer groups have reported increased yields with reduced environmental impact.
- Rwanda has emerged as a hub for biofertilizer innovation, where cooperatives under the Rwanda Organic Agriculture Movement (ROAM) transitioned to biofertilizer production. This initiative not only reduced synthetic fertilizer costs but also improved crop yields and supported economic empowerment for local communities (Rwanda Organic Agriculture Movement 2024). Such cooperative models highlight the role of community-driven approaches in scaling biofertilizer adoption.
- Innovative technologies also contribute to the success of biofertilizers. For example, biochar-enhanced carriers have improved the viability and stability of microbial inoculants, resulting in better performance in crops like onions (ScienceVolks 2024). These advancements underscore the importance of research and innovation in addressing the limitations of biofertilizers.

The outcomes of these success stories are significant. Increased yields, enhanced soil fertility, and reduced environmental degradation are recurring themes. Furthermore, the economic benefits, including reduced input costs and new revenue streams for farmers, contribute to food security and rural development. However, for broader impact, challenges like low adoption rates and limited awareness must be addressed through targeted education and robust policy frameworks. Therefore, biofertilizers are proving to be a viable and impactful alternative in African agriculture, but scaling these successes will require continued innovation and systemic support.

1.6 Regulatory Framework About Implementation of Policy in Promoting Biofertilizer

1.6.1 Policy Making and Regulatory Aspects of Promoting Biofertilizer

Promoting the use of biofertilizer in agriculture marks a significant shift towards sustainable farming practices, necessitating carefully thought-out regulations and successful implementation plans to guarantee widespread adoption and optimal impact (Raimi et al. 2017). Policy frameworks ought to concentrate on educating farmers on the benefits of biofertilizers, clearing up myths, and demonstrating their efficacy through demonstration plots and field tests, which calls for informational campaigns, farmer training sessions, and agricultural extension services that highlight the financial and ecological advantages of applying biofertilizers (Mącik et al. 2020; Ramasamy et al. 2020). Policies should also support biofertilizer technology research and development, with a focus on identifying and isolating highly effective microbial strains, refining formulation techniques, and developing biofertilizers that are suited to particular crops and climates (Samantaray et al. 2024). It was anticipated that by 2022, the biofertilizer market would have grown to USD 2.3 billion (Basiru et al. 2020). Policy initiatives should focus on bolstering the supply chain from production to distribution in order to further promote biofertilizers. This includes establishing regional production facilities, promoting local entrepreneurship in the biofertilizer manufacturing industry, and developing effective distribution networks that reach outlying rural communities (Saleh et al. 2020). Along with creating strong regulatory frameworks for biofertilizers that address registration, quality control, labelling, and safety standards, policymakers should also prioritize the adoption of integrated nutrient management strategies, which combine the use of biofertilizers with other organic and inorganic nutrient sources to maximize nutrient use efficiency and minimize environmental effects (Ka et al. 2018; Santos et al. 2024).

1.6.2 Policy Implementation Aspects of Promoting Biofertilizer

Government support and incentives (such as direct subsidies for biofertilizer purchases, tax incentives for producers, or favourable credit options for farmers adopting biofertilizer techniques) can significantly encourage the uptake of biofertilizers, especially among smallholder farmers facing financial limitations, where the implementation of quality control standards and certification processes is crucial to ensure the reliability and effectiveness of biofertilizer products, preventing the spread of inferior or contaminated formulas that could undermine farmers' trust (Okur 2018). In order to smoothly transfer technology from research settings to agricultural fields and guarantee that biofertilizer products are available, reasonably priced, and efficient, effective implementation requires a multifaceted strategy that involves cooperation between government agencies, research institutions, private sector organizations, and farmer organizations (Ataci et al. 2022). Additionally, the adoption rates and overall efficacy of biofertilizers can be greatly increased by integrating them into government-supported agricultural projects and programs, such as those that encourage organic farming or manage soil health (Ramasamy et al. 2020).

The potential of biofertilizers to increase agricultural yields, improve soil fertility, and lessen reliance on synthetic fertilizers is highlighted by recent studies. For example, adding chitin promoted the growth of *Bacillus subtilis* AF 1 and effectively inhibited the development of *Aspergillus niger*, a pathogen that causes crown rot in peanuts, and has been shown to promote the growth of plant growth-promoting microorganisms, leading to increased levels of calcium and nitrogen in the soil (Ngasotter et al. 2023). The beneficial effects of arbuscular mycorrhizal fungi found in rhizospheric soils on plant productivity further highlight the necessity of funding studies to find and describe novel microbial strains with exceptional biofertilization skills and to create creative delivery systems that guarantee the viability and efficiency of these microorganisms in the field (Oruru and Njeru 2016).

Thus, promoting the use of biofertilizers in agriculture requires a whole approach that involves support from policymakers, effective implementation strategies, research and innovation, and stakeholder participation. Governments can improve resilient, productive, and sustainable agricultural systems that benefit farmers and the environment by creating an environment that is conducive to the use of biofertilizers (Ahmadbeigi et al. 2021; Ngasotter et al. 2023).

1.7 Challenges and Limitations Associated with the Use of Biofertilizers

Since 2014, biofertilizers have gained prominence in the African agricultural landscape, with promising outcomes but also considerable challenges. The integration of biofertilizers into agricultural practices has been promoted as an eco-friendly, sustainable alternative to chemical fertilizers, helping to address issues of soil degradation, nutrient depletion, and environmental pollution. Despite these successes, the adoption of biofertilizers has faced challenges. Limited awareness among farmers, insufficient infrastructure for distribution, and a lack of tailored solutions for diverse African soils have hindered wider application (Raimi et al. 2017). Additionally, regulatory issues, including inconsistent quality control and market access, have slowed growth in biofertilizer markets across the continent (Stürmer et al. 2020).

While biofertilizers present a promising avenue for sustainable agriculture, their application is not without challenges and limitations (Mahapatra et al. 2022). The efficacy of biofertilizers is highly context-dependent, influenced by factors such as soil type, pH, moisture content, temperature, and the presence of indigenous microbial communities (Malusà et al. 2016; Ramasamy et al. 2020). For instance, the successful establishment and activity of nitrogen-fixing bacteria like *Rhizobium* spp. are significantly affected by soil pH, with acidic soils often inhibiting their growth and nodulation efficiency (Mącik et al. 2020). Similarly, the availability of organic carbon in the soil can limit the activity of heterotrophic microorganisms *Azotobacter*, *Azospirillum*, *Xanthomonas*, *Bacillus*, *Pseudomonas* spp., which play crucial roles in nutrient cycling and disease suppression (Wei et al. 2015).

Another significant challenge lies in the production, storage, and delivery of biofertilizers. Maintaining the viability and activity of microorganisms during these processes is crucial, yet often difficult to achieve (Ahmadbeigi et al. 2021). Many biofertilizer products have short shelf lives and require specific storage conditions to prevent the decline in microbial populations (higher O_2 condition and lesser CO_2). Inoculation methods, timing, and dosage can significantly impact their effectiveness. For example, seed inoculation with *Rhizobium* requires proper adhesion of the bacteria to the seed surface, which can be affected by seed coatings or the presence of other microorganisms (Ka et al. 2018). Temperature, pH, moisture, and nutrient levels are just a few of the environmental variables that interact intricately to determine how well biofertilizers work to improve plant growth and soil health (Cremonez et al. 2021; Banerjee et al. 2022). According to Okur (2018), excessive moisture can produce anaerobic conditions that are detrimental to many helpful microbes, while insufficient moisture might hinder microbial activity by limiting oxygen diffusion and nutrient delivery. The soil's carbon-to-nitrogen (C:N) ratio is particularly crucial since large variations might result in nutrient shortages or insufficient buffering ability, which will restrict microbial activity and growth (Lohani and Havukainen 2017). Furthermore, denitrification a process that affects nitrogen availability is significantly impacted by soil structure, aeration, temperature, and pH, where anaerobic digestion depends on maintaining the proper pH because values above 8 or below 5 can disrupt biological processes (Lohani and Havukainen 2017), and methanogens require optimal conditions within a pH range of 6.5–8.0 (Nsair et al. 2020). The enzymatic activities of biofertilizers, especially those that aid in nitrogen fixation, are greatly impacted by the availability of essential trace elements like molybdenum and cobalt (Mącik et al. 2020).

The efficacy of biofertilizers may be diminished by certain agricultural practices, such as the use of synthetic fertilizers and pesticides, which disrupt microbial communities and reduce the number of beneficial microorganisms that aid in nutrient cycling and pathogen suppression (Liu et al. 2017), negatively affecting the soil microbial diversity and activity (Hatfield and Walthall 2015). In extreme environments, such as arid and semi-arid regions, where the ability of biofertilizers to promote plant growth depends on their tolerance to water stress and salinity, the impact of environmental factors, including temperature, pH, and moisture, on biofertilizer efficiency is especially noticeable (Gougoulias et al. 2014). By increasing water intake, reducing transpiration losses, and inoculating plants with drought-resistant microbial strains can improve their capacity to deal with water constraint in crops like wheat. Some biofertilizers promote root growth (Ramasamy et al. 2020). For example, chitin additions can reduce reliance on synthetic phosphorus and nitrogen fertilizers and increase the growth of beneficial bacteria (Ngasotter et al. 2023). By improving nutrient absorption and encouraging osmotic regulation in plants exposed to saline circumstances, the addition of halotolerant bacteria and fungi to biofertilizers can help lessen the negative impacts of salinity (Oruru and Njeru 2016). Additionally, soil health has declined as a result of the usage of chemical fertilizers and plant protection agents, including reductions in soil humus and organic carbon (Parmar et al. 2020). Furthermore, the lack of standardized quality control measures for biofertilizer products can result in inconsistencies in their composition and efficacy (Bharti and Suryavanshi 2021).

While biofertilizers are cost-effective, eco-friendly, and a renewable source of plant nutrients, we cannot yet claim bio-inoculants as a right alternative to chemical fertilizers, but the scientific understanding of the same will pave the way for its future (Mahanty et al. 2017). Plant growth enhancement by microorganisms like

PGPR is related to their ability to act as biofertilizers by increasing the availability of nutrients (Aloo et al. 2022). Overcoming these limitations will be crucial for realizing the full potential of biofertilizers as a sustainable alternative to chemical fertilizers and promoting environmentally sound agricultural practices (Srivastava 2021). Ultimately, addressing these challenges requires a multi-faceted approach that includes, conducting rigorous research to understand the complex interactions between biofertilizers, plants, and the soil environment; developing robust production and formulation technologies to ensure the viability and efficacy of biofertilizer products; establishing standardized quality control measures to ensure product consistency; and providing farmers with education and training on the proper use and application of biofertilizers.

Looking forward, the future of biofertilizers in Africa depends on overcoming these barriers. With the rising demand for sustainable agricultural practices, there is a growing focus on research, education, and policy support to encourage wider adoption. Investments in local production, better quality control, and targeted farmer education programs will be key to scaling up the use of biofertilizers and ensuring long-term sustainability in African agriculture (Stürmer et al. 2020).

1.8 Future Direction of Study

Future research on biofertilizer technology should concentrate on developing a deeper understanding of the ways in which bacteria function to increase their efficacy in diverse environmental settings (Mącik et al. 2020). To develop more accurate and effective biofertilizer products, it is essential to investigate the complex interactions among biofertilizers, plant microbiomes, and the soil environment, where the functions of microbial communities in nutrient cycling and supporting plant growth can be elucidated through the application of sophisticated metagenomic and meta-transcriptomic approaches (Liu et al. 2017; Nosheen et al. 2021). Future studies should also look for novel microbial strains with enhanced phosphate solubilization, nitrogen fixation, and other plant-growth-promoting properties like enduring environmental problems like drought, salinity, and high temperatures to ensure their dependable performance in agricultural settings (Samantaray et al. 2024). To improve the durability and delivery systems of biofertilizers, formulation technologies such as immobilization and microencapsulation should be studied and it is also crucial to develop standardized techniques for evaluating their efficacy and safety (Santos et al. 2024).

Additionally, to maximize the benefits of biofertilizers for different crops and soil types, research must concentrate on rhizosphere and mycorrhizosphere system optimization with fertilizers intake as well as improving their application techniques through nano and micro formulations (Khan 2006). To create all-encompassing plans for controlling soil fertility, studies should also look into how biofertilizers can be used with other sustainable farming methods like crop rotation and conservation tillage, and improved mixing tactics through the application of computational fluid dynamics, which would ultimately increase process efficiency and lower internal energy consumption (Raimi et al. 2017; Wu et al. 2021). To find better candidates or tactics for inoculation in certain situations, more information about plant microbiota with regard to biotic and abiotic stressors, plant genotype, and environmental variables is essential (Compant et al. 2019). Last but not least, future studies should take into account the socioeconomic aspects of using biofertilizer, such as its availability, cost, and farmer acceptance, particularly in underdeveloped countries and developing nations.

1.9 Conclusion

Biofertilizers are organic alternatives to synthetic chemical fertilizers and help rejuvenate existing soil ecosystems with improved qualitative and quantitative crop production, along with lesser leaching and percolation inside the soil. The biofertilizers improve the nutrient-absorbing quality of the root with enhanced growth of additional root hairs. They organically improve the nutritional quality of grain produced and that helps reduce hunger among the people. The majority of the African farmers are poor and they have a small to marginal amount of land that needs improved productivity per volume against the area of cultivation. The application of synthetic fertilizers causes some unwanted reduction in the quality of the produce impacting the crop as well as the soil health. So, the global scientific community and leading agricultural organizations should implement the use of biofertilizers in crops instead of chemicals for better and more sustainable agricultural operations. Biofertilizers help in improving the soil texture and yield of plants, do not allow pathogens to flourish, are eco-friendly and cost-effective, and also protect the environment from pollutants, since they are natural fertilizers.

References

Abbey L, Abbey J, Leke-Aladekoba A, Iheshiulo EMA, Ijenyo M (2019) Biopesticides and biofertilizers: types, production, benefits, and utilization. In: Byproducts from agriculture and fisheries: adding value for food, feed, pharma, and fuels, vol 20. Wiley, pp 479–500

Adeleke R, Dames JF (2014) Kalaharituber pfeilii and associated bacterial interactions. S Afr J Bot 90:68–73

Adeleke RA, Cloete TE, Bertrand A, Khasa DP (2010) Mobilisation of potassium and phosphorus from iron ore by ectomycorrhizal fungi. World J Microbiol Biotechnol 26:1901–1913

Adesemoye AO, Ugoji EO (2009) Evaluating Pseudomonas aeruginosa as plant growth-promoting rhizobacteria in West Africa. Arch Phytopathol Plant Protect 42(2):188–200

Ahemad M, Kibret M (2014) Mechanisms and applications of plant growth promoting rhizobacteria: current perspective. J King Saud Univ Sci 26(1):1–20

Ahmadbeigi A, Mahmoudi M, Fereidooni L, Akbari M, Kasaeian A (2021) Biodiesel production from waste cooking oil: a review on production methods, recycling models, materials and catalysts. J Therm Eng 10(5):1362–1389

Ahmed HF, El-Araby MM (2012) Evaluation of the influence of nitrogen fixing, phosphate solubilizing and potash mobilizing biofertilizers on growth, yield, and fatty acid constituents of oil in peanut and sunflower. Afr J Biotechnol 11(43):10079–10088

Ahsan ML, Ali A, Ahmed I (2012) Biofertiliser a highly potent alternative to chemical fertilisers: uses and future prospects. J Chem Eng Biol Sci 6(4):10–23

Akintokun AK, Taiwo MO (2016) Comparison of single culture and the consortium of growth-promoting rhizobacteria from three tomato (Lycopersicon esculentum Mill) varieties. Adv Plants Agric Res 5(1):00167

Alam S, Seth RK (2014) Comparative study on effect of chemical and bio-fertilizer on growth, development and yield production of paddy crop (Oryza sativa). Int J Sci Res 3(9):411–414

Alencar TLD, Chaves AF, Santos CLAD, Assis RND, Mota JCA (2015) Atributos físicos de um Cambissolo cultivado e tratado com biofertilizante na Chapada do Apodi, Ceará. Rev Bras Ciênc Solo 39(3):737–749

Ali A, Liu X, Yang W, Li W, Chen J, Qiao Y, Gao Z, Yang Z (2024) Impact of bio-organic fertilizer incorporation on soil nutrients, enzymatic activity, and microbial community in wheat-maize rotation system. Agronomy 14(9):1942

Alkurtany A, Ali S, Mahdi W (2018) The efficiency of prepared biofertilizer from local isolate of Bradyrhizobium sp. on growth and yield of mungbean plant. Iraqi J Agric Sci 49(5):722–730

Almaghrabi OA, Abdelmoneim TS, Albishri HM, Moussa TA (2014) Enhancement of maize growth using some plant growth promoting rhizobacteria (PGPR) under laboratory conditions. Life Sci J 11(11):764–772

Aloo BN, Mbega ER, Makumba BA, Hertel R, Daniel R (2021a) Molecular identification and in vitro plant growth-promoting activities of culturable Potato (Solanum tuberosum L.) rhizobacteria in Tanzania. Potato Res 64:67–95

Aloo BN, Mbega ER, Tumuhairwe JB, Makumba BA (2021b) Advancement and practical applications of rhizobacterial biofertilizers for sustainable crop production in sub-Saharan Africa. Agric Food Secur 10(1):57–68

Aloo BN, Tripathi V, Makumba BA, Mbega ER (2022) Plant growth-promoting rhizobacterial biofertilizers for crop production: the past, present, and future. Front Plant Sci 13:1002448

Al-Zabee MR, Al-Maliki SM (2019) Effect of biofertilizers and chemical fertilizers on soil biological properties and potato yield. Euphrates J Agric Sci 11:1–13

Ataei P, Karimi H, Klöckner CA, Es'haghi SR, Zarei R (2022) The promotion of biofertilizer application on farms: Farmers' intentional processes. Environ Technol Innovation 28:102722

Azeem M, Haider MZ, Javed S, Saleem MH, Alatawi A (2022) Drought stress amelioration in maize (Zea mays L.) by inoculation of Bacillus spp. strains under sterile soil conditions. Agriculture 12(1):50

Babalola OO, Berner DK, Amusa NA (2007) Evaluation of some bacterial isolates as germination stimulants of Striga hermonthica. Afr J Agric Res 2(1):27–30

Bahadur I, Meena VS, Kumar S (2014) Importance and application of potassic biofertilizer in Indian agriculture. Int Res J Biol Sci 3(12):80–85

Bahrulolum H, Nooraei S, Javanshir N, Tarrahimofrad H, Mirbagheri VS, Easton AJ, Ahmadian G (2021) Green synthesis of metal nanoparticles using microorganisms and their application in the agrifood sector. J Nanobiotechnol 19(1):1–26

Bairwa P, Kumar N, Devra V, Abd-Elsalam KA (2023) Nano-biofertilizers synthesis and applications in agroecosystems. Agrochemicals 2(1):118–134

Banerjee S, Prasad N, Selvaraju S (2022) Reactor design for biogas production-a short review. J Energy Power Technol 4(1):1–22

Barnawal D, Bharti N, Pandey SS, Pandey A, Chanotiya CS, Kalra A (2017) Plant growth-promoting rhizobacteria enhance wheat salt and drought stress tolerance by altering endogenous phytohormone levels and TaCTR1/TaDREB2 expression. Physiol Plant 161(4):502–514

Basiru S, Mwanza HP, Hijri M (2020) Analysis of arbuscular mycorrhizal fungal inoculant benchmarks. Microorganisms 9(1):81

Bationo A, Hartemink A, Lungu O, Naimi M, Okoth P, Smaling E, Thiombiano L (2012) African soils: their productivity and profitability of fertilizer use. In: Improving soil fertility recommendation in Africa using decision support system for agrotechnology transfer. Springer, New York, pp 19–42

Bazargani K, Deemyad T (2024) Automation's impact on agriculture: opportunities, challenges, and economic effects. Robotics 13(2):33

Beneduzi A, Ambrosini A, Passaglia LM (2012) Plant growth-promoting rhizobacteria (PGPR): their potential as antagonists and biocontrol agents. Genet Mol Biol 35:1044–1051

Beyranvand H, Farnia A, Nakhjavan SH, Shaban M (2013) Response of yield and yield components of maize (Zea mayz L.) to different bio fertilizers. Int J Adv Biol Biomed Res 1(9):1068–1077

Bharti N, Suryavanshi M (2021) Quality control and regulations of biofertilizers: current scenario and future prospects. In: Biofertilizers. Woodhead Publishing, pp 133–141

Bharti N, Yadav D, Barnawal D, Maji D, Kalra A (2013) Exiguobacterium oxidotolerans, a halotolerant plant growth promoting rhizobacteria, improves yield and content of secondary metabolites in Bacopa monnieri (L.) Pennell under primary and secondary salt stress. World J Microbiol Biotechnol 29:379–387

Bhattacharjee R, Dey U (2014) Biofertilizer, a way towards organic agriculture: a review. Afr J Microbiol Res 8(24):2332–2343

Bot A, Benites J (2005) The importance of soil organic matter: key to drought-resistant soil and sustained food production (No. 80). Food & Agriculture Org

Breedt G, Labuschagne N, Coutinho TA (2017) Seed treatment with selected plant growth-promoting rhizobacteria increases maize yield in the field. Ann Appl Biol 171(2):229–236

Camara O, Heineman E (2006) Overview of the fertilizer situation in Africa: backround paper of the African Fertilizer Summit. June 9–13, 2006, Abuja, Nigeria. Nepad, Pretoria

Chandra R, Collis S (2021) Digital agriculture for small-scale producers: challenges and opportunities. Commun ACM 64(12):75–84

Chaudhary P, Singh S, Chaudhary A, Sharma A, Kumar G (2022) Overview of biofertilizers in crop production and stress management for sustainable agriculture. Front Plant Sci 13:930340

Chen J, Lü S, Zhang Z, Zhao X, Li X, Ning P, Liu M (2018) Environmentally friendly fertilizers: a review of materials used and their effects on the environment. Sci Total Environ 613:829–839

Chianu JN, Nkonya EM, Mairura FS, Chianu JN, Akinnifesi FK (2011) Biological nitrogen fixation and socioeconomic factors for legume production in sub-Saharan Africa: a review. Agron Sustain Dev 31:139–154

Choudhary KK (2010) Cyanobacteria for biofertilizer, bioremediation and bioactive compounds. Algal Biotechnol New Vistas 2010:260–276

Choudhary DK, Varma A, Tuteja N (2017) Plant-microbe interaction: an approach to sustainable agriculture. In: Plant-microbe interaction: an approach to sustainable agriculture. Springer, pp 1–509

Compant S, Samad A, Faist H, Sessitsch A (2019) A review on the plant microbiome: ecology, functions, and emerging trends in microbial application. J Adv Res 19:29–37

Cremonez PA, Teleken JG, Meier TRW, Alves HJ (2021) Two-stage anaerobic digestion in agroindustrial waste treatment: a review. J Environ Manag 281:111854

de Andrade LA, Santos CHB, Frezarin ET, Sales LR, Rigobelo EC (2023) Plant growth-promoting rhizobacteria for sustainable agricultural production. Microorganisms 11(4):1088

Dhawale AB, Warade SD, Bhangre KK (2011) Integrated nutrient management in Bhendi. Asian J Hortic 6(1):145–147

Dinesh Kumar R, Kumaravel R, Gopalsamy J, Sikder MNA, Sampathkumar P (2018) Microalgae as bio-fertilizers for rice growth and seed yield productivity. Waste Biomass Valoriz 9:793–800

Diniz MA, Teixera G, Carrapico F (2015) Azolla as a biofertiliser in Africa. A challenge for the future. Rev Ciênc Agrár 23(3):120–138

Durán-Lara EF, Valderrama A, Marican A (2020) Natural organic compounds for application in organic farming. Agriculture 10(2):41

Eichmann R, Richards L, Schafer P (2021) Hormones as go-betweens in plant microbiome assembly. Plant J 105:518–541

Etesami H, Mirseyed Hosseini H, Alikhani HA (2014) Bacterial biosynthesis of 1-aminocyclopro pane-1-caboxylate (ACC) deaminase, a useful trait to elongation and endophytic colonization of the roots of rice under constant flooded conditions. Physiol Mol Biol Plants 20:425–434

Figueiredo MDVB, Seldin L, de Araujo FF, Mariano RDLR (2011) Plant growth promoting rhizobacteria: fundamentals and applications. In: Plant growth and health promoting bacteria. Springer, pp 21–43

Fu L, Penton CR, Ruan Y, Shen Z, Xue C, Li R, Shen Q (2017) Inducing the rhizosphere microbiome by biofertilizer application to suppress banana Fusarium wilt disease. Soil Biol Biochem 104:39–48

Galindo FS, Pagliari PH, Fernandes GC, Rodrigues WL, Boleta EHM, Jalal A et al (2022) Improving sustainable field-grown wheat production with Azospirillum brasilense under tropical conditions: a potential tool for improving nitrogen management. Front Environ Sci 10:821628

García J, Ortiz A, Álvarez E, Belohlav V, García-Galán MJ, Díez-Montero R, Alvarez JA, Uggetti E (2018) Nutrient removal from agricultural run-off in demonstrative full-scale tubular photobioreactors for microalgae growth. Ecol Eng 120:513–521

García-Fraile P, Menéndez E, Rivas R (2015) Role of bacterial biofertilizers in agriculture and forestry. Aims Bioeng 2(3):183–205

Ghavami N, Alikhani HA, Pourbabaei AA, Besharati H (2017) Effects of two new siderophore-producing rhizobacteria on growth and iron content of maize and canola plants. J Plant Nutr 40(5):736–746

Ghimire AR, Nainawasti A, Shah TB, Dhakal S (2021) Effect of different bio fertilizers on yield of spring rice (Oryza sativa L.) CV. Hardinath-1 in Rajapur Municipality, Bardiya. SAARC J Agric 19(1):57–69

Ghosh D, Gupta A, Mohapatra S (2019) A comparative analysis of exopolysaccharide and phytohormone secretions by four drought-tolerant rhizobacterial strains and their impact on osmotic-stress mitigation in Arabidopsis thaliana. World J Microbiol Biotechnol 35:1–15

Gougoulias C, Clark JM, Shaw LJ (2014) The role of soil microbes in the global carbon cycle: tracking the below-ground microbial processing of plant-derived carbon for manipulating carbon dynamics in agricultural systems. J Sci Food Agric 94(12):2362–2371

Hadide STE, Moussa LA, El-Gharbawy HM (2019) Effect of biofertilizers rates on yield, physicochemical and technological characteristics of wheat. Middle East J Agric Res 8(3):928–936

Hafeez BMKY, Khanif YM, Saleem M (2013) Role of zinc in plant nutrition-a review. Am J Exp Agric 3(2):374–391

Harman G, Khadka R, Doni F, Uphoff N (2020) Benefits to plant health and productivity from enhancing plant microbial symbionts. Front Plant Sci 11:610065

Haro R, Benito B (2019) The role of soil fungi in K+ plant nutrition. Int J Mol Sci 20:3169

Hassan MK, McInroy JA, Kloepper JW (2019) The interactions of rhizodeposits with plant growth-promoting rhizobacteria in the rhizosphere: a review. Agriculture 9(7):142

Hassen AI, Bopape FL, Sanger LK (2016) Microbial inoculants as agents of growth promotion and abiotic stress tolerance in plants. In: Microbial inoculants in sustainable agricultural productivity: vol. 1: research perspectives. Springer, New Delhi, pp 23–36

Hatfield JL, Walthall CL (2015) Soil biological fertility: foundation for the next revolution in agriculture? Commun Soil Sci Plant Anal 46(6):753

Htwe AZ, Moh SM, Moe K, Yamakawa T (2019) Biofertilizer production for agronomic application and evaluation of its symbiotic effectiveness in soybeans. Agronomy 9:162

Hussain A, Arshad M, Zahir ZA, Asghar M (2015) Prospects of zinc solubilizing bacteria for enhancing growth of maize. Pak J Agric Sci 52(4):915–922

Igiehon NO, Babalola OO (2018) Rhizosphere microbiome modulators: contributions of nitrogen fixing bacteria towards sustainable agriculture [Review of rhizosphere microbiome modulators:

contributions of nitrogen fixing bacteria towards sustainable agriculture]. Int J Environ Res Public Health 15(4):574

Ilyas N, Mumtaz K, Akhtar N, Yasmin H, Sayyed RZ, Khan W, Enshasy HAE, Daili DJ, Elsayed EA, Ali Z (2020) Exopolysaccharides producing bacteria for the amelioration of drought stress in wheat. Sustainability 12(21):8876

Issifu M, Songoro EK, Onguso J, Ateka EM, Ngumi VW (2022) Potential of Pseudarthrobacter chlorophenolicus BF2P4-5 as a biofertilizer for the growth promotion of tomato plants. Bacteria 1(4):191–206

Jain P, Khichi DS (2014) Phosphate solubilizing microorganism (PSM): an eco-friendly biofertilizer and pollution manager. J Dyn Agric Res 1(4):23–28

Jing X, Cui Q, Li X, Yin J, Ravichandran V, Pan D et al (2020) Engineering Pseudomonas protegens Pf-5 to improve its antifungal activity and nitrogen fixation. Microb Biotechnol 13:118–133

Ka M, Kala S, Wati L (2018) Biofertilizers: potential candidate for sustainable agriculture. Int J Curr Microbiol App Sci 7(3):2839

Kaiser DE, Rosen CJ, Lamb JA (2016) Potassium for crop production. University of Minnesota Extension. FO-6794-D

Kawalekar JS (2013) Role of biofertilizers and biopesticides for sustainable agriculture. J Bio Innov 2(3):73–78

Khan AG (2006) Mycorrhizoremediation—an enhanced form of phytoremediation. J Zhejiang Univ Sci B 7:503–514

Khosro M, Yousef S (2012) Bacterial biofertilizers for sustainable crop production: a review. J Agric Biol Sci 7(5):307–316

Kumakech A, Turyagyenda FL (2024) Effect of BioNiK-Phos and Phosphorus-based fertilizers on yield of hybrid sunflower in Northern Uganda. Braz J Sci 3(4):48–54

Kumar D, Kalita P (2017) Reducing postharvest losses during storage of grain crops to strengthen food security in developing countries. Foods 6(1):8

Kumar M, Kumar K (2019) Role of Bio-fertilizers in vegetables production: a review. J Pharmacogn Phytother 8(1):328–334

Kumar S, Kumar S, Mohapatra T (2021) Interaction between macro- and micro-nutrients in plants. Front Plant Sci 12:665583

Kumari B, Mallick M, Hora A (2016) Plant growth-promoting rhizobacteria (PGPR): their potential for development of sustainable agriculture. In: Bioexploitation for sustainable agriculture. Springer, Singapore, pp 1–19

Lahiff E, Cousins B (2005) Smallholder agriculture and land reform in South Africa. IDS Bull 36(2):127–131

Li F, Chen L, Zhang J, Yin J, Huang S (2017) Bacterial community structure after long-term organic and inorganic fertilization reveals important associations between soil nutrients and specific taxa involved in nutrient transformations. Front Microbiol 8:187

Lichtfouse E, Navarrete M, Debaeke P, Souchère V, Alberola C, Ménassieu J (2009) Agronomy for sustainable agriculture: a review. Sustain Agric:1–7

Liu C, Sakimoto KK, Colón B, Silver PA, Nocera DG (2017) Ambient nitrogen reduction cycle using a hybrid inorganic–biological system. Proc Natl Acad Sci 114(25):6450

Lohani SP, Havukainen J (2017) Anaerobic digestion: factors affecting anaerobic digestion process. In: Waste bioremediation. Springer Singapore, Singapore, pp 343–359

Mącik M, Gryta A, Frąc M (2020) Biofertilizers in agriculture: an overview on concepts, strategies and effects on soil microorganisms. Adv Agron 162:31–87

Mahanty T, Bhattacharjee S, Goswami M, Bhattacharyya P, Das B, Ghosh A, Tribedi P (2017) Biofertilizers: a potential approach for sustainable agriculture development. Environ Sci Pollut Res 24:3315–3335

Mahapatra DM, Satapathy KC, Panda B (2022) Biofertilizers and nanofertilizers for sustainable agriculture: phycoprospects and challenges. Sci Total Environ 803:149990

Mahdi SS, Hassan GI, Samoon SA, Rather HA, Dar SA, Zehra B (2010) Bio-fertilizers in organic agriculture. J Phytol 2(10):42–54

Majumdar K (2015) Bio-fertilizer use in Indian agriculture. Indian J Res 4(6):377–381

Malusà E, Ciesielska J (2014) Biofertilizers: a resource for sustainable plant nutrition. Fertil Technol 1(1):282–319

Malusà E, Pinzari F, Canfora L (2016) Efficacy of biofertilizers: challenges to improve crop production. In: Microbial inoculants in sustainable agricultural productivity: vol. 2: functional applications. Springer, New Delhi, pp 17–40

Masso C, Baijukya F, Ebanyat P, Bouaziz S, Wendt J, Bekunda M, Vanlauwe B (2017) Dilemma of nitrogen management for future food security in sub-Saharan Africa–a review. Soil Res 55(6):425–434

Mazid M, Khan TA (2015) Future of bio-fertilizers in Indian agriculture: an overview. Int J Agric Food Res 3(3):10–23

Miransari M (2011) Interactions between arbuscular mycorrhizal fungi and soil bacteria. Appl Microbiol Biotechnol 89:917–930

Mishra P, Dash D (2014) Rejuvenation of biofertilizer for sustainable agriculture and economic development. Consilience 11:41–61

Mishra D, Rajvir S, Mishra U, Kumar SS (2013) Role of bio-fertilizer in organic agriculture: a review. Res J Recent Sci ISSN 2277:2502

Mujawar MI (2014) Bacteria and fungi can contribute to nutrients. J King Saud Univ 26(1):1–20

N'Cho C, Lesueur D, Yusuf AA (2015) Combined microbial inoculations as a promising approach to enhance promiscuous soybean nodulation and nitrogen content in Sudan Savana. Int J Sustain Agric Res 2(3):86–97

Nacoon S, Jogloy S, Riddech N, Mongkolthanaruk W, Kuyper T, Boonlue S (2020) Interaction between phosphate solubilizing bacteria and arbuscular mycorrhizal fungi on growth promotion and tuber inulin content of Helianthus tuberosus L. Sci Rep 10:4916

Nagy VD, Zhumakayev AR, Vörös M, Bordé Á, Szarvas A, Szűcs A, Kocsubé S, Jakab P, Monostori T, Škrbić B, Mohai E, Hatvani L, Vágvölgyi C, Kredics L (2023) Development of a multicomponent microbiological soil inoculant and its performance in sweet potato cultivation. Microorganisms 11(4):914

Naher UA, Panhwar QA, Othman R, Ismail MR, Berahim Z (2016) Biofertilizer as a supplement of chemical fertilizer for yield maximization of rice. J Agric Food Dev 2:16–22

Naserirad H, Soleymanifard A, Naseri R (2011) Effect of integrated application of bio-fertilizer on grain yield, yield components and associated traits of maize cultivars. Am Eurasian J Agric Environ Sci 10(2):271–277

Ngasotter S, Xavier KM, Meitei MM, Waikhom D, Pathak J, Singh SK (2023) Crustacean shell waste derived chitin and chitin nanomaterials for application in agriculture, food, and health–a review. Carbohydr Polym Technol Appl 6:100349

Nile SH, Thiruvengadam M, Wang Y, Samynathan R, Shariati MA, Rebezov M, Kai G (2022) Nano-priming as emerging seed priming technology for sustainable agriculture recent developments and future perspectives. J Nanobiotechnol 20(1):1–31

Niu X, Song L, Xiao Y, Ge W (2018) Drought-tolerant plant growth-promoting rhizobacteria associated with foxtail millet in a semi-arid agroecosystem and their potential in alleviating drought stress. Front Microbiol 8:2580

Nosheen S, Ajmal I, Song Y (2021) Microbes as biofertilizers, a potential approach for sustainable crop production. Sustainability 13(4):1868

Nsair A, Onen Cinar S, Alassali A, Abu Qdais H, Kuchta K (2020) Operational parameters of biogas plants: a review and evaluation study. Energies 13(15):3761

Okur N (2018) A review-bio-fertilizers-power of beneficial microorganisms in soils. Biomed J Sci Tech Res 4(4):4028–4029

Oruru MB, Njeru EM (2016) Upscaling arbuscular mycorrhizal symbiosis and related agroecosystems services in smallholder farming systems. Biomed Res Int 2016(1):4376240

Parmar T, Baweja HS, Dilta BS, Baweja PK (2020) Impact of environmental factors and mulching on growth, quality and yield of carnation (Dianthus caryophyllus L.) cv. 'Loris'. IJCS 8(5):1889–1882

Patel N, Patel Y, Mankad A (2014) Bio fertilizer: a promising tool for sustainable farming. Int J Innov Res Sci Eng Technol 3(9):56–69

Patel KK, Pandey AK, Baheliya AK, Singh V, Yadav R, Pandey S, Adesh, Yadav V (2024) A comparative study of the effects of biofertilizers and chemical fertilizers on soil physical and biological properties under chickpea crop (Cicer arietinum L.). Int J Environ Clim Chang 14(3):72–80

Prajapati V, Mankotia BS, Rana BB, Sharma A, Dogra P, Sharma S, Verma A, Sharma J, Meena D (2024) Effect of biofertilizers at different levels on nutrient content and uptake by Gobhi sarson (Brassica napus L.) under Himalayan region. J Exp Agric Int 46(12):214–221

Raimi A, Adeleke R, Roopnarain A (2017) Soil fertility challenges and Biofertiliser as a viable alternative for increasing smallholder farmer crop productivity in sub-Saharan Africa. Cogent Food Agric 3(1):1400933

Rajawat KS, Ameta KD, Kaushik RA, Dubey RB, Jain HK, Jain D, Kaushik MK (2019) Effect of integrated nutrient management on growth attributes and soil nutrient status of tomato under naturally ventilated polyhouse. Int J Curr Microbiol App Sci 8(10):512–517

Raji SG, Tzanakakis V, Dörsch P (2019) Bradyrhizobial inoculation and P application effects on haricot and mung beans in the Ethiopian Rift Valley. Plant Soil 442:271–284

Rajkumar M, Ae N, Prasad MNV, Freitas H (2010) Potential of siderophore-producing bacteria for improving heavy metal phytoextraction. Trends Biotechnol 28(3):142–149

Ramakrishnan K, Selvakumar G (2012) Effect of biofertilizers on enhancement of growth and yield on Tomato (Lycopersicum esculentum Mill.). Int J Res Bot 2(4):20–23

Ramasamy M, Geetha T, Yuvaraj M (2020) Role of biofertilizers in plant growth and soil health. IntechOpen. https://doi.org/10.5772/intechopen.87429

Rose MT, Phuong TL, Nhan DK, Cong PT, Hien NT, Kennedy IR (2014) Up to 52% N fertilizer replaced by biofertilizer in lowland rice via farmer participatory research. Agron Sustain Dev 34:857–868

Rwanda Organic Agriculture Movement (2024) KHEA multiplier journey from training to profitable biofertilizer innovation. https://khea-africa.org

Saleh S, Anshary A, Made U, Mahfudz M, Basir-Cyio M (2020) Application of mycorrhizae and beauveria in organic farming system effectively control leafminers and enhance shallot production. AGRIVITA J Agric Sci 43(1):79–88

Salomon MV, Bottini R, de Souza Filho GA, Cohen AC, Moreno D, Gil M, Piccoli P (2014) Bacteria isolated from roots and rhizosphere of Vitis vinifera retard water losses, induce abscisic acid accumulation and synthesis of defense-related terpenes in in vitro cultured grapevine. Physiol Plant 151(4):359–374

Samantaray A, Chattaraj S, Mitra D, Ganguly A, Kumar R, Gaur A, Mohapatra PKD, de los Santos-Villalobos S, Rani A, Thatoi H (2024) Advances in microbial based bio-inoculum for amelioration of soil health and sustainable crop production. Curr Res Microb Sci 7:100251

Santos F, Melkani S, Oliveira-Paiva C, Bini D, Pavuluri K, Gatiboni L et al (2024) Biofertilizer use in the United States: definition, regulation, and prospects. Appl Microbiol Biotechnol 108(1):1–16

Santosh S, Velmourougane K, Idapuganti RG, Manikandan A, Blaise D (2022) Potassium solubilizing potential of native bacterial isolates from cotton rhizosphere of rainfed vertisols. Natl Acad Sci Lett 45(3):209–212

Savci S (2012) An agricultural pollutant: chemical fertilizer. Int J Environ Sci Dev 3(1):73

Schütz L, Gattinger A, Meier M, Müller A, Boller T, Mäder P, Mathimaran N (2018) Improving crop yield and nutrient use efficiency via biofertilization—a global meta-analysis. Front Plant Sci 8:2204

ScienceVolks (2024) Challenges and opportunities in biofertilizer commercialization. https://sciencevolks.com

Shaheen AM, Fatma AR, Omiama MS, Ghoname AA (2007) The integrated use of bio-inculants and chemical nitrogen fertilizer on growth, yield and nutritive value of two okra (Abelmoschus esculentus, L.). cultivars. Aust J Basic Appl Sci 1(3):307–312

Sharma B, Yadav L, Pandey M, Shrestha J (2022) Application of biofertilizers in crop production: a review. Peruvian J Agron 6(1):13–31

Sharma B, Tiwari S, Kumawat KC, Cardinale M (2023) Nano-biofertilizers as bio-emerging strategies for sustainable agriculture development: potentiality and their limitations. Sci Total Environ 860:160476

Shridhar BS (2012) Nitrogen fixing microorganisms. Int J Microbiol Res 3(1):46–52

Singh G, Sekhon HS, Sharma P (2011) Effect of irrigation and biofertilizer on water use, nodulation, growth and yield of chickpea (Cicer arietinum L.). Arch Agron Soil Sci 57(7):715–726

Singh DP, Singh HB, Prabha R (eds) (2016) Microbial inoculants in sustainable agricultural productivity, vol 2. Springer, New Delhi, p 308

Sonawane H, Shelke D, Chambhare M, Dixit N, Math S, Sen S, Borah SN, Islam NF, Joshi SJ, Yousaf B, Rinklebe J (2022) Fungi-derived agriculturally important nanoparticles and their application in crop stress management-Prospects and environmental risks. Environ Res 212:113543

Srivastava RK (2021) Biofertilizers application in agriculture: a viable option to chemical fertilizers. Biofertil Study Impact 13:393–411

Statistics South Africa (2017) Poverty trends in South Africa an examination of absolute poverty between 2006 and 2015. Statistics South Africa, Government Printers, Pretoria

Stürmer SL et al (2020) Advancements in bio-based fertilizers. Springer

Suhag M (2016) The potential of biofertilizers to replace chemical fertilizers. Int Adv Res J Sci Eng Technol 3(5):163–167

Sun R, Zhang XX, Guo X, Wang D, Chu H (2015) Bacterial diversity in soils subjected to long-term chemical fertilization can be more stably maintained with the addition of livestock manure than wheat straw. Soil Biol Biochem 88:9–18

SustInAfrica (2023) Scaling off-farm production of organic and biofertilizers in Africa. https://sustinafrica.com

Suyal DC, Soni R, Sai S, Goel R (2016) Microbial inoculants as biofertilizer. In: Microbial inoculants in sustainable agricultural productivity: vol. 1: research perspectives. Springer, New Delhi, pp 311–318

Tahir HA, Gu Q, Wu H, Raza W, Hanif A, Wu L et al (2017) Plant growth promotion by volatile organic compounds produced by Bacillus subtilis SYST2. Front Microbiol 8:171

Tairo EV, Ndakidemi PA (2014) Macronutrients uptake in soybean as affected by Bradyrhizobium japonicum inoculation and phosphorus (P) supplements. Am J Plant Sci 5(4):1–9

Timmusk S, Abd El-Daim IA, Copolovici L, Tanilas T, Kännaste A, Behers L et al (2014) Drought-tolerance of wheat improved by rhizosphere bacteria from harsh environments: enhanced biomass production and reduced emissions of stress volatiles. PLoS One 9(5):e96086

Tripathi S, Srivastava P, Devi RS, Bhadouria R (2020) Influence of synthetic fertilizers and pesticides on soil health and soil microbiology. In: Agrochemicals detection, treatment and remediation. Butterworth-Heinemann, pp 25–54

Trivedi P, Leach JE, Tringe SG, Sa T, Singh BK (2020) Plant–microbiome interactions: from community assembly to plant health. Nat Rev Microbiol 18(11):607–621

Vacheron J, Desbrosses G, Bouffaud ML, Touraine B, Moënne-Loccoz Y, Muller D et al (2013) Plant growth-promoting rhizobacteria and root system functioning. Front Plant Sci 4:356

Velázquez E, Rojas M, Lorite MJ, Rivas R, Zurdo-Piñeiro JL, Heydrich M, Bedmar EJ (2008) Genetic diversity of endophytic bacteria which could be found in the apoplastic sap of the medullary parenchyma of the stem of healthy sugarcane plants. J Basic Microbiol 48(2):118–124

Vessey JK (2003) Plant growth promoting rhizobacteria as biofertilizers. Plant Soil 255:571–586

Vishwakarma K, Kumar N, Shandilya C, Mohapatra S, Bhayana S, Varma A (2020) Revisiting plant–microbe interactions and microbial consortia application for enhancing sustainable agriculture: a review. Front Microbiol 11:560406

Vurukonda SSKP, Vardharajula S, Shrivastava M, SkZ A (2016) Enhancement of drought stress tolerance in crops by plant growth promoting rhizobacteria. Microbiol Res 184:13–24

Wagner SC (2012) Biological nitrogen fixation. Nat Educ Knowl 3(10):15

Walling E, Vaneeckhaute C (2020) Greenhouse gas emissions from inorganic and organic fertilizer production and use: a review of emission factors and their variability. J Environ Manag 276:111211

Wei H, Chen X, Xiao G, Guenet B, Vicca S, Shen W (2015) Are variations in heterotrophic soil respiration related to changes in substrate availability and microbial biomass carbon in the subtropical forests? Sci Rep 5(1):18370

Wu H, Li Z (2022) Nano-enabled agriculture: how do nanoparticles cross barriers in plants? Plant Commun 3(6):100346

Wu D, Peng X, Li L, Yang P, Peng Y, Liu H, Wang X (2021) Commercial biogas plants: review on operational parameters and guide for performance optimization. Fuel 303:121282

Yadav KK, Sarkar S (2019) Biofertilizers, impact on soil fertility and crop productivity under sustainable agriculture. Environ Ecol 37(1):89–93

Youseif SH, Abd El-Megeed FH, Saleh SA (2017) Improvement of faba bean yield using Rhizobium/Agrobacterium inoculant in low-fertility sandy soil. Agronomy 7(1):2

Chapter 2
Biofertilizers: A Green Approach to Enhance Crop Fertility

Bharat Bhushan Rana, Vinod Sharma, Deekshith H N, Chinmaya Sahoo, and Ankit Saini

Abstract The rising global population has intensified the need for sustainable agricultural practices to ensure food security without degrading the environment. Chemical fertilizers, while effective in enhancing crop productivity, have contributed to soil degradation, water pollution, and ecological imbalance. Biofertilizers are the formulations containing live or dormant microorganisms emerged as eco-friendly alternatives that enhance nutrient availability and improve soil health. These beneficial microbes, including nitrogen-fixers (e.g., *Rhizobium*, *Azospirillum*, *Azotobacter*), phosphate and potassium solubilizing bacteria, mycorrhizal fungi, and plant growth-promoting rhizobacteria (PGPR), play crucial roles in nutrient cycling, stress tolerance, and crop productivity. Symbiotic and free-living nitrogen-fixers significantly contribute to atmospheric nitrogen conversion into plant-usable forms, while phosphate and potassium solubilizers mobilize essential nutrients locked in soil. Additionally, biofertilizers improve plant resilience to environmental stresses such as drought and salinity, promote root growth, and increase crop yields by 10–50%. Biofertilizers also offer a sustainable solution to micronutrient deficiencies like sulfur and zinc, improving plant nutrition and reducing dependence on synthetic inputs. Their use aligns with the principles of organic farming and climate-smart agriculture, reducing greenhouse gas emissions and conserving nonrenewable resources such as phosphate rock. Scientific evidence supports the efficacy of biofertilizers in enhancing nitrogen and phosphorus use efficiency, improving crop yield, and maintaining long-term soil fertility. This chapter highlights the various

B. B. Rana · V. Sharma · D. H. N
Chaudhary Sarwan Kumar Himachal Pradesh Krishi Vishvavidyalaya, Palampur, Himachal Pradesh, India

C. Sahoo
Dr. Rajendra Prasad Central Agricultural University, Pusa, Bihar, India

A. Saini (✉)
Chaudhary Sarwan Kumar Himachal Pradesh Krishi Vishvavidyalaya, Palampur, Himachal Pradesh, India

Dr. KSG Akal College of Agriculture, Eternal University, Sirmaur, Himachal Pradesh, India

A. D. Nciizah et al. (eds.), *Biofertilizers for Sustainable Agriculture in Africa*, Sustainability Sciences in Asia and Africa, https://doi.org/10.1007/978-981-95-6009-7_2

types of biofertilizers, their mechanisms of action, and their vital role in sustainable agriculture as a green alternative to chemical fertilizers, ultimately contributing to food security and environmental sustainability.

Keywords Biofertilizers · BNF · Yield · Environment · Soil fertility

2.1 Introduction

The global demand for food is escalating due to a rapidly growing population, necessitating increased agricultural productivity without compromising environmental sustainability. According to FAO estimates, agricultural product consumption is expected to increase by 60% by 2030 (Mia and Shamsuddin 2010). Improving productivity while ensuring environmental safety is a significant challenge in the twenty-first century (Berg 2009). Farmers are exploring various crop nutrition tactics to address the increasing food demand. Fertilizers have been widely used to enhance agricultural yields from cultivable land. Chemical fertilizers can make nations independent of food but harm the environment and life. Chemical fertilizers are expensive, degrade soil quality, reduce water retention and fertility, upset nutrient balance, and pollute water (Sprent and Sprent 1990). In this context, biofertilizers are microbial inoculants that promote plant growth by increasing availability of nutrients and emerged as an eco-friendly and sustainable alternatives to chemical fertilizers. Biofertilizers maintain soil structure and crop diversity and are environmentally friendly, economically feasible, safe, and easy to use. They are good alternatives to chemical fertilizers (Deepali and Gangwar 2010; Thomas and Singh 2019).

The adoption of biofertilizers is particularly relevant in the era of climate smart agriculture and organic farming, where the goals include maximizing productivity while minimizing ecological footprints. They have increased plant growth and yield by 10–40% (Kawalekar 2013). The use of biofertilizers can be a potential approach to enhance soil microbial activity that stimulates the native soil microbiota therefore improving nutrient accessibility and promotes the decomposition of organic matter (Yadav et al. 2021). When applied to the plant material, crop surface, or soil, these bioinoculants colonize the rhizosphere and the plant core, promoting plant development (Raghuwanshi 2012). The ability of biofertilizers to promote a high-level microbial diversity in soil can contribute to improved crop productivity for sustainable agriculture practices. Supplementing soil with nutrients and safeguarding plants from pests and diseases boosts agricultural output.

2.2 Biofertilizers

Biofertilizers are formulations containing live or latent microbial strains that, when applied to seeds, plant surfaces, or soil, enhance the availability and uptake of essential nutrients, thereby improving soil fertility and crop productivity. These beneficial microorganisms, including nitrogen-fixing bacteria (*Rhizobium*, *Azotobacter*, *Azospirillum*), phosphate-solubilizing bacteria (PSB), potassium-solubilizing bacteria (KSB), and mycorrhizal fungi, play a pivotal role in nutrient cycling and sustainable crop nutrition (Kumawat et al. 2017). Nitrogen-fixing biofertilizers convert atmospheric nitrogen into ammonia, making it accessible to plants, while PSB release organic acids and phosphatases to mobilize insoluble phosphates in the soil (Janati et al. 2021). Biological fertilizers contain microorganisms. When applied to plant surfaces, they increase thc primary nutrient availability and plant growth. Plant-based fertilizers increase nutrient availability by fixing nitrogen, solubilizing phosphorus, and producing growth-regulating chemicals. Biofertilizer is inherently viable and can form a mutually advantageous relationship with plant roots. Microorganisms can efficiently transform complex organic materials into simpler materials for easier absorption by plants. This preserves the natural soil habitat. Development increases by 20–30%, while artificial nitrogen and phosphorus substitutes are reduced by 25%, thereby promoting plant development. It can protect against drought and certain soilborne diseases. Biofertilizers are formulations of beneficial microorganisms that, when applied to seeds, roots, or soil, promote the availability and functionality of these microorganisms, thereby improving soil health (Bhattacharjee and Dey 2014). Biofertilizers can be applied to seed, root, soil, or as foliar spray to enhance the microbial activity through their multiplication, which then facilitates the nutrient mobilization to the target plants. This process significantly improved the soil fertility and subsequently enhanced the crop health and productivity (Chakraborty and Akhtar 2021).

2.3 Types of Biofertilizers

According to Alengebawy et al. (2023), global biofertilizers are generally classified into two main types based on their source and raw materials: organic residue-based biofertilizers and microorganism-based biofertilizers. Organic residue-based types include materials such as green manure, crop residues, farmyard manure, and treated sewage sludge. In contrast, microorganism-based biofertilizers, as noted by Adetunji et al. (2023), contain beneficial microbes such as bacteria, fungi, and algae. Biofertilizers influence plant growth either directly or indirectly. Direct mechanisms act on plants by facilitating processes such as nitrogen fixation, phosphate solubilization, micronutrient solubilization, and phytohormone production (Rawat et al. 2021). Indirect mechanisms support plant health by protecting them against harmful pathogens through the release of lytic enzymes, antibiotics, siderophores, and the production of cyanide. Various types of biofertilizers are classified based on microorganisms such as bacteria and fungi and function of the biofertilizers.

2.4 Nitrogen Fixing Biofertilizers

Plants require nitrogen for shoot system growth, reproduction, chlorophyll production, and grain size (Sandhu et al. 2021). Plants cannot use dinitrogen, which has three bonds and accounts for 78% of the atmosphere. Although atmospheric nitrogen (78%) is abundant, it exists as dinitrogen (N_2) with a triple bond, making it unavailable to plants (Ohyama 2010). Diazotrophs convert N_2 into ammonia (NH_3) through biological nitrogen fixation, which is then transformed into nitrite and nitrate by ammonia-oxidizing and nitrifying bacteria, respectively (Sun et al. 2021; Rana et al. 2025). Unused nitrate undergoes denitrification in deeper soil, returning to the atmosphere as N_2 gas, completing the nitrogen cycle. Nitrogen-fixing microbes such as *Azotobacter* and *Bacillus* promote maize and forest crop growth, while *Bradyrhizobium japonicum* enhances biomass, nodulation, and N fixation in soybean (Meena et al. 2018). *Azotobacter chroococcum* boosts plant height and chlorophyll in maize, and *Bradyrhizobium* spp. aid mung bean yield by fixing nitrogen and producing IAA and siderophores (Mahmood et al. 2021). These microbes are classified as symbiotic, free-living, or associative nitrogen fixers based on their plant interactions (Fig. 2.1).

Diazotrophs must first convert dinitrogen into a dissolved, beneficial form of ammonia through a mechanism known as nitrogen fixation. Bacteria oxidize ammonia to nitrite and nitrate (Roy et al. 2020). Denitrification in deeper soil layers

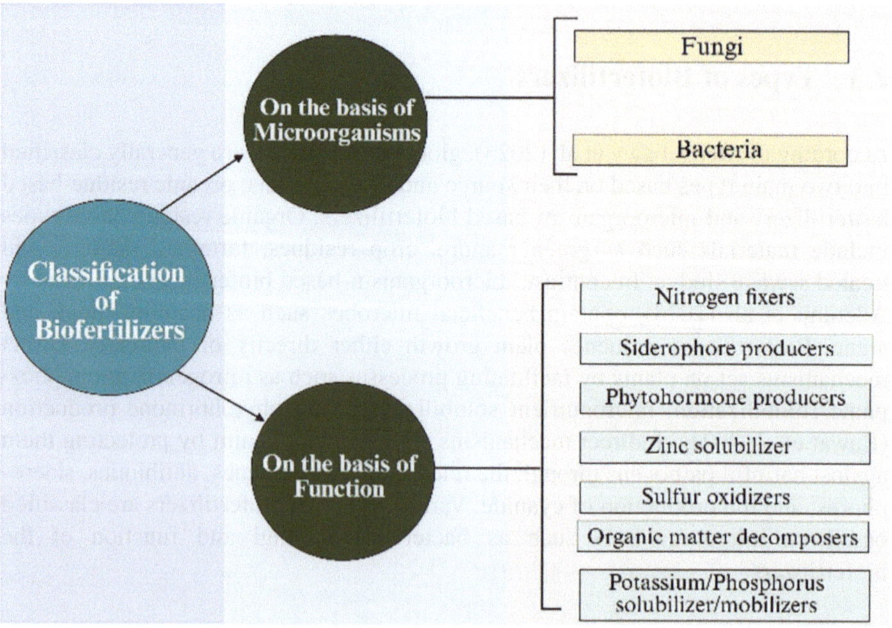

Fig 2.1 Classification of biofertilizers

converts excess nitrate into atmospheric nitrogen, which is emitted as dinitrogen gas. This is a typical nitrogen cycle (Mahanty et al. 2017).

A specific consortium of microorganisms is necessary to assimilate nitrogen and render it accessible to plants. These are biological nitrogen-fixers. According to Reed et al. (2011), this enables physiological access to neutral N_2. Crop production has increased by 10–50%, and 300–400 kg N/ha is produced annually by nitrogen fixation. Nitrogen fixation in plants accounts for 25% of the total nitrogen. Chemicals generated by the roots of crops help in soil bacterial development and nitrogen fixation. They reduce environmental chemical pollution by replacing synthetic fertilizers to varying degrees.

2.5 Symbiotic Nitrogen-Fixing Microorganisms

The plant is a macrosymbiont, whereas prokaryotic bacteria are microsymbionts. One of the best-studied mutualistic relationships between plant root nodules and nitrogen-fixing bacteria is that of *Rhizobium* legumes. The symbiotic association between *Rhizobium* and legumes is among the most extensively studied mutualistic relationships, involving nitrogen-fixing microorganisms within plant root nodules (Clúa et al. 2018). Mutualistic relationships begin when the plant releases flavonoids and isoflavonoids into the rhizosphere, which are detected and recognized by *Rhizobium*.

2.5.1 Rhizobium

Rhizobium is an endosymbiotic nitrogen-fixing bacterium that exists freely in the soil and possesses the ability to fix atmospheric nitrogen in a symbiotic relationship. The root nodules function as small fermenters for organic nitrogen stabilization, enabling the conversion of atmospheric nitrogen into ammonia. The infection process begins with the differentiation of root hairs, followed by the formation of an infection thread that extends into the root hair cell. This thread then releases the bacteria into the cytoplasmic region. Subsequently, the bacterial cells undergo terminal differentiation into bacteroids, which further develop into symbiosomes—the specialized structures where nitrogen fixation occurs (Tsyganova et al. 2017; Parwan et al. 2025). Its association with leguminous hosts is highly specific, enabling it to fix nitrogen exclusively in particular plant species. *Rhizobium* variants increase growth and yield parameters compared to the control, according to Datta et al. (2015). The bacterium Rh increased crop height, the development of seeds, leaf chlorophyll, and nitrogen (Sara et al. 2013). Rice grain yield increased by 8% to 22% and straw production increased by 4% to 19% when rice seeds were inoculated with different *Rhizobium* strains at incremental nitrogen levels (Sammauria et al. 2020).

2.5.2 *Azolla-Anabaena*

Azolla is a small, free-floating aquatic fern renowned for its rapid growth and its symbiotic relationship with the nitrogen fixing cyanobacterium *Anabaena azollae*. This symbiosis enables Azolla to fix atmospheric nitrogen, making it a valuable biofertilizer, particularly in rice cultivation systems (Yadav et al. 2014; Bocchi and Malgioglio 2010). Rice receives 20–40 kg of nitrogen/ha from one Azolla crop in 20–25 days (Setiawati et al. 2018).

2.5.3 *Cyanobacteria*

Nitrogen-fixing Cyanobacteria fix nitrogen worldwide. Nostoc, Anabaena, Oscillatoria, Aulosira, and Lyngbya are cyanobacteria (Sharma et al. 2011). Cyanobacterial biofertilizers are applicable across several biomes and settings, including terrestrial, rainforest, and desert ecosystems (Chittora et al. 2020; Bothe et al. 2010). They sequester 20 to 30 kilogram N per ha in wet rice fields and boost crop output by 10–15% at 10 kg per ha. Cyanobacteria in rice fields increased plant nitrogen availability (Singh et al. 2016a, b; Mishra and Pabbi 2004). In another study, infecting rice with field-grown cyanobacteria improved soil and plant health (Hasan 2020). Strains could release bioactive compounds and boost plant yield. Another study found that infecting rice with field-grown cyanobacteria improved soil and plant health.

2.5.4 *Free-Living Nitrogen Fixing*

2.5.4.1 *Azotobacter*

It is an aerobic, heterotrophic, nitrogen-fixing bacteria commonly found in the rhizosphere of various crop plants, particularly in neutral to alkaline soils. Unlike symbiotic diazotrophs, *Azotobacter* can fix atmospheric nitrogen independently, without requiring a host plant. Among the species, *Azotobacter chroococcum* is the most extensively studied for agricultural applications. These bacteria possess nitrogenase enzymes that catalyze the reduction of atmospheric nitrogen (N_2) to ammonia (NH_3), thus enriching the soil nitrogen pool (Wani et al. 2016a, b). *Azotobacter* can synthesize vitamins such as thiamine and riboflavin (Revillas et al. 2000). *Azotobacter* inoculation increased rapeseed and mustard seed output, according to Dutta and Singh (2002).

Azotobacter chroococcum serves as an effective biofertilizer due to its ability to fix up to 10 mg N per gram of carbon source under in vitro conditions (Chaudhary et al. 2022). It produces plant growth-promoting substances such as indole acetic acid, gibberellic acid, naphthalene acetic acid, and B-complex vitamins. Additionally, *Azotobacter* suppresses root pathogens, stimulates root development, enhances mineral uptake, and contributes to improved soil fertility.

2.5.5 Associative Nitrogen-Fixing Bacteria

2.5.5.1 *Azospirillum*

It is a genus of free-living, microaerophilic, nitrogen-fixing bacteria predominantly found in the rhizosphere of various crop plants. Among its species, *Azospirillum brasilense* and *Azospirillum lipoferum* are extensively used for their plant growth-promoting capabilities. These microbe fixes a lot of soil nitrogen, making it vital. It fixes 20–40 kg N/ha in the rhizosphere for various crops (Isawa et al. 2010; Choudhary et al. 2024). *Azospirillum* is a gram-negative, non-nodulating, aerobic, associative nitrogen-fixing bacterium that forms associations with plants possessing a C_4 dicarboxylic pathway of photosynthesis, such as sugarcane, sorghum, maize, bajra, as well as C_3 cereals like wheat, rice, barley (Chaudhary et al. 2022).

2.5.6 Phosphate Solubilizing and Mobilizing Biofertilizers

Plants need phosphorus, which makes up 0.2% of its dry mass, to grow. Plants can carry phosphorus the least under diverse soil conditions. Various bacterial and fungal species engage in the phosphate solubilization process (Antoun 2012). They also support chelation, organic acid production, and ion exchange processes. PSB makes phosphate cation plant-soluble by chelating it with hydroxyl (gluconic) and carboxyl (ketogluconic) groups (Itelima et al. 2018). Microorganisms employ both the organic method and the proton-extrusion process to solubilize phosphate (Park et al. 2009; Patel and Goswami 2020). Phosphorus-mobilizing bacteria can mobilize the immobile forms of phosphorus (Suthar et al. 2017). Arbuscular mycorrhiza infiltrate roots, increase circumference, stimulate metabolism, and consume minerals. Some phosphate-solubilizing bacteria are also called mobilizers (Chang and Yang 2009). According to Asoegwu et al. (2020), under the right circumstances, they may solubilize or mobilize approximately 30 to 50 kg/ha of P_2O_5, which can boost crop yield by 10% to 20%. Phosphate-solubilizing bacteria (PSB) and fungi (PSF) include genera such as *Bacillus*, *Rhizobium*, *Aerobacter*, *Burkholderia*, *Aspergillus*, and *Penicillium*. For instance, inoculation with *Alcaligenes* sp. has been shown to enhance plant growth through phosphorus solubilization and indole-3-acetic acid (IAA) production. Similarly, *Rhizobium leguminosarum* and *Pseudomonas moraviensis* improved wheat growth and yield by exhibiting both IAA production and phosphorus solubilization activity. Phosphorus in soils often exists in forms unavailable to plants, including soluble phosphorus bound in inaccessible complexes and organic phosphorus compounds requiring mineralization. Phosphorus-mobilizing microorganisms (PMB), such as *Bacillus*, *Pseudomonas*, and *Rhizobium*, play a key role in making these forms available. Arbuscular mycorrhizal fungi (AMF) offer the added advantage of transporting both inorganic and organic forms of phosphorus to host plants. Common AMF genera include *Glomus*, *Acaulospora*, *Entrophospora*, and *Paraglomus*, while typical ectomycorrhizal fungi include *Laccaria*, *Amanita*, and *Boletus* species. Additionally, certain fungal endophytes, such as *Serendipita*, can increase potassium content in crops like maize and confer tolerance to salinity stress (Souza 2015).

2.6 Potassium Solubilizing and Mobilizing Biofertilizers

After phosphorus and nitrogen, potassium (K) represents the second vital nutrient in plants. Using the potassium-solubilizing bacterium *Bacillus pseudomycoides*, tea plants cultivated in mica waste-treated land absorbed more potassium (Pramanik et al. 2019). *Aspergillus niger* as well as *Penicillium niger* can solubilize along with mobilize potassium from various inorganic and organic sources (Xiafang and Weiyi 2002).

2.6.1 Sulfur-Oxidizing Biofertilizers

The micronutrient sulfur (S) is needed for plants. Soil microbiological and physical properties are enhanced with sulfur. It is commonly known that sulfur regulates soil pH. Sulfur increases plant micronutrient uptake and phosphate and nitrogen fertilizers (El-Halfawi et al. 2010). Sulfur-oxidizing *Thiobacillus* bacteria are the most important microorganisms. Sulfur-oxidizing bacteria like *Thiobacillus* sp. can convert sulfur into crop-available sulfates, improving the health of plants (Riaz et al. 2020; Vidyalakshmi et al. 2009). Reduced sulfur compounds pollute heavily. Sulfur-oxidizing bacteria reduce sulfur pollution, protecting the environment (Pokorna and Zabranska 2015). K in soil predominantly exists in immobilized forms within silicate minerals such as illite, orthoclase, biotite, and feldspar. Its solubilization is facilitated by both bacteria and fungi, with acidification being the primary mechanism for converting unavailable forms of K into plant-available forms. Additional mechanisms include siderophore production, exchange reactions, and complexation. Examples of potassium-solubilizing bacteria include *Bacillus mucilaginosus*, *B. circulans*, *B. edaphicus*, *Acidithiobacillus ferrooxidans*, *Frateuria aurantia*, *Herbaspirillum* spp., and *Clostridium* spp., while potassium-solubilizing fungi include *Aspergillus* spp. (Das and Pradhan 2016).

2.6.2 Zinc-Solubilizing Biofertilizers

Plant tissues require little zinc for development and reproduction. Zinc deficiency in the soil is prevalent, resulting from inadequate fertilizer application, intensive agricultural practices, and compromised soil health. Wheat with zinc deficiency causes chlorosis and reduces its development. Wheat with zinc deficiency can also cause zinc deficiency (Kamran et al. 2017). Zinc deficiency is the sixth leading cause of death in underdeveloped countries. Thus, agricultural Zn shortages must be addressed to address other issues (Graham 2008; Kumar et al. 2019; Hussain et al. 2018). Many rhizobacterial taxa, including *Pseudomonas*, *Bacillus*, Mycorrhiza, and *Saccharomyces*, increase soil Zn availability. Zinc-solubilizing biofertilizers contain bacteria capable of producing organic acids that convert insoluble zinc into the plant-available Zn^{2+} form, thereby improving zinc uptake. Representative zinc-solubilizing bacteria and fungi include *Bacillus subtilis*, *Serratia*, *Pseudomonas striata*, *Burkholderia cenocepacia*, *Aspergillus niger*, *A. nomius*, and *A. oryzae*. These microorganisms have been reported to enhance soil enzyme activities and increase zinc availability in crops (Haroon et al. 2022).

2.6.3 Plant Growth-Promoting Rhizobacteria

PGPR are diverse group of beneficial soil bacteria that colonize the rhizosphere and enhance plant growth through various mechanism. They infiltrate roots and stimulate growth (Beneduzi et al. 2012). PGPRs are soil-inhabiting bacteria that enhance the growth and improve the tolerance of plants toward stress conditions (Majeed et al. 2018). Plant growth-promoting rhizobacteria (PGPR) enhance plant development through multiple mechanisms, including nitrogen fixation, macro- and micronutrient mineralization, phytohormone production, exopolysaccharide secretion, and the synthesis of siderophores, hydrogen cyanide, and antibiotics that suppress phytopathogens. For example, *Rhizobium lupini* has been shown to promote alfalfa growth and improve nutrient uptake efficiency (Steinberga et al. 2008) (Table 2.1).

2.7 Role of Biofertilizers in Crop Production

1. *A sustainable alternative to chemical fertilizers*
   Biofertilizers fix nitrogen, solubilize or mineralize phosphate and potassium, release plant growth regulators, produce antibiotics, and biodegrade organic matter to maintain nutrient-rich soils (Bhardwaj et al. 2014). Biofertilizers made from intentionally raised beneficial microbes can progressively restore soil bio-

Table 2.1 Plant growth-promoting rhizobacteria (PGPR) and their beneficial effects on crop performance

PGPR species/combination	Reported crop effects	References
Rhizobium leguminosarum	Stimulates growth in rapeseed and lettuce	Noel et al. (1996)
Pseudomonas putida	Promotes tomato development; enhances early rapeseed growth	Liu et al. (2019) and Glick et al. (1997)
Azospirillum brasilense, A. irakense	Improves growth of cereals such as maize and wheat	Hungria et al. (2010) and Coelho et al. (2022)
Pseudomonas fluorescens	Enhances growth, nutritional status, and productivity in pearl millet; improves vigor in banana leaves	Jogaiah et al. (2010), Babu et al. (2015), and Rajendran and Saravanan (2015)
Azotobacter spp., *Azospirillum* spp.	Increases canola yield	Yasari and Patwardhan (2007)
Pseudomonas alcaligenes, Mycobacterium phlei, Bacillus polymyxa	Boosts N, P, and K uptake in maize	Egamberdiyeva (2007)
Pseudomonas, Azotobacter, Azospirillum spp.	Promotes chickpea growth and yield	Rokhzadi et al. (2011)
Rhizobium leguminosarum, Pseudomonas spp.	Improves P uptake and wheat yield	Li et al. (2022) and De Freitas and Germida (1992)
P. putida, P. fluorescens, A. brasilense, A. lipoferum	Increases maize germination, growth, and productivity	Coelho et al. (2022)
P. putida, P. fluorescens, A. lipoferum, A. brasilense (greenhouse maize)	Enhances germination, development, and yield under controlled conditions	Bashan et al. (2004) and Cassán et al. (2014)

logical activity impaired by chemical fertilizer abuse. Biofertilizers reduce ecological disruption. Biofertilizers can replace environmentally harmful synthetic fertilizers. Sustainable agriculture considers biofertilizers an alternative to toxic chemical fertilizers (Nosheen et al. 2021). Air nitrogen is delivered to plants by biofertilizers. The nitrogen cycle showed that nitrogen-fixing bacteria are essential for soil nitrogen fixation. Nitrogen fertilizers increase NO_2 emissions. This makes biofertilizers a global warming mitigation tool (Naher et al. 2015).

2. *Improvement of Plant Resilience to Environmental Stress*
   Biofertilizers have been shown to increase the ability of plants to withstand environmental issues. Inorganic and microbiological variables decrease agricultural yields. According to Ansari et al., inoculation of plants with arbuscular fungi boosts their growth under saline stress. (2013). Nitrogen-fixing bacteria and AM fungus work together to help drought-tolerant legume plants. *Pseudomonas* sp. enhances the antioxidant and photosynthetic pigment levels in basal plants during periods of water constraint, and in salty and drought conditions, rice plants were inoculated with arbuscular mycorrhiza to investigate their impact on photosynthetic efficiency and antioxidant response (Ruiz-Sánchez et al. 2010).

Table 2.2 Nutrient contribution and crop associations of major biofertilizers

Biofertilizer type	Nitrogen fixed/ phosphorus mobilized	Commonly associated crops	References
Blue-green algae (BGA)	~25 kg N/ha/ year	Rice, banana	Venkataraman and Neelakantan (1967) and Singh et al. (2017)
Azolla	~900 kg N/ha/ year	Rice	Lumpkin and Plucknett (1982) and Watanabe et al. (1992)
Phosphate-solubilizing bacteria & fungi	Mobilize 50–60% of soil-bound phosphorus	All crop types	Rodríguez and Fraga (1999) and Sharma et al. (2013)
Rhizobium	50–300 kg N/ha/ year	Black gram, green gram, lentil, Bengal gram, cowpea, fodder legumes	Graham and Vance (2003) and Dakora and Keya (1997)
Azotobacter	0.026–20 kg N/ ha/year	Vegetables, plantation crops, cotton, rice, wheat, barley, ragi, mustard, safflower, sunflower, niger, tobacco, fruits, spices, condiments, ornamental flowers	Wani et al. (2016a, b) and Mrkovacki and Milic (2001)
Azospirillum	10–20 kg N/ha/ year	Vegetables, sugarcane, maize, rice, wheat, pearl millet, fodder crops, oilseeds	Okon and Labandera-Gonzalez (1994) and Bashan and de Bashan (2010)

3. *Improvement of Soil Fertility*
   Crop biofertilizers captured the soil nutrients (Table 2.2). Combining N fertilizer and *A. brasilense* boosted grain output by 29%. Observations of *Rhizobium* variants have revealed plant-microbe interactions and how many factors affect plant development and crop output. These studies showed that biofertilizers and rhizobial secondary metabolites boost grain crop growth and yield.
4. *Enhancement of Nitrogen and Phosphorus Use Efficiency*
   Soil phosphorus increased the efficiency of AMF, phosphorus solubilizers, and nitrogen fixers. Nitrogen and phosphorus are essential yet limiting nutrients for living. Most soils in agriculture contain both inorganic as well as organic phosphorus, but plants cannot access most of it. Due to the ease of phosphorus immobilization, its low solubility, and limited accessibility to plants, vegetation may still lack sufficient phosphorus despite the application of phosphorus fertilizers to the soil. Primary phosphorus source phosphate ore is a not renewable resource, and its extraction and trade contribute to world consumption of energy, which is environmentally harmful and inefficient. Rock phosphate is degrading rapidly, especially over cadmium pollution, as economical and excellent sources disappear.

2.8 Research Supporting the Use of Biofertilizers in Crop Production

The most well-known technique for inoculating legumes with rhizobia is the use of microbial inoculants. It has significantly transformed legume cultivation techniques, diminishing the requirement for nitrogen fertilizer to achieve good production. In Brazil, Argentina, and the US, BNF and legume inoculation agents with rhizobia and chemical compounds like lipo-chito oligosaccharides (LCOs) increased crop yields. To promote plant growth, *Azospirillum* bacteria synthesize phytohormones, particularly Indole-3-acetic acid (IAA) (Halpern et al. 2015). In clay and sandy Cerrado soils, Ferreira et al. (2013) examined maize's reaction towards *Azospirillum brasilense* infection and nutritional treatments. Maize infected with *Azospirillum* brasilense responded significantly; however, soil type affected it. When combined with A. brasilense, N fertilizer increased grain production by 29%. Marks et al. (2013) successfully inoculated maize and soybean fields with *Azospirillum brasilense* and *Bradyrhizobium* sp., respectively, using rhizobial signaling molecules. Research indicates that the combination of rhizobial secondary metabolites with biofertilizers enhances the growth and yield of grain crops. Kaur and Reddy (2014) found two P-solubilizing isolates of bacteria in various agroclimatic regions that are essential for enhancing plant development and improving soil fertility.

2.9 Challenges in the Use of Biofertilizers

Biofertilizers' shelf life is limited by carrier shortages. According to Singh et al. (2016a, b), fermentation-induced mutation is one of the main drawbacks of biofertilizers. The primary obstacles to the application and advancement of biofertilizers are their uniformity, low quality, high levels of contamination, unsuitable formulations, and dependability in field settings. Biofertilizers that exhibit efficacy in laboratory and greenhouse settings may not demonstrate the same performance in field conditions due to their susceptibility to biotic and abiotic stressors. Crops are cultivated under diverse conditions, encompassing temperature, precipitation, soil composition, biodiversity, and crop variety. The discrepancies lead to variable efficacy of biofertilizers. Furthermore, biofertilizers take longer to work than synthetic fertilizers since the inoculum needs time to colonize the root and build up its concentration (Roy 2021). Isolates that work in different crops, soils, and climates must be chosen to address these issues (Basu et al. 2021). Instead of replacing regular fertilizers, biofertilizers should be used with them to reduce application (Roy 2021). Biofertilizer shelf life is crucial throughout research and marketing. Biofertilizers' live bacterial cells last 6 months at 20–25 °C, necessitating careful storage and shipping and increasing expenses (Mitter et al. 2021). Product licensing and requests for patents are limited. No clear legal definition of "plant biostimulant" or "plant biofertilizer" hinders global coordination. Biofertilizer registration is daunting, time-consuming, and complicated in most countries (Basu et al. 2021).

2.10 Conclusion

In conclusion, biofertilizers improve soil health and crop output while providing a natural, eco-friendly alternative to chemical fertilizers. Due to population expansion, sustainable agriculture techniques are needed, and biofertilizers improve soil fertility, plant growth, and environmental impact. Nitrogen fixation, phosphorus solubilization, potassium mobilization, and plant growth regulator production enable plants to flourish in nutrient-deficient soils under drought and salt stress. Biofertilizers restore the soil biological activity impaired by synthetic fertilizers by using beneficial microbes, lowering chemical consumption, and minimizing soil erosion, water contamination, and greenhouse gas emissions. Many studies have demonstrated that biofertilizers increase crop output by 10–40%. Biofertilizers have several benefits; however, their widespread implementation is difficult. Their short shelf life, high susceptibility to environmental stress, uneven performance across climates and soil types, and complicated commercial production and registration are the major obstacles. The efficiency of biofertilizers depends on the environmental circumstances and takes longer to work than conventional fertilizers. Despite these challenges, biofertilizers can improve agricultural sustainability by reducing the ecological footprint of farming and improving soil health. Biotechnology research and advances may solve many of these challenges, making biofertilizers more feasible and available globally. Biofertilizers can improve food security, environmental resilience, and agricultural sustainability.

References

Adetunji CO, Akram M, Adetuyi BO, Khan FS, Rashid A, Anwar H et al (2023) Biofertilizer derived from cyanobacterial: recent advances. Next-Generation Algae 1:295–320

Alengebawy A, Ran Y, Ghimire N, Osman AI, Ai P (2023) Rice straw for energy and value-added products in China: a review. Environ Chem Lett 21(5):2729–2760

Ansari MW, Trivedi DK, Sahoo RK, Gill SS, Tuteja N (2013) A critical review on fungi mediated plant responses with special emphasis to *Piriformospora indica* on improved production and protection of crops. Plant Physiol Biochem 70:403–410

Antoun H (2012) Beneficial microorganisms for the sustainable use of phosphates in agriculture. Procedia Eng 46:62–67

Asoegwu CR, Awuchi CG, Nelson KCT, Orji CG, Nwosu OU, Egbufor UC, Awuchi CG (2020) A review on the role of biofertilizers in reducing soil pollution and increasing soil nutrients. Himal J Agric 1(1):34–38

Babu S, Jogaiah S, Ito S, Nagaraj AK, Tran LSP (2015) Improvement of growth, fruit yield, and nutrition of tomato plants by rhizosphere bacteria and their formulations. Biocatal Agric Biotechnol 4(4):409–415

Bashan Y, de Bashan LE (2010) How the plant growth-promoting bacterium *Azospirillum* promotes plant growth—a critical assessment and advancements. Adv Agron 108:77–136

Bashan Y, Holguin G, de Bashan LE (2004) Azospirillum–plant relationships: physiological, molecular, agricultural, and environmental advances (1997–2003). Can J Microbiol 50(8):521–577

Basu A, Prasad P, Das SN, Kalam S, Sayyed RZ, Reddy MS, El Enshasy H (2021) Plant growth promoting rhizobacteria (PGPR) as green bioinoculants: recent developments, constraints, and prospects. Sustainability 13(3):1140

Beneduzi A, Ambrosini A, Passaglia LM (2012) Plant growth-promoting rhizobacteria (PGPR): their potential as antagonists and biocontrol agents. Genet Mol Biol 35:1044–1051

Berg G (2009) Plant–microbe interactions promoting plant growth and health: perspectives for controlled use of microorganisms in agriculture. Appl Microbiol Biotechnol 84(1):11–18

Bhardwaj D, Ansari MW, Sahoo RK, Tuteja N (2014) Biofertilizers function as key player in sustainable agriculture by improving soil fertility, plant tolerance and crop productivity. Microb Cell Fact 13:1–10

Bhattacharjee R, Dey U (2014) Biofertilizer, a way towards organic agriculture: a review. Afr. J. Microbiol. Res. 8(24):2332–2343

Bocchi S, Malgioglio A (2010) Azolla-Anabaena as a biofertilizer for rice paddy fields in the Po Valley, a temperate rice area in Northern Italy. Int J Agron 2010(1):152158

Bothe H, Schmitz O, Yates MG, Newton WE (2010) Nitrogen fixation and hydrogen metabolism in cyanobacteria. Microbiol Mol Biol Rev 74(4):529–551

Cassán F, Vanderleyden J, Spaepen S (2014) Physiological and agronomical aspects of phytohormone production by model plant-growth-promoting rhizobacteria (*Azospirillum* spp.). Plant Growth Regul 73(2):207–220

Chakraborty T, Akhtar N (2021) Biofertilizers: characteristic features and applications. In: Biofertilizers: study and impact, pp 429–489

Chang CH, Yang SS (2009) Thermo-tolerant phosphate-solubilizing microbes for multi-functional biofertilizer preparation. Bioresour Technol 100(4):1648–1658

Chaudhary P, Singh S, Chaudhary A, Sharma A, Kumar G (2022) Overview of biofertilizers in crop production and stress management for sustainable agriculture. Front Plant Sci 13:930340

Chittora D, Meena M, Barupal T, Swapnil P, Sharma K (2020) Cyanobacteria as a source of biofertilizers for sustainable agriculture. Biochem Biophys Rep 22:100737

Choudhary R, Kumar R, Rana BB, Choudhary M, Gurjar SN, Choudhary J, Choudhary K (2024) Effects of cow-based preparations and mulching on weed management and nodulation in chickpea under intercropping system. J Exp Agric Int 46(11):571–579

Clúa J, Roda C, Zanetti ME, Blanco FA (2018) Compatibility between legumes and rhizobia for the establishment of a successful nitrogen-fixing symbiosis. Genes 9(3):125

Coelho MR, Marriel IE, Jenkins SN, Lacerda LA, Moreira FM (2022) Long-term impact of *Azospirillum brasilense* inoculation on wheat productivity. Field Crops Res 276:108380

Dakora FD, Keya SO (1997) Contribution of legume nitrogen fixation to sustainable agriculture in Sub-Saharan Africa. Soil Biol Biochem 29(5–6):809–817

Das I, Pradhan M (2016) Potassium-solubilizing microorganisms and their role in enhancing soil fertility and health. In: Potassium solubilizing microorganisms for sustainable agriculture. Springer, New Delhi, pp 281–291

Datta A, Singh RK, Tabassum S (2015) Isolation, characterization and growth of *Rhizobium* strains under optimum conditions for effective biofertilizer production. Int. J. Pharm. Sci. Rev. Res 32(1):199–208

De Freitas JR, Germida JJ (1992) Growth promotion of winter wheat by fluorescent pseudomonads under growth chamber conditions. Soil Biol Biochem 24(12):1137–1146

Deepali, G. K., & Gangwar, K. (2010). Biofertilizers: an ecofriendly way to replace chemical fertilizers. Available online. Accessed on 7 Feb 2021

Dutta RK, & Agrawal M (2002) Effect of tree plantations on the soil characteristics and microbial activity of coal mine spoil land. Tropical ecology, 43(2):315–324

Egamberdiyeva D (2007) The effect of plant growth promoting bacteria on growth and nutrient uptake of maize in two different soils. Appl Soil Ecol 36(2–3):184–189

El-Halfawi MH, Ibrahim SA, Kandil H, Niculită M, Rusu C (2010) Influence of elemental sulphur, organic matter, sulfur oxidizing bacteria and cabronite alone or in combination on cowpea plants and the used soil. Factori °i Procese Pedogenetice din Zona Temperatã 9 S. nouã, 13–29.

Ferreira AS, Pires RR, Rabelo PG, Oliveira RC, Luz JMQ, Brito CH (2013) Implications of *Azospirillum brasilense i*noculation and nutrient addition on maize in soils of the Brazilian Cerrado under greenhouse and field conditions. Appl Soil Ecol 72:103–108

Glick BR, Karaturovic DM, Newell PC (1997) A novel procedure for rapid isolation of plant growth-promoting rhizobacteria. Can J Microbiol 43(9):896–901

Graham RD (2008) Micronutrient deficiencies in crops and their global significance. In: Micronutrient deficiencies in global crop production. Springer, Dordrecht, pp 41–61

Graham PH, Vance CP (2003) Legumes: importance and constraints to greater use. Plant Physiol 131(3):872–877

Halpern M, Bar-Tal A, Ofek M, Minz D, Muller T, Yermiyahu U (2015) The use of biostimulants for enhancing nutrient uptake. Adv Agron 130:141–174

Haroon M, Khan ST, Malik A (2022) Zinc-solubilizing bacteria: an option to increase zinc uptake by plants. In: Microbial biofertilizers and micronutrient availability: the role of zinc in agriculture and human health, pp 207–238

Hasan MA (2020) Cyanobacteria from rice field and comparative study of their performances as biofertilizer on rice plants. J. Glob. Biosci 9:8078–8087

Hungria M, Campo RJ, Souza EM, Pedrosa FO (2010) Inoculation with selected strains of *Azospirillum brasilense* and *A. lipoferum* improves yields of maize and wheat in Brazil. Plant Soil 331:413–425

Hussain A, Zahir ZA, Asghar HN, Ahmad M, Jamil M, Naveed M, Zaman Akhtar MFU (2018) Zinc solubilizing bacteria for zinc biofortification in cereals: a step toward sustainable nutritional security. In: Role of rhizospheric microbes in soil: volume 2: nutrient management and crop improvement, pp 203–227

Isawa T, Yasuda M, Awazaki H, Minamisawa K, Shinozaki S, Nakashita H (2010) *Azospirillum* sp. strain B510 enhances rice growth and yield. Microbes Environ 25(1):58–61

Itelima JU, Bang WJ, Onyimba IA, Sila MD, Egbere OJ (2018) Bio-fertilizers as key player in enhancing soil fertility and crop productivity: a review. Direct Res. J. Agric Food Sci 6(3):73–83

Janati W, Benmrid B, Elhaissoufi W, Zeroual Y, Nasielski J, Bargaz A (2021) Will phosphate biosolubilization stimulate biological nitrogen fixation in grain legumes? Front Agron 3:637196

Jogaiah S, Kumar PK, Shetty HS (2010) Evaluation of plant growth-promoting rhizobacteria for their efficacy in pearl millet against downy mildew disease. Crop Prot 29(6):564–572

Kamran S, Shahid I, Baig DN, Rizwan M, Malik KA, & Mehnaz S (2017) Contribution of zinc solubilizing bacteria in growth promotion and zinc content of wheat. Frontiers in microbiology, 8:2593

Kaur G, Reddy MS (2014) Influence of P-solubilizing bacteria on crop yield and soil fertility at multilocational sites. Eur J Soil Biol 61:35–40

Kawalekar JS (2013) Role of biofertilizers and biopesticides for sustainable agriculture. J Bio Innov 2(3):73–78

Kumar A, Dewangan S, Lawate P, Bahadur I, Prajapati S (2019) Zinc-solubilizing bacteria: a boon for sustainable agriculture. In: Plant growth promoting rhizobacteria for sustainable stress management: volume 1: rhizobacteria in abiotic stress management, pp 139–155

Kumawat N, Kumar R, Kumar S, Meena VS (2017) Nutrient solubilizing microbes (NSMs): its role in sustainable crop production. In: Agriculturally important microbes for sustainable agriculture: volume 2: applications in crop production and protection, pp 25–61

Li H, Jiang X, Ma Y, Wang G, Dong J (2022) Phosphate-solubilizing bacteria enhance phosphorus uptake and wheat growth under phosphorus-deficient conditions. Appl Soil Ecol 169:104226

Liu H, Carvalhais LC, Crawford M, Singh E, Dennis PG, Pieterse CM, Schenk PM (2019) Inner plant values: diversity, colonization and benefits from endophytic bacteria. Front Microbiol 10:271

Lumpkin, T. A., & Plucknett, D. L. (1982). *Azolla* as a green manure: use and management in crop production. Westview Tropical Agriculture Series.

Mahanty T, Bhattacharjee S, Goswami M, Bhattacharyya P, Das B, Ghosh A, & Tribedi P (2017) Biofertilizers: A potential approach for sustainable agriculture development. Environmental Science and Pollution Research, 24(4):3315–3335

Mahmood A, Shahzad T, Hussain S, Ali Q, Ali HM, Yasin S et al (2021) Evaluation of symbiotic association between various rhizobia, capable of producing plant-growth-promoting biomolecules, and mung bean for sustainable production. Sustainability 13(24):13832

Majeed A, Muhammad Z, Ahmad H (2018) Plant growth promoting bacteria: role in soil improvement, abiotic and biotic stress management of crops. Plant Cell Rep 37(12):1599–1609

Marks BB, Megías M, Nogueira MA, Hungria M (2013) Biotechnological potential of rhizobial metabolites to enhance the performance of *Bradyrhizobium* spp. and *Azospirillum brasilense* inoculants with soybean and maize. AMB Express 3:1–10

Meena RS, Vijayakumar V, Yadav GS, Mitran T (2018) Response and interaction of *Bradyrhizobium japonicum* and arbuscular mycorrhizal fungi in the soybean rhizosphere. Plant Growth Regul 84:207–223

Mia MB, Shamsuddin Z (2010) *Rhizobium* as a crop enhancer and biofertilizer for increased cereal production. Afr. J. Biotechnol. 9:6001–6009

Mishra U, Pabbi S (2004) Cyanobacteria: a potential biofertilizer for rice. *Resonance* 9:6–10

Mitter EK, Tosi M, Obregón D, Dunfield KE, Germida JJ (2021) Rethinking crop nutrition in times of modern microbiology: innovative biofertilizer technologies. Front Sustain Food Syst 5:29

Mrkovacki N, Milic V (2001) Use of *Azotobacter chroococcum* as potentially useful in agricultural application. Ann Microbiol 51(2):145–158

Naher UA, Othman R, Panhwar QA, Ismail MR (2015) Biofertilizer for sustainable rice production and reduction of environmental pollution. In: Crop production and global environmental issues, pp 283–291

Noel TC, Sheng C, Yost CK, Pharis RP, Hynes MF (1996) Rhizobium leguminosarum as a plant growth-promoting rhizobacterium: direct growth promotion of canola and lettuce. Can J Microbiol 42(3):279–283

Nosheen S, Ajmal I, Song Y (2021) Microbes as biofertilizers, a potential approach for sustainable crop production. Sustainability 13(4):1868

Ohyama T (2010) Nitrogen as a major essential element of plants. Nitrogen Assim. Plants 37:1–17

Okon Y, Labandera-Gonzalez CA (1994) Agronomic applications of *Azospirillum*: an evaluation of 20 years of worldwide field inoculation. Soil Biol Biochem 26(12):1591–1601

Park KH, Lee CY, Son HJ (2009) Mechanism of insoluble phosphate solubilization by *Pseudomonas fluorescens* RAF15 isolated from ginseng rhizosphere and its plant growth-promoting activities. Lett Appl Microbiol 49(2):222–228

Parwan S, Banyal DK, Sharma S (2025) Epidemiological factors affecting purple blotch (Alternaria porri) progression in garlic. Himachal J Agric Res 51(1):137–143

Patel D, Goswami D (2020) Phosphorus solubilization and mobilization: mechanisms, current developments, and future challenge. In: Advances in plant microbiome and sustainable agriculture: functional annotation and future challenges, pp 1–20

Pokorna D, Zabranska J (2015) Sulfur-oxidizing bacteria in environmental technology. Biotechnol Adv 33(6):1246–1259

Pramanik P, Goswami AJ, Ghosh S, Kalita C (2019) An indigenous strain of potassium-solubilizing bacteria *Bacillus pseudomycoides* enhanced potassium uptake in tea plants by increasing potassium availability in the mica waste-treated soil of North-east India. J Appl Microbiol 126(1):215–222

Raghuwanshi R (2012) Opportunities and challenges to sustainable agriculture in India. *Nebio* 3(2):78–86

Rajendran L, Saravanan R (2015) Effect of biofertilizers on growth and yield of banana (*Musa* spp.). Plant Arch 15(2):839–842

Rana BB, Sharma V, Manuja S, Sharma SK, Singh A, Sharma R, Chauhan G (2025) Evaluation of herbicide efficacy on growth indices of maize cultivated under conservation agriculture system. Plant Arch 25(1):2808–2817

Rawat P, Das S, Shankhdhar D, Shankhdhar SC (2021) Phosphate-solubilizing microorganisms: mechanism and their role in phosphate solubilization and uptake. J Soil Sci Plant Nutr 21(1):49–68

Reed SC, Cleveland CC, Townsend AR (2011) Functional ecology of free-living nitrogen fixation: a contemporary perspective. Annu Rev Ecol Evol Syst 42(1):489–512

Revillas JJ, Rodelas B, Pozo C, Martínez-Toledo MV, González-López J (2000) Production of B-group vitamins by two *Azotobacter* strains with phenolic compounds as sole carbon source under diazotrophic and adiazotrophic conditions. J Appl Microbiol 89(3):486–493

Riaz U, Mehdi SM, Iqbal S, Khalid HI, Qadir AA, Anum W et al (2020) Bio-fertilizers: eco-friendly approach for plant and soil environment. In: Bioremediation and biotechnology: sustainable approaches to pollution degradation, pp 189–213

Rodríguez H, Fraga R (1999) Phosphate solubilizing bacteria and their role in plant growth promotion. Biotechnol Adv 17(4–5):319–339

Rokhzadi A, Asgharzadeh A, Darvish F, Nour-Mohammadi G, Majidi E (2011) Influence of plant growth-promoting rhizobacteria on dry matter accumulation and yield of chickpea (*Cicer arietinum* L.) under field conditions. Ann Biol Res 2(2):352–357

Roy A (2021) Biofertilizers for agricultural sustainability: current status and future challenges. In: Current trends in microbial biotechnology for sustainable agriculture. Springer, pp 525–553

Roy S, Liu W, Nandety RS, Crook A, Mysore KS, Pislariu CI et al (2020) Celebrating 20 years of genetic discoveries in legume nodulation and symbiotic nitrogen fixation. Plant Cell 32(1):15–41

Ruiz-Sánchez M, Aroca R, Muñoz Y, Polón R, Ruiz-Lozano JM (2010) The arbuscular mycorrhizal symbiosis enhances the photosynthetic efficiency and the antioxidative response of rice plants subjected to drought stress. J Plant Physiol 167(11):862–869

Sara S, Morad M, & Reza CM (2013) Effects of seed inoculation by Rhizobium strains on chlorophyll contentand protein percentage in common bean cultivars (Phaseolus vulgaris L.)

Sammauria R, Kumawat S, Kumawat P, Singh J, Jatwa TK (2020) Microbial inoculants: potential tool for sustainability of agricultural production systems. Arch Microbiol 202(4):677–693

Sandhu N, Sethi M, Kumar A, Dang D, Singh J, Chhuneja P (2021) Biochemical and genetic approaches improving nitrogen use efficiency in cereal crops: a review. Front Plant Sci 12:657629

Setiawati MR, Damayani M, Herdiyantoro D, Suryatmana P, Anggraini D, Khumairah FH (2018, February) The application dosage of Azolla pinnata in fresh and powder form as organic fertilizer on soil chemical properties, growth and yield of rice plant. In: AIP conference proceedings (vol. 1927, no. 1). AIP Publishing

Sharma NK, Tiwari SP, Tripathi K, Rai AK (2011) Sustainability and cyanobacteria (blue-green algae): facts and challenges. J Appl Phycol 23:1059–1081

Sharma SB, Sayyed RZ, Trivedi MH, Gobi TA (2013) Phosphate solubilizing microbes: sustainable approach for managing phosphorus deficiency in agricultural soils. Springer Plus 2:587

Singh M, Dotaniya ML, Mishra A, Dotaniya CK, Regar KL, Lata M (2016a) Role of biofertilizers in conservation agriculture. In: Conservation agriculture: an approach to combat climate change in Indian Himalaya, pp 113–134

Singh JS, Kumar A, Rai AN, Cyanobacteria DS (2016b) A precious bio-resource in agriculture, ecosystem, and environmental sustainability. Front Microbiol 7:529. https://doi.org/10.3389/fmicb

Singh AL, Bora R, Ram R (2017) Blue-green algae in rice cultivation and soil fertility. Agric Rev 38(4):299–305

Souza T (2015) Handbook of arbuscular mycorrhizal fungi. Springer, Cham, p 153

Sprent JI, Sprent P (1990) Nitrogen fixing organisms: pure and applied aspects, vol 256. Chapman and hall, London

Steinberga V, Alsina I, Ansevica A, Dubova L, Liepina L (2008) The evaluation of effectiveness of *Rhizobium lupini* strains. Lat J Agron/Agronomija Vestis 10

Sun WENLI, Shahrajabian MH, Cheng QI (2021) Nitrogen fixation and diazotrophs–a review. Rom Biotechnol Lett 26(4):2834–2845

Suthar H, Hingurao K, Vaghashiya J, Parmar J (2017) Fermentation: a process for biofertilizer production. In: Microorganisms for green revolution: volume 1: microbes for sustainable crop production, pp 229–252

Thomas L, Singh I (2019) Microbial biofertilizers: types and applications. In: Biofertilizers for sustainable agriculture and environment, pp 1–19

Tsyganova AV, Kitaeva AB, Tsyganov VE (2017) Cell differentiation in nitrogen-fixing nodules hosting symbiosomes. Funct Plant Biol 45(2):47–57

Venkataraman GS, Neelakantan S (1967) The cultivation, blue-green algae, and its application in agriculture. Indian Counc Agric Res Bull 128

Vidyalakshmi R, Paranthaman R, Bhakyaraj R (2009) Sulphur oxidizing bacteria and pulse nutrition- a review. World J Agric Sci 5(3):270–278

Wani SA, Chand S, Wani MA, Ramzan M, Hakeem KR (2016a) *Azotobacter chroococcum*–a potential biofertilizer in agriculture: an overview. In: Soil science: agricultural and environmental prospectives, pp 333–348

Wani SP, Rupela OP, Lee KK (2016b) Sustainable agriculture in the semi-arid tropics through biological nitrogen fixation in grain legumes. Plant Soil 174(1):29–49

Watanabe I, Ventura W, Mascarina GB (1992) Mineral nitrogen production from *Azolla* in rice fields and its use by rice. Soil Sci Plant Nutr 38(4):781–790

Xiafang S, Weiyi H (2002) Mechanism of potassium release from feldspar affected by the sprain Nbt of silicate bacterium. Acta Pedologica Sinica 39(6):863–871

Yadav RK, Abraham G, Singh YV, Singh PK (2014, June) Advancements in the utilization of Azolla-Anabaena system in relation to sustainable agricultural practices. Proc. Indian Natl. Sci. Acad 80(2):301–316

Yadav AN, Kour D, Kaur T, Devi R, Yadav A, Dikilitas M et al (2021) Biodiversity, and biotechnological contribution of beneficial soil microbiomes for nutrient cycling, plant growth improvement and nutrient uptake. Biocatal Agric Biotechnol 33:102009

Yasari E, Patwardhan AM (2007) Physiological analysis of the growth and development of canola (*Brassica napus* L.) as affected by Azotobacter and Azospirillum inoculation. Am Eurasian J Agric Environ Sci 2(6):678–685

Chapter 3
Liquid Biofertilizers: Types, Production Methods and Potential Use for Vegetable Production in Hydroponic Systems

Farai Solomon Zirebwa, Rangarirai Lucia Mhindu, Loice Mhengera, Blessing Mirika Ndau, and Tirivashe Phillip Masere

Abstract Food insecurity is ever increasing in Southern Africa due to climatic, economic as well as other idiosyncratic shocks. Previously, this food insecurity and hidden hunger was associated with rural communities but is now affecting the urban space as well. Efforts to build the resilience of the vulnerable people must now extend to include urban communities. One way of doing that is through urban agriculture. However, urban crop production is limited by shortage of land and unavailability of water for irrigation. To offset these pertinent limitations, soilless crop production in the form of hydroponic systems emerges as a worthy alternative with immense potential. Despite this, the adoption of hydroponics is very low because most smallholder farmers do not afford liquid inorganic fertilizers. The other pertinent challenge is that these fertilizers are not easily accessible. This chapter aimed to establish the potential of using liquid bio-fertilizers for crop production under hydroponics systems. This was achieved through a systematic literature review focusing on different types of liquid biofertilizers, technologies that can be used for their production, characterization of microbial communities and potential for use under hydroponic systems. The chapter revealed that the main methods that can be used for bio or organic fertilizer production are composting and vermicomposting. The chapter suggested a myriad of waste substrates that can be used for production of organic fertilizers, and these include agricultural waste, kitchen waste, bio solids, rumen waste, slaughter house waste, market waste, foliage of vegetation as well as effluent water. Nutrients from these substrates can be extracted as compost teas (aerated or non-aerated), vermicomposting teas (aerated or non-aerated), fermentation (submerged, solid state, co-culture, bokashi) and algal tea. The chapter recommends that for optimal liquid biofertilizers formulations to be developed, it is

F. S. Zirebwa · R. L. Mhindu · L. Mhengera · B. M. Ndau · T. P. Masere (✉)
Department of Land and Water Resources Management, Faculty of Agriculture, Environment and Natural Resources Management, Midlands State University, Gweru, Zimbabwe
e-mail: maseretp@staff.msu.ac.zw

A. D. Nciizah et al. (eds.), *Biofertilizers for Sustainable Agriculture in Africa*, Sustainability Sciences in Asia and Africa, https://doi.org/10.1007/978-981-95-6009-7_3

imperative that the production processes be modified to ensure optimum nutrient availability, presence of beneficial microorganisms and absence of harmful pathogens.

Keywords Biofertilizers · Urban agriculture · Hydroponics · Compost teas

3.1 Introduction

The overreliance on cereal imports, high population growth rates, coupled with stagnated crop productivity in both irrigated and rain-fed agriculture, has confined sub-Saharan Africa (SSA) to perennial food insecurity (van Ittersum et al. 2016). As if this is not enough, climate change and variability adds an additional layer of challenges to the already grave situation in SSA. As noted by Lottering et al. (2020) and Ouko and Odiwuor (2023), SSA has experienced numerous episodes of crop failure that are triggered by the increase in the incidence and severity of extreme climatic events, particularly droughts and other natural hazards. The situation will remain precarious given the rampant land degradation associated with unsustainable soil management practices (Zingore et al. 2015; Masere 2022) and ineptitude of governments to promote sustainable agricultural practices. Whilst the rural folk who generally rely heavily on rain-fed crop production are the hardest hit, the urban populace of most SSA countries are not spared either from food insecurity. For example, in Zimbabwe, only about 20% of the urban populace are food secure (ZimVac 2019). Thus, efforts to build the resilience and adaptive capacity of the vulnerable people against economic, climate change and other eccentric shocks must now extend to include urban communities.

Sustainable agricultural intensification (SAI) offers a credible entry point for adapting to the effects of climate change and building climate-resilience, sustainably increasing productivity and incomes, and mitigates greenhouse gas emissions, thereby decreasing the environmental footprint of agriculture. SAI is also relevant in SSA given high rates of land degradation. One way of building the resilience of urban population is through the use of locally available resources, which is a more sustainable way of improving livelihoods. SAI is hinged on eco-friendly production techniques (Ammar et al. 2023) that are essential for management of biological waste (Kurniawati et al. 2023). There are several crop production technologies that have been developed to cater for this. One of these methods is soilless production or hydroponics which emerged due to the adverse effects of climate change especially water shortages (Dasgan et al. 2023a).

The hydroponic system has numerous advantages over traditional soil cultivation including that it is environmentally compatible and conducive in urban areas because of space constraints. Further, if the system is vertical, there can be more production per unit area all year round. Crop yield and quality under hydroponic systems are usually better compared to the traditional soil cultivation since the root zone environment is easily regulated (Ammar et al. 2023; Dasgan et al. 2023a; Dere et al. 2019). Plant nutrition management which is a precursor for the success of plant production is ensured under soilless production systems (Dasgan et al. 2023a). To maintain the high yield and quality, there is corresponding high application of inorganic mineral fertilizers (Dasgan et al. 2023b; Dere et al. 2019). However, intensive usage of inorganic mineral fertilizers in hydroponic systems is expensive and not environmentally friendly (Dasgan et al. 2023b; Daniel et al. 2022; Kadhum et al. 2021; Dere et al. 2019). The success of the soilless crop production systems is restricted by the high cost of fertilizers that are required for the preparation of nutrient solutions. On small scale production, profitability is therefore severely compromised.

To ameliorate the challenges mentioned earlier, there is need to consider use of liquid organic or biofertilizers. Partial or full implementation of liquid biofertilizers is essential for reducing production costs and improving environmental sustainability (Dasgan et al. 2023a, b; Kadhum et al. 2021). However, the potential replacement of inorganic mineral fertilizers with biofertilizers has a myriad of challenges. The nutrients may be in proportions and quantities that do not match plant requirements. Nutrients in biofertilizers are usually present in low concentration and are bound to organic molecules and hence not readily available for plant uptake. Nutrient release is also slow since the organic material has to be transformed into inorganic form which is available to plants. Some of the main challenges regarding the application of organic fertilizers to plants is the emergence of diseases (Dantroliya et al. 2022), contamination by heavy metals and pathogens (Kurniawati et al. 2023).

The production of biofertilizers is therefore supposed to be done in ways that offset these challenges. As such, there are various composting and extraction methods in the production of the liquid biofertilizers. This chapter focuses on the different types of biofertilizers, production methods and potential use for vegetable production under hydroponic systems. The objectives of the chapter are as follows:

1. To give an overview of liquid biofertilizers used in hydroponic systems.
2. To categorize different technologies that are used for the production of liquid biofertilizers.
3. To identify the factors affecting the adoption of liquid biofertilizers in hydroponics systems.

3.2 Overview of Liquid Fertilizers Used in Hydroponic Systems

As already discussed, hydroponic systems involve growing of plants in soilless conditions where mineral nutrients are fed directly to the roots (Phibunwatthanawong and Riddech 2019). The technology offers an important platform for environmentally friendly agricultural and food production sector, particularly in areas characterized by severe soil degradation and limited water availability (Sambo et al. 2019).

Moreover, hydroponics has numerous advantages over traditional farming including high productivity through the increased number of plants per unit area, less labour-intensive and potentially harvesting multi-crops throughout the year, without chaotic discharges of either pesticides or fertilizers to the environment (Margaret et al. 2022; Velazquez-Gonzalez et al. 2022). There are six main types of hydroponic systems, namely, nutrient film technique, Deep Water Culture, Drip system, Aeroponics, Ebb and Flow and Aquaponics (Santosh and Gaikwad 2023; Velazquez-Gonzalez et al. 2022). In all these hydroponic systems, liquid fertilizer is a major constituent as it is one of the most critical factor in determining quality and quantity of plant yield (Upendri and Karunarathna 2021). Liquid fertilizer types are discussed in detail in the next section.

3.2.1 Types of Liquid Fertilizers

Broadly there are two liquid fertilizer types and these are chemical nutrient solutions and soluble organic fertilizers.

3.2.1.1 Chemical Nutrient Solutions

Hydroponic systems usually make use of nutrient solutions derived from chemical fertilizers and from inorganic compounds supplied by using different chemical combinations (Velazquez-Gonzalez et al. 2022). The chemical nutrient solutions provide the essential micro and macro plant nutrients. They are formulated depending on the nutritional requirements of the crops being grown. The most common chemical nutrient solutions for most crops are the Hoagland (Hoagland and Arnon 1938) and the Steiner's universal nutrient solution (Steiner 1961). The typical nutrient compositions of the solutions include the following macronutrients: nitrogen (NO_3^- or NH_4^+), phosphorous ($H_2PO_4^-$), potassium (K), calcium (Ca), magnesium (Mg) and sulphur (S). The micronutrients are iron (Fe), manganese (Mn), zinc (Zn), copper (Cu), boron (B) and molybdenum (Mo). The concentrations depend on formulations, but the standard Hoagland solution consists of nitrogen (N) 210 ppm, phosphorous (P) 31 ppm, potassium (K) 235 ppm, calcium (Ca) 200 ppm, magnesium (Mg) 48 ppm, sulphur (S), 64 ppm, iron (Fe) 1-5 ppm, manganese (Mn) 0.5 ppm, boron (B) 0.5 ppm, zinc (Zn) 0.05 ppm and molybdenum (Mo) 0.01 ppm (Resh 2022). The Steiner's universal nutrient solution focuses on maintaining a balance between cations and anions through optimization of electrical conductivity and pH.

However, these chemical nutrient solutions present a plethora of challenges, particularly when higher concentrations of inorganic chemical fertilizers are used. The challenges range from cost implications, human health, plant growth and environmental. The cost of these chemical nutrient solutions is prohibitive especially for small-scale farmers. A premixed solution costs up to \$20 per litre (Resh 2022).

Several studies (Phibunwatthanawong and Riddech 2019; Sambo et al. 2019; Upendri and Karunarathna 2021; Szekely and Jijakli 2022; Velazquez-Gonzalez et al. 2022) found chemical nutrient solutions to pose human health risk due to accumulation of toxic chemicals in vegetables like lettuce. Further, the manufacturing of fertilizers like ammonia (from which 97% of mineral N fertilizers are derived), phosphates and potash is not only costly but also expends a great deal lot of energy, gas and fuel from activities like rock mining and extraction (Szekely and Jijakli 2022). Furthermore, nitric acid production during the making of N fertilizers used in hydroponic system generates large emissions of nitrous oxide, which has a global warming potential that is 298 times greater than that of carbon dioxide (Basosi et al. 2014). The aforementioned challenges show that use of chemical mineral fertilizers in hydroponics systems is not only unsustainable but has far-reaching effects to human health and the environment hence the need for shifting to liquid organic fertilizers.

3.2.1.2 Liquid Organic Fertilizers

Organic fertilizers are naturally derived fertilizers containing essential elements that help the growth and development of plants (Torres and Somera 2023). The use of liquid organic fertilizer as a substitute to chemical nutrient solutions is a promising solution for hydroponic farming systems. Liquid organic fertilizers are produced by simple fermentation processes of organic substrates and industrial wastes. Examples of liquid biofertilizers include plant extracts, compost watery extracts, compost leachate, compost teas, animal-based liquid manures (including digested animal slurry, blood meal and fish hydrolysates) and manure teas (Govere et al. 2011; Torres and Somera 2023). At the end of the fermentation process, phytohormones such as auxin and cytokinin, organic acids and plant growth promoters are present in the liquid organic fertilizer (Arancon et al. 2019; Upendri and Karunarathna 2021). The nutrient composition of liquid biofertilizers vary considerably depending on the substrates and brewing process (Ingham 2005). Typical nutrient compositions are outlined in Table 3.1. Despite the many advantages of using liquid organic fertilizers over chemical nutrient solutions in hydroponics, liquid organic fertilizers do not have an established optimum operation parameter and face quality variations (physicochemical characteristics of various organic materials) as well as nutrient management issues (Torres and Somera 2023). Similarly, Upendri and Karunarathna

Table 3.1 Typical nutrient composition of liquid biofertilizers. (Ingham 2005)

Fertilizer	N (%)	P (%)	K (%)	Key benefits
Compost tea	~0.03	~0.01	~0.04	Microbial benefits + general nutrition
Fish emulsion	2–5	1–3	0.5–2	High N, quick response
Seaweed extract	<1	<0.5	1–5	Hormones + micronutrients
Vermiwash	~0.005	~0.002	~0.007	Enzymes + beneficial microbes
Cow urine fertilizer	~0.006	~0.004	~0.005	Boosts microbial soil activity

(2021) noted that different types of liquid organic fertilizers and crops perform differently thereby making it difficult to disclose the efficacy of these organic fertilizers in hydroponic systems.

3.2.2 Production of Liquid Biofertilizers

There are various methods that can be used for the production of biofertilizers. The two broad categories of the production methods are the composting and non-composting. These are discussed in the next section.

3.2.2.1 Composting Methods

Thermophilic composting and vermicomposting are two popular methods for organic waste management and nutrient extraction for reuse in agriculture (Ning et al. 2021). Thermophilic composting involves aerobic decomposition of organic waste materials. It generates high temperatures which speed up breakdown of organic matter and nutrient release. Vermicomposting involves the use of earthworms for decomposing organic wastes (Gupta et al. 2020). Earthworms degrade waste by ingesting it, passing it though their digestive tracts, thereby converting it into nutrient-rich vermicompost (vermicasts) (Huang and Xia 2018).

Thermophilic Composting Systems

Thermophilic composting systems differ in the size or scale of operation and vary in their temperature profiles. Composting systems may be mesophilic whereby temperatures in the compost mass do not exceed 45 °C, or thermophilic, with internal temperature exceeding 60 °C. Thermophilic composting is preferred where waste sanitization is desired, and this is achieved as temperature within the compost mass exceed 55 °C and remain elevated for a minimum of 3 days. High temperatures achieved during thermophilic composting accelerate the decomposition process, resulting in faster production of compost compared to other methods. Thus, the heat generated during thermophilic composting effectively kills many pathogens, weed seeds and harmful organisms present in the organic waste, reducing the risk of disease transmission.

The optimization of the compost moisture content, C:N ratio, temperature and turning frequency is vital to improve nutrient extraction (Yeh et al. 2020). In-vessel composting permits varying levels of automation and may be operated at an industrial scale or at municipality level. It relies on the use of an enclosed or contained environment that allows close control of the process. In-vessel composting systems include vertical flow, horizontal flow and agitated bins. Open systems may be small compost heaps or long windrows designed to allow aerobic biodegradation of wastes.

Vermicomposting

Vermicomposting involves the use of earthworms for decomposing organic wastes (Gupta et al. 2020). Earthworms degrade waste by ingesting it, passing it though their digestive tracts, thereby converting it into nutrient-rich vermicompost (vermicasts) (Huang and Xia 2018). Vermicomposting structures include bins, windrows, towers or bags. Bins, usually plastic or wooden, are designed for ventilation, drainage and access for worms and tend to be used for small-scale composting. In a continuous flow-vermicomposting system, organic waste material is added to the top of the bin, and the worms continuously process the material as it moves downward with the finished compost harvested from the bottom. Vermicomposting bags are usually simple and portable and made of material that permits aeration. Large-scale vermicomposting structures are usually in the form of windrows long piles of organic waste aerated periodically. Vermicomposting towers consist of multiple trays or compartments organized in a vertical stacking arrangement that is compact enough to hold organic waste. In this system, worms gradually move up through the trays as they degrade the waste. Vermicomposting is used in combination with aerobic or anaerobic composting whereby worms are introduced in the composting mass following some pretreatment and conditioning that allows the worms to survive in a vermicomposting bed or vessel.

Each type of vermicomposting system has its own advantages and suitability depending on factors such as available space, amount of waste and individual needs. The choice of system depends on the specific circumstances and requirements of the user. Several studies have investigated the nutrient dynamics during vermicomposting and have shown that vermicompost can have higher nutrient concentrations compared to traditional compost (Dume et al. 2022). The vermicompost contains nutrients in plant-available forms as well as beneficial biomolecules and microbial communities, which can enhance soil fertility and plant growth. For instance, kitchen waste compost innoculated with *Bacillus safensis* YM1 compost suppressed pests, stimulated the growth of lettuce and enhanced bacterial communities in soils (Zhang et al. 2023).

It is important to note that the effectiveness of nutrient extraction and the nutrient content of compost and vermicompost can vary depending on elements like feedstock configuration, process considerations and environmental circumstances (Table 3.2). Upendri and Karunarathna (2021) reported the use of a wide range of substrates to generate liquid organic fertilizers including grey water, sewage effluent with no detrimental effects to plants, whilst Dhananjani and Pakeerathan (2022) observed the highest N, P and K nutrient content from vegetable waste extracts subjected to fermentation for 2 weeks.

Waste Substrates Used for Composting and Vermicomposting

Waste materials range from municipal solid waste, household canteen or kitchen waste, agricultural waste, biosolids, domestic waste water as well as waste water generated as effluent by manufacturing companies. A detailed outline of the substrates and their treatment processes and crops that were fertilized using the liquid fertilizer is outlined in Table 3.3. Animal manures including rumen waste, slaughter

Table 3.2 Nutrient composition of various vermicomposting teas

Vermicompost tea (VCT) Type	N (mg/L)	P (mg/L)	K (mg/L)	Ca (mg/L)	Mg (mg/L)	Micronutrients	pH	Authors
Cow manure	2.31%	1862	2482	–	–	Cu: 0.09 mg/L, Zn: 25.56 mg/L, Fe: 12.07 mg/L	–	Atiyeh et al. (2000)
Leaf meal	0.45%	311	820	–	–	Fe: 1.27 mg/L, Zn: 7.96 mg/L	–	Atiyeh et al. (2000)
Cow manure + leaf meal	–	–	–	–	Mn: 0.51 mg/L	–	–	Atiyeh et al. (2000)
Bovine manure (dry weight)	18,000	23,000	7000	169,000	11,000	Fe: 1683 μg/g, Mn: 828 μg/g, Zn: 552 μg/g, Cu: 91 μg/g, B: 56 μg/g	6.9	Congilosi et al. (2022)
20% VCT from cow manure	58.9	3.7	2.9	75.1	5.5	–	8.4	Souffront et al. (2022)
10% VCT from cow manure	35.1	2.0	1.3	38.2	2.7	–	7.9	Souffront et al. (2022)
Food waste	–	–	–	–	–	High extraction efficiency for P and Ca; active microbial populations	–	Pant et al. (2012)
Chicken manure-based thermophilic compost tea	–	–	–	–	–	Greater NO_3^-N, K, Mg, Na, Zn, Cu than other compost teas	–	Vukobratović et al. (2014)
With additives (e.g., oat bran, sunflower press cake)	57	405	13	–	–	High microbial counts; enhanced plant growth parameters	–	Pant et al. (2012)

Table 3.3 Organic waste materials and techniques used for production of nutrient solutions

#	Author(s)	Product(s)	Substrates	Treatment	Duration	Country	Crop
1	Nishana and Rahman (2021)	Hydroponics nutrient solution	Common weeds (fresh); *Phyllanthus niruri*; unrefined sugar, kitchen waste (fruit veg peels)	Fermentation	7 days	India	Spear mint; chilies
2	Takemura et al. (2020)	Nitrified biogas digestate filtrate	Organic fraction of municipal solid waste	AD digestate notified & filtered		Japan	Chrysanthemum
3	Wongkiew et al. (2023)	Vermicompost leachate	Food waste derived from cafeteria vegetable waste	Vermicompost leachate supplementation; co-composting	1.5mths	Thailand	Lettuce
4	Kechasov et al. (2021)	Nutrient solution—liquid biogas digestate	Pig waste biogas digestate	Pig manure digestion; + nitrification bioreactor		Norway	Tomato
5	Phibunwatthanawong and Riddech (2019)	Liquid organic fertilizers	Molasses, distillery slop and sugarcane leaves	Fermentation (plastic cans) and ambient temp stirred weekly	30 days	Thailand	Lettuce
6	Pant et al. (2011)	Vermicompost tea biofertilizer	700 g vermicompost 19 L water with 50 ml molasses and 25 g soil mender	Vermicompost extraction in mesh bags	24–72 h		Lettuce
7	Lopez (2019)	Hydroponic solution	Bio digester effluents, commercial org fertilizers, compost tea and urine	Dilution	n/a	Italy	Lettuce
8	Heintze (2023)	Nutrient solution	Blood meal, bone meal and potato peel (N-, P- and K-rich residues, resp.)	Mixed with water & digested anaerobically	123 or 54 days	Germany	Lettuce
9		Review paper	n/a	n/a	n/a	n/a	n/a
10	Bergstrand et al. (2020)	Bioponics substrate	Green & catering waste, animal manure and slaughterhouse waste ($^1/_3$each) biogas digestate	Biochips in water biogas digestate concentrate and bioponics substrate	n/a	Switzerland	n/a

(continued)

Table 3.3 (continued)

#	Author(s)	Product(s)	Substrates	Treatment	Duration	Country	Crop
11		Review paper	n/a	n/a	n/a	n/a	n/a
12	Upendri and Karunarathna (2021)	Review paper	n/a	n/a	n/a	n/a	n/a
13	Pelayo et al. (2020)	Hydroponic solution	Crop residues 85.5%; food industry waste; 12.5% and iron chloride 2% (biogas digestate)	Biogas reactors (80 days), nitrification reactor	80 days		Rape
14	Dhananjani and Pakeerathan (2022)	1. Liquid fertilizer (AD) 2. Vermi tea	1. Canteen veg waste-pumpkin /potato; cow dung, Gliricidia leaves, brown sugar; 2. Vermicompost: (teak+Jack leaves):	Fermentation + agitation every other day—"Vermi tea"	14 days	Sri Lanka	Spinach
15	Zarei et al. (2018)	Filtrate/ concentrate	Goat manure; banana peel	20 L + 5 kg boiled to 5 L conc		Pakistan	
16	Salas-Sanjuán et al. (2023)	Aerated/ non-aerated aqueous extract	Vegetable waste vermicompost	Aeration of aqueous vermicompost mixture	12–16 days	Spain	Lettuce
17	Deepthi et al. (2021)	Vermiwash hydroponic solution	Cattle dung				Amaranthus
18	Malayil and Chanakyna (2020)	Biogas digester residue	Bio waste	AD	n/a	n/a	Vegetables; mushrooms
19	Al-najjar et al. (2016)	Solid & liquid digester effluents	Cow manure	Fermentation/ AD			
20	Mauerer et al. (2023)	Nutrient solution	Human urine			Germany	Lettuce
21	Wongkiew et al. (2021)	Nutrient film technique bioponics	Chicken manure	200–400 g/18 plants bioponic NFT			Lettuce

22	Ruiz and Salas (2022)	Vermicompost bio fertilizer	Composted greenhouse horticultural crop residues	Inoculated vermicompost solutions (PGPB) aerated [blower pump, 3.5 L min^{-1}, 0.012 MPa]	16 days	n/a	Tomatoes
23	Fragalà et al. (2022)	Alkaline hydrolysates	Municipal bio waste	Fermented -anaerobic/aerobic			Hibiscus
24	Tikasz et al. (2019)	Hydroponic nutrient solution	Chicken, cow and Turkey manure extract solutions (10, 25 and 50 g/L manure)	Manure in 30 L tap water, aerated with air pump & filtered (1-mm sieve)	48 hr	Canada	Lettuce; kale
25	Franco (2025)	Aquaponics	Food waste	Black soldier fly and aquaponics	n/a		
26	Panuccio (2021)	Solid digestate medium	Crop residues and other vegetable by-products	Biogas AD	n/a	Sweden	Tomato
27	Bidabadi et al. (2016)	Vermicompost leachate	Cattle manure	VL brewed	48 h	Israel	Stevia

house waste, chicken, goat, cattle as well as rabbit manure are used as additives that activate the decomposition process by providing microorganisms that constitute the decomposer community. Such organic materials as banana stalks, banana peels, moringa, coffee waste, sugarcane waste, slaughterhouse waste and the foliage of specific vegetation have also been demonstrated to be important feedstock in the production of organic fertilizers. Waste from the manufacturing industry that use organic raw materials have also provided feed stocks that can be processed into bio-fertilizers. Research has also been done to explore the use of various digestates from bio-energy production systems. Waste materials from fish farming, green feed industry and other manufacturing industries were successfully used as bio fertilizers.

3.2.3 Nutrient Extraction Techniques

Nutrients that are produced from composting methods can be extracted as compost teas or vermicomposting teas.

3.2.3.1 Compost Teas (Cts)

Compost teas (CTs) have emerged as eco-friendly alternatives for enhancing plant growth and controlling plant diseases (González-Hernández et al. 2022). They are liquid extracts produced by steeping mature compost in water under controlled conditions, resulting in a nutrient-rich solution containing beneficial microorganisms, soluble nutrients and organic matter. These components contribute to their positive effects on plant growth promotion and pathogen suppression. CTs offer several advantages over solid compost application. They can reduce transportation costs and provide larger area coverage. Studies have shown that CTs improve plant nutrient status, decrease disease incidence, promote plant growth and reduce reliance on chemical pesticides and fertilizers (Raza et al. 2023).

To optimize the preparation and application of compost tea, laboratory-scale experiments have been conducted. Researchers found that the optimal compost-to-water ratio for brewing compost tea is from 1:3 to 1:10 (Garg and Rakshit 2024; Pilla et al. 2023; Hatam et al. 2015; Martin and Brathwaite 2012). The brewing time of 5 days resulted in the highest nutritive value. However, it was observed that the nitrogen content of compost tea decreased immediately after production and at storage temperatures above 30 °C. Field experiments demonstrated the positive effects of compost tea application on potato yield enhancement and disease suppression (Raza et al. 2023). The quality of compost teas can be influenced by various factors, including compost stability, water quality, brewing time, aeration and additives. These factors should be carefully considered to ensure the efficacy of the compost tea. Compost teas are considered a cost-effective technology for converting organic waste into organic fertilizers (El-Haddad et al. 2014). The brewing of compost teas can be done under aerated and non-aerated conditions.

Aerated Compost Tea (ACT)
Aerated compost tea (ACT) is a type of compost tea that involves deliberate aeration during the brewing process. Similar to non-aerated compost tea (NCT), the compost is placed in a mesh or nylon bag and steeped in water. However, in ACT production, an air pump is used to provide aeration by bubbling air through the mixture. Additionally, a sugar source is added to serve as a microbial food source (Dearborn et al. 2011). The deliberate aeration in ACT promotes the growth of beneficial aerobic microorganisms, including bacteria, fungi, nematodes and protozoa. These aerobic microorganisms help break down organic matter more efficiently, releasing additional nutrients and enhancing the microbial diversity of the tea. ACT can be brewed for a shorter period of time compared to non-aerated compost teas, typically ranging from 12 h to 3 days, due to the presence of oxygen. The addition of a microbial food source further enhances the growth and activity of beneficial microorganisms. ACT is commonly used as a soil drench or foliar spray in agricultural systems. It is believed to have disease suppressive effects and can contribute to improved plant health and productivity (Dearborn et al. 2011).

Non-aerated Compost Tea (NCT)
Non-aerated compost tea (NCT) is a subset of compost tea production methods that involves steeping stable compost in water without the addition of aeration or sugar additives (Dearborn et al. 2011). In NCT production, the stable compost is typically placed in a mesh or nylon bag and steeped in water for a specific period, with occasional stirring to ensure proper mixing. NCT is brewed under low oxygen conditions, leading to anaerobic fermentation. The lack of aeration restricts the growth of oxygen-requiring microorganisms. The average brewing period for NCT is around 14 days, which differs from the longer brewing times of aerated compost teas (Dearborn et al. 2011). The shorter brewing period in NCT helps minimize the dominance of anaerobic microorganisms, as they require more time to thrive in the absence of oxygen. Although non-aerated compost teas may have a different microbial composition compared to aerated compost teas, they still contain beneficial microorganisms and soluble nutrients extracted from the compost. While they may not be as microbe-dense as aerated compost teas, NCT can provide nutritional benefits to plants when applied as a foliar spray or soil drench (Dearborn et al. 2011).

3.2.3.2 Vermicomposting Tea

Vermicomposting tea is a liquid biofertilizer that is derived from the process of vermicomposting, which involves the decomposition of organic waste materials by earthworms and microorganisms. The production of vermicomposting tea can be achieved through either an aerated or non-aerated process. Several factors can influence the quality of vermicompost and subsequently the quality of the tea extracted from it. These factors include the type and composition of the organic waste used, the species of earthworms involved, the choice of bedding material, temperature conditions and the pH levels of the vermicompost (Zaini and Syafi 2021). Both aerated and non-aerated vermicompost teas offer valuable benefits for plants and soil health. The choice between the two methods depends on factors such as available resources, desired microbial populations and specific plant requirements. Vermicomposting tea, regardless of the method used, offers advantages such as ease of application, enrichment of the medium with beneficial microorganisms and the provision of essential macro and micronutrients. However, a challenge associated with vermicomposting tea is maintaining its quality during storage (Arosha and Sarvananda 2022). The nutrient composition of various vermicomposting teas is outlined in Table 3.2.

Aerated Vermicomposting Tea (AVT)

Aerated vermicompost tea is brewed by introducing oxygen into the vermicompost tea mixture during the brewing process. This aeration process creates an aerobic environment that supports the growth of beneficial aerobic microorganisms derived from the vermicompost. These microorganisms play a vital role in breaking down organic matter and releasing soluble nutrients, resulting in a nutrient-rich and microbe-dense tea. Aerated vermicomposting tea (AVT) is created using a tea brewer and requires regular aeration to maintain its quality. Additional substances such as molasses or brown sugar can be added to enhance its quality. AVT is more stable and yields better crop production outcomes compared to non-aerated tea (Arosha and Sarvananda 2022). A study by Natchavatthon and Thammared (2019) found that adding molasses improved the nutrient content and electrical conductivity of vermicompost tea, while oxygen addition raised pH and electrical conductivity but lowered nutrient content. When applied to plants as a foliar spray or soil drench, aerated vermicompost tea provides a concentrated source of nutrients, beneficial microorganisms, enzymes and bioactive compounds, thereby enhancing plant growth, improving soil fertility and boosting plant defences against diseases and pests (Pant et al. 2011).

Non-aerated Vermicomposting Tea (NVT)

On the other hand, non-aerated vermicomposting tea (NVT) as the name suggests, is brewed without the addition of aeration. The shorter brewing time of NVT helps minimize the growth of anaerobic microorganisms. Non-aerated vermicompost tea is obtained by soaking the vermicompost in water, similar to steeping tea, facilitating the dissolution of nutrients and beneficial microorganisms into the water (Pant et al. 2011). While it may have a different microbial composition compared to aerated tea, NVT still contains beneficial microorganisms, enzymes and soluble nutrients extracted from the vermicompost. It can contribute to improving soil fertility, enhancing nutrient availability and stimulating plant growth, albeit with potentially lower microbe density compared to AVT. Non-aerated vermicompost tea can be a convenient option for those who prefer a simpler brewing process or have limited access to aeration equipment (Ingham 2005).

3.2.4 Non-compost-Based Methods of Producing Liquid Biofertilizer or Tea

Examples of non-compost-based biofertilizer production methods include fermentation (various forms) and algal tea.

3.2.4.1 Fermentation Methods

Fermentation is a valuable process used in the production of liquid biofertilizers. It involves the breakdown of organic matter by microorganisms in the absence of oxygen (Uparivong 2012). The resulting liquid biofertilizer is rich in essential nutrients and can be applied to hydroponic systems or used as a foliar spray to enhance plant growth and soil fertility. Fermentation employs microbial inoculants to ferment the organic matter and optimize the biofertilizer production process (Yusuf and Nazif (2023). Uzor (2018) described a fermentation method for producing liquid biofertilizer that involved the cultivation and isolation of nitrogen-fixing and phosphate-solubilizing bacteria. This fermentation process optimized the efficiency of these microorganisms, providing a safer and more effective alternative to chemical fertilizers. There are different ways of fermentation that can be used, and these include submerged, solid state, co-culture and Bokashi.

Submerged Fermentation
Submerged fermentation involves culturing microorganisms in a liquid medium without direct contact with air or solid substrates. This method provides an optimal environment for microbial growth and metabolite production. The use of submerged fermentation enables efficient nutrient uptake by microorganisms and uniform distribution of nutrients throughout the liquid medium. This promotes the release of enzymes and metabolites that convert complex organic compounds into simpler, bioavailable forms. The scalability of submerged fermentation allows for the production of large volumes of liquid biofertilizers, ensuring a consistent supply for agricultural applications (Uparivong 2012).

The optimization of submerged liquid fermentation processes is crucial for the production of high-quality liquid fertilizers. Strategies such as selecting an appropriate fermentation model, optimizing medium components, fine-tuning process parameters and designing efficient bioreactors can be employed to enhance biomass production and the production of specific metabolites with plant beneficial properties (Choubey et al. 2018).

Solid-State Fermentation
Solid-state fermentation (SSF), on the other hand, involves culturing microorganisms on solid substrates with limited moisture content. It is advantageous due to its simplicity, low cost and utilization of agricultural and industrial by-products as substrates. This process releases enzymes and metabolites, transforming complex compounds into nutrient-rich liquid biofertilizers that contain beneficial microorganisms.

SSF utilizes agricultural and industrial by-products as substrates, contributing to sustainability and improved nutrient availability when applied to plants or soil (Thomas et al. 2013).

Thomas et al. (2013) explored the use of solid-state fermentation for producing biofertilizer from agro-wastes. The study utilized watermelon, papaya, pineapple, citrus orange and banana wastes, subjecting them to solid-state fermentation to produce biofertilizers. The study found promising physical characteristics and nutrient content in the biofertilizers derived from watermelon, papaya and banana wastes.

Co-culture Fermentation

Co-culture fermentation involves the simultaneous growth of multiple microbial strains to enhance synergistic effects in biofertilizer production. It is an emerging method for producing liquid biofertilizers. By cultivating different microorganisms together, it enhances nutrient content and effectiveness. The selected microorganisms complement each other's functions, promoting nutrient cycling and plant growth. This approach improves nutrient availability and uptake, resulting in balanced biofertilizers. Careful selection and optimization of strains, along with environmental control, are essential for successful co-culture fermentation. It offers a sustainable solution for agriculture by maximizing nutrient utilization and fertility.

Bokashi Fermentation

Bokashi fermentation is an innovative method for producing liquid fertilizer from organic waste materials, offering a sustainable solution for nutrient recycling and soil enrichment. Bokashi fermentation involves the controlled breakdown of organic waste through anaerobic microbial activity. This process is initiated by introducing a consortium of beneficial microorganisms, including lactic acid bacteria and yeasts, to the waste material. These microorganisms metabolize the organic matter, resulting in the conversion of the waste into a nutrient-rich liquid fertilizer, also known as Bokashi tea or Bokashi leachate (Gebiola et al. 2023).

During Bokashi fermentation, the introduced microorganisms play a vital role in breaking down complex organic compounds and solubilizing nutrients. This process enhances the availability of essential nutrients such as nitrogen, phosphorus, potassium and micronutrients, as well as beneficial microorganisms crucial for plant growth (Gebiola et al. 2023). The resulting liquid fertilizer obtained through Bokashi fermentation exhibits improved nutrient content, making it highly beneficial for enhancing soil fertility and promoting plant productivity (Gebiola et al. 2023).

To carry out Bokashi fermentation, organic waste materials such as kitchen scraps, plant residues or agricultural by-products are collected and placed in an airtight container or Bokashi bin. A microbial inoculant, typically in the form of Bokashi bran or starter, is then added to the waste material (Gebiola et al. 2023). This inoculant contains a mixture of beneficial microorganisms, including lactic acid bacteria and yeasts, which kick start the fermentation process. The waste material and inoculant are thoroughly mixed to ensure uniform distribution of the microorganisms. The container is sealed to create an anaerobic environment necessary for the fermentation process to occur.

Over a period of several weeks, the microorganisms in the Bokashi bin break down the organic waste through anaerobic respiration. This process involves the microbial conversion of complex organic compounds into simpler forms, releasing nutrients and metabolites in the process. The microorganisms produce organic acids and enzymes that help to solubilize and break down the organic matter. These microbial activities result in the production of a liquid by product, known as Bokashi tea or Bokashi leachate, which accumulates at the bottom of the container (Gebiola et al. 2023).

3.2.4.2 Algal Tea

The production of algal tea as a liquid fertilizer involves the cultivation and extraction of specific strains of beneficial algae, such as *Chlorella*, *Spirulina* or seaweed, to create a concentrated nutrient-rich liquid solution. The process begins with the careful cultivation of these algae in controlled environments such as ponds, tanks or bioreactors. These algae are selected for their high nutrient content, including essential macronutrients like nitrogen, phosphorus, potassium as well as trace minerals. During cultivation, the algae are provided with optimal conditions, including appropriate temperature, light exposure and nutrient availability, to ensure their robust growth and nutrient accumulation. Once the algae reach their desired growth stage, they are harvested and processed to extract the valuable compounds and nutrients. Techniques like pressing, drying or enzymatic treatments are employed to release the liquid fraction containing the desired nutrients and bioactive substances. The resulting liquid, known as algal tea, is a concentrated liquid fertilizer that contains a wide range of essential nutrients, growth-promoting substances and beneficial microorganisms derived from the cultivated algae. Algal tea can be further processed or diluted with water before being applied to plants.

The application of algal tea as a liquid fertilizer offers numerous benefits for plant growth and soil fertility. Firstly, algal tea provides a rich source of readily available nutrients, including macronutrients and micronutrients, which can support healthy plant growth, improve nutrient absorption and enhance crop productivity. Additionally, algal tea contains bioactive substances such as plant hormones, enzymes and organic acids that can stimulate root development, improve soil structure and enhance plant resilience against various stress conditions. Moreover, the use of algal tea as a liquid fertilizer promotes sustainable agriculture by utilizing renewable resources and reducing reliance on synthetic fertilizers. Algae cultivation for liquid fertilizer production can also contribute to environmental benefits such as carbon dioxide sequestration and wastewater treatment, depending on the cultivation system and algae species used. To ensure the production of high-quality algal tea with optimal nutrient content, it is important to follow established protocols and guidelines for algal cultivation, harvesting and extraction.

3.2.5 Adoption of Liquid Biofertilizers in Hydroponics Vegetable Production

Use of liquid biofertilizers in vegetable production is now preferred to counteract challenges being faced globally from climate change, limited arable land and water scarcity. As the global environment is facing a demand for high-quality fresh food due to the growing global population, scalable and geographically adaptable alternative agricultural systems such as hydroponics are necessary to preserve food security without compromising environmental sustainability. The demand for organic food has also stimulated increasing concern in developing organic hydroponic or soilless production systems (Di Gioia and Rosskopf 2021; Bergstrand 2022). Use of liquid biofertilizers enhances the efficiency of hydroponic systems. This results in leafy vegetables and fruit of better quality, characterized by higher levels of antioxidant compounds and lower nitrate levels (Bergstrand 2022). When using liquid biofertilizers for vegetable production in hydroponic system, nutrients are directly delivered to plants through nutrient solutions, ensuring efficient uptake and utilization. They offer a natural means to improve nutrient availability, root health and overall plant vitality as liquid organic fertilizers consist of essential plant nutrients and beneficial microorganisms, which recycle organic matter. This then maximizes crop yields which is crucial to meet the growing global food demand.

Growing of vegetables with hydroponics that utilize liquid fertilizer has been on the rise. A study done by Dasgan et al. (2023b) found out that the growth of capia (*Capsicum annuum L.*) utilizing hydroponics was increasing among users that have adopted the system. They determined the effect of replacing 20% of mineral fertilizer with biofertilizers, and there was an improvement on fruit quality and growth parameters, such as fruit weight, diameter, volume, the electric conductivity of the fruit juice and total soluble solids. Liquid fertilizer use in hydroponics has also been found to be effective in basil production. Dasgan et al. (2022) showed that biofertilizers could affect the leaf yield, nitrate amount and antioxidant content of basil (*Ocimum basilicum* L.) in a floating culture.

Baum et al. (2015) demonstrated that the presence of mycorrhizae in biofertilizers led to increased levels of ascorbic acid in chili peppers, beta carotene in potatoes and various beneficial compounds such as carotenoids, phenolics, anthocyanins, chlorophyll and essential mineral nutrients in lettuce. These findings indicate that biofertilizers contribute significantly to improving vegetable quality. Similarly, Michałojć et al. (2015) examined the impact of mycorrhiza under two electrical conductivity (EC) conditions—2600 and 1900 $\mu S\ cm^{-1}$ on the quality of tomatoes cultivated in soilless systems. Their results showed that tomatoes treated with mycorrhiza contained notably higher concentrations of total soluble solids compared to those without mycorrhizal treatment.

Churilova and Midmore (2019) assessed the use of vermiliquer (vermicompost leachate) as a complete liquid fertilizer for hydroponic cultivation of Pak Choi. Their study found that both vermicast and vermiliquer provide plant-available nutrients, with vermiliquer being more suitable for hydroponic use. However, the optimal pH required for worm activity and vermiliquer generation rendered the solution too alkaline for direct application in hydroponic systems. Despite this limitation, under carefully managed hydroponic conditions, Pak Choi (Brassica chinensis) grown with pH-buffered vermiliquer harvested after 8–10 weeks of vermicomposting exhibited growth and yields comparable to those produced using conventional inorganic hydroponic fertilizers.

Further research by Muthu et al. (2023) compared organic liquid fertilizer derived from fermented vegetable waste with commercial inorganic fertilizers. Their study concluded that a 30-day fermentation process of vegetable waste yields an organic liquid fertilizer that is as effective as its commercial counterparts. The process enhances nutrient availability and concentration, thereby increasing the agricultural value of vegetable waste and promoting a circular, bio-based economy. Moreover, organic liquid fertilizers not only reduce fertilizer costs but also offer a sustainable alternative for hydroponic nutrient solution remediation. As such, their application in both traditional agriculture and hydroponics is recommended to reduce landfill waste, curb environmental degradation and boost crop productivity.

Conversely, Alneyadi et al. (2024) found that the effectiveness of organic liquid nutrient solutions (NS) in hydroponics significantly depends on the fertilizer source. While organically grown lettuce using organic NS had lower nitrate content and elevated levels of chlorophyll, carotene, phenolics and flavonoids—suggesting enhanced antioxidant properties and nutritional value—these plants also showed lower concentrations of calcium, potassium and magnesium, but higher phosphorus levels in their leaves. These findings highlight the sustainability and health benefits of organic agriculture but also point to the challenges of using organic NS in hydroponic systems due to potential nutrient imbalances. Further information on the adoption of liquid bio fertilizers in hydroponic vegetable production is summarized in Table 3.4.

Table 3.4 Studies on the adoption of liquid biofertilizers in hydroponic vegetable production

Liquid biofertilizer type	Vegetables grown	Yield observations	Author(s)
Vermiwash (0.5%) + ½ Albert's solution	Green bean (*Phaseolus vulgaris L.*)	Highest yield at third picking; improved number of pods and pod weight	Madushani and Karunarathna (2023)
Vegetable waste solution (0.5%) + ½ Albert's solution	Green bean (*Phaseolus vulgaris L.*)	Highest yield at first and second picking; enhanced plant height and leaf area	Madushani and Karunarathna (2023)
FoodLift (food waste-derived)	Lettuce (*Lactuca sativa*), cucumber (*Cucumis sativus*), cherry tomato (*Solanum lycopersicum*)	Lettuce: 156 g/plant; cucumber: 187 g/plant; cherry tomato: Lower yield due to low phosphorus content	Siddiqui et al. (2023)

(continued)

Table 3.4 (continued)

Liquid biofertilizer type	Vegetables grown	Yield observations	Author(s)
CLF (commercial liquid fertilizer)	Lettuce (*Lactuca sativa*), cucumber (*Cucumis sativus*), cherry tomato (*Solanum lycopersicum*)	Lettuce: 161 g/plant; cucumber: 243 g/plant; cherry tomato: Higher yield compared to FoodLift	Siddiqui et al. (2023)
Liquid organic fertilizer from molasses, distillery slop, sugarcane leaves	Green Cos lettuce (*Lactuca sativa var. longifolia*)	Total nitrogen: 0.14–0.33%; total phosphorus (P_2O_5): 0.002–0.017%; total potassium (K_2O): 0.81–11.8%; effective in enhancing lettuce growth	Phibunwatthanawong and Riddech (2019)
Liquid organic fertilizer from dairy cow manure, laying hen manure and *Indigofera zollingeriana*	Lettuce (*Lactuca sativa*)	Total nitrogen: 3.11%; P_2O_5: 2.12%; K_2O: 1.94%; optimal lettuce growth observed with 75% AB Mix +25% LOF and 50% AB Mix +50% LOF combinations	Shaik et al. (2022)
Fish-based and fish-cum-plant-based organic fertilizers	Lettuce (*Lactuca sativa*)	Lower yield compared to inorganic fertilizers; however, fish-based (OF_1) and fish-cum-plant-based (OF_2 and OF_6) fertilizers performed better among organic options	Samarakoon et al. (2019)
Liquid biofertilizer (Impro Tanotec) + foliar fertilizer (Growmore)	Water spinach (*Ipomoea reptans*)	Best results with 3.5 ml/l biofertilizer and 3 g/l foliar fertilizer; significant improvements in plant height, leaf number and fresh weight per plant	Wang et al. (2017)

3.3 Conclusions

Soilless crop production is on the increase due to irrigation water challenges, shortage of agricultural land as well as environmental concerns. The essential plant macro and micronutrients are fed directly to the roots. Usually chemical mineral solutions are used to supply the nutrients. These are formulated depending on nutritional requirements of crops being grown. Although the efficacy of the chemical nutrient solutions cannot be doubted, they are expensive and not environmentally friendly. The challenges associated with the use of chemical nutrient solutions triggered the shift towards liquid biofertilizers. The nutrient compositions of these liquid biofertilizers however depend on the type of substrates and the production methods. Organic fertilizers must be carefully selected to produce a balanced nutrient solution specifically designed for hydroponic systems to optimize crop production. Liquid fertilizer especially in hydroponics production of vegetable production is increasingly becoming popular; hence further research is therefore required to improve the efficacy of such systems and to understand their significance for food security, nutrition and overall sustainability given the rise in global issues associated with them.

References

Al-najjar D, Jamjom R, & Ayaydeh H (2016) Application of anaerobic digester effluent as a fertilizer for vegetation. Palestine Polytechnic University Thesis

Alneyadi K, Almheiri M, Tzortzakis N, Di Gioia F, Ahmed Z (2024) Organic-based nutrient solutions for sustainable vegetable production in a zero-runoff soilless growing system. J Agric Food Res 15:101035

Ammar EE, Rady HA, Khattab AM, Mohamed B, Amer H, Mohamed SA, Elodamy NI, Farga A, Aioub AAA (2023) A comprehensive overview of eco-friendly bio fertilizers extracted from living organisms. Environ Sci Pollut Res 2023(30):113119–113137. https://doi.org/10.1007/s11356-023-30260-x

Arancon NQ, Owens JD, Converse C (2019) The effects of vermicompost tea on the growth and yield of lettuce and tomato in a non-circulating hydroponics system. J Plant Nutr 42(19):2447–2458

Arosha KPL, Sarvananda L (2022) Vermicomposting tea: a potential liquid biofertilizer. Front Life Sci Res 1(1). https://doi.org/10.56397/FLSR.2022.10.01

Atiyeh RM, Subler S, Edwards CA, Bachman G, Metzger JD, Shuster W (2000) Effects of vermicomposts and composts on plant growth in horticultural container media and soil. Pedobiologia 44(5):579–590. https://doi.org/10.1078/S0031-4056(04)70073-6

Basosi R, Spinelli D, Fierro A, Jez S (2014) Mineral nitrogen fertilizers: environmental impact of production and use. NOVA Science Publishers, New York, NY, pp 1–43

Baum C, El-Tohamy W, Gruda N (2015) Increasing the productivity and product quality of vegetable crops using arbuscular mycorrhizal fungi: a review. Sci Hortic 187:131–141

Bergstrand KJ, Asp H, Hultberg M (2020) Utilizing anaerobic digestates as nutrient solutions in hydroponic production systems. Sustainability. 12(23):10076

Bergstrand KJ (2022) Organic fertilizers in greenhouse production systems–a review. Sci Hortic 295:110855

Bidabadi S, Afazel M, Dehghani Poodeh S (2016) The effect of vermicompost leachate on morphological, physiological and biochemical indices of Stevia rebaudiana Bertoni in a soilless culture system. Int J Recycl Org Waste Agricult. 5(3):251–262. https://doi.org/10.1007/s40093-016-0135-5

Choubey G, Dhargave T, Joshi Y, Karadbhajne V, Jambhekar H (2018) A pilot scale process for the production of high shelf life multi-functional liquid biofertilizers. Int J Bio-Technol Res (IJBTR) 8(5):1–10

Churilova EV, Midmore DJ (2019) Vermiliquer (Vermicompost Leachate) as a complete liquid fertilizer for hydroponically-grown pak choi (*Brassica chinensis* L.) in the tropics. Horticulturae 5:26

Daniel AI, Fadaka AO, Gokul A, Bakare OO, Aina O, Fisher S, Burt AF, Mavumengwana V, Keyster M, Klein A (2022) Biofertilizer: the future of food security and food safety. Microorganisms 10:1220. https://doi.org/10.3390/microorganisms10061220

Dantroliya S, Joshi C, Mohapatra A, Shah D, Bhargava P, Bhanusha S, Pandita R, Joshi C, Joshi M (2022) Creating wealth from waste: an approach for converting organic waste in to value-added products using microbial consortia. Environ Technol Innov 25(2022):102092

Dasgan HY, Aldiyab A, Elgudayem F, Ikiz B, Gruda NS (2022) Effect of biofertilizers on leaf yield, nitrate amount, mineral content and antioxidants of basil (*Ocimum basilicum* L.) in a floating culture. Sci Rep 12:20917

Dasgan HY, Yilmaz M, Dere S, Ikiz B, Gruda NS (2023a) Bio-fertilizers reduced the need for mineral fertilizers in soilless-grown Capia pepper. Horticulturae 9:188. https://doi.org/10.3390/horticulturae9020188

Dasgan HY, Kacmaz S, Arpaci BB, Ikiz B, Gruda NS (2023b) Biofertilizers improve the leaf quality of hydroponically grown baby spinach (*Spinacia oleracea* L.). Agronomy 13:575. https://doi.org/10.3390/agronomy13020575

Dearborn Y (2011) Compost tea: literature review on production, application and plant disease management. San Francisco Department of Environment

Dere S, Coban A, Akhoundnejad Y, Ozsoy S, Dasgan HY (2019) Use of mycorrhiza to reduce mineral fertilizers in soilless melon (*Cucumis melo* L.) cultivation. Not Bot Horti Agrobo 47(4). https://doi.org/10.15835/nbha47411738

Deepthi MP, Nivethitha S, Saminathan K, Narendhirakannan RT, Karmegam N, & Kathireswari P (2021) Effect of vermiwash prepared from livestock biowaste as vermiponics medium on the growth and biochemical indices of Amaranthus viridis L. Environmental Technology & Innovation, 21:101300. https://doi.org/10.1016/j.eti.2020.101300

Dhananjani BA, Pakeerathan K (2022) Organic nutrient solutions for hydroponic spinach (*Basella alba*) production in urban agriculture. J Agric Sci – Sri Lanka 18(1):01–13. https://doi.org/10.4038/jas.v18i1.10095

Di Gioia F, Rosskopf E (2021) Organic hydroponics: a US reality challenging the traditional concept of "organic" and "soilless" cultivation. In: III international symposium on soilless culture and hydroponics: innovation and advanced technology for circular horticulture. 1321: 275–282

Dume B, Pizzeghello D, Francioso O, Mwandira M, Musonge P (2022) Vermicomposting technology as a process able to reduce the content of potentially toxic elements in sewage sludge. Agronomy 12(9):2049. https://doi.org/10.3390/agronomy12092049

El-Haddad ME, Zayed MS, El-Sayed GAM, Hassanein MK, Abd El-Satar AM (2014) Evaluation of compost, vermicompost and their teas produced from rice straw as affected by addition of different supplements. Ann Agric Sci 59(2):243–251

Fragalà F, Castello I, Puglisi I, Padoan E, Baglieri A, Montoneri E, Vitale A (2022) New insights into municipal biowaste derived products as promoters of seed germination and potential antifungal compounds for sustainable agriculture. Chem Biol Technol Agric. 9(1):33. https://doi.org/10.1186/s40538-022-00333-z

Franco A, Pucciarelli V, Hosseini SA, Schmitt E, Bovera F, Scieuzo C, Falabella P (2025) Bioconversion of Meat and Fish-Based Former Foodstuffs by Black Soldier Fly Larvae: A Sustainable Pathway for Reducing Food Waste, Enhancing Nutrient Recovery, with a Circular Economy Approach. Insects. 16(5):508. https://doi.org/10.3390/insects16050508

Garg J, Rakshit A (2024) Compost tea: an emerging nature-based supplement strengthening options for durable agriculture. J Soil Sci Plant Nutr 24:8075–8098. https://doi.org/10.1007/s42729-024-02100-5

Gebiola M, Rodriguez MV, Garcia A, Garnica A, Tomberlin JK, Hopkins FM, Mauck KE (2023) Bokashi fermentation of brewery's spent grains positively affects larval performance of the black soldier fly *Hermetia illucens* while reducing gaseous nitrogen losses. Waste Manag 171:411–420. https://doi.org/10.1016/j.wasman.2023.09.033

Congilosi JL, Wallace JS, Neher TP, Howe A, Soupir ML, Aga DS (2022) Co-occurrence of antimicrobials and metals as potential drivers of antimicrobial resistance in swine farms. Frontiers in Environmental Science 10:1018739. https://doi.org/10.3389/fenvs.2022.1018739

González-Hernández AI, Pérez-Sánchez R, Plaza J, Morales-Corts MR (2022) Compost tea as a sustainable alternative to promote plant growth and resistance against Rhizoctonia solani in potato plants. Sci Hortic 300:111090. https://doi.org/10.1016/j.scienta.2022.111090

Govere S, Madziwa B, Mahlatini P (2011) The nutrient content of organic liquid fertilizers in Zimbabwe. Int J Mod Eng Res 1:196–202

Gupta S, Garg VK, Yadav S (2020) Vermicomposting of the organic waste: environmental benefits and implications for sustainable agriculture. J Environ Manage 261:110275. https://doi.org/10.1016/j.jenvman.2019.110275

Hatam S, Bahman F-N, Hasan A, Gholam A, Sadegh S (2015) An overview of the benefits of compost tea on plant and soil structure. Adv Bioresearch 6(61):154–158. https://doi.org/10.15515/abr.0976-4585.6.1.154158

Heintze S (2023) Evaluation of a low-tech approach to mobilize nutrients from organic residues for the production of a hydroponic nutrient solution. Master Thesis submitted to Institute of Agricultural Sciences in the tropics Management of Crop Water Stress in the tropics and subtropics. University of Hohenheim

Hoagland DR, Arnon DI (1938) The water-culture method for growing plants without soil (Circular 347). California Agricultural Experiment Station

Huang K, Xia H (2018) Role of earthworms' mucus in vermicomposting system: biodegradation tests based on humification and microbial activity. Sci Total Environ 610:703–708

Ingham ER (2005) The Compost Tea Brewing Manual. 5th ed. Corvallis (OR): Soil Foodweb Incorporated

Kadhum AA, ALobaidy BJ, Al-joboory W (2021) The effect of bio and mineral fertilizers on growth and yield of wheat (*Triticum aestivum* L.). IOP Conf Series: Earth Environ Sci 761(2021):012004. https://doi.org/10.1088/1755-1315/761/1/012004

Kechasov D, Verheul MJ, Paponov M, Panosyan A, & Paponov IA (2021) Organic Waste-Based Fertilizer in Hydroponics Increases Tomato Fruit Size but Reduces Fruit Quality. Frontiers in Plant Science, 12, 680030. https://doi.org/10.3389/fpls.2021.680030

Kurniawati A, Toth G, Ylivainio K, Toth Z (2023) Opportunities and challenges of bio-based fertilizers utilization for improving soil health. Org Agr 2023(13):335–350. https://doi.org/10.1007/s13165-023-00432-7

Lopez FA (2019) Alternative nutrient solutions in simplified hydroponics. Free University of Bozen-Bolzano

Lottering S, Mafongoya P, Lottering R (2020) Drought and its impacts on small-scale farmers in sub-Saharan Africa: a review. S Afr Geogr J 103(3):319–341. https://doi.org/10.1080/03736245.2020.1795914

Madushani GHK, Karunarathna B (2023) Impact of organic liquid fertilizers on growth and yield performance of Green Bean (Phaseolus vulgaris L.) in non-circulating hydroponic system. AGRIEAST J Agric Sci. 17(1):25–41

Malayil S, Chanakya HN (2020) Leaf Litter Biogas Digester Residue—A Nutrient Supplement for Mushroom Cultivation. In: Ghosh S, Sen R, Chanakya H, Pariatamby A (eds) Bioresource Utilization and Bioprocess. Springer, Singapore. https://doi.org/10.1007/978-981-15-1607-8_18

Margaret GS, Becky AN, Billy M, Baptist TJ, Patrick NA, Ernest MR (2022) Prospects for soilless farming in Africa: a review on the aids of plant growth-promoting rhizobacteria inoculants in hydroponics. Int J Biosci 21(5):273–282

Martin CCGS, Brathwaite RAI (2012) Compost and compost tea: principles and prospects as substrates and soil-borne disease management strategies in soil-less vegetable production. Biol Agric Hortic 28(1). https://doi.org/10.1080/01448765.2012.67151

Masere TP (2022) Evaluation of the role of small-scale farmers in soil and water conservation management in the context of climate change. In: Ondrasek G, Zhang L (eds) Resource management in agroecosystems. IntechOpen. https://doi.org/10.5772/intechopen.108889

Mauerer M, Rocksch T, Dannehl D, Schuch I, Mewis I, Förster N, Ulrichs C, Schmidt U (2021). Replacing Mineral Fertilizer with Nitrified Human Urine in Hydroponic Lettuce (Lactuca sativa L.) Production. Sustainability, 15(13), 10684. https://doi.org/10.3390/su151310684

Michałojć Z, Jarosz Z, Pitura K, Dzida K (2015) Effect of mycorrhizal colonization and nutrient solutions concentration on the yielding and chemical composition of tomato grown in rockwool and straw medium. Acta Sci Pol Hortorum Cultus 14:15–27

Muthu HD, Izhar TNT, Zakarya IA, Saad FNM, Ngaa MH (2023) Comparative study between organic liquid fertilizer and commercial liquid fertilizer and their growth performances on mustard greens. IOP Conf Ser Earth Environ Sci 1135(1):012002

Natchavatthon K, Thammared C (2019) Nutrient dynamic of vermicompost tea after adding molasses and oxygen. Int J Environ Rural Dev 10(1):6–7

Ning J-Y, Zhu X-D, Liu H-G, Yu G-H (2021) Coupling thermophilic composting and vermicomposting processes to remove Cr from biogas residues and produce high value-added biofertilizers. Bioresour Technol 329:124869. https://doi.org/10.1016/j.biortech.2021.124869

Nishana F, & Rahman R (2021) Sustainable Hydroponics using Fermented Plant Juice Nutrition. International Journal of Engineering Research & Technology, 10(6)

Ouko KO, Odiwuor MO (2023) Contributing factors to the looming food crisis in sub-Saharan Africa: opportunities for policy insight. Cogent Soc Sci 9(1):1–14. https://doi.org/10.1080/23311886.2023.2173716

Pant A, Radovich TJK, Hue NV, Arancon NQ (2011) Effects of vermicompost tea (aqueous extract) on pak choi yield, quality, and soil biological properties. Compost Sci Util. 19(4):172–181

Pant AP, Radovich TJK, Hue NV, Paull RE (2012) Biochemical properties of compost tea associated with compost quality and effects on pak choi growth. Scientia Horticulturae 148:138–146. https://doi.org/10.1016/j.scienta.2012.09.019

Panuccio MR, Mallamaci C, Attinà E, Muscolo A (2021) Using digestate as fertilizer for a sustainable tomato cultivation. Sustainability. 13(3):1574. https://doi.org/10.3390/su13031574

Pelayo L, Hultberg M, Bergstrand KJ, Larsson-Jönsson H, Caspersen S, & Asp H (2020) Biogas Digestate in Vegetable Hydroponic Production: pH Dynamics and pH Management by Controlled Nitrification. Waste and Biomass Valorization, 11:2645–2650. https://doi.org/10.1007/s12649-020-00965-y

Phibunwatthanawong T, Riddech N (2019) Liquid organic fertilizer production for growing vegetables under hydroponic condition. Int J Recycl Org Waste Agric 8:369–380. https://doi.org/10.1007/s40093-019-0257-7

Pilla N, Tranchida-Lombardo V, Gabrielli P, Aguzzi A, Caputo M, Lucarini M, Durazzo A, Zaccardelli M (2023) Effect of compost tea in horticulture. Horticulturae 9:984

Raza A, Ali Z, Ali SA (2023) Standardization of protocol for the formulation of compost tea and its efficacy study on potato. Pure Appl Biol 12(2):1044–1055. https://doi.org/10.19045/bspab.2023.120107

Resh HM (2022) Hydroponic food production: a definitive guidebook for the advanced home gardener and the commercial hydroponic grower, 8th edn. CRC Press

Ruiz JL, Salas Sanjuán MDC (2022) The use of plant growth promoting bacteria for biofertigation; effects on concentrations of nutrients in inoculated aqueous vermicompost extract and on the yield and quality of tomatoes. Biol Agric Hortic. 38(1):1–19. https://doi.org/10.1080/01448765.2021.2019652

Salas-Sanjuán MC, Ruiz-Zubiate JL, Valenzuela JL, & Campos AX (2023) Nutrient Solution from Aqueous Extracts as an Alternative to Fertigation in Hydroponic. Horticulturae, 9(12):1281. https://doi.org/10.3390/horticulturae9121281

Samarakoon UC, Weerasinghe P, Weerakkody WAP (2019) Evaluation of fish-based organic fertilizers for hydroponic lettuce production. Trop Agric Res. 30(2):112–123

Sambo P, Nicoletto C, Giro A, Pii Y, Valentinuzzi F, Mimmo T, Lugli P, Orzes G, Mazzetto F, Astolfi S, Terzano R, Cesco S (2019) Hydroponic solutions for soilless production systems: issues and opportunities in a smart agriculture perspective. Front Plant Sci 10:923. https://doi.org/10.3389/fpls.2019.00923

Santosh DT, Gaikwad DJ (2023) Advances in hydroponic systems: types and management. In: Maitra S, Gaikwad DJ, Santosh DT (eds) Advances in agricultural technology. Griffon, Ottawa, pp 16–28. ISBN: 978-17-77795-91-7

Shaik A, Singh H, Singh S, Montague T, Sanchez J (2022) Liquid organic fertilizer effects on growth and biomass of lettuce grown in a soilless production system. HortScience. 57(3):447–452

Siddiqui Z, Hagare D, Liu MH, Panatta O, Hussain T, Memon S, et al. (2023) A food-waste-derived organic liquid fertiliser for sustainable hydroponic cultivation of lettuce, cucumber and cherry tomato. Foods. 12(4):719

Steiner AA (1961) A universal method for preparing nutrient solutions of a certain desired composition. Plant Soil 15(2):134–154

Szekely I, Jijakli MH (2022) Bioponics as a promising approach to sustainable agriculture: a review of the main methods for producing organic nutrient solution for hydroponics. Water 14:3975. https://doi.org/10.3390/w14233975

Souffront DK, Salazar-Amoretti D, Jayachandran K (2022) Influence of vermicompost tea on secondary metabolite production in tomato crop. Scientia Horticulturae 301:111135. https://doi.org/10.1016/j.scienta.2022.111135

Takemura K, Endo R, Shibuya T, & Kitaya Y (2020) Application of Biogas Digestate as a Nutrient Solution for the Hydroponic Culture of Chrysanthemum morifolium Ramat with Rockwool Substrate. Waste and Biomass Valorization, 11, 2645–2650. https://doi.org/10.1007/s12649-018-00576-8

Thomas L, Larroche C, Pandey A (2013) Current developments in solid-state fermentation. Biochem Eng J. 81:146–161

Tikasz P, MacPherson S, Adamchuk V, Lefsrud M (2019) Aerated chicken, cow, and turkey manure extracts deferentially affect lettuce and kale yield in hydroponics. Int J Recycl Org Waste Agric. 8(3):241–251. https://doi.org/10.1007/s40093-019-0261-y

Torres EC, Somera CGG (2023) How organic fertilizers can be used as a plant nutrient source in hydroponics: a review. Appl Sci Eng Prog 16(4):6359

Uparivong S (2012) Use of bio clean and liquid biofertilizers in the temples for enhancing a natural environment. Int J Environ Rural Dev 3(2):200–210

Upendri HFL, Karunarathna B (2021) Organic nutrient solution for hydroponic system. Acad Lett:Article 1893. https://doi.org/10.20935/AL1893

Uzor EF (2018) Production of liquid biofertilizer and biochemical characterization of component nitrogen-fixing and phosphate-solubilizing bacteria species. (Unpublished bachelor's thesis). Godfrey Okoye University, Enugu

van Ittersum MK, van Bussel LG, Wolf J, Grassini P, van Wart J, Guilpart N, Claessens L, de Groot H, Wiebe K, Mason-D'Croz D, Yang H, Boogaard H, van Oort PA, van Loon MP, Saito K, Adimo O, Adjei-Nsiah S, Agali A, Bala A, Chikowo R, Cassman KG (2016) Can sub-Saharan Africa feed itself? Proc Natl Acad Sci USA 113(52):14964–14969. https://doi.org/10.1073/pnas.1610359113

Velazquez-Gonzalez RS, Garcia-Garcia AL, Ventura-Zapata E, Barceinas-Sanchez JDO, Sosa-Savedra JC (2022) A review on hydroponics and the technologies associated for medium and small-scale operations. Agriculture 12:646. https://doi.org/10.3390/agriculture12050646

Vukobratović M, Vukobratović Z, Lončarić Z, Kerovac D (2014) Heavy metals in animal manure and effects of composting on it. Acta Horticulturae 1034:75–82. https://doi.org/10.17660/ActaHortic.2014.1034.75

Wang L, Chen X, Guo W, Li Y, Yan H, Xue X (2017) Yield and nutritional quality of water spinach (Ipomoea aquatica) as influenced by hydroponic nutrient solutions with different pH adjustments. International Journal of Agriculture & Biology. 19(4):635–642

Wongkiew S, Koottatep T, Polprasert C, Prombutara P, Jinsart W, & Khanal SK (2021) Bioponic system for nitrogen and phosphorus recovery from chicken manure: Evaluation of manure loading and microbial communities. Waste Management, 120:67–76

Wongkiew S, Polprasert C, Noophan PL, et al. (2023) Effects of vermicompost leachate (VCL) supplementation on nitrogen, phosphorus and microbiome in a food-waste bioponic system. J Environ Manag 339:117860. https://doi.org/10.1016/j.jenvman.2023.117860

Yeh CK, Lin C, Shen HC, Cheruiyot NK, Camarillo ME, Wang CL (2020) Optimizing food waste composting parameters and evaluating heat generation. Appl Sci 2020(10):2284. https://doi.org/10.3390/app10072284

Yusuf H, Nazif DM (2023) Production of liquid biofertilizer from cassava peels and chicken droppings using microbial inoculants. Int Res J Modern Eng Technol Sci 05(07):2362–2368

Zaini MSM, Syafi WA (2021) Recycling of waste tea leaves via vermicomposting process and the effect on water spinach growth. Kem Ind 70(7–8):387–392

Zhang X, Khalid M, Wang R, Chi Y, Zhang D, Chu S, Yang X, Zhou P (2023) Enhancing lettuce growth and rhizosphere microbial community with Bacillus safensis YM1 compost in soilless cultivation: an agricultural approach for kitchen waste utilization. Sci Hortic 21:112345. https://doi.org/10.1016/j.scienta.2023.112345

Zarei M, Jahandideh MA, Moridi A (2018) Comparison of vermiwash and vermicompost tea properties produced from different organic beds under greenhouse conditions. International Journal of Recycling of Organic Waste in Agriculture 7:25–32. https://doi.org/10.1007/s40093-017-0186-2

ZimVac (2019) Urban livelihoods assessment report. Food and Nutrition Council, Harare

Zingore S, Mutegi J, Agesa B, Desta LT, Kihara J (2015) Soil degradation in sub-Saharan Africa and crop production options for soil rehabilitation. Better Crops with Plant Food 99(1):24–26

Chapter 4
Potential of Digestate as an Emerging Biomass-Based Biofertilizer for Sustainable Agriculture in Africa

Haripriya Rama, Tetelo Tolo, Gemina Abigail Boitumelo Letsoalo, and Ashira Roopnarain

Abstract Worldwide, the demand for energy and food has increased due to the growth in global population and economy. In Africa, energy and food security are of great concern particularly to resource-poor, smallholder farmers, and communities. Anaerobic digestion (AD) is a microbial-driven technology that can be used to convert organic wastes such as food waste, livestock manure and agricultural residues that are often found on farms, to renewable energy in the form of biogas, which can be used for cooking, heating, and lighting. Although AD technology has been widely established in countries like China and India, among others, its adoption in African countries remains stagnant. Valorizing biogas as well as the effluent released from AD reactors, which is termed "digestate," may promote the economic feasibility and adoption of the AD technology in Africa. Digestate is oftentimes nutrient-rich and contains microorganisms involved in the decomposition of organic wastes; hence it has the potential to be used as a biomass-based biofertilizer. While traditional biofertilizers are typically composed of plant growth-promoting microbes, digestate biomass-based biofertilizers contain various types of microorganisms, nutrients, and organic matter. The digestate can therefore potentially be beneficial for soil health and agricultural productivity, and serve as an alternative to expensive chemical fertilizers. This chapter provides an overview of the biological and chemical properties of digestate. Additionally, the influence of microorganisms associated with the digestate in soil health restoration is discussed. Lastly, the potential of

H. Rama (✉) · T. Tolo · G. A. B. Letsoalo
Microbiology and Environmental Biotechnology Research Group, Soil Science Division, Agricultural Research Council-Natural Resources and Engineering, Pretoria, South Africa
e-mail: ramah@arc.agric.za

A. Roopnarain
Microbiology and Environmental Biotechnology Research Group, Soil Science Division, Agricultural Research Council-Natural Resources and Engineering, Pretoria, South Africa

Department of Environmental Sciences, College of Agriculture and Environmental Sciences, University of South Africa-Florida Campus, Johannesburg, South Africa

A. D. Nciizah et al. (eds.), *Biofertilizers for Sustainable Agriculture in Africa*, Sustainability Sciences in Asia and Africa, https://doi.org/10.1007/978-981-95-6009-7_4

digestate as biomass-based biofertilizers is revealed and recommendations to promote its adoption and use in sustainable agriculture in Africa is provided.

Keywords Digestate · Biomass-based · Biofertilizer · Agriculture · Soil health · Africa

4.1 Introduction

Africa has a growing population of about 1.53 billion people (i.e., 18.3% of the world population) that is expected to reach 2.47 billion in 2050 (Worldometers 2024). Majority of the African population (55.5%) inhabit rural regions. Approximately, 90% of agricultural production in sub-Saharan Africa (SSA) arises from smallholder farming systems that are primarily located in deep rural regions or towns (Raimi et al. 2017). However, a gradual decline in food crop productivity on farmlands in SSA has been observed, which was fundamentally initiated from colonialism and the continued political focus on resource extraction and exportation of non-value-added products (cash crops) to the global north (Bjornlund et al. 2022). The lack of political will to drive agricultural policies that can foster growth within local economies and support smallholder farming systems has contributed to the persisting decline in food crop productivity and security in SSA (Bjornlund et al. 2022). The dual economies experienced in developing countries of SSA has further provoked food insecurity, whereby higher productivity is observed in the modern sectors compared to informal sectors that are the source of labor supply (Raimi et al. 2017). To feed the growing population, it is imperative to enhance food crop productivity by fostering sustainable farming systems that optimally maximize on financial resources and capacitate humans, while harnessing environmental resources including organic matter, nutrients, water, and energy (Roopnarain et al. 2022).

Food security is impacted by the current global energy crisis as most agriculture and food production systems rely on energy to conduct various activities such as automated irrigation, operation of farming machinery, packaging, transportation, and distribution (Rubagumya et al. 2023). In most countries of SSA, only one in three people have access to modern energy (Rubagumya et al. 2023). Accessible fuel sources that are typically used for energy provision include non-renewable fossil fuels such as coal, firewood, and kerosene, which pollute the environment upon combustion, and result in the emission of various greenhouse gases (GHGs) (e.g., methane, carbon dioxide, nitrous oxide, and water vapor) (Bakari et al. 2024). The release of GHGs into the atmosphere as well as utilization of expensive agrochemicals (e.g., bulk inorganic fertilizers and pesticides) further exacerbates the climate crisis (Mgadi et al. 2024). The rapidly changing climatic conditions subsequently pose environmental risks to agro-ecosystems that are susceptible to soil infertility, soil erosion, droughts, and disease, among other threats (Raimi et al. 2021).

In addition to fossil fuels and agro-chemicals, organic waste is decomposed by microorganisms under anoxic conditions, contributing to emission of GHGs due to conventional waste management approaches, i.e., disposal into landfills and open dumping (Rama et al. 2023). Approximately, 48.7% of the waste produced in SSA is organic, including food waste, municipal solid waste, and wastewater as well as human and livestock manure (Rubagumya et al. 2023). Biomass derived from organic agricultural waste can be valorized as a sustainable input source for waste to energy and biofertilizer conversion technologies (Fertahi et al. 2023). African countries generate various types of organic residues from crops such as cassava, maize, sorghum, and banana; however, these waste streams are predominantly underutilized for bioenergy generation (Bakari et al. 2024).

Over the years, anaerobic digestion (AD) technology has emerged as an important opportunity to harness renewable energy in the form of biogas via microbial-driven decomposition of organic waste (Rama et al. 2023). During the AD process, hydrolytic, acidogenic, acetogenic, and methanogenic microorganisms convert complex organic matter that is fed to anaerobic digesters into biogas, which is primarily composed of combustible methane. Biogas technology can therefore be utilized by smallholder farmers as a sustainable renewable energy source for cooking, heating, and lighting by harnessing methane that would have essentially been released into the atmosphere (Rama et al. 2024). India and China have approximately 5 and 42 million operating smallholder farm digesters, respectively, that have greatly improved the livelihood of their farmers by contributing to poverty reduction (Rama et al. 2024). Despite the many benefits of implementing biogas technology on smallholder farming systems, anaerobic digesters have not been widely adopted in Africa (Tolessa 2024). To promote the adoption of biogas technology in Africa, it is essential to valorize organic waste available on smallholder farms for generation of biogas energy as well as digestate production, which has the potential to be used as a soil ameliorant or fertilizer (Roopnarain et al. 2023). This process can encourage a circular economy, as shown in Fig. 4.1.

Various types of fertilizers have been classified including organic and inorganic fertilizers as well as biofertilizers for improving crop productivity (Raimi et al. 2017; Fertahi et al. 2023). Inorganic fertilizers are chemicals that are synthesized and applied in bulk. Inorganic fertilizers are unsustainable due to the associated cost implications and detrimental effects on the environment. On the other hand, organic fertilizers comprise plant or animal materials that are oftentimes environmentally friendly; however, there are some challenges with its use by African farmers, which include cost, management constraints, and availability (Raimi et al. 2021). Biofertilizers are the third type of fertilizer that comprise microorganisms with capabilities of improving soil fertility as well as promoting plant growth and development (Fertahi et al. 2023). At the intersection of the different types of fertilizers lies nutrient-rich digestate that comprises recalcitrant organic matter from the AD process as well as microorganisms with the vast untapped potential to improve agricultural soil productivity (Karimi et al. 2022; Roopnarain et al. 2023). Digestate has previously been designated as a biomass-based biofertilizer (Fertahi et al. 2023). Biomass-based biofertilizer is derived from biomass that was previously processed

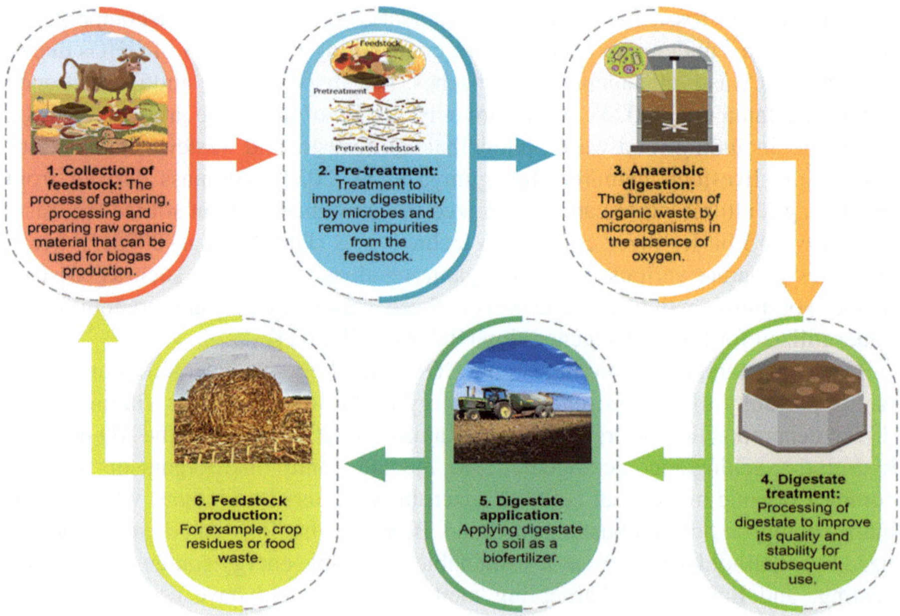

Fig. 4.1 Circular economy waste management strategy through anaerobic digestion: From feedstock collection to feedstock production. (Source: Author's own work)

chemically, biologically, or thermochemically. Examples of other biomass-based biofertilizers include biochar, compost, and biopolymer-coated fertilizers (Fertahi et al. 2023). To valorize digestate as a biomass-based biofertilizer, it is necessary that the properties of digestate are environmentally benign and that it improves soil fertility and quality for the production of high quality foods that are safe for human and animal consumption (Sharma et al. 2023). Digestate may potentially promote sustainable agriculture as an alternative to expensive and polluting agro-chemicals conventionally used by African farmers (Sharma et al. 2023). This chapter aims to provide an overview of the biological and chemical properties of digestates as well as their potential to be valorized as a biomass-based biofertilizer for sustainable agriculture in African countries.

4.2 Properties of Digestate

Chemical and biological properties provide insight on how substances interact and function within living organisms. Chemical properties describe the characteristics of a substance that become evident during chemical reactions, such as reactivity, flammability, pH, and toxicity, whereas biological properties focus on how chemical substances function within biological systems (Siheri et al. 2017). Digestate is the by-product of AD, a biological process that involves the breakdown of organic waste in oxygen-free environments. The type of feedstock used and the specific AD process parameters influence the characteristics of digestate. Utilizing digestate protects the environment and preserves the process's commercial viability. This section explores the digestate's chemical and biological properties, focusing on its nutrient content, microbial composition, and implications for agricultural use.

4.2.1 Chemical Composition of Digestate Fractions

Digestate consists of valuable quantities of nitrogen (N), phosphorus (P), and potassium (K) that are essential macronutrients for plant growth (Celletti et al. 2021; Asli et al. 2022). It can also be classified based on its dry matter (DM) content. It is normally produced "whole" in a slurry consistency. The initial phase in digestate processing is mechanical separation into two fractions: solid and liquid digestate. According to Makádi et al. (2012), solid digestate typically has a DM content of more than 15%, and liquid digestate typically has a DM content of less than 15%. To stay clear of mobility constraints and make soil-digestate integration easier, digestate is generally fractioned into solid and liquid portions (Kovačić et al. 2022). The solid fraction is rich in organic matter, particularly fibrous materials, encompassing lignin and cellulose, and some nutrients, including P (Celletti et al. 2021; Asli et al. 2022). On the other hand, the liquid fraction typically contains more readily available forms of N, P, K, organic compounds, as well as dissolved mineral salts (Vaneeckhaute et al. 2017; Celletti et al. 2021). The nutrient composition of digestate is influenced by initial feedstock with food-based digestate typically exhibiting higher nutrient concentrations than crop and animal manure-based materials (O'Connor et al. 2022). Due to its high N composition in the form of plant available ammonium-N, the liquid digestate can be used as a supplement for mineral N fertilizers (Sigurnjak et al. 2017; Tambone et al. 2017). Conversely, the solid digestate is abundant in P and organic carbon and can be utilized as a P fertilizer or soil-enhancing agent (Egene et al. 2021). However, the fertilizer value of digestate in soil is affected by the nutritional status, crop nutrient requirements, and soil organic carbon content (García-López et al. 2023).

The nutrients in digestate can primarily contribute to long-term soil fertility and act as a slow-release nutrient source. The potential for obtaining soluble nutrients from liquid digestate using extraction techniques is substantial (Vasudevan et al. 2010). These extraction techniques include liquid-liquid extraction, which separates components based on their solubility in different liquids, and Soxhlet extraction, which is a traditional method that involves continuous extraction of a solid sample using a solvent, allowing for thorough extraction of soluble compounds. Methods like liquid-liquid extraction and Soxhlet extraction are accessible and can be implemented at a smaller scale, allowing local farmers to produce organic fertilizers from agricultural waste. By leveraging these techniques, communities can enhance soil fertility, improve crop yields, and promote environmental sustainability through the recycling of nutrients (Vasudevan et al. 2010; Celletti et al. 2021). Chemical treatments employ reagents to precipitate specific nutrients like P, while thermal treatments, such as incineration or pyrolysis, convert waste into nutrient-rich biochar or ash (Kovačić et al. 2022; Korba et al. 2024). These techniques offer sustainable and cost-effective solutions for nutrient recovery, particularly relevant for enhancing agricultural productivity in Africa. The long-term use of digestate as a fertilizer on agricultural land is contingent on excellent management practices (Czekała et al. 2023). According to Massé et al. (2011), digestate has a more balanced nutritional content than pure livestock waste, which decreases the need for nutrient supplementation using conventional fertilizers when addressing crop nutrient requirements.

4.2.2 Digestate pH and Electrical Conductivity

Electrical conductivity (EC) serves as an indicator of the nutrient concentration within the digestate. The EC values can typically range between 0.5 and 30.5 $mS \cdot cm^{-1}$ (milliSiemens/centimeter), depending on the type of substrates added to the anaerobic digesters (Table 4.1). The ionic strength (IS) of digestate also plays a role in determining EC, with significant correlations observed between these parameters (Snyders et al. 2015). Food waste digestate often exhibits high sodium (Na) and chloride (Cl) ions, contributing to elevated EC readings (Matjuda et al. 2023). The presence of these ions can affect plant growth if not managed properly; hence understanding the ionic composition is crucial for effective application in agriculture (Vasudevan et al. 2010; Snyders et al. 2015). Besides EC, variations in pH can also occur based on the type of organic material digested (Table 4.1) and the conditions under which AD occurs, such as temperature and retention time. The pH of digestate from AD of different feedstocks ranged between 7.4 and 8.5 (Roopnarain et al. 2022). The recommended pH range for digestate is between 6.5 and 9. Digestate that has a slightly alkaline pH is beneficial for soil health as it can enhance nutrient availability and microbial activity (Matjuda et al. 2023). In addition, higher pH levels can indicate the presence of ammonium ions, which are readily available for plant uptake (Celletti et al. 2021). The implications of pH and EC extend beyond mere measurements; these factors are vital for predicting how digestate will perform as a fertilizer in agricultural settings. High EC may indicate a rich nutrient source but could also pose salinity risks if applied excessively. Conversely, a balanced pH is essential for optimizing nutrient uptake by plants. Monitoring pH and EC provides essential insights into the quality and usability of digestate as a biomass-based biofertilizer. These parameters not only inform agricultural practices but also contribute to sustainable waste management strategies by promoting the recycling of nutrients back into the soil ecosystem. As research advances, further understanding of these properties will enhance the efficacy and safety of using digestate in various agricultural applications.

Table 4.1 Chemical properties of digestate derived from anaerobic digestion of various feedstock

Feedstock	Dry matter (DM)	pH	Electrical conductivity	Total carbon (C)	Total nitrogen (N)	Phosphorus	Potassium	C/N ratio	References
Beef manure, hay, and silage corn	92.0(g kg^{-1})	8.40	8.38(mS cm^{-1})	–	27.8(mg g^{-1})	9.21(mg g^{-1})	26.93(mg g^{-1})	–	Sharifi et al. (2019)
Cattle slurries (solid digestate)	290(g kg^{-1})	7.8	1.7(mS cm^{-1})	–	27.3(g kg^{-1})	0.72(g kg^{-1})	16(g kg^{-1})	12	Horta and Carneiro (2020)
Food waste	1.4–7.88(%)	7.9–8.3	–	0.44–3.56(% dry weight (DW))	0.06–1.24 (% DW)	0.008–0.126 (% DW)	0.03–0.64 (% DW)	2.87–8.8	Barampouti et al. (2020)
Pig slurry with 1.0% sludge from a slaughterhouse wastewater treatment plant and 6.5% biodiesel wastewaters	1.9(%)	8.3	30.5(mS cm^{-1})	0.47(% fresh weight (FW)-organic)	0.38(% FW)	–	–	1.2	Alburquerque et al. (2012)
Poultry manure	–	8.2	19.8(mS cm^{-1})	28.8(%-organic)	4.5(g L^{-1})	1.2(g L^{-1})	1.82(g L^{-1})	–	Mortola et al. (2019)
Beef slaughtering waste (sludge from wastewater, stomach content and blood)	–	7.3	2.5(mS cm^{-1})	40.3(% DW-organic)	5.0(% DW)	2.4(% DW)	0.2(% DW)	8.0	Ciurli et al. (2025)

(continued)

Table 4.1 (continued)

Feedstock	Dry matter (DM)	pH	Electrical conductivity	Total carbon (C)	Total nitrogen (N)	Phosphorus	Potassium	C/N ratio	References
Typha latifolia grass (cattail), rumen fluid, sewage sludge	–	2.04	–	2.139($g\ L^{-1}$ – organic)	3.635($g\ L^{-1}$)	1.88($g\ L^{-1}$)	2.756($g\ L^{-1}$)		Petrovič et al. (2021)
Silage maize	6.88(%)	7.8	29.2($mS\ cm^{-1}$)	40.9(% DW)	8.79(% DW)	1.01(% DW)	4.82(% DW)	4.65	Häfner et al. (2022)
Cattle manure	–	7.29 ± 0.006	0.557($mS\ cm^{-1}$)	19.945 ± 0.977 (% DW)	1.337 ± 0.011 (% DW)	0.827 ($g\ kg^{-1}$ DM)	6.03 ($g\ kg^{-1}$ DM)	14.92 ± 0.605	Srivastava et al. (2024)
Swine manure	–	8.1	–	–	2.93($g\ L^{-1}$)	0.93($g\ L^{-1}$)	1.37($g\ L^{-1}$)	–	Loria et al. (2007)

4.2.3 Biological Properties of Digestate

The biological properties of digestate highlight its potential as a biomass-based biofertilizer and valuable resource for agricultural practices (Ahmed et al. 2023). Its rich microbial community and high nutrient content can significantly enhance soil health and fertility when managed appropriately (García-López et al. 2023). However, understanding the dynamics of microbial interactions and nutrient availability is essential for maximizing the agronomic benefits of digestate while minimizing adverse effects on soil ecosystems (García-López et al. 2023; Ahmed et al. 2023). The potential role of digestate microorganisms in biofertilization is significant (Table 4.2).

By introducing microbes residing in the digestate into the soil, farmers can enhance microbial diversity and activity, which are critical for nutrient cycling and plant growth (Rowland et al. 2018). For instance, certain bacteria such as Proteobacteria in digestate can produce enzymes that degrade complex organic compounds, making nutrients more accessible to plants (Nigam 2013). Additionally, some fungi in digestate can form symbiotic relationships with plant roots, further improving nutrient uptake and plant health. However, the effectiveness of digestate as biomass-based biofertilizers can vary based on factors such as the carbon-to-nitrogen ratio of the feedstock used in AD and the specific microbial composition of the digestate (Nigam 2013; Rowland et al. 2018). The AD process increases the

Table 4.2 Microorganisms with potential agricultural roles isolated from digestate in Africa

Microorganism(s)	Potential agricultural role(s)	Feedstock	Source location	References
Clostridiales and *Bacteroidales*	Metabolism of organic macromolecules	Cow dung	Kenya	Muturi et al. (2021)
Pseudomonas, *Klebsiella*, *Clostridium*, *Bacillus*, *Bacteroides*, *Penicillium*, *Shigella*, *Salmonella* and *Aspergillus*	Phosphate solubilization and nitrogen fixation	Cow dung and chicken droppings	Nigeria	Alfa et al. (2014)
Bacteroidota, *Proteobacteria*, *Patescibacteria*, and *Acidovorax*	Metabolism of organic macromolecules	Paperboard sludge cake / sewage sludge	Egypt	Hafez et al. (2024)
Pseudomonas sp., *Proteus* sp., *Bacillus* sp. and *Streptococcus* sp.	Phosphate solubilization, nitrogen fixation, and ammonia and indole acetic acid production	Cattle rumen	Nigeria	Oshoma et al. (2025)
Acinetobacter, *Bacillus*, *Pseudomonas*, and *Micrococcus*	Phosphate solubilization and indole acetic acid production	Cow dung	South Africa	Mukhuba et al. (2018)
Aspergillus, *Thermomyces*, *Chaetomium*, and *Microascus*	Organic matter decomposition	Water hyacinth and cow dung	South Africa	Obi et al. (2025)

biological stability of organic matter in digestate, which is characterized by a higher concentration of recalcitrant organic matter fractions, such as lignin and humic substances. Recalcitrant organic matter is less readily decomposed by microorganisms (Nikolaidou et al. 2024). Resistance of recalcitrant substrates to microbial decomposition enhances the long-term benefits of digestate as a soil amendment (García-López et al. 2023; Nikolaidou et al. 2024). The presence of recalcitrant organic matter in digestate not only enhances its stability as a soil amendment but can also foster diverse microbial communities and contribute significantly to maintaining and improving soil health over time (Vaneeckhaute et al. 2017).

4.3 Influence of Digestate on Soil Microbial Activity

Digestate application has been linked to increased soil microbial activity, when combined with carbon-rich amendments like biochar or humic acids (Garg et al. 2005). Based on literature, digestate organic matter can improve soil structure, infiltration rate, and water-holding capacity, resulting in increased productivity and decreased fertilizer costs (Chaudhry et al. 2012). However, more research into potential negative effects on soil biota, particularly microorganisms that play critical roles in the soil ecosystem, is required to assess the sustainability of large-scale digestate land application in agricultural fields (Celletti et al. 2021). It is now widely known that organic additions benefit soil microbial communities by increasing soil microbial biomass and modifying prokaryotic and fungal community structures (García-López et al. 2023). The impact of digestate on soil microbial communities depends on both the soil type and the characteristics of the digestate (Zeng et al. 2016; van Midden et al. 2023). Digestate can potentially serve as a valuable amendment that can enhance microbial activity and support soil health through improved nutrient dynamics and biodiversity (O'Connor et al. 2022). Its potential application in agriculture not only promotes sustainable practices but also contributes to the circular economy by recycling organic waste into productive resources. Continued research into the interactions between digestate and soil microbial activity is essential for optimizing its use in agricultural systems.

Nevertheless, the incorporation of digestate can improve soil structure and function by enhancing microbial activity that contributes to nutrient mineralization, disease suppression, and pollutant degradation (Garg et al. 2005). This biological activity is essential for maintaining ecosystem functions and promoting sustainable agricultural practices. However, the enhanced activity may be temporary, with lasting effects varying based on the digestate's chemical composition, soil respiration, microbial biomass, and carbon levels (Chen et al. 2012). While immediate benefits such as increased microbial activity are observed post-application, studies suggest that these effects may not be sustained over time without ongoing amendments or management practices (Roopnarain et al. 2023; Nikolaidou et al. 2024). In a recent study, digestate was shown to alter soil microbial communities' diversity and composition. In particular, microbial diversity was shown to remain stable while species richness increased as a result of application of digestate as well as its mixture with conventional fertilizers (Nikolaidou et al. 2024). This suggests that digestate supports a broad range of microbial life, crucial for maintaining soil fertility and structure, by enriching the soil with organic matter and nutrients (Nikolaidou et al. 2024). Although digestate can initially boost microbial activity in soils, its prolonged use may lead to shifts in microbial community structure and function. In addition, high doses of digestate were shown to decrease certain microbial activities compared to control treatments (García-López et al. 2023). Therefore, further research is required to understand the long-term impacts of digestate on soil health and microbial dynamics as well as careful management to optimize benefits while mitigating potential drawbacks is necessary.

4.4 Management of Digestate

Challenges such as the presence of contaminants ranging from heavy metals to pathogens pose risks to agricultural soils and necessitate careful digestate management strategies (US EPA 2021). Moreover, the emission of odorous compounds from the digestate can limit its utilization; thus, techniques like composting and direct injection are being explored to mitigate these issues (Zeng et al. 2016; Riva et al. 2016). Mismanagement of digestate may lead to nitrate leaching and ammonia volatilization, contributing to water pollution and air quality degradation, respectively. However, when applied correctly, digestate can reduce these negative impacts. For instance, studies have shown that subsurface injection of digestate can decrease ammonia emissions by as much as 77% compared to surface applications (Riva et al. 2016). This method not only minimizes odor but also enhances nutrient retention in the soil.

To ensure its agronomic benefits, digestate must meet specific quality standards. A digestate of high quality is that which has minimal physicochemical impurities and pathogens (Viancelli et al. 2023). For example, the Department of Agriculture, Forestry, and Fisheries (DAFF) (Republic of South Africa) requires registration of digestate derived from abattoir waste for use as a soil amendment in which the quality and characteristics of the digestate are to be reported and approved (DAFF (RSA) 2018). Each country's individual legislation determines the maximum limitations of these criteria. Regulation of the use of digestate in less developed African countries remains in infancy. For instance, Mmereki et al. (2024) reported that Liberia and Nigeria are delayed in terms of the judicial framework for the use of digestate. On the other hand, South Africa, has "an act to reform the law regulating waste management in order to protect health and the environment by providing reasonable measures for the prevention of pollution and ecological degradation and for securing ecologically sustainable development" (South Africa 2008). Although not explicitly stated, this act may imply the regulation of waste management, including digestate. According to Fertahi et al. (2023), the majority of African countries are keen on valorizing their by-products in order to establish efficient biorefinery processes. Therefore, standardizing comprehensive regulatory frameworks is important to classify and ensure uniformity in the use of digestate across Africa.

The quality of the final digestate is governed by the temporal, operational, and spatial differences together with the type of feedstock (Hammerschmiedt et al. 2022). Nonetheless, proper digestate management is required to minimize the negative effects on an agricultural and environmental scale and to establish an approach that promotes a circular economy (García-López et al. 2023). Provision of digestate-use regulations could facilitate greater adoption of digestate in agriculture, thereby promoting nutrient recycling and reducing reliance on synthetic fertilizers (Möller and Müller 2012; Alburquerque et al. 2012). Further research is essential to optimize digestate application methods and inform policy adjustments that support the sustainable use of this valuable resource in farming systems. As the agricultural sector seeks to balance productivity with environmental stewardship, digestate represents a viable solution for enhancing soil fertility and promoting sustainable practices across Africa and beyond. This holistic approach to soil management not only improves nutrient availability but also promotes overall soil health.

4.5 Sustainable Agronomic Benefits of Digestate

Sustainable agriculture is described by Jeranyama et al. (2020), as a tool that drives environmental health, economic profitability, as well as social and economic equity. Smallholder farmers in many African countries are prone to various challenges that pose a threat to their crop yield (Touch et al. 2024). In addition, land deterioration, such as soil erosion, water scarcity, and loss of biodiversity, leads to significant productivity limitations, making it challenging for smallholder farmers to fulfil the growing demand for food supply (Touch et al. 2024). Therefore, Setsoafia et al. (2022) advises that smallholder farmers apply sustainable agricultural practices in order to counteract the negative externalities of crop intensification.

4.5.1 Soil Fertility Benefits of Digestate

Soil fertility improvement is a critical aspect of sustainable agriculture, and the use of digestate has emerged as a promising solution. The application of digestate can potentially enhance soil fertility and crop productivity by providing essential nutrients such as N, P, K, calcium (Ca), and magnesium (Mg) to the plant (Fig. 4.2). Some studies indicate that digestate can effectively replace traditional mineral fertilizers without compromising yield (Alburquerque et al. 2012; Riva et al. 2016). Mineral fertilizers have been widely used for many farming practices in Africa. However, according to Tongwane et al. (2016), they have contributed to over 57%

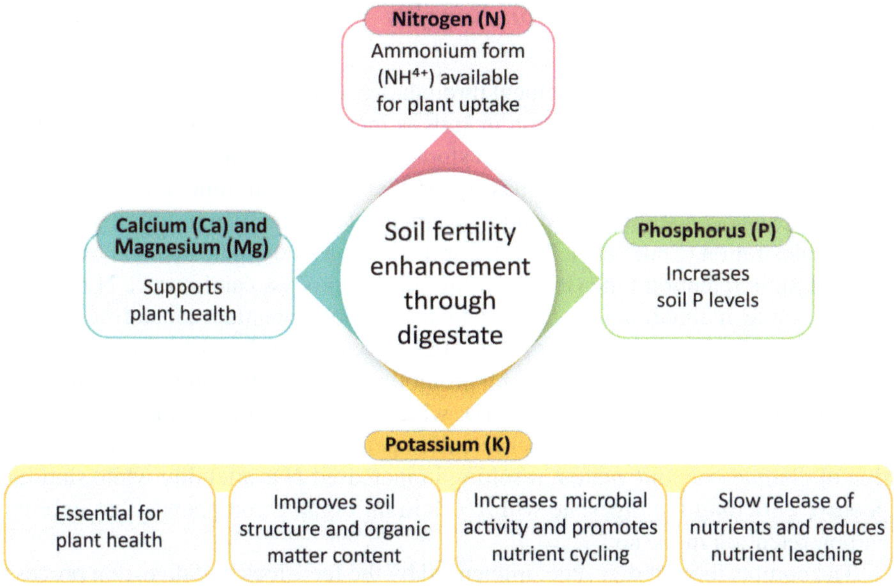

Fig. 4.2 Soil fertility enhancement through digestate application. (Source: Author's own work)

of GHGs in South Africa. Hence, the use of digestate can mitigate environmental issues associated with conventional fertilization practices (Alburquerque et al. 2012). For example, Riva et al. (2016) demonstrated that both digestate and its liquid fraction could substitute urea in maize silage production, achieving comparable agronomic performance while reducing odor and ammonia emissions when applied through subsurface injection methods.

Since digestate is packed with essential nutrients and is a vigorous soil amendment, its valorization as a biofertilizer is optimized by recovering nutrients such as N, K, and P, alleviating the increased demand for synthetic chemical fertilizers in Africa (Matjuda et al. 2023). These nutrients can be combined in different ratios to produce fertilizers that meet the specific nutritional needs of different crops (Chojnacka and Moustakas 2024), thus reducing the environmental expenses caused by synthetic fertilizers (Matjuda et al. 2023). Unlike synthetic fertilizers, which are harmful to the environment, digestate provides a balanced and varied nutrient profile. Moreover, according to Shaji et al. (2021), digestate gradually releases nutrients into the soil, promoting balanced nutrient levels for optimal plant growth. Due to its slow-release nature, digestate contributes to reduced environmental impact, and this, therefore, results in a lower risk of run-offs. Synthetic fertilizers, on the other hand, are emission intensive and usually leach into water streams, thus leading to eutrophication and harmful algal blooms (Akinnawo 2023). In addition, digestate is a cost-effective solution for smallholder farmers, as it offers a bargain offer of organic matter and nutrients in one product.

4.5.2 Carbon Sequestration Benefits

Apart from soil fertility enhancement through digestate application, applying digestate for the sequestration of carbon in soils is increasingly recognized as a significant avenue for enhancing soil carbon storage while providing agronomic benefits (Nyang'au et al. 2022). Research has shown that the carbon mineralization dynamics of digestate can significantly influence their effectiveness in promoting soil carbon sequestration (Smith et al. 2005). For instance, a study indicated that extending the hydraulic retention time (HRT) during the AD process can improve N mineralization while maintaining or enhancing the carbon retention capabilities of the digestate applied to soil (Egene et al. 2021). Extending the HRT during the AD process enhances N mineralization by promoting a stable environment for microorganisms, particularly nitrifying bacteria, which convert N sources into bioavailable forms (Egene et al. 2021). A longer retention period also facilitates the decomposition of complex organic matter, resulting in increased N availability while simultaneously enriching the digestate with recalcitrant organic compounds that improve carbon retention in the soil.

The composition of digestate, influenced by the feedstock and digestion process, plays a crucial role in its carbon sequestration potential. Digestate derived from a two-step AD process has been found to exhibit lower carbon mineralization rates compared to those from single-step processes, suggesting that prolonged digestion reduces the amount of decomposable organic matter, thereby enhancing long-term carbon stability in soils (Uludag-Demirer and Demirer 2022). Additionally, studies have demonstrated that applying digestate to agricultural soils can lead to increased soil organic carbon levels, contributing to climate change mitigation efforts by reducing atmospheric CO_2 levels (Møller et al. 2022; Uludag-Demirer and Demirer 2022). Moreover, the use of digestate as a soil amendment not only supports carbon sequestration but also reduces reliance on synthetic fertilizers, which are associated with significant GHG emissions during production and application (Zilio et al. 2022). This dual benefit aligns with sustainable agricultural practices aimed at improving soil health while minimizing environmental impacts (Møller et al. 2022; Uludag-Demirer and Demirer 2022). Furthermore, optimizing AD process parameters, such as HRT, and incorporating secondary digestion steps can enhance carbon retention in soils when the resulting digestate is applied as a biomass-based biofertilizer (Romio et al. 2021).

4.6 Digestate Application and Adoption

Digestate application is a complex process that is shaped by environmental, agricultural, and economic factors. Studies have shown that South Africa has normalized making digestate from slaughterhouse waste and applying it as a biofertilizer on agricultural land (Jankauskienė et al. 2024). However, effective application of digestate requires suitable equipment and technology to ensure precision and efficient application to soil with efforts to enhance soil nutrient content, reduce waste, promote sustainable agriculture, and minimize nutrient run-off (Jin et al. 2022). Moreover, the application techniques of digestate are reliant on solid-liquid separation (Dhull et al. 2024). Therefore, it is imperative to understand the solid and liquid digestate properties in order to optimize its use.

4.6.1 Strategies for Applying Digestate

Digestate can be effectively applied using various techniques. Direct application to soil is a conventional and cost-effective approach that provides an effective means of waste disposal and nutrient recovery for agricultural systems (Barłóg et al. 2020). For instance, solid digestate can be applied directly to the soil and incorporated into the top layer. Based on the study by Slepetiene et al. (2023), the application of solid digestate heightened the soil organic carbon content in the 0–10 cm soil top layer significantly, but did not reach the 20–30 cm layer. Nonetheless, liquid digestate is generally preferred over solid digestate as it allows nutrients to be easily accessible to plants. Although the quality of digestate is primarily dependent on the feedstock used, its effective use can be optimized through various application methods. For example, liquid digestate can be applied through injection in irrigation water to enhance efficient use of nutrients, provided that a proper solid-liquid separation system is in place to prevent clogging (Finzi et al. 2021). Additionally, co-composting is an effective technique for solid digestate as it allows the aerobic degradation of organic waste mixtures, primarily aiming at obtaining compost that can be used as fertilizer or soil amendment (Giagnoni et al. 2020). Composting of digestate effectively reduces potential negative consequences of digestate application to soils (Weldon et al. 2023).

Other digestate application methods that can significantly affect its effectiveness include broadcasting, bandspreading, or injection (van Midden et al. 2023). Broadcasting involves uniform distribution of digestate across the entire soil surface (Yan et al. 2023). Bandspreading, on the other hand, applies digestate in narrow, concentrated bands near crop rows or roots thus reducing waste and increasing nutrient efficiency (Nicholson et al. 2017). However, Korba et al. (2024) state that it is being replaced by precision spreading equipment, which offers precise application by a band spreader or by direct injection into the soil, thereby reducing ammonia emissions that would occur through volatilization of field-applied digestate. Some applicators inject digestate into the soil through a straw layer to minimize ammonia losses, although this method can result in spatial separation of the straw and digestate. Among the different methods proposed to reduce GHGs, the direct injection of digestate into the soil has been proposed as a successful sustainable agricultural practice (Orzi et al. 2018).

4.6.2 Integration with Other Fertilization Practices

Digestate can be integrated effectively with other fertilization practices to optimize nutrient management, improve soil health, and enhance crop productivity. According to Nikolaidou et al. (2024), combining chemical fertilizers with digestate enhances microbial biomass and carbon availability, leading to accelerated nitrite immobilization and improved soil health. Similarly, a study by Ehmann et al. (2018) revealed that the combination of liquid digestate with mineral fertilizer led to a higher quantitative biomass yield in all types of cropping systems compared to mineral fertilizer alone. On the other hand, a study by Show et al. (2023) revealed that digestate made from combining cow dung and water hyacinth digestate with a 3:1:1 ratio of NPK resulted in a higher yield of grain with increased panicle length and test weight compared to other combinations. Overall, although there is limited information pertaining to the effect of digestate on plant growth, it is revealed that integrating digestate with other fertilizers can balance nutrient levels (Ehmann et al. 2018; Jamison et al. 2021; Show et al. 2023).

4.6.3 Challenges of Digestate Adoption

Despite its many benefits, digestate presents challenges that hinder its widespread adoption (Piadeh et al. 2024). Improving smallholder farmer perception of digestate use as a biomass-based biofertilizer can be achieved by developing and offering high-quality digestate products that match mineral fertilizers in both agronomic efficiency and physical attributes (Salas et al. 2024). Although regulations and standards for digestate management are in place for some African countries and are aimed at addressing the pollution concerns and educating end-users on digestate storage and applications, the regulatory frameworks of other African countries are not up-to-date, thus creating uncertainty for industry stakeholders (Logan and Visvanathan 2019). In addition, while digestate is a promising alternative to synthetic fertilizers, poor management poses a risk to public health (Nag et al. 2020). A major source of pathogens is digestate derived from digesters fed with animal manure, municipal sewage wastewater, and abattoir waste. The types of pathogens mainly include bacteria, viruses, parasites, and fungi. Therefore, pathogens in feedstock could potentially contaminate the environment if not effectively inactivated during the process of AD (Lin et al. 2022).

Caution needs to be taken when adopting the technology, to ensure that the full potential of the digestate is harnessed and to prevent inadvertent negative implications on animals, humans, and the environment. Unlike commercial biofertilizers, biomass-based fertilizers vary with regard to their biological, physical, and chemical properties. This variability has been widely reported in literature (Zirkler et al. 2014; Coelho et al. 2018; Häfner et al. 2022; Romio et al. 2024) and is a consequence of several factors including the feedstock used for AD, digestion process, digestate storage method, and post-treatment of digestate (Roopnarain et al. 2022). Digestate variability contributes to challenges related to its adoption as a biofertilizer, since this variability makes it difficult to establish standard practices for its management and use, including guidance on rate of fertilization (Coelho et al. 2018). It is therefore strongly recommended that the use of digestate as a biomass-based fertilizer be preceded by a thorough evaluation of its physicochemical, nutritional, and microbial properties (Coelho et al. 2018). The results of these analyses will aid in making informed decisions on the optimal use of the digestate thereby reaping maximal benefits and casting a positive light on the technology (Roopnarain et al. 2022).

Moreover, as digestate variability is dictated by several factors, such factors can be manipulated to "engineer" a desired resident microbiome in the digestate, as dictated by the downstream application (Roopnarain et al. 2023). For example, when digestate is targeted for direct land application as a soil ameliorant, thermophilic AD and digestate post-treatment methods such as composting may be considered as these processes manipulate the resistome profile of the microbes residing in the digestate, by way of decreasing the abundance of antibiotic resistance genes (ARGs) (Sun et al. 2016; Gurmessa et al. 2021). Such strategies may aid in improving the quality, safety, and value of digestate, thereby changing its outlook as a valuable product rather than a waste by-product of the AD process.

Apart from the impact of chemical and microbial constituents of digestate on its adoption as a biomass-based biofertilizer, the method of application also directly impacts the magnitude of benefits associated with its use. A substantial amount of nutrients can be inadvertently lost, through volatilization and leaching, as a result of poor application of digestate to soil. Suitable methods, equipment, and technology are required for the optimal application of digestate to soil to ensure maximal soil nutrient content, minimize nutrient run-off, reduce digestate wastage, and promote sustainable agriculture (Jin et al. 2022). However, when selecting the ideal method for digestate application, cost-effectiveness needs to be maintained, particularly when considering application on resource-constrained African farms. A careful balance needs to therefore be established between cost and level of technicality of the application methods, while minimizing negative environmental impacts. The special requirements of small-scale farmers may actually open up a niche for specialized equipment that are less technologically advanced but may effectively apply digestate on small plots of land while reducing GHG emissions and maximizing nutrient and microbe integration into the soil, at a low cost. This emphasizes the need for a regional context when generating strategies for improved adoption of digestate as a biomass-based biofertilizer. A solution that applies to African farmers also needs to be established in relation to digestate storage prior to application, to prevent negative environmental impacts as a result of GHG emissions from storage ponds (Amon et al. 2006), which may be exacerbated in areas that experience elevated temperatures, such as Africa.

Overall, in order to overcome existing barriers to the adoption of biomass-based biofertilizers such as digestate in Africa, it is crucial to implement innovative strategies. To promote digestate commercialization, a combination of incentives, quality regulation, and social awareness is necessary to foster a culture of sustainability within the circular economy (Lamolinara et al. 2022). For instance, alternative strategies could include policy incentives that optimize waste valorization and penalties that aim at reducing the intensive use of non-renewable materials for fertilizer production. However, the effectiveness of every action plan varies from country to country, depending on its socioeconomic characteristics (Kurniawati et al. 2023). Collaborative efforts between industry stakeholders, researchers, and policymakers are vital for driving progress. By addressing these barriers, the biofertilizer industry can unlock digestate's full potential (Kurniawati et al. 2023).

4.7 Conclusions and Recommendations

With the escalating population in Africa and globally, the shift towards agricultural practices that are sustainable and eco-friendly are fundamental to ensure food security. Biomass-based biofertilizers, such as digestate obtained from the AD process, offer a multitude of benefits over conventional chemical fertilizers and may contribute significantly in the trajectory towards sustainable agriculture. Apart from boosting plant production by availing nutrients to crops, these biomass-based biofertilizers are also inexpensive in comparison to chemical fertilizers as it is a by-product of the AD process that would be required to be disposed of anyway. Valorizing this AD by-product not only contributes to improving the economic feasibility of AD but also aids in waste management. To improve adoption of biomass-based fertilizers, future research and development ought to shed light upon existing knowledge gaps and challenges associated with the effective use of digestate in Africa. The development of novel digestate treatment technologies is necessary to enhance its quality and safety. Delving deep into the effects of digestate on soil microbiology will help further understand its agronomic value. Additionally, it is necessary to conduct further research into application techniques and strategies that are feasible for African farmers, required to optimize digestate's fertilization potential. Furthermore, studying the impact of digestate on the overall environment will drive efficient policy and decision-making. Thus, exploring the potential of digestate as a potential biomass-based biofertilizer in the African context merits further investigation.

Acknowledgments The authors are grateful to the Agricultural Research Council for providing the facilities to conduct this research. This work is based on the research supported in part by the National Research Foundation of South Africa (NRF; Grant numbers 121924, 128307 and AJCR240708241418). Opinions expressed and conclusions reached are those of the authors and not necessarily endorsed by the NRF.

References

Ahmed T, Noman M, Qi Y et al (2023) Fertilization of microbial composts: a technology for improving stress resilience in plants. Plants 12:3550. https://doi.org/10.3390/plants12203550

Akinnawo SO (2023) Eutrophication: causes, consequences, physical, chemical and biological techniques for mitigation strategies. Environ Challenges 12:100733. https://doi.org/10.1016/j.envc.2023.100733

Alburquerque JA, De La Fuente C, Campoy M et al (2012) Agricultural use of digestate for horticultural crop production and improvement of soil properties. Eur J Agron 43:119–128. https://doi.org/10.1016/j.eja.2012.06.001

Alfa MI, Adie DB, Igboro SB et al (2014) Assessment of biofertilizer quality and health implications of anaerobic digestion effluent of cow dung and chicken droppings. Renew Energy 63:681–686. https://doi.org/10.1016/j.renene.2013.09.049

Amon B, Kryvoruchko V, Amon T, Zechmeister-Boltenstern S (2006) Methane, nitrous oxide and ammonia emissions during storage and after application of dairy cattle slurry and influence of slurry treatment. Agric Ecosyst Environ 112:153–162. https://doi.org/10.1016/j.agee.2005.08.030

Asli S, Massalha N, Diab M, Hugerat M (2022) Phytotoxic effects of treated wastewater used for irrigation on root hydraulic conductivity and plant growth of maize seedlings (Zea mays L. PR 32w86). Irrig Sci 40:817–828. https://doi.org/10.1007/s00271-022-00793-z

Bakari R, Asha R, Hossein M et al (2024) Converting food waste to biofuel: a sustainable energy solution for Sub-Saharan Africa. Sustain Chem Environ 7:100126. https://doi.org/10.1016/j.scenv.2024.100126

Barampouti EM, Mai S, Malamis D et al (2020) Exploring technological alternatives of nutrient recovery from digestate as a secondary resource. Renew Sust Energ Rev 134:110379. https://doi.org/10.1016/j.rser.2020.110379

Barłóg P, Hlisnikovský L, Kunzová E (2020) Effect of digestate on soil organic carbon and plant-available nutrient content compared to cattle slurry and mineral fertilization. Agronomy 10:379. https://doi.org/10.3390/agronomy10030379

Bjornlund V, Bjornlund H, Van Rooyen A (2022) Why food insecurity persists in sub-Saharan Africa: a review of existing evidence. Food Sec 14:845–864. https://doi.org/10.1007/s12571-022-01256-1

Celletti S, Lanz M, Bergamo A et al (2021) Evaluating the aqueous phase from hydrothermal carbonization of cow manure digestate as possible fertilizer solution for plant growth. Front Plant Sci 12:687434. https://doi.org/10.3389/fpls.2021.687434

Chaudhry V, Rehman A, Mishra A et al (2012) Changes in bacterial community structure of agricultural land due to long-term organic and chemical amendments. Microb Ecol 64:450–460. https://doi.org/10.1007/s00248-012-0025-y

Chen R, Blagodatskaya E, Senbayram M et al (2012) Decomposition of biogas residues in soil and their effects on microbial growth kinetics and enzyme activities. Biomass Bioenergy 45:221–229. https://doi.org/10.1016/j.biombioe.2012.06.014

Chojnacka K, Moustakas K (2024) Anaerobic digestate management for carbon neutrality and fertilizer use: a review of current practices and future opportunities. Biomass Bioenergy 180:106991. https://doi.org/10.1016/j.biombioe.2023.106991

Ciurli A, Di Biase G, Rossi M et al (2025) Dried anaerobic digestate from slaughterhouse by-products: emerging cues for a bio-based fertilization. Waste Biomass Valor 16:927–943. https://doi.org/10.1007/s12649-024-02737-4

Coelho JJ, Prieto ML, Dowling S et al (2018) Physical-chemical traits, phytotoxicity and pathogen detection in liquid anaerobic digestates. Waste Manag 78:8–15. https://doi.org/10.1016/j.wasman.2018.05.017

Czekała W, Nowak M, Piechota G (2023) Sustainable management and recycling of anaerobic digestate solid fraction by composting: a review. Bioresour Technol 375:128813. https://doi.org/10.1016/j.biortech.2023.128813

DAFF (RSA) (2018) Guidelines for registration of digestate derived from abattoir waste used as a soil conditioner or amendment – issued by the registrar: ACT 36 of 1947. Department of Agriculture, Forestry and Fisheries, Republic of South Africa. https://www.greenagri.org.za/assets/documents-/Tips-Tools/Useful-Tips/Livestock/AIC-GUIDELINE-FOR-THE-REGISTRATION-REQUIREMNTS-OF-THE-DIGESTATE-SEPTEMBER-2018.pdf. Accessed 28 Feb 2025

Dhull P, Mozhiarasi V, Kumar S et al (2024) Current and prognostic overview of digestate management and processing practices, regulations and standards. Biocatal Agric Biotechnol 61:103345. https://doi.org/10.1016/j.bcab.2024.103345

Egene CE, Sigurnjak I, Regelink IC et al (2021) Solid fraction of separated digestate as soil improver: implications for soil fertility and carbon sequestration. J Soils Sediments 21:678–688. https://doi.org/10.1007/s11368-020-02792-z

Ehmann A, Thumm U, Lewandowski I (2018) Fertilizing potential of separated biogas digestates in annual and perennial biomass production systems. Front Sustain Food Syst 2:12. https://doi.org/10.3389/fsufs.2018.00012

Fertahi S, Elalami D, Tayibi S et al (2023) The current status and challenges of biomass biorefineries in Africa: a critical review and future perspectives for bioeconomy development. Sci Total Environ 870:162001. https://doi.org/10.1016/j.scitotenv.2023.162001

Finzi A, Guido V, Riva E et al (2021) Performance and sizing of filtration equipment to replace mineral fertilizer with digestate in drip and sprinkler fertigation. J Clean Prod 317:128431. https://doi.org/10.1016/j.jclepro.2021.128431

García-López AM, Delgado A, Anjos O, Horta C (2023) Digestate not only affects nutrient availability but also soil quality indicators. Agronomy 13:1308. https://doi.org/10.3390/agronomy13051308

Garg RN, Pathak H, Das DK, Tomar RK (2005) Use of Flyash and biogas slurry for improving wheat yield and physical properties of soil. Environ Monit Assess 107:1–9. https://doi.org/10.1007/s10661-005-2021-x

Giagnoni L, Martellini T, Scodellini R et al (2020) Co-composting: an opportunity to produce compost with designated tailor-made properties. In: Hettiarachchi H, Caucci S, Schwärzel K (eds) Organic waste composting through nexus thinking. Springer, Cham, pp 185–211

Gurmessa B, Milanovic V, Foppa Pedretti E et al (2021) Post-digestate composting shifts microbial composition and degrades antimicrobial resistance genes. Bioresour Technol 340:125662. https://doi.org/10.1016/j.biortech.2021.125662

Hafez RM, Tawfik A, Hassan GK et al (2024) Synergism of floated paperboard sludge cake / sewage sludge for maximizing biomethane yield and biochar recovery from digestate: a step towards circular economy. Chemosphere 362:142639. https://doi.org/10.1016/j.chemosphere.2024.142639

Häfner F, Hartung J, Möller K (2022) Digestate composition affecting N fertiliser value and C mineralisation. Waste Biomass Valor 13:3445–3462. https://doi.org/10.1007/s12649-022-01723-y

Hammerschmiedt T, Kintl A, Holatko J et al (2022) Assessment of digestates prepared from maize, legumes, and their mixed culture as soil amendments: effects on plant biomass and soil properties. Front Plant Sci 13:1017191. https://doi.org/10.3389/fpls.2022.1017191

Horta C, Carneiro JP (2020) Assessment of fertilising properties of a solid digestate in comparison with undigested cattle slurry applied to an acidic soil. OJSS 10:307–325. https://doi.org/10.4236/ojss.2020.108017

Jamison J, Khanal SK, Nguyen NH, Deenik JL (2021) Assessing the effects of digestates and combinations of digestates and fertilizer on yield and nutrient use of Brassica juncea (Kai Choy). Agronomy 11:509. https://doi.org/10.3390/agronomy11030509

Jankauskienė J, Laužikė K, Kaupaitė S (2024) The use of anaerobic digestate for greenhouse horticulture. Agronomy 14:2437. https://doi.org/10.3390/agronomy14102437

Jeranyama P, Shrestha A, Neupane N (2020) Sustainable food systems. In: The role of ecosystem services in sustainable food systems. Elsevier, pp 1–16

Jin K, Ran Y, Alengebawy A et al (2022) Agro-environmental sustainability of using digestate fertilizer for solanaceous and leafy vegetables cultivation: insights on fertilizer efficiency and risk assessment. J Environ Manag 320:115895. https://doi.org/10.1016/j.jenvman.2022.115895
Karimi B, Sadet-Bourgeteau S, Cannavacciuolo M et al (2022) Impact of biogas digestates on soil microbiota in agriculture: a review. Environ Chem Lett 20:3265–3288. https://doi.org/10.1007/s10311-022-01451-8
Korba J, Šařec P, Novák V et al (2024) Digestate application methods and rates with regard to greenhouse gas emissions and crop conditions. Agronomy 14:336. https://doi.org/10.3390/agronomy14020336
Kovačić Đ, Lončarić Z, Jović J et al (2022) Digestate management and processing practices: a review. Appl Sci 12:9216. https://doi.org/10.3390/app12189216
Kurniawati A, Stankovics P, Hilmi YS et al (2023) Understanding the future of bio-based fertilisers: the EU'S policy and implementation. Sustain Chem Clim Action 3:100033. https://doi.org/10.1016/j.scca.2023.100033
Lamolinara B, Pérez-Martínez A, Guardado-Yordi E et al (2022) Anaerobic digestate management, environmental impacts, and techno-economic challenges. Waste Manag 140:14–30. https://doi.org/10.1016/j.wasman.2021.12.035
Lin M, Wang A, Ren L et al (2022) Challenges of pathogen inactivation in animal manure through anaerobic digestion: a short review. Bioengineered 13:1149–1161. https://doi.org/10.1080/21655979.2021.2017717
Logan M, Visvanathan C (2019) Management strategies for anaerobic digestate of organic fraction of municipal solid waste: current status and future prospects. Waste Manag Res 37:27–39. https://doi.org/10.1177/0734242X18816793
Loria ER, Sawyer JE, Barker DW et al (2007) Use of anaerobically digested swine manure as a nitrogen source in corn production. Agron J 99:1119–1129. https://doi.org/10.2134/agronj2006.0251
Makádi M, Tomcsik A, Orosz V (2012) Digestate: a new nutrient source – review. In: Kumar S (ed) Biogas. InTech
Massé DI, Talbot G, Gilbert Y (2011) On farm biogas production: a method to reduce GHG emissions and develop more sustainable livestock operations. Anim Feed Sci Technol 166–167:436–445. https://doi.org/10.1016/j.anifeedsci.2011.04.075
Matjuda DS, Tekere M, Thaela-Chimuka M-J (2023) Characterization of the physicochemical composition of anaerobically digested (digestate) high throughput red meat abattoir waste in South Africa and the determination of its quality as a potential biofertilizer. Heliyon 9:e21647. https://doi.org/10.1016/j.heliyon.2023.e21647
Mgadi K, Ndaba B, Roopnarain A et al (2024) Nanoparticle applications in agriculture: overview and response of plant-associated microorganisms. Front Microbiol 15:1354440. https://doi.org/10.3389/fmicb.2024.1354440
Mmereki D, David VE, Wreh Brownell AH (2024) The management and prevention of food losses and waste in low- and middle-income countries: a mini-review in the Africa region. Waste Manag Res 42:287–307. https://doi.org/10.1177/0734242X231184444
Möller K, Müller T (2012) Effects of anaerobic digestion on digestate nutrient availability and crop growth: a review. Eng Life Sci 12:242–257. https://doi.org/10.1002/elsc.201100085
Møller HB, Sørensen P, Olesen JE et al (2022) Agricultural biogas production—climate and environmental impacts. Sustainability 14:1849. https://doi.org/10.3390/su14031849
Mortola N, Romaniuk R, Cosentino V et al (2019) Potential use of a poultry manure digestate as a biofertiliser: evaluation of soil properties and Lactuca sativa growth. Pedosphere 29:60–69. https://doi.org/10.1016/S1002-0160(18)60057-8
Mukhuba M, Roopnarain A, Adeleke R et al (2018) Comparative assessment of bio-fertiliser quality of cow dung and anaerobic digestion effluent. Cogent Food Agric 4:1435019. https://doi.org/10.1080/23311932.2018.1435019

Muturi SM, Muthui LW, Njogu PM et al (2021) Metagenomics survey unravels diversity of biogas microbiomes with potential to enhance productivity in Kenya. PLoS One 16:e0244755. https://doi.org/10.1371/journal.pone.0244755

Nag R, Whyte P, Markey BK et al (2020) Ranking hazards pertaining to human health concerns from land application of anaerobic digestate. Sci Total Environ 710:136297. https://doi.org/10.1016/j.scitotenv.2019.136297

Nicholson F, Bhogal A, Cardenas L et al (2017) Nitrogen losses to the environment following food-based digestate and compost applications to agricultural land. Environ Pollut 228:504–516. https://doi.org/10.1016/j.envpol.2017.05.023

Nigam P (2013) Microbial enzymes with special characteristics for biotechnological applications. Biomolecules 3:597–611. https://doi.org/10.3390/biom3030597

Nikolaidou C, Mola M, Papakostas S et al (2024) The effect of anaerobic digestate as an organic soil fertilizer on the diversity and structure of the indigenous soil microbial and nematode communities. Environ Sci Pollut Res. https://doi.org/10.1007/s11356-024-32850-9

Nyang'au JO, Møller HB, Sørensen P (2022) Nitrogen dynamics and carbon sequestration in soil following application of digestates from one- and two-step anaerobic digestion. Sci Total Environ 851:158177. https://doi.org/10.1016/j.scitotenv.2022.158177

O'Connor J, Mickan BS, Rinklebe J et al (2022) Environmental implications, potential value, and future of food-waste anaerobic digestate management: a review. J Environ Manage 318:115519. https://doi.org/10.1016/j.jenvman.2022.115519

Obi LU, Roopnarain A, Tekere M et al (2025) Dynamics and insights into the unique ecological guild of fungi in bacteria-bioaugmented anaerobic digesters. J Fungi 11:56. https://doi.org/10.3390/jof11010056

Orzi V, Riva C, Scaglia B et al (2018) Anaerobic digestion coupled with digestate injection reduced odour emissions from soil during manure distribution. Sci Total Environ 621:168–176. https://doi.org/10.1016/j.scitotenv.2017.11.249

Oshoma CE, Ajieh MU, Onigho OO et al (2025) Assessment of the effects of temperature and time on the bacteriological quality and heavy metal of digestate as a safe biofertiliser. Geomicrobiol J 42:205–212. https://doi.org/10.1080/01490451.2024.2448162

Petrovič A, Simonič M, Čuček L (2021) Nutrient recovery from the digestate obtained by rumen fluid enhanced anaerobic co-digestion of sewage sludge and cattail: precipitation by MgCl2 and ion exchange using zeolite. J Environ Manag 290:112593. https://doi.org/10.1016/j.jenvman.2021.112593

Piadeh F, Offie I, Behzadian K et al (2024) A critical review for the impact of anaerobic digestion on the sustainable development goals. J Environ Manag 349:119458. https://doi.org/10.1016/j.jenvman.2023.119458

Raimi A, Adeleke R, Roopnarain A (2017) Soil fertility challenges and Biofertiliser as a viable alternative for increasing smallholder farmer crop productivity in sub-Saharan Africa. Cogent Food Agric 3:1400933. https://doi.org/10.1080/23311932.2017.1400933

Raimi A, Roopnarain A, Adeleke R (2021) Biofertilizer production in Africa: current status, factors impeding adoption and strategies for success. Sci Afr 11:e00694. https://doi.org/10.1016/j.sciaf.2021.e00694

Rama H, Akindolire M, Obi L et al (2023) Anaerobic digestion: climate change mitigation through sustainable organic waste valorization. In: Handbook of nature-based solutions to mitigation and adaptation to climate change. Springer, pp 1–19

Rama H, Ndaba B, Dhlamini MS et al (2024) Elucidating key microbial drivers for methane production during cold adaptation and psychrophilic anaerobic digestion of cattle manure and food waste. Fermentation 10:370. https://doi.org/10.3390/fermentation10070370

Riva C, Orzi V, Carozzi M et al (2016) Short-term experiments in using digestate products as substitutes for mineral (N) fertilizer: agronomic performance, odours, and ammonia emission impacts. Sci Total Environ 547:206–214. https://doi.org/10.1016/j.scitotenv.2015.12.156

Romio C, Kofoed MVW, Møller HB (2021) Digestate post-treatment strategies for additional biogas recovery: a review. Sustainability 13:9295. https://doi.org/10.3390/su13169295

Romio C, Ward AJ, Møller HB (2024) Characterization and valorization of biogas digestate and derived organic fertilizer products from separation processes. Front Sustain Food Syst 8:1415508. https://doi.org/10.3389/fsufs.2024.1415508

Roopnarain A, Ndaba B, Rama H et al (2022) Liquid gold: harnessing the potential of digestate to enhance smallholder farmer food security and livelihood. In: Mupambwa HA, Nciizah AD, Nyambo P, Muchara B, Gabriel NN (eds) Food security for African smallholder farmers. Springer, Singapore, pp 313–341

Roopnarain A, Akindolire MA, Rama H, Ndaba B (2023) Casting light on the micro-organisms in digestate: diversity and untapped potential. Fermentation 9:160. https://doi.org/10.3390/fermentation9020160

Rowland I, Gibson G, Heinken A et al (2018) Gut microbiota functions: metabolism of nutrients and other food components. Eur J Nutr 57:1–24. https://doi.org/10.1007/s00394-017-1445-8

Rubagumya I, Komakech AJ, Kabenge I, Kiggundu N (2023) Potential of organic waste to energy and bio-fertilizer production in sub-Saharan Africa: a review. Waste Dispos Sustain Energy 5:259–267. https://doi.org/10.1007/s42768-022-00131-1

Salas MÁ, Sica P, Rydgård M et al (2024) Current challenges on the widespread adoption of new bio-based fertilizers: insights to move forward toward more circular food systems. Front Sustain Food Syst 8:1386680. https://doi.org/10.3389/fsufs.2024.1386680

Setsoafia ED, Ma W, Renwick A (2022) Effects of sustainable agricultural practices on farm income and food security in northern Ghana. Agric Econ 10:9. https://doi.org/10.1186/s40100-022-00216-9

Shaji H, Chandran V, Mathew L (2021) Organic fertilizers as a route to controlled release of nutrients. In: Controlled release fertilizers for sustainable agriculture. Elsevier, pp 231–245

Sharifi M, Baker S, Hojabri L, Hajiaghaei-Kamrani M (2019) Short-term nitrogen dynamics in a soil amended with anaerobic digestate. Can J Soil Sci 99:173–181. https://doi.org/10.1139/cjss-2018-0060

Sharma P, Bano A, Verma K et al (2023) Food waste digestate as biofertilizer and their direct applications in agriculture. Bioresour Technol Rep 23:101515. https://doi.org/10.1016/j.biteb.2023.101515

Show BK, Panja S, GhoshThakur R et al (2023) Optimisation of anaerobic digestate and chemical fertiliser application to enhance rice yield—a machine-learning approach. Sustainability 15:13706. https://doi.org/10.3390/su151813706

Sigurnjak I, De Waele J, Michels E et al (2017) Nitrogen release and mineralization potential of derivatives from nutrient recovery processes as substitutes for fossil fuel-based nitrogen fertilizers. Soil Use Manag 33:437–446. https://doi.org/10.1111/sum.12366

Siheri W, Alenezi S, Tusiimire J, Watson DG (2017) The chemical and biological properties of propolis. In: Alvarez-Suarez JM (ed) Bee products – chemical and biological properties. Springer, Cham, pp 137–178

Slepetiene A, Ceseviciene J, Amaleviciute-Volunge K et al (2023) Solid and liquid phases of anaerobic digestate for sustainable use of agricultural soil. Sustainability 15:1345. https://doi.org/10.3390/su15021345

Smith P, Andrén O, Karlsson T et al (2005) Carbon sequestration potential in European croplands has been overestimated. Glob Chang Biol 11:2153–2163. https://doi.org/10.1111/j.1365-2486.2005.01052.x

Snyders CA, Bradshaw SM, Akdogan G, Eksteen JJ (2015) Determination of the equilibrium and film diffusion constants of the platinum cyanide anions during the elution from activated carbon. Miner Eng 80:57–68. https://doi.org/10.1016/j.mineng.2015.07.001

South Africa (2008) No. 59 of 2008: National Environmental Management: Waste Act, 2008. In: Government Gazette. https://www.dffe.gov.za/sites/default/files/legislations/nema_amendment_act59.pdf. Accessed 28 Feb 2025

Srivastava S, Singh P, Barbora L et al (2024) Performance assessment of innovative waste management system developed for the production of bio-fertilizer. Sustain Chem Environ 7:100148. https://doi.org/10.1016/j.scenv.2024.100148

Sun W, Qian X, Gu J et al (2016) Mechanism and effect of temperature on variations in antibiotic resistance genes during anaerobic digestion of dairy manure. Sci Rep 6:30237. https://doi.org/10.1038/srep30237

Tambone F, Orzi V, D'Imporzano G, Adani F (2017) Solid and liquid fractionation of digestate: mass balance, chemical characterization, and agronomic and environmental value. Bioresour Technol 243:1251–1256. https://doi.org/10.1016/j.biortech.2017.07.130

Tolessa A (2024) Current status and future prospects of small-scale household biodigesters in sub-Saharan Africa. J Energy 2024:1–19. https://doi.org/10.1155/2024/5596028

Tongwane M, Mdlambuzi T, Moeletsi M et al (2016) Greenhouse gas emissions from different crop production and management practices in South Africa. Environ Dev 19:23–35. https://doi.org/10.1016/j.envdev.2016.06.004

Touch V, Tan DKY, Cook BR et al (2024) Smallholder farmers' challenges and opportunities: implications for agricultural production, environment and food security. J Environ Manag 370:122536. https://doi.org/10.1016/j.jenvman.2024.122536

Uludag-Demirer S, Demirer GN (2022) Post-anaerobic treatability and residual biogas potential of digestate. Biomass Conv Bioref 12:1695–1702. https://doi.org/10.1007/s13399-021-01290-7

US EPA (2021) Emerging issues in food waste management plastic contamination. U.S. Environmental Protection Agency

van Midden C, Harris J, Shaw L et al (2023) The impact of anaerobic digestate on soil life: a review. Appl Soil Ecol 191:105066. https://doi.org/10.1016/j.apsoil.2023.105066

Vaneeckhaute C, Lebuf V, Michels E et al (2017) Nutrient recovery from digestate: systematic technology review and product classification. Waste Biomass Valor 8:21–40. https://doi.org/10.1007/s12649-016-9642-x

Vasudevan P, Thapliyal A, Srivastava R et al (2010) Fertigation potential of domestic wastewater for tree plantations. J Sci Ind Res 69:146–150

Viancelli A, Schneider TM, Demczuk T et al (2023) Unlocking the value of biomass: exploring microbial strategies for biogas and volatile fatty acids generation. Bioresour Technol Rep 23:101552. https://doi.org/10.1016/j.biteb.2023.101552

Weldon S, Rivier P-A, Joner EJ et al (2023) Co-composting of digestate and garden waste with biochar: effect on greenhouse gas production and fertilizer value of the matured compost. Environ Technol 44:4261–4271. https://doi.org/10.1080/09593330.2022.2089057

Worldometers (2024) Africa population. https://www.worldometers.info/world-population/africa-population/. Accessed 7 Dec 2024

Yan M, Tian H, Song S et al (2023) Effects of digestate-encapsulated biochar on plant growth, soil microbiome and nitrogen leaching. J Environ Manag 334:117481. https://doi.org/10.1016/j.jenvman.2023.117481

Zeng Y, De Guardia A, Dabert P (2016) Improving composting as a post-treatment of anaerobic digestate. Bioresour Technol 201:293–303. https://doi.org/10.1016/j.biortech.2015.11.013

Zilio M, Pigoli A, Rizzi B et al (2022) Using highly stabilized digestate and digestate-derived ammonium sulphate to replace synthetic fertilizers: the effects on soil, environment, and crop production. Sci Total Environ 815:152919. https://doi.org/10.1016/j.scitotenv.2022.152919

Zirkler D, Peters A, Kaupenjohann M (2014) Elemental composition of biogas residues: variability and alteration during anaerobic digestion. Biomass Bioenergy 67:89–98. https://doi.org/10.1016/j.biombioe.2014.04.021

Chapter 5
Biofertilizers as a Nature-Based Solution to Improve Soil Health, Yield, and Quality in African Medicinal Plant Systems

Patrick Nyambo, H. Motsi, M. M. Mabitsela, N. T. Mkhize, and C. V. Mashamaite

Abstract Calls for the cultivation of medicinal plants in Africa are growing louder due to their cultural, therapeutic, and economic value. However, due to the declining soil quality and environmental pollution (caused partly by the overuse of chemical fertilizers), overharvesting of wild populations and use of unsustainable agricultural practices negatively impact medicinal plants' quality and yield. As such, biofertilizers can be an eco-friendly alternative by exploiting beneficial microorganisms like Rhizobacteria, *Azotobacter*, and *Azospirillum* in order to enhance nutrient availability, improve soil health, and stimulate plant growth and the biosynthesis of secondary metabolites critical for therapeutic efficacy. Nevertheless, their adoption by African smallholder farmers is still low, partly due to a lack of knowledge, limited access, and economic and policy barriers. In this chapter, we explore possible solutions to these challenges, such as farmer education and changes to policies, and mix them with various sustainable climate-smart agriculture. Incorporating biofertilizers in African medicinal plant production systems is significant for progressing towards attaining sustainability and climatic resilience.

Keywords Biofertilizers · Medicinal plants · Soil health · Secondary metabolites · Sustainable agriculture

P. Nyambo (✉) · N. T. Mkhize · C. V. Mashamaite
Agricultural Research Council—Vegetables, Industrial and Medicinal Plants, Pretoria, South Africa
e-mail: nyambop@arc.agric.za

H. Motsi
Department of Soil Science, Faculty of AgriSciences, Stellenbosch University, Matieland, South Africa

M. M. Mabitsela
Department of Agriculture, University of Zululand, KwaDlangezwa, South Africa

A. D. Nciizah et al. (eds.), *Biofertilizers for Sustainable Agriculture in Africa*, Sustainability Sciences in Asia and Africa, https://doi.org/10.1007/978-981-95-6009-7_5

5.1 Introduction

Medicinal plants are an integral part in traditional healthcare, with over 80% of the African population relying on plant-based medicines for primary healthcare (Ikhoyameh et al. 2024). Approximately 40% of the pharmaceuticals are made from medicinal plants (WHO 2013). The therapeutic significance of medicinal plants is well-documented and attributed to the diverse range of species found globally. In addition to their medicinal value, these plants are widely used in the cosmetic industry and contribute to income generation at both national and household levels. However, declining soil fertility, climate change-induced temperature and rainfall variability, and unsustainable agricultural practices such as excessive chemical fertilizer use have hindered the adoption and expansion of medicinal plant cultivation. To attain sustainable agricultural systems, the use and promotion of nature-based agricultural practices are essential. Among these, biofertilizers can be a sustainable and eco-friendly strategy to improve soil quality and agricultural productivity and support the resilient cultivation of medicinal plants.

Raimi et al. (2021) define biofertilizers as biologically active formulations comprising beneficial living microbes that are applied in soil to enhance plant growth and development by increasing the supply and availability of nutrients and overall soil health. Nitrogen-fixing bacteria, phosphate-solubilizing bacteria, and plant growth-promoting rhizobacteria are examples of the biofertilizers. Specifically, biofertilizers improve nutrient cycling, stimulate root development, and improve soil structure, thereby supporting sustainable agriculture. Furthermore, they play a role in secondary metabolite biosynthesis, e.g., as alkaloids, flavonoids, and essential oils, that determines pharmacological efficacy (Rizvi et al. 2022; Çakmakçı et al. 2023). The application of microbial consortia has been shown to significantly enhance phytochemical content in species such as *Curcuma longa*, *Stevia rebaudiana*, and *Withania somnifera* under field and greenhouse conditions (Sontsa-Donhoung et al. 2022; Prakash et al. 2022). Such outcomes are critical in African farming contexts, where resource-constrained farmers must optimize yields while preserving ecological integrity.

Although evidence of biofertilizers' potential exists, adoption among African farmers is still low due to a lack of knowledge, limited access, and economic and policy barriers. Moreover, their efficacy is influenced by biophysical factors like climate, soil type, crop genotype, and microbial compatibility. This chapter aims to explore and explain the benefits of biofertilizers in medicinal plant production across African farming systems, by synthesizing current research on their modes of action and impacts on secondary metabolite profiles. It further discusses the possibility of including biofertilizers into broader sustainable farming systems as a pathway to resilient, productive, and health-promoting medicinal plant systems in Africa.

5.2 The Role of Biofertilizers in Enhancing Soil Health

5.2.1 Biofertilizers' Mechanisms of Action in Soil

Biofertilizers rely on living microorganisms such as bacteria, fungi, protozoa, and nematodes to enhance overall soil and plant growth through multiple mechanisms of action such as biological nitrogen fixation, plant growth-promoting rhizobacteria, phosphate, and potassium solubilizing (Table 5.1) (Singh et al. 2021). Biofertilizers contain strains of different species, such as *Rhizobium*, *Azotobacter*, *Azospirillum*, *Bacillus*, and cyanobacteria, as either a single strain or a combination of various strains (Dimitrova et al. 2024). They can also be applied as powders, granules, or liquids (Saif et al. 2021).

Table 5.1 The mechanisms of biofertilizers in soil

Type of biofertilizer	Mechanisms of action	Effects observed in soil	References
UGmax (commercialized biofertilizer containing *Lactobacillus* spp., *Azotobacter*)	Phosphate solubilizing	Increased phosphorus, organic carbon, total nitrogen content, and microbial biomass carbon content	Długosz and Piotrowska-Długosz (2023)
Bacteria strains of *Pseudomonas aeruginosa* and *Bacillus firmus*	Plant growth-promoting *rhizobacteria*	Application of biofertilizer increased the heavy metal bioavailability and the activities of soil enzymes. Soil pH were positively correlated with soil Zn	Haroun et al. (2023)
Biochar	Nitrogen fixation	The application of solid biofertilizers and organic ameliorants improved biological properties and increased soil chemical properties (total N, available P, organic C, and cation exchange capacity)	Fitriatin et al. (2021)
Azotobacter and *Azospirillum*	Nitrogen fixation	Application of *Azotobacter* and *Azospirillum* bacteria to the soil led to a significant increase in electrical conductivity, cation exchange capacity, bulk density, and organic matter content	Ajeel and Al-Hakeim (2024)
Bacillus megaterium, *B. circulans*, and *Pseudomonas putida*	Potassium solubilizing	Lower soil respiration was observed compared to commercial NPK fertilizers, reducing CO_2 emissions	Mátyás et al. (2020)
Enterobacter hormaechei and fungal *Aspergillus terreus*	Potassium solubilizing	Potassium content was improved in the soil	Prajapati et al. (2013)

5.2.1.1 Nitrogen-Fixing Biofertilizers

The atmospheric nitrogen gas which makes up approximately 78% of the atmosphere is biologically fixed into ammonia by microorganisms such as *Rhizobium*, *Azotobacter*, and *Azospirillum*, making it usable by plants (Lawal 2021). *Rhizobium* forms a symbiotic relationship with leguminous plants (cow peas, beans etc.) to convert atmospheric nitrogen into ammonia (Allito et al. 2015). Through this mechanism of action, *Rhizobium* releases essential nitrogen compounds to both the soil and the host plant, thereby promoting plant growth and contributing to long-term soil fertility (Allito et al. 2020; Yamamoto et al. 2023). Findings revealed that the effectiveness of *Rhizobium* biofertilizer is influenced by several factors, including soil organic carbon (SOC) content, with higher SOC levels generally supporting larger rhizobial populations (Swanepoel et al. 2011). Patel et al. (2024) reported the influence of *Rhizobium* to the enhancement of soil macro and micronutrients through their metabolic activities, enriching the soil physical properties and improving the soil structure and capacity to retain water and nutrients, which are essential for crop productivity. On the other hand, *Azotobacter* biofertilizer is a free-living microorganism with the ability to release nitrogen element in the soil without the need for a host plant. Mohite and Patil (2022) and Ali and Kamal (2025) reported an improvement in the physical and biological properties of soil, including enhanced soil structure and stimulation of beneficial soil microorganisms that contribute to nutrient cycling at the application of *Azotobacter*. *Azotobacter* does not only convert nitrogen gas into ammonia but also produce plant growth-promoting substances, including phytohormones such as auxins, gibberellins, and cytokinins, which enhance root development and crop productivity (Glick 2012). Additionally, inoculation with *Azospirillum* increased microbial biomass, positively influencing soil respiration and enzyme activity, both of which indicate the health status of soil (Sharma et al. 2013). *Azospirillum* integration with arbuscular mycorrhizae was found to improve soil characteristics, enhancing nutrient cycles and overall microbial diversity, which is essential for maintaining soil health (Sharma et al. 2013).

5.2.1.2 Phosphate Solubilization Biofertilizers

Solubilizing inorganic phosphate is done through the secretion of organic acids such as citric and gluconic acids. These acids are essential in lowering the pH of the rhizosphere and contributing to the mobilization of fixed phosphorus by dissolving insoluble phosphate compounds (Selvi 2023). De-Andrade et al. (2021) reported that application of phosphoric acids does not only serve as a source of phosphorus during fertigation but also lowers pH in the soil thereby reducing calcium-induced precipitation. *Rhizobium* biofertilizers were also found to not only fix nitrogen, but also contributes to the availability of phosphorus in crops like maize and pigeon pea grown under phosphorus limited soils (Yamamoto et al. 2022, 2023). Additionally, Li et al. (2018) found that single strain of *Burkholderia* enhanced phosphate solubilization through the secretion of organic acids, which help in the acidification of the soil environment. Meanwhile *Pseudomonas fluorescens* produce organic acids that increase the bioavailability of phosphorus in soil, thus facilitating its absorption by plants (Sandhya et al. 2010). Moreover, the secretion of phytohormones such as indole-3 acetic acid (IAA) promotes root elongation and branching, further enhancing nutrient uptake from the soil (Hu et al. 2017).

5.2.1.3 Potassium-Solubilizing Biofertilizers

Potassium-solubilizing bacteria (KSB) are reported to increase potassium availability in soils as well as contribute to crop growth under potassium-limited soils (Ahmad et al. 2016).

Potassium-solubilizing microorganisms such as *Bacillus mucilaginosus*, *Bacillus subtilis*, and *Pseudomonas aeruginosa* have an ability to convert insoluble forms of potassium such as mica and feldspar into soluble form (Shanware et al. 2014). Enzymes such as phosphatases and cellulases contribute to the organic matter decomposition, indirectly releasing potassium from organic pools (Sen et al. 2016; Wu et al. 2023). Another crucial mechanism is chelation, which secretes chelating agents that bind aluminum and silicon ions in soil minerals, smoothening the release of potassium (Saheewala et al. 2023). The production of capsular polysaccharides by potassium-solubilizing biofertilizer has also been involved in the dissolution of minerals, releasing potassium into a form accessible by plants (Velázquez et al. 2016). This solubilization process does not only enhance potassium availability but also contributes to overall soil nutrients, thereby promoting plant growth under conditions where chemical fertilizers are limited or costly.

5.2.2 *Impact of Biofertilizers on Soil Fertility and Microbial Activity*

The application of biofertilizers plays a significant role in enhancing long-term soil fertility and promoting microbial activity, thus contributing to sustainable agriculture (Li et al. 2023). The biofertilizers containing nitrogen-fixing bacteria and mineral solubilizing have been reported to improve nutrient availability and plant uptake efficiency (Saloum 2025). For instance, (Yadav et al. 2023) applied biofertilizers and observed the significant increase in essential nutrient availability and higher yield compared to synthetic fertilizers. The presence of plant growth-promoting rhizobacteria and mycorrhizal fungi has also been reported to enhance soil microbial activity, improving soil structure and nutrient cycling and enhancing the tolerance of crops against pathogens and environmental sensors (Zheng et al. 2021). Dinesh et al. (2010) incorporated biofertilizers and organic manures and observed an increase in microbial biomass, soil respiration, nitrogen mineralization, and enzyme activities compared to synthetic fertilizers. Additionally, Dimitrova et al.

(2024) applied biofertilizers containing *Priestia megatherium* and found an increased microbial diversity.

5.2.3 Comparative Analysis of Chemical and Fertilizers and Biofertilizers

Chemical fertilizers contain essential nutrients including nitrogen, phosphorus, and potassium which are rapidly absorbed by plants and soils (Bhardwaj et al. 2014). However, the production and application of chemical fertilizers poses threat to the environmental due to greenhouse gas emissions, soil degradation, soil acidification, nutrient leaching, and reduced fertility (Etesami et al. 2017). The mechanisms of chemical fertilizer are straightforward as the nutrients are released directly to the soil (Altobaishi et al. 2023). In contrast, biofertilizers contain active microorganisms which promote soil health by gradually releasing nutrients into the soil (Mącik et al. 2020). They are beneficial to farmers especially smallholder farmers as they are cost-effective and offer eco-friendly effects to the environment resulting in long-term soil fertility. The biofertilizers release nutrients through multiple mechanisms of action such as nitrogen fixation and nutrient solubilization.

5.3 Biofertilizers and Medicinal Plant Growth

5.3.1 Effects of Biofertilizers on Medicinal Plant Growth and Yield

Evidence is available which supports the beneficial effects of biofertilizers on growth and yield of medicinal plants across agricultural systems. Biofertilizers enhance attributes such as plant height, branches, leaves, stem diameter, leaf area, biomass, and yield-related characteristics compared to controls (Table 5.2). These effects are modulated by factors including biofertilizer type, application methods (foliar or root), co-inoculation, timing, and agronomic practices like crop rotation, multi-cropping, fertilization, cover cropping, plant density, and water management. Paramanik and Chikkaswamy (2014) conducted a pot study on *Hemigraphis colorata* and *Ocimum basilicum* using biofertilizers like *Azospirillum*, vesicular arbuscular mycorrhizae (VAM), and phosphate-solubilizing bacteria (PSB), alone or combined, and observed improvements in plant height, number of leaves, leaf area, and stem diameter relative to control. Similarly, Dawood et al. (2019) examined *Linum usitatissimum L.* using *mycorrhiza*, *milk whey*, and *yeast extract*, and noted enhancements in plant growth and yield traits. Numerous studies have also reported positive effects of biofertilizers on growth and yield traits of medicinal plants such as *Salvia officinalis* (Nadjafi et al. 2014), *Thymus vulgaris* (Nadjafi et al. 2014),

Table 5.2 Impact of plant growth-promoting rhizobacteria on biochemical properties and yield of selected medicinal plants

Medicinal plants	Biofertilizers	Study place	Parameters and observed effects	Source
Waffle plant (*Hemigraphis colorata*)	*Azospirillum, VAM*	Pot	PH (30.4%–45.5%), NB (60%), LA (12.7%–16.4%), SD (40%)	Paramanik and Chikkaswamy (2014)
Basil (*Ocimum basilicum*)	*PSB, Rhizobium*	Pot	PH (33.3%–100%), NB (40%–133.3%), LA (26.5%–29.4%), SD (50%)	Paramanik and Chikkaswamy (2014)
Sage (*Salvia officinalis*)	*PSB, Azospirillum, Azotobacter*	Field	PH (7.1%), PB (4.3%)	Nadjafi et al. (2014)
Thyme (*Thymus vulgaris*)	*PSB, Azospirillum, Azotobacter*	Field	PH (3.2%), PB (6.7%)	Nadjafi et al. (2014)
Flax (*Linum usitatissimum L.*)	*Mycorrhiza, milk whey, yeast extract*	Field	SB (74.1%–146.8%), RB (37.1%–180%), EOC (7.8%–17.8%), yield (48.6%–102.5%)	Dawood et al. (2019)
Candyleaf (*Stevia rebaudiana*)	*Bacillus safensis, VAM*	Field	PH (94.5%–130.3%), NL (81.2%–98.5%), PB (73.4%–184.2%)	Prakash et al. (2022)
Ashwagandha (*Withania somnifera L. Dunal.*)	*Azospirillum, VAM*	Field	PH (12.7%–20.7%), NB (18.7%–27.2%), RL (13.4%–21.8%), RB (27.3%–45.6%)	Pramodkumar et al. (2018)
Atractylodes lancea	*AMF*	Greenhouse	PH (18.9%–23.5%), PB (22.1%–29.7%), RB (26.4%–34.2%)	Liang and Tang (2018)
Wormwood (*Artemisia annua*)	*Rhizophagus irregularis*	Greenhouse	PH (19.9%), LA (23.6), SB (33.2%), RB (42.4%), PB (34.4)	Domokos et al. (2018)
Pennyroyal (*Mentha pulegium* L.)	*Pseudomonas fluorescens, Azospirillum lipoferum*	Greenhouse	PH (34.6%), SB (52.2%), RB (53.6%), LA (36.4%)	Asghari et al. (2020)
Turmeric (*Curcuma longa*)	*Pseudomonas* sp., *Bacillus* sp.	Greenhouse	PH (34.5%), NB (43.1%), RhB (74.9%), LA (41.5%)	Sontsa-Donhoung et al. (2022)
Begonia malabarica	*Glomus mosseae, Bacillus coagulans, Trichoderma viride*	Greenhouse	PH (0%–77%), NL (27.2%–100%), NB (12.5%–62.5%), PB (259.3%)	Selvaraj et al. (2008)

PH plant height, *NB* number of branches, *NL* number of leaves, *LA* leaf area, *SD* stem diameter, *PB* plant biomass, *RB* root biomass, *SB* shoot biomass, *RL* root length, *RB* root biomass, *RhB* rhizome biomass

Stevia rebaudiana (Prakash et al. 2022), *Withania somnifera L. Dunal* (Pramodkumar et al. 2018), *Atractylodes lancea* (Liang and Tang 2018), and *Curcuma longa* (Sontsa-Donhoung et al. 2022).

5.3.2 Mechanisms Associated with Biofertilizer Effects on Medicinal Plant Growth and Yield

The enhancement of plant growth and yield through the application of biofertilizers can be attributed to a range of factors, which can be classified into direct and indirect mechanisms. Direct mechanisms include nitrogen fixation, phosphorus solubilization, and the production of phytohormones, siderophores, volatile organic compounds (VOCs), and enzymes (Prakash et al. 2022). Biofertilizers contain bacteria and fungi microorganisms, that aid in nitrogen fixation by converting atmospheric nitrogen into forms plants can utilize, such as ammonium (Nadjafi et al. 2014). Among these microbes are *Azospirillum*, *Rhizobium*, and *Azotobacter* (Paramanik and Chikkaswamy 2014). This process increases nitrogen availability to plants, essential for growth and productivity. Additionally, it reduces the need for inorganic nitrogen fertilizers, which have energy-intensive production and negative environmental impacts (Pramodkumar et al. 2018; Wei et al. 2023). Therefore, incorporating biofertilizers can support sustainable agricultural practices. Phosphorus solubilizers facilitate the solubilization of phosphorus, which is typically immobilized in a form that is inaccessible to plants within the soil (Rizvi et al. 2022). Various bacteria, including *Bacillus*, *Pseudomonas*, and *Aspergillus*, are recognized for their ability to solubilize phosphorus through the production of organic acids and phosphatases (Calvo et al. 2014), thereby positively impacting plant growth and yield.

Biofertilizers such as *Rhizobium*, *Azotobacter*, and PSB are known to produce and modify plant hormones like ethylene, cytokinins, gibberellins, and auxins (Zafar et al. 2012; Emmanuel and Babalola 2020). These hormones can improve various plant growth traits, including the length of roots, the initiation of roots, the formation of root hairs, nutrient uptake, and resistance to stress, ultimately leading to increased yields and better crop quality (Prakash et al. 2022; Karapetyan 2024). Biofertilizers, such as *Azotobacter* and *Pseudomonas fluorescent*, generate siderophores, which can boost plant growth by enhancing the availability of iron. These microorganisms secrete siderophores into the soil, which attach to iron and create a form that plant roots can absorb (Stamenković et al. 2018; Sontsa-Donhoung et al. 2022; Çakmakçı et al. 2023). This mechanism is especially advantageous in soils with limited iron content or where iron exists in insoluble forms.

Biofertilizers indirectly benefit plants by helping them cope with abiotic stress and enhancing resistance to harmful microbes. These growth-promoting biofertilizers assist plants in managing abiotic stress conditions, minimizing yield reductions (Wei et al. 2023). Drought and salinity remain prominent agricultural stresses, worsened by climate change and high-input agriculture. Biofertilizers modify physiological and enzymatic activities, particularly plant antioxidative responses (Lubna et al. 2018). For example, biofertilizers that address drought stress help plants by altering root growth, enhancing water and nutrient absorption through increased root penetration (Domokos et al. 2018; Asghari et al. 2020). Besides indole-3-acetic acid (IAA), abscisic acid (ABA) is another crucial plant hormone involved in plant response to drought conditions (Calvo et al. 2014). In terms of salinity, biofertilizers play a role in maintaining ion equilibrium and minimizing the absorption of harmful ions such as sodium (Na^+), while promoting the uptake of vital nutrients like potassium (K^+) (Nieves-Cordones et al. 2016; González-Delgado et al. 2023). Additionally, they assist plants in osmotic regulation by generating osmoprotectants, which help sustain cell turgor and shield cellular structures from the adverse effects of elevated salt levels (Lubna et al. 2018; Asghari et al. 2020). Microorganisms in biofertilizers can produce antibiotics that kill or suppress disease-causing organisms, protecting plants from infections (Xie et al. 2016). By colonizing root surfaces and surrounding soil, these beneficial microbes outcompete harmful pathogens, reducing their numbers and preventing plant infections (Zafar et al. 2012). Microorganisms in biofertilizers secrete hydrolytic enzymes that degrade pathogenic microbe cell walls (Zhang et al. 2023). Enzymes like chitinases, glucanases, and proteases break down pathogens' structural components, leading to their destruction.

Another important aspect of biofertilizers is their ability to reduce heavy metal contamination, thereby enhancing plant growth. For example, in a study conducted by Wei et al. (2023) on soil contaminated with heavy metals, the application of biofertilizers resulted in a reduction of Cd and Hg concentrations to below detectable levels, which subsequently increased the biomass of *Salvia miltiorrhiza*.

5.4 Enhancing Medicinal Compound Production Through Biofertilizers

5.4.1 Significance of Biofertilizer Application on Secondary Metabolites of Medicinal Plants

The use of biofertilizers in the commercial cultivation of medicinal plants is an effective management strategy that boosts biomass production and enhances the synthesis of secondary metabolites (Wei et al. 2023). Biofertilizers' capacity to improve secondary metabolite production positions them as a valuable asset for the commercial growth and value of medicinal plants. This method can result in increased yields of important medicinal compounds, thereby facilitating the production process more efficient and sustainable (Espinoza et al. 2019). Bioactive compounds from medicinal plants play a crucial role in enhancing and preserving their value in the healthcare, food, and cosmetics sectors (Prakash et al. 2022). In plants, secondary metabolites serve several important functions, such as providing a carbon and energy source, protecting plants against pathogens, coloring of flowers for pollinator attraction, and shielding plants from detrimental UV rays (Lubna et al. 2018; Dawood et al. 2019; Asghari et al. 2020; Çakmakçı et al. 2023). These benefits are amplified by incorporating biofertilizers. Thus, utilizing biofertilizers in the commercial cultivation of medicinal plants is an effective management strategy that boosts biomass production and enhances the synthesis of secondary metabolites in these plant species (Wei et al. 2023).

5.4.2 Effects of Biofertilizer Application on Secondary Metabolites of Medicinal Plants

Research has been carried out to examine how biofertilizers influence the buildup of secondary metabolites in medicinal plants, aiming to enhance their medicinal and commercial value (Dawood et al. 2019; Espinoza et al. 2019). The establishment of biofertilizers follows a traditional "pharmacological" method, where secondary metabolites are evaluated under controlled conditions, following a systematic process to identify promising candidates, transitioning from laboratory settings to more practical environments (Rizvi et al. 2022). Table 5.3 illustrates examples of how various biofertilizers impact different secondary metabolites in various medicinal plants under diverse cultivation conditions. Çakmakçı et al. (2023) demonstrated that plant growth-promoting rhizobacteria (PGPR) increased the concentrations of thymol, carvacrol, linalool, borneol, as well as the essential oil content and yield of *Origanum onites* in comparison to the control group in a greenhouse study. Similarly, a field study conducted by Sontsa-Donhoung et al. (2022) revealed that *Pseudomonas* sp. and *Bacillus* sp. enhanced the levels of phenolics, curcumin, and polyphenol oxidase, while also increasing antioxidant and guaiacol peroxidase activity relative to the control. Additional research has reported positive effects of secondary metabolites under various biofertilizers on medicinal plants, including *Thymus vulgaris* (Nadjafi et al. 2014), *Callisia fragrans* (Karapetyan 2024), *Atractylodes lancea* (Liang and Tang 2018), *Artemisia annua* (Domokos et al. 2018), *Mentha pulegium L.* (Asghari et al. 2020), and *Dracocephalum moldavica* (Ghanbarzadeh et al. 2020). These effects are modulated by factors like inoculant type, application methods (foliar or root), co-inoculation, timing, and agronomic practices like crop rotation, multi-cropping, fertilization, cover cropping, plant density, and water management.

Table 5.3 Impact of plant growth-promoting rhizobacteria on biochemical properties and yield of selected medicinal plants

Medicinal plants	Biofertilizers	Study place	Parameters and observed effects	Source
Basket plant (*Callisia fragrans*)	*Azospirillum*, *Azotobacter*, *Bacillus* spp.	Field	Flavonoids (70%–100%), anthocyanins (50%)	Karapetyan (2024)
Sage (*Salvia officinalis*)	*PSB*, *Azospirillum*, *Azotobacter*	Field	α-Thujone (0%), camphor (0%)	Nadjafi et al. (2014)
Thyme (*Thymus vulgaris*)	*PSB*, *Azospirillum*, *Azotobacter*	Field	Thymol (18.9%), γ-terpinene (13.2%)	Nadjafi et al. (2014)
Oregano (*Origanum onites*)	*PGPR*	Greenhouse	Thymol (14.5%–16.9%), carvacrol (66.1%–72.2%), linalool (1.38%–3.68%), borneol (32.2%–65%), EOC (0.5%–40.1%), EOY (5.9%–71.9%)	Çakmakçı et al. (2023)
Flax (*Linum usitatissimum L.*)	*Mycorrhiza*, *milk whey*, *yeast extract*	Field	Protein (8%–18.2%), phenolic (18.3%–42.7%), flavonoid (15%–34.2%)	Dawood et al. (2019)
Candyleaf (*Stevia rebaudiana*)	*Bacillus safensis*, *VAM*	Field	Protein (153.1%–33.8%), flavonoid (56.9%–16.2%), antioxidant activity (50.3%–12.6%), carbohydrate (25.6%–103.8%)	Prakash et al. (2022)
Ashwagandha (*Withania somnifera L. Dunal.*)	*Azospirillum*, *VAM*	Field	Withanolide (29.2%–44.3%), alkaloid (16.1%–26.1%)	Pramodkumar et al. (2018)
Atractylodes lancea	*AMF*	Greenhouse	EOC (35.7%–41.3%), atractylon (32.4%–35.6%), atractylodin (28.3%–31.2%), hinesol (26.4%–34.2%), β-eudesmol (20.3%–25.6%)	Liang and Tang (2018)
Wormwood (*Artemisia annua*)	*Rhizophagus irregularis*	Greenhouse	Artemisinin (32.3%), guaiacol peroxidase activity (46.5%)	Domokos et al. (2018)
Pennyroyal (*Mentha pulegium L.*)	*Pseudomonas fluorescens*, *Azospirillum lipoferum*	Greenhouse	Phenolics (35.2%), flavonoids (38.1%), EOC (33.6%), antioxidant (40.1%), proline (41.7%)	Asghari et al. (2020)

(continued)

Table 5.3 (continued)

Medicinal plants	Biofertilizers	Study place	Parameters and observed effects	Source
Dracocephalum moldavica	*Micrococcus yunnanensis*, *Claroideoglomus etunicatum*	Greenhouse	Phenolics (53.9%), flavonoids (50.9%), EOC (73.2%), antioxidant (47.7%), proline (41.4%)	Ghanbarzadeh et al. (2020)
Turmeric (*Curcuma longa*)	*Pseudomonas* sp., *Bacillus* sp.	Greenhouse	Phenolics (44.6%), curcumin (55.8%), antioxidant (36.8%), guaiacol peroxidase activity (68.8%), polyphenol oxidase activity (62.8%)	Sontsa-Donhoung et al. (2022)

VAM vesicular arbuscular mycorrhizae, *PSB* phosphate-solubilizing bacteria, *PGPR* plant growth-promoting rhizobacteria, *EOC* essential oil content, *EOY* essential oil yield

5.5 Challenges and Limitations in the Adoption of Biofertilizer Use for Medicinal Plants

5.5.1 Barriers to Adoption by Smallholder Farmers

The limited development and utilization of biofertilizers among African smallholder farmers can be attributed to several challenges. Foremost among these is the lack of awareness among many African farmers regarding biofertilizers, including their nature, application methods, and specific benefits, particularly in the context of medicinal plant cultivation. These farmers are accustomed to conventional inorganic and organic fertilizers, and this lack of familiarity impedes the adoption of biofertilizers (Stamenković et al. 2018).

Although some farmers may be aware of biofertilizers, they lack the technical expertise required for their effective utilization. For example, research conducted by Singla and Rampal (2021) found that 58% of participants expressed a lack of understanding regarding biofertilize utilization. This lack of understanding can hinder their adoption of these sustainable options, as they might be unaware of the advantages or the correct application methods. Successful use of biofertilizers often demands specific knowledge about soil health and crop management. Thus, farmers without this technical know-how may struggle to implement biofertilizer strategies successfully (Emmanuel and Babalola 2020). The situation is worsened by the fact that in some regions researchers and extension officers lack sufficient knowledge about the advantages of biofertilizers, which has adversely impacted farmers' awareness of this technology. According to Singla and Rampal (2021), 41% of participants in their research reported not receiving sufficient guidance from extension services on biofertilizers. This deficiency in support can result in farmers lacking the essential information and help needed to effectively adopt biofertilizer practices in medicinal plant cultivation.

There is a paucity of research concerning biofertilizer formulations specifically designed for medicinal plants. This results in a lack of data and evidence to persuade farmers of their benefits, leading to skepticism regarding their efficacy and consequently limiting the adoption of biofertilizers as an alternative. Also, in rural areas, biofertilizers may not be readily accessible, and those available in the market often suffer from poor quality and lack quality assurance, leading to unsatisfactory results and skepticism (Zaidi et al. 2015). This deficiency in quality undermines farmers' confidence and presents challenges for commercialization (Stamenković et al. 2018).

5.5.2 *Economic and Logistical Challenges*

The expense of transitioning to biofertilizers can discourage many farmers, particularly if they are unsure about the financial returns compared to conventional fertilizers. Farmers frequently express concern over the possibility of reduced yields and profits when opting for biofertilizers instead of traditional synthetic fertilizers. This worry about potential economic losses serves as a major obstacle to their adoption (Stamenković et al. 2018). Furthermore, the absence of government subsidies to alleviate the high costs of biofertilizers for resource-poor farmers restricts their use in cultivating medicinal plants.

The absence of dependable distribution systems in Africa results in unpredictable supply chains mechanisms, particularly during planting periods when the fertilizers are needed most. The market for biofertilizers is frequently weak and unstable in Africa, impacting both their market potential and availability (Zaidi et al. 2015). Limited marketing channels for biofertilizers restrict farmers' awareness of these products and their advantages. Additionally, farmers often encounter difficulties in accessing markets for their medicinal plants, which can discourage investment in biofertilizers (Nwafor et al. 2021).

There is scarce of essential infrastructure and technologies vital for producing biofertilizers in Africa. Key technologies, including formulation methods, fermenters, storage solutions, and sterilization equipment, which require significant capital investment and are predominantly sourced outside the continent, are largely unavailable in most African countries (Raimi et al. 2021). This limits biofertilizer availability, slowing down their application in medicinal plants. Storage conditions significantly influence biofertilizer use and advancement in Africa. Many biofertilizers are vulnerable to heat and light, requiring cold storage facilities and quick application, conditions often lacking among small-scale farmers in Africa. Storing biofertilizers at lower temperatures extends shelf life (Satinder 2012). However, maintaining storage at 4 °C with refrigerators is challenging in Africa, where farmers face financial constraints, unreliable electricity, and temperatures of 32–40 °C (Raimi et al. 2021). These conditions are not conducive to effective biofertilizer storage and may have led to inferior products on the continent.

Also, inadequate regulation of biofertilizer products leads to markets filled with inferior formulations. The low quality of available biofertilizer undermines farmers' trust and hinders commercialization. Also, complex regulatory frameworks can create obstacles for new products entering the market (Zaidi et al. 2015). Without appropriate government policies encouraging biofertilizer use, farmers may be less motivated to adopt these alternatives (Emmanuel and Babalola 2020). Additionally, biofertilizers are often not tailored to specific local crops or ecological conditions, which can impact how effectively they are used.

5.5.3 Potential Solutions and Policy Recommendations

5.5.3.1 Awareness and Knowledge

Despite the relatively low adoption of biofertilizers in Africa due to numerous constrains, significant efforts are required to promote their application across the continent. Given that farmers are not yet well-familiar with biofertilizers, it is crucial to begin by raising awareness and training programs of these fertilizers as viable alternatives to the inorganic and organic fertilizers with which farmers are predominantly familiar.

Educational initiatives, such as workshops and training sessions, can be implemented to inform farmers about the benefits of biofertilizers. Conducting field trials allows farmers to witness the advantages of biofertilizers firsthand, which can be highly convincing. Collaborating with agricultural extension services can enhance the dissemination of information. Extension agents can offer direct support and advice to farmers. Presenting information in local languages and adapting it to the region's specific agricultural practices and conditions can make it more accessible and understandable for farmers. Utilizing social media and mobile apps can help reach a wider audience. Engaging farmers through informative content, videos, and interactive sessions can provide them with valuable insights. Creating brochures, pamphlets, and videos that detail the benefits and application methods of biofertilizers can serve as helpful resources for farmers.

5.5.3.2 Financial Incentives

Securing funding is deemed essential for the commercialization and widespread use of biofertilizers for smallholder farmers in the cultivation of medicinal plants. Financial assistance or government subsidies for purchasing biofertilizers can help ease the economic strain on farmers. African governments should also enhance their efforts in financing biofertilizer startups and technology, as well as intentionally create policies to foster their growth. Additionally, governments and nongovernmental organizations (NGOs) should allocate resources to training programs that focus on the application of biofertilizers in sustainable medicinal plant cultivation.

Policies such as subsidies or incentive programs, like voucher systems, that lower the financial burden could encourage more smallholder farmers to adopt these practices.

5.5.3.3 Research and Development

Promoting research into the local production of biofertilizers can enhance their availability and lower expenses. For example, identifying effective microbes with distinct characteristics, such as high sensitivity to drying and the capacity to thrive in extreme temperatures, could result in innovative formulations (Zhang et al. 2023). Consequently, formulation technologies might be employed to create a specialized product tailored for a specific field function within a particular environment.

The effectiveness of biofertilizers is influenced by crop species, soil nutrient complexity, and climate. Research on suitable biofertilizers requires agronomists who understand crop-nutrient-climate interactions in the African context (Aloo et al. 2022). A key research phase involves developing efficient microbial strains with enhanced colonization potential and field success. Creating formulations using local strains with high field efficiency is essential. Biofertilizer application should be optimized alongside agronomic practices like crop rotation, amendments, tillage, and irrigation to enhance effectiveness. Understanding inoculum dynamics under various practices could help develop suitable biofertilizers for different crop stages. Advances in formulations with extended shelf lives and reliable performance will accelerate this technology's commercialization.

5.5.3.4 Logistics and Infrastructure

Despite the potential of biofertilizers, the marketing channels in Africa have been inadequate and ineffective, posing difficulties in promoting these products to various regions, particularly among rural farmers (Raimi et al. 2021). Therefore, enhancing infrastructure, marketing channels, and developing strategies are crucial for advancing the African biofertilizer market, which will, in turn, increase awareness, acceptance, and usage among African farmers. Investing in both current and future infrastructure and technologies is essential for producing and distributing the appropriate biofertilizer formulations needed by farmers. An effective delivery system should enhance the product's cost-effectiveness, applicability, and efficiency. Additionally, encouraging decentralized production units can help overcome logistical challenges, ensuring better access to fresher products. Thus, investing in such a high-quality delivery system is vital.

5.5.3.5 Develop Standards and Certifications

The growing interest in biofertilizers, driven by sustainable and organic farming, calls for strong standards and certifications. Establishing quality control systems for biofertilizer products is crucial to ensure farmers receive effective formulations, a vital step in encouraging their use. The regulatory framework for biofertilizers is evolving (Rizvi et al. 2022). Detailed quality concepts, manuals, and regulations are needed to guide biofertilizer production and use. This framework should cover the entire lifecycle from production to shelf life and application. Managing biofertilizer quality during production is critical and must be consistently upheld for farmers. Quality control documentation is needed, addressing infrastructure, transportation, formulation, and shelf life. Institutions must conduct regular monitoring with periodic checks to ensure product quality meets specified standards. Private organizations, farmers, manufacturers, and government institutions in Africa should collaborate on policies that foster commercialization and realization of the economic benefits of biofertilizers on the continent. Moreover, although biofertilizers are typically sourced from natural environments, it is important to assess any possible environmental hazards, particularly when additives are present (Zaidi et al. 2015; Stamenković et al. 2018). Toxicological assessments might be necessary for certain strains that are recognized as opportunistic pathogens.

5.6 Emerging Biofertilizer Technologies

Innovative technologies are set to drive substantial progress in the use of biofertilizers for medicinal plants. The focus of new biofertilizer technologies should be on improving the effectiveness and targeting of microbial formulations, which can enhance the growth and yield of medicinal crops. Over time, these innovative methods will improve the comprehensive understanding of biofertilizers and their impact on plant nutrient uptake and other biofertilizer characteristics, aiding in their practical use. The subsequent sections detail key innovations in this area Africa can adopt.

5.6.1 Next-Generation Microbial Consortia

This technology utilizes a blend of various beneficial microorganisms that can be mixed to boost plant growth (Khumairah et al. 2022). These microbial consortia offer multiple plant growth-promoting characteristics, such as nitrogen fixation and nutrient solubilization, making them more efficient than single-strain biofertilizers (Aloo et al. 2022). For instance, combining nitrogen-fixing bacteria like *Rhizobium* and *Azospirillum* with phosphate-solubilizing bacteria such as *Bacillus* and *Pseudomonas* can create synergistic effects that improve plant growth and nutrient uptake. The integration of different microbes can result in healthier plants and higher crop yields (Singh et al. 2020).

5.6.2 Nano-biofertilizers

The incorporation of nanotechnology enhances the delivery and efficacy of biofertilizers, boosting microbial survival and nutrient utilization efficiency, which is vital for growing medicinal plants. Nanoparticles can improve the delivery and performance of biofertilizers by enabling precise nutrient release and better interaction with roots. This cutting-edge method is regarded as a promising direction for future biofertilizer research and establishment in sustainable agriculture (Rizvi et al. 2022).

5.6.3 Encapsulation Techniques

Innovative methods such as alginate bead encapsulation shield microbes from environmental challenges, facilitating a slow release of nutrients and enhancing root colonization, which is crucial for the well-being of medicinal plants (Chaudhary et al. 2020; Aloo et al. 2022). This approach involves encasing beneficial microorganisms within a protective layer, which can boost their stability and longevity (Yuliani et al. 2024). Microencapsulation simplifies the handling and application of biofertilizers, ensuring that the microbes remain active until they are introduced to the soil. The goal of this technology is to improve the efficacy of biofertilizers by delivering microorganisms to plants in optimal condition (Zhang et al. 2023). However, most research in this area has been confined to laboratory settings, and efforts should be made to standardize this technology for industrial and field use including medicinal plants.

5.6.4 Genetically Modified Organisms (GMOs)

Research into GMOs is a burgeoning field. These biofertilizers are designed to enhance plant growth and nutrient absorption more effectively (Bumunang and Babalola 2014). By modifying certain genes, scientists can develop strains that excel in nitrogen fixation, phosphorus solubilization, or the production of growth hormones. This technology offers the possibility of creating highly efficient biofertilizers that are customized for particular crops and soil types (Aloo et al. 2022). The aim is to produce biofertilizers that can greatly boost crop yields while remaining environmentally safe (Haskett et al. 2021). Nonetheless, there are hurdles in achieving public acceptance and obtaining regulatory approval for GMOs.

5.6.5 Liquid Biofertilizers;

Liquid biofertilizers are becoming increasingly popular because they are easy to apply and are quickly absorbed by plants. These formulations can be administered via irrigation systems or as foliar sprays, offering plants an immediate supply of beneficial microorganisms and nutrients. Additionally, liquid biofertilizers can be more efficient in delivering nutrients directly to the roots of plants (Singh et al. 2020).

5.6.6 Carrier-Enriched Biofertilizers

These biofertilizers are formulated with carrier materials that enhance microorganism durability and effectiveness during storage and application. This method ensures beneficial microbes remain viable until soil application (Khumairah et al. 2022). They consist of live or dormant beneficial microorganisms combined with carriers like peat or lignite, which improves moisture retention and supports microbial proliferation, enhancing plant nutrient availability and crop yield (Haskett et al. 2021). Additionally, materials like gum and pea fiber improve microbial entrapment and survival. The microstructural characteristics of these carriers play a crucial role in biofertilizer efficiency during storage and use, as shown in the study (Yuliani et al. 2024).

5.6.7 Technologically Integrated Approaches

There is need to incorporate current advanced technologies, including artificial intelligence, data analytics, and machine learning, into the creation of biofertilizers. These methods can assist in forecasting how biofertilizers will perform in different environmental settings, thereby facilitating more strategic use in agriculture. This integration is part of a larger shift towards sustainable agricultural practices that utilize technology to achieve better results.

5.7 Conclusion

The current chapter presented the potential use of biofertilizers as environmentally friendly and sustainable alternative to synthetic chemical fertilizers for improving productivity and phytochemical composition of medicinal plants in African farming systems. This is due to their important role in improving soil health and microbial diversity, as well as their numerous benefits in fixing nitrogen, production of phytohormone (e.g., IAA), and potassium and phosphate solubilization. Furthermore, the role of biofertilizers in promoting plant growth and yield as well as in enhancing bioactive secondary metabolites crucial for medicinal effectiveness was emphasized. Despite their well-documented potential, adoption of these biofertilizers by smallholder farmers remains low. This has been ascribed to lack of awareness by farmers, infrastructural and financial barriers, as well as inadequate regulatory frameworks. As such, it is important to address these challenges through interventions like targeted education, supportive policies, improved logistics, and investment in research. In this chapter, we concluded that integrating advanced technologies, such as microbial consortia, nano-enabled biofertilizers, encapsulation techniques, and AI-driven decision support systems, can significantly augment the efficacy of biofertilizers in promoting the sustainable cultivation of African medicinal plants.

References

Abdel Aal AM, Assiri ME, Al-Farga A, Moustafa YM, Hammam AA, Haddad SA and Abdelkarim NS (2023) Exploration of the benefits of biofertilizers for attaining food security in Egypt's agriculture. Agronomy, 13(10):2477. https://doi.org/10.3390/agronomy13102477

Ahmad M, Nadeem SM, Naveed M. and Zahir ZA (2016) Potassium-solubilizing bacteria and their application in agriculture. Potassium solubilizing microorganisms for sustainable agriculture, 293–313. https://doi.org/10.1007/978-81-322-2776-2_21

Ajeel MR, Al-Hakeim MS (2024) The influence of biological and organic fertilization and boron spraying on some soil characteristics, Basrah Journal of Agricultural Sciences, 38(1):312–323. https://doi.org/10.1088/1755/1371/8/082022

Ali HAA, Kamal JAK (2025) April. The Effect of Organic Manure Type and Molecularly Identified Azotobacter Bacteria on some Soil Properties. In IOP Conference Series: Earth and Environmental Science, 1487(1):012207 IOP Publishing. https://doi.org/10.1088/1755-1315/1487/1/012207

Allito BB, Ewusi-Mensah N. and Logah V (2020) Legume-rhizobium strain specificity enhances nutrition and nitrogen fixation in faba bean (Vicia faba L.). Agronomy, 10(6):826. https://doi.org/10.3390/agronomy10060826

Allito BB, Nana EM, Alemneh AA (2015) Rhizobia strain and legume genome interaction effects on nitrogen fixation and yield of grain legume: a review. Molecular Soil Biology, 6. https://doi.org/10.5376/msb.2015.06.0004

Altobaishi SF, Almana FA, Abd-ElGawad AM, Al-Yafrsi MA, Elhindi KM (2023) Ivy Geranium (Pelargonium peltatum (L.) L'Hér.) Plant growth and flowering as affected by mineral or biofertilizer with or without compost amendment. Agriculture, 13(5):1106. https://doi.org/10.3390/agriculture13051106

Aloo BN, Tripathi V, Makumba BA, Mbega ER (2022) Plant growth-promoting rhizobacterial biofertilizers for crop production: the past, present, and future. Front Plant Sci 13:1–15

Asghari B, Khademian R, Sedaghati B (2020) Plant growth promoting rhizobacteria (PGPR) confer drought resistance and stimulate biosynthesis of secondary metabolites in pennyroyal (*Mentha pulegium* L.) under water shortage condition. Sci Hortic 263:109132

Bhardwaj D, Ansari MW, Sahoo RK, Tuteja N (2014) Biofertilizers function as key player in sustainable agriculture by improving soil fertility, plant tolerance and crop productivity. Microbial cell factories, 13(1):66. https://doi.org/10.1186/1475-2859-13-66

Bumunang EW, Babalola OO (2014) Characterization of rhizobacteria from field grown genetically modified (GM) and non-GM maize. Braz Arch Biol Technol 57:1–8

Çakmakçı R, Haliloglu K, Türkoğlu A, Özkan G, Kutlu M, Varmazyari A, Molnar Z, Jamshidi B, Pour-Aboughadareh A, Bocianowski J (2023) Effect of different plant growth-promoting rhizobacteria on biological soil properties, growth, yield and quality of Oregano (*Origanum onites* L.). Agronomy 13:1–17

Calvo P, Nelson L, Kloepper JW (2014) Agricultural uses of plant biostimulants. Plant Soil 383:3–41

Chaudhary T, Dixit M, Gera R, Shukla AK, Prakash A, Gupta G, Shukla P (2020) Techniques for improving formulations of bioinoculants. 3 Biotech 10:1–9

Chen Y, Ye J. and Kong Q (2020) Potassium-solubilizing activity of Bacillus aryabhattai SK1-7 and its growthpromoting effect on Populus alba L. Forests, 11(12):1348. https://doi.org/10.3390/f11121348

Dawood MG, Sadak MS, Bakry BA, El Karamany MF (2019) Comparative studies on the role of benzoic, t-cinnamic, and salicylic acids on growth, some biochemical aspects, and yield of three flax cultivars grown under sandy soil conditions. Bull Natl Res Cent 43:112

de Andrade ARS, Franco ES, Guerra HOC, Jadoski SO, da Silva EG, Cruz AFDS, Ferreira MEDS, da Silva ET, Ramalho e Soares H (2021) Chemical fertilization through fertirrigation in the evaluation of the biochemical characteristics of the watermelon fruits. https://doi.org/10.32404/rean.v8i3.5561

Dimitrova K, Kaiyrbekov T. and Balabanova D (2024) Assessing the impact of biofertilizer on soil microbial dynamics and metabolic activity in a controlled maize pot-grown experiment. Eurasian Journal of Soil Science, 13(3):202–209. https://doi.org/10.1016/j.biortech.2010.01.108, https://doi.org/10.18393/ejss.1439846

Dinesh R, Srinivasan V, Hamza S, Manjusha A (2010) Short-term incorporation of organic manures and biofertilizers influences biochemical and microbial characteristics of soils under an annual crop [Turmeric (*Curcuma longa* L.)]. Bioresour Technol 101(12):4697–4702

Długosz J. and Piotrowska-Długosz A (2023) Different response of carbon and P-related soil properties toward microbial fertilizer application. Agronomy, 13(11):2751. https://doi.org/10.3390/agronomy13112751

Domokos E, Jakab-Farkas L, Darkó B, Bíró-Janka B, Mara G, Albert C, Balog A (2018) Increase in artemisia annua plant biomass artemisinin content and guaiacol peroxidase activity using the arbuscular mycorrhizal fungus *Rhizophagus irregularis*. Front Plant Sci 9:1–9

Emmanuel OC, Babalola OO (2020) Productivity and quality of horticultural crops through co-inoculation of arbuscular mycorrhizal fungi and plant growth promoting bacteria. Microbiol Res 239:126569

Espinoza F, Vidal S, Rautenbach F, Lewu F, Nchu F (2019) Effects of *Beauveria bassiana* (Hypocreales) on plant growth and secondary metabolites of extracts of hydroponically cultivated chive (*Allium schoenoprasum* L. [Amaryllidaceae]). Heliyon 5:e03038

Etesami H, Emami S. and Alikhani HA (2017) Potassium solubilizing bacteria (KSB): Mechanisms, promotion of plant growth, and future prospects A review. Journal of soil science and plant nutrition, 17(4):897–911. https://doi.org/10.4067/S0718-95162017000400005

Fitriatin BN, Amanda AP, Kamaluddin NN, Khumairah FH, Sofyan ET, Yuniarti A. and Turmuktini T (2021) Some soil biological and chemical properties as affected by biofertilizers and organic ameliorants application on paddy rice. Eurasian Journal of Soil Science, 10(2):105–110. https://doi.org/10.18393/ejss.829695

García YB, Peñaranda MCP, Marrero G, Cabrera AR, Pérez M, Pino AS, Vega VRM and Alarcón Mok MJ (2024) Effects of a New Rhizobium-Based Biofertilizer (Fertiriz) on Growth and the Yield of Cowpea (Vigna unguiculata (L.) Walp). Journal of Environmental Analysis & Ecology Studies. https://doi.org/10.31031/EAES.2024.12.000778

Ghanbarzadeh Z, Mohsenzadeh S, Rowshan V, Zarei M (2020) Mitigation of water deficit stress in *Dracocephalum moldavica* by symbiotic association with soil microorganisms. Sci Hortic 272:109549

Glick BR (2012) Plant growth-promoting bacteria: mechanisms and applications. Scientifica, 2012(1):963401. https://doi.org/10.6064/2012/963401

González-Delgado M, Minjares-Fuentes R, Mota-Ituarte M, Pedroza-Sandoval A, Comas-Serra F, Quezada-Rivera JJ, Sáenz-Esqueda Á, Femenia A (2023) Joint water and salinity stresses increase the bioactive compounds of Aloe vera (*Aloe barbadensis* Miller) gel enhancing its related functional properties. Agric Water Manag 285:108374

Haroun M, Xie S, Awadelkareem W, Wang J and Qian X (2023) Influence of biofertilizer on heavy metal bioremediation and enzyme activities in the soil to revealing the potential for sustainable soil restoration. Scientific Reports, 13(1):20684

Haskett TL, Tkacz A, Poole PS (2021) Engineering rhizobacteria for sustainable agriculture. ISME J 15:949–964

Hu J, Wei Z, Weidner S, Friman VP, Xu YC, Shen QR, Jousset A (2017) Probiotic Pseudomonas communities enhance plant growth and nutrient assimilation via diversity-mediated ecosystem functioning. Soil Biology and Biochemistry, 113:122–129

Ikhoyameh M, Okete WE, Ogboye RM, Owoyemi OK, Gbadebo OS (2024) Integrating traditional medicine into the African healthcare system post-Traditional Medicine Global Summit: challenges and recommendations. Pan African Medical Journal, 47(1)

Karapetyan A (2024) The role of biofertilizer on the growing efficiency of *Callisia fragrans* cultivated under open-air hydroponic conditions of the Ararat Valley. Bioact Compd Health Dis 7:211–220

Khumairah FH, Azwari F, Setiawati MR, Fitriatin BN, Simarmata T (2022) Effect of different carrier composition on viability of halotolerant N-fixing rhizobacteria, water content and pH level in nitrogen biofertilizer for saline ecosystem. Kultivasi 21:1–9

Lawal I (2021) Nitrogen fixing bacteria and their application for heavy metal removal: a mini review. Journal of Biochemistry, Microbiology and Biotechnology, 9(2):43–47. https://doi.org/10.54987/jobimb.v9i2.617

Li GX, Wu XQ, Ye JR, Yang HC (2018) Characteristics of organic acid secretion associated with the interaction between Burkholderia multivorans WS-FJ9 and poplar root system. BioMed Research International, 2018(1):9619724. https://doi.org/10.1155/2018/9619724

Li L, Tong L. and Lv Y (2023) Influence of bio-fertilizer type and amount jointly on microbial community composition, crop production and soil health. Agronomy, 13(7):1775. https://doi.org/10.3390/agronomy13071775

Liang X, Tang M (2018) Effects of three arbuscular mycorrhizal fungi (AMF) species on the growth, physiology, and major components of essential oil of *Atractylodes lancea*. Chin J Ecol 37:1871

Lodi LA, Klaic R, Bortoletto-Santos R, Ribeiro C. and Farinas CS (2022) Unveiling the solubilization of potassium mineral rocks in organic acids for application as K-fertilizer. Applied biochemistry and biotechnology, 194(6):2431–2447. https://doi.org/10.1007/s12010-022-03826-7

Lubna AS, Hamayun M, Khan AL, Waqas M, Khan MA, Jan R, Lee IJ, Hussain A (2018) Salt tolerance of *Glycine max*. L induced by endophytic fungus *Aspergillus flavus* CSH1, via regulating its endogenous hormones and antioxidative system. Plant Physiol Biochem 128:13–23

Nadjafi F, Mahdavi Damghani M, Tabrizi L, Nejad ES (2014) Effect of biofertilizers on growth, yield and essential oil content of Thyme (*Thymus vulgaris* L.) and Sage (*Salvia officinalis* L.). J Essent Oil-Bear Plants 17:237–250

Mącik M, Gryta A. and Frąc M (2020) Biofertilizers in agriculture: An overview on concepts, strategies and effects on soil microorganisms. Advances in agronomy, 162:31–87. https://doi.org/10.1016/bs.agron.2020.02.001

Mátyás BG, Lowy DA, Singla A, Melendez JR, Sándor Z (2020) Comparison of effects exerted by biofertilizers, NPK fertilizers, and cultivation methods on soil respiration in Chernozem soil. https://doi.org/10.17163/lgr.n32.2020.01

Mohite BV, Patil SV (2022) Isolation and Identification of Nonsymbiotic Azotobacter and Symbiotic Azotobacter Paspali–Paspalum notatum. Practical handbook on agricultural microbiology, 25–33

Nieves-Cordones M, Martínez V, Benito B, Rubio F (2016) Comparison between Arabidopsis and rice for main pathways of K+ and Na+ uptake by roots. Front Plant Sci 7:1–14

Nwafor I, Nwafor C, Manduna I (2021) Constraints to cultivation of medicinal plants by smallholder farmers in South Africa. Horticulturae 7:1–15

Paramanik RC, Chikkaswamy BK (2014) Effects of VAM and Biofertilizers on some Medicinal Plants. Int J Curr Microbiol Appl Sci 3:1016–1027

Prajapati K, Sharma MC, Modi HA (2013) Growth promoting effect of potassium solubilizing microorganisms on Abelmoscus esculantus. International Journal of Agricultural Science 3(1): 181–188

Patel Krishna Kumar, Alok Kumar Pandey, Ajay Kumar Baheliya, Veerendra Singh, Rishikesh Yadav, Shubham Pandey, Adesh, Vikash Yadav (2024) A Comparative Study of the Effects of Biofertilizers and Chemical Fertilizers on Soil Physical and Biological Properties under Chickpea Crop (Cicer Arietinum L.). International Journal of Environment and Climate Change 14 (3):72–80. https://doi.org/10.9734/ijecc/2024/v14i34020

Prakash J, Egamberdieva D, Arora NK (2022) A novel *Bacillus safensis*-based formulation along with mycorrhiza inoculation for controlling *Alternaria alternata* and simultaneously improving growth, nutrient uptake, and steviol glycosides in *Stevia rebaudiana* under field conditions. Plants 11:1857

Pramodkumar T, Shiragur M, Chandrakant B (2018) Effect of bio-inoculants on growth, dry root yield and quality in Ashwagandha (*Withania somnifera* L. Dunal.). Int J Curr Microbiol App Sci 7:2203–2212

Raimi A, Roopnarain A, Adeleke R (2021) Biofertilizer production in Africa: current status, factors impeding adoption and strategies for success. Sci Afr 11:e00694

Rizvi A, Ahmed B, Khan MS, El-Beltagi HS, Umar S, Lee J (2022) Bioprospecting plant growth promoting rhizobacteria for enhancing the biological properties and phytochemical composition of medicinally important crops. Molecules 27:1–31

Saheewala H, Sanadhya S, Upadhyay SK, Mohanty SR, Jain D (2023) Polyphasic characterization of indigenous potassium-solubilizing bacteria and its efficacy studies on maize. Agronomy, 13(7):1919. https://doi.org/10.3390/agronomy13071919

Saif S, Abid Z, Ashiq MF, Altaf M. and Ashraf RS (2021) Biofertilizer formulations. Biofertilizers: Study and Impact, 211–256. https://doi.org/10.1002/9781119724995.ch7

Saloum A (2025) April. Effect of Mineral and Bio-Fertilizers on the Growth and Productivity of Sunflower (Helianthus annuus L.). In IOP Conference Series: Earth and Environmental Science 1487, (1):012092. IOP Publishing. https://doi.org/10.1088/1755-1315/1487/1/012092

Sandhya VSKZ, Ali SZ, Grover M, Reddy G. and Venkateswarlu B (2010) Effect of plant growth promoting Pseudomonas spp. on compatible solutes, antioxidant status and plant growth of maize under drought stress. Plant growth regulation, 62:21–30. https://doi.org/10.1007/s10725-010-9479-4

Selvi BK (2023) Preparation of Novel Rhizoazophos Biofertilizer Consortium and Evaluation of Their Shelf Life in Rice Bran, Coir Pith and Sugarcane Trash. https://doi.org/10.53550/EEC.2023.v29i03.033

Sen A, Padhan D. and Poi SC (2016) Isolation and characterization of mineral potassium solubilizing bacteria from rhizosphere soils. Journal of Applied and Natural Science, 8(2):705

Shah I, Sarim KM, Sikka VK, Dudeja SS, Gahlot DK (2024) Developed Rhizobium Strains Enhance Soil Fertility and Yield of Legume Crops in Haryana, India. Journal of Basic Microbiology, 64(11):e2400327. https://doi.org/10.1002/jobm.202400327

Shanware AS, Kalkar SA, Trivedi MM (2014) Potassium solublisers: occurrence, mechanism and their role as competent biofertilizers. Int J Curr Microbiol App Sci, 3(9):622–629

Sharma RC, Sarkar S, Das D, Banik P (2013) Impact assessment of arbuscular mycorrhiza Azospirillum and chemical fertilizer application on soil health and ecology. Communications in soil science and plant analysis 44(6):1116–1126. https://doi.org/10.1080/00103624.2012.750335

Singh A, Singh B, Chinna GS, Chahal HS, Devi R (2023) Revitalization of potassium solubilizing microbes in food production system: AN overview. Agricultural Reviews, 44(3):311–319

Singh D, Thapa S, Geat N, Mehriya ML, Rajawat MVS (2021) Biofertilizers: Mechanisms and application. In Biofertilizers, 151–166. Woodhead Publishing. https://doi.org/10.1016/B978-0-12-821667-5.00024-5

Satinder KB (2012) Shelf-life of biofertilizers: an accord between formulations and genetics. J Biofertil Biopestic 03:1–2

Swanepoel PA, Botha PR, Truter WF, Surridge-Talbot AK (2011) The effect of soil carbon on symbiotic nitrogen fixation and symbiotic Rhizobium populations in soil with Trifolium repens as host plant. African Journal of Range & Forage Science, 28(3):121–127. https://doi.org/10.2989/10220119.2011.642096

Selvaraj T, Rajeshkumar S, Nisha MC, Wondimu L, Tesso M (2008) Effect of *Glomus mosseae* and plant growth promoting rhizomicroorganisms (PGPR) on growth, nutrients and content of secondary metabolites in *Begonia malabarica* Lam. Maejo Int J Sci Technol 2:516–525

Singh J, Singh AV, Upadhayay VK, Khan A (2020) Comparative evaluation of developed carrier based bio-formulations bearing multifarious PGP properties and their effect on shelf life under different storage conditions. Environ Ecol 38:96–103

Singla R, Rampal VK (2021) Constraints faced by farmers in adoption of polyhouse technology. Indian J Ext Educ 57:22–27

Sontsa-Donhoung AM, Bahdjolbe M, Hawaou and Nwaga D. (2022) Selecting endophytes for rhizome production, curcumin content, biocontrol potential, and antioxidant activities of turmeric (*Curcuma longa*). Biomed Res Int 2022:8321734

Stamenković S, Beškoski V, Karabegović I, Lazić M, Nikolić N (2018) Microbial fertilizers: a comprehensive review of current findings and future perspectives. Span J Agric Res 16:1–18

Velázquez E, Silva LR, Ramírez-Bahena MH, Peix A (2016) Diversity of potassium-solubilizing microorganisms and their interactions with plants. Potassium solubilizing microorganisms for sustainable agriculture, 99–110

Wangiyana W, Arianti DF, Ngawit IK (2023) October. Improvement of growth and yield of several varieties of soybean following paddy rice using Rhizobium biofertilizer with reduced NPK doses. In IOP Conference Series: Earth and Environmental Science 1253, (1):012054. IOP Publishing. https://doi.org/10.1088/1755-1315/1253/1/012054

Wei X, Bai X, Cao P, Wang G, Han J, Zhang Z (2023) Bacillus and microalgae biofertilizers improved quality and biomass of *Salvia miltiorrhiza* by altering microbial communities. Chin Herb Med 15:45–56

World Health Organization (WHO) (2013) WHO traditional medicine strategy: 2014–2023, World Health Organization, Geneva.

Wu X, Zhao Z, Zhao Z, Zhang Y, Li M, Yu Q (2023) Analysis of the potassium-solubilizing Priestia megaterium strain NK851 and its potassium feldspar-binding proteins. International Journal of Molecular Sciences, 24(18):14226. https://doi.org/10.3390/ijms241814226

Xie J, Shi H, Du Z, Wang T, Liu X, Chen S (2016) Comparative genomic and functional analysis reveal conservation of plant growth promoting traits in *Paenibacillus polymyxa* and its closely related species. Sci Rep 6:1–12

Yamamoto S, Okazaki S, Monica ND, Ohkama-Ohtsu N, Tanaka H, Sugihara S (2022) Rhizobium inoculation improved the rhizosphere P dynamics and P uptake capacity of pigeon pea plants grown in strongly weathered soil only under P fertilized conditions. https://doi.org/10.3390/agronomy12123149

Yamamoto S, Okazaki S, Monica ND, Ohkama-Ohtsu N, Tanaka H, Sugihara S (2023) Effect of rhizobium inoculation on rhizosphere phosphorous dynamics and fertilised phosphorous use efficiency in a maize–pigeon pea intercropping system in weathered tropical soil. Journal of Sustainable Agriculture and Environment, 2(3):357–368. https://doi.org/10.1002/sae2.12067

Yadav V, Pandey AK, Meena AK (2023) Role of Biofertilizer on Soil Fertility and Crop Production. Vigyan Varta 4(12):280–284

Yuliani S, Pratiwi E, Ratnaningsih. (2024) Potential carriers for biofertilizers: microstructural and entrapment properties. IOP Conf Ser Earth Environ Sci 1377:012004

Zafar M, Abbasi MK, Khan MA, Khaliq A, Sultan T, Aslam M (2012) Effect of plant growth-promoting rhizobacteria on growth, nodulation and nutrient accumulation of lentil under controlled conditions. Pedosphere 22:848–859
Zaidi A, Ahmad E, Khan MS, Saif S, Rizvi A (2015) Role of plant growth promoting rhizobacteria in sustainable production of vegetables: current perspective. Sci Hortic 193:231–239
Zheng H, Yang T, Bao Y, He P, Yang K, Mei X, Wei Z, Xu Y, Shen Q, Banerjee S (2021) Network analysis and subsequent culturing reveal keystone taxa involved in microbial litter decomposition dynamics. Soil Biology and Biochemistry, 157:108230. https://doi.org/10.1016/j.soilbio.2021.108230
Zhang S, Han S, Yu X, Gao J, Hu S, Borjigin N, Borjigin Q, Guo J, Bai J, Zhang B, Huang Z, Lei Y (2023) Successful formulation and application of low-temperature bacterial agents for corn stover degradation. Agronomy 13:1032

Chapter 6
Enhancing Resilience and Sustainability in African Agricultural Systems Through Biochar–Biofertilizer Synergies

Patrick Nyambo, H. Motsi, M. M. Mabitsela, Adornis D. Nciizah, Hupenyu Allan Mupambwa, and B. Muchara

Abstract Productivity in smallholder farming systems is low due to the high levels of soil degradation, nutrient depletion, and climate variability, all of which threaten food security and rural livelihoods. Corrective measures like amending soils with biochar and biofertilizers have been suggested for their ability to enhance soil health, increase crop productivity, and improve resilience of the farming systems. This chapter integrates evidence on the synergistic potential of biochar and biofertilizer co-application in African farming systems. When co-applied, the two amendments interact in ways that amplify microbial activity, nutrient retention, and plant resilience, resulting in improved yields and carbon sequestration. This chapter

P. Nyambo (✉)
Agricultural Research Council—Vegetables, Industrial and Medicinal Plants, Pretoria, South Africa
e-mail: nyambop@arc.agric.za

H. Motsi
Department of Soil Science, Faculty of AgriSciences, Stellenbosch University, Matieland, South Africa

M. M. Mabitsela
Department of Agriculture, University of Zululand, KwaDlangezwa, South Africa

A. D. Nciizah
Department of Agriculture and Animal Health, University of South Africa, Florida Park, Roodepoort, South Africa

Agricultural Research Council of South Africa– Natural Resources & Engineering (ARC–NRE), Pretoria, South Africa

H. A. Mupambwa
Sam Nujoma Marine and Coastal Resources Research Centre, Sam Nujoma Campus, University of Namibia, Henties Bay, Namibia

B. Muchara
University of South Africa (UNISA), Graduate School of Business Leadership, Pretoria, South Africa

A. D. Nciizah et al. (eds.), *Biofertilizers for Sustainable Agriculture in Africa*, Sustainability Sciences in Asia and Africa, https://doi.org/10.1007/978-981-95-6009-7_6

explores the mechanisms underpinning this synergy, including the effects on soil structure, microbial habitats, water retention, and greenhouse gas mitigation. Additionally, this chapter assesses the economic feasibility, scalability, and sustainability of adopting biochar–biofertilizer systems across diverse African agroecosystems.

6.1 Introduction

Agricultural production in Africa is often low because of challenges, including high levels of soil degradation, erratic and unpredictable rainfall, and climate risks (Nyambo et al. 2023a, 2024). The level of severity of the challenges varies across the region, but they all lead to a decline in food security, ecosystem services, and rural livelihoods (Mashamaite et al. 2024). For instance, the challenges result in changes in alterations in the growing seasons, lowering crop growth and yield. Additionally, ecosystem functions like the natural water and nutrient cycles are disrupted (Weiskopf et al. 2020). To optimize African agricultural systems, it is vital to look at agricultural practices that can be tailored to suit the ecological and socioeconomic dynamics of smallholder farming systems across the region.

Previous calls for the adoption of biochar and biofertilizers as cheaper and sustainable agricultural practices that can potentially minimize the effects of land degradation and climate change-associated challenges were presented by Nyambo et al. (2023b), Katakula et al. (2022), and Dzvene et al. (2019). Biochar is a carbon-rich by-product of pyrolyzing any organic material in the absence or under low oxygen concentration (Laird et al. 2010). Biochar application enhances soil quality by improving soil structure, enriching microbial communities, and promoting efficient nutrient cycling. These benefits are largely attributed to biochar's high porosity and large surface area, which create favorable microhabitats for soil microorganisms, enhance soil aeration, and contribute to improved soil aggregation and stability. In contrast, biofertilizers use viable microbes in liquid, powder, or granule formulations that play a role in facilitating and enhancing soil nutrient availability and uptake by plants. Through the action of living microorganisms such as bacteria, fungi, protozoa, and nematodes, biofertilizers improve soil health and plant growth through actions like nitrogen fixation, phosphate, and potassium solubilizing (Singh et al. 2021). These microbial inoculants make it easier for plants to thrive by making nutrients more available and helping them deal with stress. They also cut down on the requirement for chemical fertilizers.

Co-applying biochar and biofertilizers can potentially have synergistic effects that make the benefits of each input even greater. For example, biochar can adsorb nutrients and provide a habitat for microbes in the biofertilizers, thus increasing their survival, activity, and effectiveness (Shikha et al. 2023). This combined effect is reported to increase nutrient release, cycling and use, microbial population and diversity, and ultimately plant growth and yields (Fwanyanga et al. 2022). Despite these promising outcomes, widespread adoption remains limited due to knowledge gaps, cost barriers, and limited awareness among farmers and policymakers. This chapter aims to synthesize the evidence on the synergies between biochar and biofertilizer, highlight the mechanisms underpinning their synergistic effects, and discuss the opportunities and challenges for mainstreaming these technologies in African agricultural systems.

6.2 Biochar Properties and Benefits

Biochar is made through the pyrolysis process, when an organic material (feedstock) is burnt under higher temperatures and low oxygen conditions (Nyambo et al. 2023b). Through this pyrolysis process, the feedstock is converted to a recalcitrant carbon, which prolongs for several hundreds to thousands of years in soils. Therefore, its properties are primarily dependent on the type of feedstock utilized and the temperature conditions, and an understanding of how biochar varies based on these factors is critical for its intended application. Most of the biochar feedstocks consist of agricultural and forestry residues, or municipal, green, and food waste, while a wide range between 300 and 900 °C is widely used in preparation of biochar (Touray et al. 2014; Tan et al. 2018; Huang et al. 2021).

6.2.1 Biochar Chemical Properties

Biochar yield, which indicates the amount of biochar produced from a given biomass input, is a crucial metric that determines the quantity of biochar produced from a specific amount of biomass input (Song and Guo 2012; Antonangelo et al. 2019). Both feedstock and temperature influence biochar yield. Feedstocks high in lignin typically produce more biochar than nonwood materials due to their greater resistance to thermal degradation during pyrolysis. Higher pyrolysis temperatures reduce biochar yield, as more material is lost at elevated temperatures, thereby decreasing the final amount of biochar (Billa et al. 2019; Sinha 2021; Ghorbani et al. 2024). Volatile matter and moisture content are mainly affected by temperature. Increasing temperature converts volatile materials into gas molecules that vaporize, degrades lignin and cellulose, and dehydrates hydroxyl groups (Choudhary et al. 2019; Sarfaraz et al. 2020). Consequently, biochar produced at high temperatures contains aromatic compounds like benzenoid aromatics and alkyl furans, while labile materials like hemiacetal, carboxyl, and carbonyl are reduced (Murtaza et al. 2022). Biochar with a high lignin content, particularly woody biochar, has a lower ash content compared to those with less lignin. This explains why biochar derived from animal manure is nutrient-dense, as ash content is rich in nutrients. Additionally, higher temperatures remove volatile matter, resulting in increased ash content (Ghorbani et al. 2024).

Understanding biochar chemistry is essential as it dictates biochar's interaction with various organic and inorganic soil compounds. Key chemical characteristics include pH, total carbon (C), electrical conductivity, cation exchange capacity, and specific nutrients and trace elements. Biochar's C is primarily present in polycondensed aromatic structures, which enhance stability and enable long-term C sequestration (Antonangelo et al. 2019; Guo et al. 2021). These structures are more prevalent in wood feedstock compared to nonwood feedstock and increase with rising temperatures. Ash content is crucial for nutrient availability in biochar. Less lignin feedstocks like manure have a higher ash and nutrient content than wood feedstocks (Sarfaraz et al. 2020). Legumes typically have more nitrogen than cereals and vegetables, while poultry manure is richer in nutrients compared to cattle, sheep, or goat manure (Dzvene et al. 2019; Ghorbani et al. 2019). Nutrient concentration generally increases with pyrolysis temperature due to rising ash content, except for nitrogen and sulfur, which volatilize beyond 200 and 375 °C. Potassium and phosphorus volatilize at 700 and 800 °C, respectively, whereas magnesium, calcium, and manganese volatilize above 1000 °C (Deluca et al. 2015). Biochar derived from animal manure has a higher pH than that from crop residues because animals consume nutrient-rich cations that influence pH (Dzvene et al. 2019). Conversely, crops are often fertilized with low-mineral fertilizers, leading to lower pH values in crop residue biochar. High temperatures raise the concentration of mineral elements (Ca, Na, Mg), while reducing acidic groups, which increases pH. Biochar exhibits high cation exchange capacity due to a large surface area, negative surface charge, and high charge density (Břendová et al. 2017). Furthermore, CEC increases with aging due to the accumulation of carboxylic functional groups after oxygen exposure (Sinha 2021). While CEC initially rises with temperature, it decreases at very high temperatures, likely due to the loss of surface functional groups from increased pyrolysis temperatures (Sarfaraz et al. 2020).

6.2.2 Biochar Physical Properties

Specific surface area (SSA), total pore volume (TPV), pore width (PW), and micropore volume (uPV) of biochar are crucial physical parameters influenced by feedstock type and pyrolysis temperature (Mashamaite et al. 2024). These properties enhance soil quality by improving nutrient and water retention, providing microbial habitats, enhancing soil aeration, and modifying soil structure. High SSA improves nutrient adsorption and retention, essential for plant growth (Pituello et al. 2015; Tan et al. 2018), and offers refuge for soil microbes, boosting soil biological activity (Jaafar et al. 2015). The complex pore structure of biochar promotes soil aeration, water movement, and structure, while reducing compaction (Sinha 2021). Different feedstocks yield varying SSA, TPV, PW, and uPV due to differences in ratios of cellulose, hemicellulose, lignin, and volatile fractions (Břendová et al. 2017). Wood-based biochars, with high levels of stable components like lignin, cellulose, and hemicellulose, resist distortion during thermal decomposition, resulting in fewer pores and lower SSA (Choudhary et al. 2019). Biomass with a high volatile content tends to produce biochar with more developed pore structures, as the removal of volatiles creates micro or meso pores (Ghorbani et al. 2019).

At higher temperatures, SSA, TPV, PW, and uPV tend to increase, with increase in temperature. This phenomenon can be attributed to the continuous decomposition of organic materials (hemicelluloses, cellulose, and lignin), coupled with the establishment of vascular bundles as temperature rises, leading to an enhancement in the SSA and pore structure of biochar (Choudhary et al. 2019; Huang et al. 2021). In addition, at higher temperatures, large quantities of volatile material are removed, which enhances highly developed biochar pore structures and establishment of micro or meso pores. For instance, Tan et al. (2018) showed that rice husk biochar produced at 300, 400, 600, and 800 °C had SSA of 2.105, 4.985, 13.511, and 303.731 m^2 g^{-1}, respectively, TPV of 0.00957, 0.0146, 0.0158, and 0.161 m^2 g^{-1}, respectively, and uPV of 0.000832, 0.00196, 0.00535, and 0.122, respectively, m^2 g^{-1}. Touray et al. (2014), using goat manure, showed that 400, 500, 600, 700, and 800 °C yield SSA of 3.27, 1.68, 13,92, 39.08, and 93.49, respectively, TPV of 0.0013, 0.0013, 0.0078, 0.0199, and 0.049, respectively, and pore diameter of 15.9, 30.95, 22.41, 20.37, and 20.96, respectively.

6.2.3 Biofertilizer Types and Functions

The term "biofertilizer" has numerous definitions in the literature, and it is frequently used interchangeably with organic fertilizers, with other terms that have distinct interpretations, such as organic manure, animal manure, green manure, and plant extracts (Mącik et al. 2020). The agreed definition is that biofertilizers are formulates containing viable and live microorganisms that, when introduced to seeds, plant surfaces, or soil, establishes itself in the rhizosphere or within the plant, and enhances growth by improving the host plant's access to or supply of essential nutrients (Sarfraz et al. 2019; Li et al. 2023). This biofertilizer definition distinguishes it from organic fertilizers, which are composed of organic materials to enhance soil fertility directly or indirectly (Shikha et al. 2023). Biofertilizers are becoming increasingly popular, particularly in the context of organic farming and sustainable agricultural practices. These natural fertilizers play a crucial role in organic cultivation methods, offering numerous advantages to farmers and the environment, mainly by reducing overreliance of inorganic fertilizers.

Biofertilizers are categorized by their roles and mechanisms: nitrogen fixers, potassium solubilizers, phosphorus solubilizers, and others for minority elements such as sulfur, zinc, and silicates (Schütz et al. 2018; Fasusi et al. 2021). These classifications are based on the distinct functions each type performs in enhancing soil fertility and plant growth (Ahmad et al. 2015). Bacteria, fungi, and cyanobacteria are the main microorganisms in these biofertilizers, often forming symbiotic relationships with plants. Their effectiveness depends on soil physical and chemical properties, organic content, phosphorus levels, and agricultural practices.

6.2.4 Nitrogen-Fixing

Nitrogen-fixing microorganisms enhance soil nitrogen levels by capturing nitrogen from the atmosphere and converting it into a form that plants can utilize. These organisms are categorized as symbiotic, free-living, and associative symbiotic. Symbiotic nitrogen-fixers comprise microorganisms such as *Rhizobium*, *Sinorhizobium*, *Frankia*, *Azorhizobium*, *Mesorhizobium*, *Anabaena azollae*, *Bradyrhizobium*, *Frankia*, *Allorhizobium*, and *Trichodesmium* (Sarfraz et al. 2019; Daniel et al. 2022; Shikha et al. 2023). Bacteria, particularly those in the Rhizobiaceae family, engage in symbiotic associations with leguminous plants such as common bean, pea, faba bean, lentil, cowpea, pigeon pea, and soybean (Ahmad et al. 2015). Leguminous plants develop specialized root structures known as nodules, which serve as sites for the process of nitrogen fixation (Shikha et al. 2023). Within nodules, bacteria convert atmospheric nitrogen into ammonium, which is subsequently transferred to plant cells, where it is utilized in metabolic processes.

Free-living nitrogen-fixing microorganisms comprise *Azotobacter*, *Azospirillum*, *Anabaena*, *Clostridium*, *Aulosira*, *Bejerinkia*, *Nostoc*, *Klebsiella*, *Stigonema*, *Desulfovibrio*, *Rhodospirillum*, and *Rhodopseudomonas* (Schütz et al. 2018; Fasusi et al. 2021). These organisms are further categorized as free-living non-photosynthetic and photosynthetic nitrogen-fixing bacteria. Free-living non-photosynthetic bacteria predominantly include *Azotobacter* and *Azospirillum*, which are aerobic microorganisms that inhabit neutral and alkaline soils (Balasubramanian and Karthickumar 2017). These organisms play a crucial role in nitrogen fixation and provide benefits to non-leguminous crops, including maize, coconuts, tobacco, wheat, oats, beets, rice, sunflowers, tea, barley, and coffee (Li et al. 2023). Moreover, these microorganisms synthesize compounds that promote plant growth, such as auxins, nicotinic acid, cytokinins, pantothenic acid, gibberellins, and B vitamins (Mącik et al. 2020). Conversely, free-living photosynthetic bacteria comprise cyanobacteria or blue-green algae. These organisms are autonomous, oxygen-producing microorganisms typically found in dams, ponds, and flowing water bodies, where they convert nitrogen to nitrogenous and ammonium compounds in these aquatic environments (Sifton et al. 2023). Furthermore, they secrete various plant growth compounds, including vitamins, polypeptides, amino acids, and other substances that possess antifungal and antibacterial characteristics. The incorporation of these microorganisms in agricultural systems offers numerous benefits in nutrient mobility, heavy metal detoxification, and protection of plants against pathogenic organisms.

Associative symbiotic microbes consist of *Azospirillum* spp., *Herbaspirillum* spp., *Alcaligenes*, *Enterobacter*, and *Azoarcus* spp. *Acetobacter diazotrophicus* (Schütz et al. 2018). They are also aerobic, free-living bacteria, but with less intimate association with roots. They are non-nodule forming and prevalent in the soil of temperate, subtropical, and tropical ecosystems. These bacteria establish a rhizosphere association by inhabiting both the external and internal portions of root systems in agricultural and wild plants. The *Azospirillum* sp. strains are mainly used in nonlegume crops like cotton, coffee, millets, oil palm, sugarcane, tea, oil seeds, lime, coconut, and banana (Schütz et al. 2018).

6.2.5 Phosphorus-Mobilizing and Solubilizing

Phosphorus mobilizers possess the capability of enhancing phosphorus absorption by plants. These organisms accomplish this by mobilizing phosphorus from the soil to the root cortex rather than by solubilizing P-containing compounds (Xu et al. 2022). Microorganisms associated with this process include *Arbuscular mycorrhiza*, *Acaulospora* spp., *Scutellospora* spp., *Glomus* spp., *Gigaspora* spp., and *Sclerocystis* spp. Among these P-mobilizing organisms, mycorrhizal fungi are considered the most significant (Fox et al. 2014). These fungi establish a symbiotic relationship with vascular plants, including several crops such as wheat, maize, potato, and rice. In this symbiosis, the fungi colonize the roots of their plant partners, creating a pathway for nutrient absorption (Sifton et al. 2023). Mycorrhizal fungi can be further categorized into endomycorrhizal and ectomycorrhizal types.

Soils and plant rhizospheres harbor abundant populations of phosphate-solubilizing bacteria. These include strains of *Penicillium* spp., *Flavobacterium*, *Pseudomonas striata*, *Trichoderma* spp., *Bacillus subtilis*, *B. polymyxa*, *Agrobacterium*, *Aspergillus awamori*, *Bacillus circulans*, *Micrococcus*, *Erwinia*, *Enterobacter*, and *Klebsiella* (Schütz et al. 2018; Pastore et al. 2020). These microorganisms comprise both aerobic and anaerobic bacteria, with the former being more prevalent in waterlogged soils (Schütz et al. 2018). The mechanism of solubilization occurs through microbial secretion of organic acids such as fumaric, 2-ketoglutonic, citric, succinic, gluconic, and malic acids. Microbial metabolism produces these organic acids, which can solubilize mineral phosphorus through two mechanisms. They can either directly dissolve phosphorus via anion exchange, where acid anions replace phosphate, or they can form chelates with cations such as Al, Fe, and Ca that are associated with P (Pastore et al. 2020). In addition to the secretion of these organic acids, they decrease soil pH, break down bound phosphates, and transform insoluble phosphorus into soluble forms in the soil.

6.2.6 Potassium-Solubilizing

Potassium-solubilizing bacteria include species such as *Paenibacillus glucanolyticus*, *Bacillus edaphicus*, *B. edaphicus*, *B. mucilaginosus*, *Mucilaginosus*, *B. circulanscan*, and *Arthrobacter* spp (Liu et al. 2017a, b; Udomkun et al. 2024). Potassium (K) solubilization involves multiple mechanisms. One process converts inaccessible K into a bioavailable form via metal–organic complexes with Si^{4+} ions, enabling K to enter the soil solution (Biswas and Basak 2014). Another mechanism is the decomposition of feldspar due to decreased pH levels. Furthermore, H^+ ions from organic acids release mineral structural cations. These microorganisms produce several organic acids, including coumaric, oxalic, syringic, malic succinic, citric, and tartaric acid. Consequently, these microbes facilitate soil fertility through enhancement of potassium uptake by plants, thereby improving their growth and development (Liu et al. 2017a, b). Research has demonstrated that potassium-solubilizing microbes provide beneficial effects on the development of various crops, including tomato, Sudan grass, pepper, tobacco, wheat, chili, rape, sorghum, and cotton.

6.2.7 Others (Sulfur-Oxidizing, Zinc-Solubilizing, Silicates-Solubilizing)

In addition to biofertilizers significant for major nutrients such as nitrogen, phosphorus, and potassium, there are others required for improving nonmajor nutrients such as sulfur (sulfur-oxidizing), zinc (zinc-solubilizing), and silicates (silicates-solubilizing). Microorganisms such as *Thiobacillus* spp., *Bacillus*, *Xanthobacter*, *Pseudomonas*, and *Alcaligenes* are responsible for oxidizing sulfur into a form that can be utilized by plants, thereby enhancing plant productivity (Fox et al. 2014; Arabi et al. 2018). The incorporation of sulfur-oxidizing microorganisms has been proposed for developing biofertilizers intended for various crops such as garlic, grape, onion, cauliflower, oats, and ginger, particularly when cultivated in alkaline soil environments. Furthermore, these microorganisms, specifically bacteria, play a significant role in safeguarding the natural environment by organically removing sulfur-based pollutants, which occurs through the transformation of sulfur from its reduced form via oxidation (Fox et al. 2014).

The utilization of zinc-solubilizing microorganisms as a complementary method for zinc supplementation is gaining prominence. Microbial inoculants containing *Mycorrhiza*, *Saccharomyces* spp., *Pseudomonas* spp., *Thiobacillus thioxidans*, and *Azospirillum* spp., along with numerous other bacteria, enhance zinc accessibility to plants (Kamran et al. 2017; Upadhayay et al. 2022). These microorganisms render zinc soluble through chelated ligands and oxidoreductive mechanisms. Furthermore, these microbes secrete substances such as vitamins, antifungal compounds, antibiotics, and phytochromes, which confer various benefits to the plant (Kamran et al. 2017). Zinc-solubilizing bacteria have demonstrated efficacy in crops such as maize and rice, additionally increasing resistance to pests and diseases, thereby improving overall plant health.

From the numerous microorganisms found in soil, only a limited number possess the capability to solubilize silicon. Microbes such as *Bacillus mucilaginosus* var. *siliceous*, *Proteus mirabilis*, *Pseudomonas*, and *Bacillus caldolyticus* have been identified as the most appropriate organisms for solubilizing Si from its silicates, primarily Al_2SiO_5 (Biswas and Basak 2014; Pastore et al. 2020). During their growth period, these microorganisms produce various organic materials that can facilitate weathering processes, additionally liberating K from K-containing minerals (Liu et al. 2017a, b). The solubilization of silica minerals by microorganisms is considered a valuable source of Si for plant uptake (Pastore et al. 2020). It has been reportedly to be significant to crops such as legumes, sugarcane, pumpkin, gourd cereals, leafy greens, cucumber, and onions.

6.2.8 Mechanisms of Interaction: Biochar–Biofertilizer Synergy

The integration of biochar and biofertilizer in agricultural systems exemplifies and complements sustainable agriculture, as both are essential components in these natural systems. However, the mechanisms governing this synergy are primarily determined by how biochar effectively improves soil conditions for the enhancement of microbes in the biofertilizers. As elucidated in the preceding biochar section, the primary factors influencing biochar are feedstock type and pyrolysis temperature. These factors additionally become primary factors affecting how biochar interacts with the biofertilizers. These factors are determinants in affecting biochar properties such as pH, carbon content, porosity, surface area, and nutrient composition, which ultimately influence the synergy between biochar and biofertilizers. Moreover, other factors such as soil texture, biochar to biofertilizer mixing ratio, crop type, and climatic variables, including temperature and precipitation, may also impact the mechanism of synergy between biochar and biofertilizers.

The capacity of biochar to modify soil acidity levels may influence the abundance of microbial life present in the soil (Pastore et al. 2020). Biochar elevates soil pH, particularly in highly acidic soils, to neutral-alkaline conditions, which are favorable for soil microbes (Fox et al. 2014). However, this increase may be influenced by the type of microorganisms in biofertilizers, with bacterial populations increasing beyond pH 7, while fungal populations may decrease at higher pH levels pH (Sarfraz et al. 2019; Sifton et al. 2023).

Biochar's high carbon content supplies energy to soil microorganisms in biofertilizer formulations, mainly through bioavailable labile carbon, such as carbohydrates, as its stable, recalcitrant carbon remains inaccessible to microbes (Dzvene et al. 2019; Ghorbani et al. 2019). Biochar can also induce native SOM mineralization via a positive priming effect, compensating the nonbioavailable recalcitrant carbon (Sarfraz et al. 2019). Additionally, biochar generates dissolved organic carbon (DOC), composed of low molecular weight organic carbon, which is easily consumed by microbes (Xu et al. 2022). The DOC has recently been proposed as a significant pathway for carbon sequestration, wherein following microbial utilization, the resultant necromass forms stable carbon (Cotrufo et al. 2015).

The addition of biochar to soil for crop cultivation increases the concentration of essential nutrients, specifically nitrogen, phosphorus, and potassium. Jatuwong et al.'s (2024) findings indicate that incorporating biochar alone or in combination with AMF into soil significantly enhanced the overall nitrogen, phosphorus, and potassium levels across various soil types. This occurs through several mechanisms associated with its ability to increase cation exchange capacity (CEC) and improve nutrient retention. This prevents the loss of nutrients produced through microbial activities in biofertilizers. These microorganisms contribute to the decomposition of organic materials, which increases the availability and accessibility of nutrients for plant uptake. Furthermore, biochar can improve nutrient absorption when used in conjunction with certain microorganisms such as AMF (Shikha et al. 2023). This improvement is attributed to AMF's ability to form symbiotic associations with plant root systems, which further enhances the plants' capacity to extract nutrients from the soil (Jatuwong et al. 2024). It has also been posited that the availability of pollutants (polychlorinated biphenyls, polyaromatic hydrocarbons), herbicides and pesticides, and heavy metals (Mn, Zn, Fe, and Cu) can be decreased by biochar due to its strong adsorption capacity for such compounds (Cruz-Méndez et al. 2021). These toxic chemicals are detrimental to microbes, and their decline due to biochar facilitates an increase in microbial activity.

Biochar functions as a supportive habitat for valuable microorganisms, which are essential in facilitating nutrient cycles and the process of mineralization. A plausible explanation for this effect is that biochar possesses a porous and aggregated structure, as well as a large surface area, which serves as a protective environment for soil microorganisms. This structure likely influenced the growth and proliferation of microbes in the rhizosphere (Fox et al. 2014; Li et al. 2023). The various oxygen-containing functional groups on biochar's surface enhance its adsorption and dissolution of simple organic compounds and ammonium ions, fostering microbial growth. This leads to beneficial changes on the biochar's surface in the soil, affecting the microbial community composition and activity (Shikha et al. 2023). Biochar's high internal porosity protects beneficial bacteria from predators such as protozoa and nematodes, potentially enhancing rhizobia survival and subsequently improving crop productivity (Ahmad et al. 2015). These microorganisms are shielded from environmental challenges such as drought or high salt content by biochar, which also facilitates their growth, enhancing soil quality and promoting plant development (Arabi et al. 2018). This has been observed to be highly dependent on feedstock type and temperature of biochar, as these factors lead to biochar with different surface areas and pore structures (Xu et al. 2022).

6.3 Soil Health Improvement Through Biochar–Biofertilizer Application

6.3.1 Enhancing Soil Structure and Water Retention

Biochar and biofertilizers may play a significant role in soil structure through enhancing soil aggregation. Biochar is known to enhance soil aggregation through its large surface area, porous structure, and diverse recalcitrant material (Ghorbani et al. 2019). Conversely, microorganisms in biofertilizers participate in soil aggregation through their enzyme secretions, which bind small aggregates into more complex structural large aggregates. Furthermore, the porous nature of biochar and its labile carbon components such as volatile fatty acids provides an improved habitat and energy availability for soil microorganisms to thrive and multiply, which ultimately contributes to the formation of soil aggregates through binding processes (Murtaza et al. 2022). This formation of larger soil aggregates contributes to reducing thc loss of nutrients from agriculture systems. Additionally, both biochar and biofertilizers enhance plant root growth and elongation in different areas in the soil, and the roots improve soil structure (Sifton et al. 2023). This is specifically mediated by microbes that form hyphae and are in symbiotic relationships with plant roots (Schütz et al. 2018).

An improvement in soil structure paves the way for enhancing other critical soil properties such as water retention. Water retention refers to the capacity of soil to absorb and store moisture within its composition, allowing plant roots access to it. This ability is essential for preserving soil moisture, facilitating plant development, and achieving better yields, particularly in regions susceptible to drought or inconsistent precipitation patterns. Biochar's porous structure can improve water retention, while its combination with biofertilizer may enhance this via improved aggregation (Waqar et al. 2022). However, most of the research on water retention has predominantly focus on biochar, while fewer studies have examined the combination of biochar and biofertilizers. One study conducted under water stress conditions by Moosavi et al. (2020) reported that water retention improvement was the highest in the treatment combining biochar and biofertilizer compared to other treatments. They further observed that due to improved water retention conditions, plant growth and yield parameters were elevated under this treatment.

6.3.2 *Nutrient Cycling and Bioavailability*

6.3.2.1 Impact on Soil Microbial Communities

The health and productivity of soil are primarily determined by its organic matter content and microbial communities. Organic matter serves as a carbon and energy source for microbial activity and contributes to maintaining soil structure that facilitates nutrient and water retention, crucial for overall soil health and fertility for plant growth (Arabi et al. 2018). In the combination of biochar and biofertilizer, biochar provides both the carbon and energy required by microorganisms, including those in the biofertilizers and those already present in the soil (Sifton et al. 2023). This combination introduces microbial abundance and biodiversity through an increase in the microbial biome into the soil by incorporating microorganisms that may not have been present previously. Furthermore, the addition of new microbes ensures the introduction of microorganisms that are specific in their interactions with the particular crop, thereby providing assurance of improved soil health. Study et al.

(2016) observed that combining biofertilizer with biochar resulted in a greater benefit, with biomass increasing by 1438% for the biofertilizer and biochar combination compared to the 1178% increase observed when biochar was used alone. They attribute this to the introduction of new microbial diversity that was introduced by the biofertilizer and biochar combination. Another point of consideration is the application of this combination in poor soils, such as sandy soils (Study et al. 2016). In such soils, there is a limited amount of soil organic matter sufficient to support microbial life, thus associated with poor soil health. The introduction of biochar and biofertilizer combination introduces additional microorganisms and biochar as organic matter to support the microbes (Ahmad et al. 2015). This can also be observed in other poor soils affected by acidity, salinity, heavy metal contamination, and compaction, where soil health can be ameliorated through the application of biochar, which creates a favorable environment for microorganisms in biofertilizers and native to the soil to persist.

6.4 Crop Productivity and Yield Enhancement with Biochar–Biofertilizer Integration

The integration of biochar and biofertilizers has been shown to significantly improve crop productivity and yield by enhancing soil conditions, nutrient availability, and microbial activity. Numerous studies have demonstrated that the co-application of biochar with biofertilizers is more effective than using either input alone.

A meta-analysis by Omidvar et al. (2022) found that biochar alone increased yield by 25.3%, while inorganic fertilizers alone boosted yield by 21.9%. However, when biochar was combined with inorganic fertilizers, yield improvements increased by an additional 10%. The highest yield improvement (179.6%) was observed when biochar was co-applied with both organic and inorganic fertilizers compared to unfertilized controls. The meta-analysis further showed that fertilizer rates of 50 kg N ha^{-1} co-applied with biochar resulted in the greatest yield increase, while rates above 300 kg N ha^{-1} did not result in further yield gains.

Crop-specific responses also varied. Root and tuber crop and vegetables exhibited the highest yield increase of up to 50.6%, while no significant improvement was observed for sorghum. Significant rice and rapeseed yields increases of up to 25% and 20%, respectively, were observed when biochar was applied with organic fertilizers (Zhang et al. 2024) (Khan et al. 2024). Similarly, Rivelli and Libutti (2022) reported a 116% increase in Swiss chard yield with the co-application of biochar and vermicompost compared to ammonium nitrate alone.

The benefits of biochar and fertilizer co-application extend over time and are influenced by soil characteristics. Yield improvements persist for years, with the highest gains (36.0%) observed 1–2 years after biochar application. The effects are particularly strong in very acidic soils ($pH \leq 5$), where yield increases by 41.9%. These improvements are attributed to enhanced nutrient retention, reduced leaching, and better pH buffering, contributing to more sustainable soil fertility management. Biochar's performance also depends on nutrient status. In phosphorous-deficient soils, co-application with phosphorus-solubilizing microbes (PSM) significantly increased yield, while no notable improvement was seen in P-rich soils, thus highlighting the context-specific nature of biochar–biofertilizer outcomes.

Cropping systems also play a key role in the effectiveness of biochar in improving crop yield. For instance, a study by Liu et al. (2017a, b) showed that biochar application in maize intercropped with soybean or peanut increased yield and nutrient uptake, enhancing legume N fixation and N transfer to maize. These findings suggest that biochar amendments can significantly enhance the productivity and sustainability of legume-based intercropping systems. For instance, biochar application at 40 t ha^{-1} increased maize yields by up to 12.1% and peanut yields by 14.9% in intercropping systems. It also amplified nitrogen transfer from peanut to maize, with biochar treatments increasing nitrogen transfer by up to 71.84 mg per box in root-box experiments. Additionally, biochar improved nitrogen fixation in legumes, with peanut nitrogen derived from the atmosphere increasing from 38.68% to 43.15% in divided root boxes and from 49.60% to 57.30% in open boxes.

The co-application of biochar and biofertilizer also contributes to sustainable nutrient management by reducing the reliance on chemical fertilizers by enhancing the slow release of nutrients, thus providing sustained nutrient availability and improving soil fertility over the long term (Zhao et al. 2024). When biochar is used with organic amendments like Ghanajeevamruth, it significantly boosts in nutrient levels in the soil, including nitrogen, phosphorus, and potassium, which are essential for plant growth (Archana et al. 2025). These findings demonstrate the potential of biochar to boost crop yields, optimize nutrient uptake, and enhance soil health, making it a valuable tool for sustainable agriculture, but the effect may depend on prevailing soil conditions. While the co-application of biochar and organic fertilizers offers numerous benefits, it is important to consider the specific soil type and crop requirements to optimize results. Additionally, the long-term effects and economic feasibility of this strategy should be evaluated to ensure its sustainability and practicality in diverse agricultural settings.

6.5 Climate Resilience and Carbon Sequestration

6.5.1 Biochar–Biofertilizer's Role in Mitigating Climate Change

The total quantity of carbon held within a specific system or reservoir over a defined timeframe is known as carbon storage. This phenomenon can occur in various environments; however, in terrestrial ecosystems, carbon is predominantly stored in living plants, deadwood, litter, roots, and soil, with the soil possessing the largest reservoir of storage capacity (Liu et al. 2018). The addition of biochar alone can exert both direct and indirect influences on increasing carbon storage. Biochar directly increases soil carbon storage directly Biochars's high carbon content directly contributes to soil carbon storage and indirectly optimising the soil for growth of plants and microbes which add on to the carbon storage (Murtaza et al. 2022). On the other hand, addition of biofertilizer alone may have an indirect influence on both living plants, nonliving plant material, and soil. Thus, a combination of biochar and biofertilizer may have both indirect and direct influence on carbon storage.

In aboveground plant carbon storage, the combination of biochar and biofertilizers directly and indirectly enhances conditions that improve plant health, thereby increasing its photosynthetic capacity. Plants acquire carbon as CO_2 from the atmosphere during photosynthesis, and enhancing the plant's ability to absorb more CO_2 occurs through improving its photosynthetic apparatus (Sokol et al. 2019). The association of biochar and biofertilizer provides nutrients such as N, P, K, Fe, Mn, Zn, and Cu, which are essential for the synthesis of chlorophyll, enzymes (primarily RuBisCO), and numerous biomolecules that participate in photosynthesis. Furthermore, the microorganisms in the biofertilizer establish symbiotic associations with plant roots. For instance, members of the Rhizobiaceae bacterial family (*Allorhizobium*, *Rhizobium*, *Frankia*, *Anabaena azollae*, etc.) form symbiotic relationships with various legumes, including common bean, and soybean (Shikha et al. 2023). These leguminous plants have nodules, which function as nitrogen fixation sites where bacteria transform atmospheric nitrogen into forms absorbable by plants (Ahmad et al. 2015). Other soil properties, including chemical properties (pH, EC, organic carbon, and organic matter) and physical properties (water-holding capacity, bulk density, porosity, infiltration, and hydraulic conductivity), are indirectly enhanced by the biochar–biofertilizer association to promote plant growth and its carbon assimilation capacity (Jatuwong et al. 2024). In instances where plant growth is impeded by challenges such as heavy metal contamination, soil acidity, and salinity, the biochar–biofertilizer association mitigates these challenges, enabling plants to grow under normal conditions (Sarfraz et al. 2019; Srivastava et al. 2023).

The association of biochar and biofertilizer may exert a direct influence on carbon storage in soil. A study conducted by Shikha et al. (2023) observed that after the harvest of groundnuts at the end of the growing season, soil carbon stocks were the highest in the biochar and biofertilizer combination treatment, measuring 12.26 t ha^{-1}, compared to the control (10.07 t ha^{-1}), biochar-only (12.04 t ha^{-1}), and biofertilizer-only (11.23 t ha^{-1}) treatments. In another study conducted by Allaire et al. (2015), following 2 years of switch grass harvesting, the highest carbon stocks were observed in the treatment combining biochar and biofertilizer, which yielded 9.72 t ha^{-1}. Other treatments, including the control, N fertilizer, biochar only, and biofertilizer only, resulted in carbon stocks of 0.54, 9.52, 6.63, and 0.009 t ha^{-1}, respectively. The reason for such high carbon storage in the biochar and biofertilizer combination could be the high carbon content of biochar that will become stable in the soil and other associated indirect effects associated with the combination (Shikha et al. 2023). These indirect effects may be that after senescence and shedding of plants in soils amended with the biochar and biofertilizer combination, they add large amounts of carbon on the ground surface as litter or deadwood that will be subsequently stored belowground in the soil after decomposition. Additional biochar and biofertilizer can influence soil organic matter mineralization via priming effects. Introducing biochar into soil increases microbial populations in the rhizosphere, enhancing microbial activity and promoting the priming effect by facilitating the introduction of fresh organic matter (Sarfraz et al. 2019).

6.5.2 Carbon Storage Potential of Biochar–Biofertilizer Synergy

Greenhouse gases (GHGs) remain a concern, which are produced by burning fossil fuels and contribute to significant global environmental and health issues. The primary GHGs of concern comprise CO_2, CH_4, and NO_x (Sun et al. 2021). Globally, various strategies have been employed to decrease GHG emissions in all potential sectors that emit these GHGs, including agriculture. The application of biochar to agriculture is recognized for its capacity to reduce GHG emissions. Multiple research studies have examined the impact of biochar on emissions from soil, and the results indicate that biochar can help decrease them (Zhang et al. 2021).

Nevertheless, a few studies have investigated how the combination of biochar and biofertilizer influences GHG emissions. The utilization of biochar and biofertilizer can reduce the necessity for synthetic fertilizers, thereby mitigating GHG emissions associated with fertilizer production, primarily N_2O (Sun et al. 2021). For instance, the capacity of biochar and biofertilizer to sequester and retain ammonium in soil diminishes the availability of nitrogen for denitrification, consequently leading to decreased N_2O emissions into the atmosphere. Also, biochar affects the quantity and function of soil microorganisms involved in nitrogen transformation, leading to reduced N_2O emissions (Li et al. 2023). Research shows that biochar application changes soil microbial community structures, decreasing N_2O emissions from agricultural fields (Zhang et al. 2021). In contrast, Sun et al. (2021) observed that combining biochar and biofertilizer might foster nitrifier growth, potentially increasing their populations and N_2O emissions. They linked this effect to the enhanced root biomass from biochar and biofertilizer use, which produced more root exudates as labile carbon sources. These compounds offer readily available energy for soil microorganisms, promoting denitrification and increasing N_2O emissions. In regards to CO_2, Shikha et al. (2023) observed that the treatment with biochar and biofertilizer exhibited the highest GHG emissions saved for SOC accumulation, which was 6.6 kg CO_2eq ha^{-1}, while other treatments such as the control, biochar only, and biofertilizer only resulted in GHG emission savings of −1.61, 5.58, and 3.7 kg CO_2eq ha^{-1}, respectively. Also, Ahmad et al. (2023) observed that the interaction between biochar particle size and biofertilizer significantly affected CO_2 production across all three incubation periods (day 3, day 6, and day 10). The study further observed that the combination was either equivalent to or emitted less

CO_2 than the control or other treatments, resulting from the application of different biofertilizers (Biozote and Rhizogold) in the combinations. Although a few studies indicate that combining biochar and biofertilizer may reduce GHG emissions, significant research gaps remain in this area. Understanding these microbial processes is vital for optimizing biochar application to mitigate GHG emissions effectively.

6.6 Sustainability of Biochar–Biofertilizer Systems in African Agriculture

6.6.1 Local Production and Cost-Effectiveness of Biochar

The integration of biochar and biofertilizers into African agriculture offers a promising pathway to enhancing soil fertility while promoting low-cost and locally adaptable solutions. Biochar can be produced from a wide range of locally available biomass residues, including crop residues, animal manure, forestry waste, and invasive species, making it a viable solution for smallholder farmers. In coastal countries like Namibia, research is also currently underway on the potential of using marine biomass in producing biochar (Katakula et al. 2020). The technologies for producing biochar range from low-tech kilns and traditional earth pits to more advanced pyrolysis systems. Biochar production kilns commonly used in low- to middle-income countries can generally be classified into several types, including flame curtain kilns, drum kilns, pyrolytic top-lit updraft (TLUD) cookstoves, retort kilns, brick or masonry kilns, and other locally adapted designs (Namaswa et al. 2023). These flexible methods enable communities to adopt biochar production at a scale suited to their technical capacity and resource availability. From a cost-effectiveness perspective, although initial investments may be needed for setup and training, the long-term benefits such as reduced fertilizer inputs, improved crop yields, and enhanced soil health contribute to positive cost–benefit ratios. The reuse of waste biomass also creates circular economy opportunities within farming systems.

Biochar–biofertilizer systems contribute significantly to environmental sustainability by addressing key challenges in African agriculture: soil degradation, nutrient depletion, greenhouse gas emissions, and waste management (Afshar and Mofatteh 2024). Biochar improves soil structure, increases water retention, and enhances cation exchange capacity, which is crucial for drought-prone and nutrient-poor soils prevalent in sub-Saharan Africa. Its carbon-rich and stable composition also makes biochar a potent tool for long-term carbon sequestration, contributing to climate change mitigation. In arid countries that lack a lot of biomass production, biochar has also been integrated into vermicomposting as a way of further enhancing the carbon sequestration potential of these organic nutrient amendments (Katakula et al. 2021).

6.6.2 Economic Feasibility in Resource-Constrained Settings

The economic feasibility of integrating biochar and biofertilizer into African agricultural systems is closely tied to the structural realities of smallholder and subsistence farming, which dominate the continent's agrarian landscape. These farming systems are typically characterized by low-input, low-output cycles, limited access to capital, and high vulnerability to climate shocks and soil degradation. As such, any technological innovation aiming for widespread adoption must not only be agronomically effective but also financially viable, accessible, and locally adaptable.

Figure 6.1 illustrates the economic ecosystem surrounding the feasibility of biochar–biofertilizer adoption. Key economic drivers such as affordability, financing, and cost-effectiveness are intricately linked with enabling factors like public–private partnerships and government incentives.

A major focus on economic feasibility must take into account the comparative advantage of biochar, which lies in its cost-saving potential. When locally produced from readily available feedstocks such as crop residues, animal manure, or invasive biomass (e.g., *Prosopis* and water hyacinth), biochar presents an affordable and renewable alternative to imported chemical fertilizers. These fertilizers remain costly due to high logistics costs, exchange rate volatility, and the global price of nitrogen, phosphate, and potash (Katakula et al. 2020; Afshar and Mofatteh 2024). The use of biochar can also reduce input costs over time by improving soil fertility, water retention, and nutrient efficiency, leading to more consistent yields with reduced reliance on synthetic fertilizers (Omidvar et al. 2022).

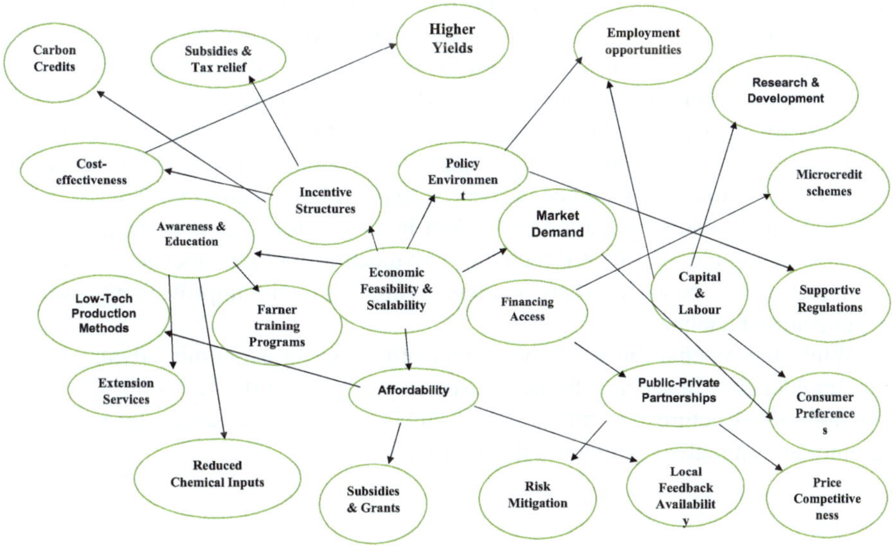

Fig. 6.1 Interconnected economic factors affecting biochar–biofertilizer adoption. (Source: Authors' illustration, 2025)

Another economic benefit arises from the durability of biochar in the soil. Unlike most chemical fertilizers that need to be applied annually or even seasonally, biochar persists for several years due to its recalcitrant carbon structure (Cotrufo et al. 2015). This means that initial investments in biochar production or procurement can yield returns over multiple cropping seasons, enhancing its cost-effectiveness in the long term.

There are notable cost barriers to entry, particularly for farmers without immediate access to pyrolysis kilns or capital to purchase ready-made biochar or formulated biofertilizers. Capital and labor requirements represent additional considerations critical to the economic feasibility of biochar–biofertilizer systems. Although biochar production is labor-intensive, it aligns well with local employment opportunities, particularly during off-peak farming periods, thus effectively integrating into existing agricultural labor dynamics (Afshar and Mofatteh 2024).

The labor intensity of biomass collection and processing, as well as the time required for biochar application, may also deter adoption unless offset by immediate yield improvements or market incentives. This integration provides supplementary income streams for rural communities, enhancing their overall economic resilience and sustainability. Furthermore, biofertilizers are often less available commercially in many African regions, with limited distribution networks and awareness among farmers about their benefits and application methods (Mącik et al. 2020).

Initial adoption and scalability often require targeted financial support mechanisms. Access to finance, including microcredit schemes and input subsidy programs, is essential to overcome initial investment barriers faced by farmers. Initiatives such as subsidized loans and input financing schemes provided by governments and NGOs can significantly improve adoption rates (Zhao et al. 2024). Evidence from agricultural financing models in East Africa demonstrates that accessible financing directly correlates with increased uptake of innovative agricultural technologies, including biochar and biofertilizers (Archana et al. 2025). As such, access to microfinance and input credit schemes could play a pivotal role in addressing these constraints. For example, input loan programs that bundle biochar and biofertilizer with crop insurance or market access arrangements could derisk adoption for smallholders. Government-backed input subsidies such as those used for lime or urea could also be redirected or expanded to include biochar–biofertilizer combinations, especially where evidence supports yield improvement and soil regeneration benefits.

While the biochar–biofertilizer system offers strong economic advantages for smallholder and subsistence farmers, targeted policy support, community-level production infrastructure, and financial inclusion strategies are essential to make these technologies a realistic and attractive option.

6.6.3 Scalability and Investment Considerations

While biochar–biofertilizer systems offer tangible economic benefits at a small scale, transitioning to large-scale adoption introduces additional complexity, requiring significant investments and structured incentive mechanisms. Economic drivers for scaling up include the potential for substantial yield improvements, long-term soil fertility enhancement, and carbon sequestration benefits, all of which contribute to broader economic resilience and sustainability (Liu et al. 2017a, b).

At a larger scale, the adoption of biochar–biofertilizer systems hinges on overcoming economic and institutional barriers. While smallholder systems may rely on local production and on-farm benefits, scaling up requires investment in biochar supply chains, improved pyrolysis technologies, and distribution logistics. Public–private partnerships (PPPs) offer a viable pathway for such investments, particularly where research institutions, agribusinesses, and local governments collaborate to derisk adoption and incentivize private sector involvement (Omidvar et al. 2022; Zhao et al. 2024). PPPs can effectively pool resources from governments, research institutions, private investors, and international donors to mitigate risks, promote innovative solutions, and scale successful interventions. Examples of effective PPPs include joint ventures between research institutions and agribusinesses that have successfully introduced biochar-based technologies in other regions, providing models that could be replicated in the African context (Omidvar et al. 2022).

Investment frameworks should consider long-term economic returns from improved soil health, reduced input costs, and enhanced resilience to climate shocks. Investment incentives such as subsidies, tax relief, and carbon credit schemes offer additional pathways to encourage large-scale adoption. Government-led incentive structures play a pivotal role by reducing the perceived financial risk of adopting new technologies. Policies that offer direct financial incentives or indirect benefits through tax deductions and environmental credits significantly enhance investment attractiveness (Zhao et al. 2024). Furthermore, international carbon markets provide a potential avenue for revenue generation through carbon sequestration credits, presenting another compelling financial incentive to adopt biochar systems at scale. Moreover, land tenure security and supportive agricultural policies are essential to encourage farmers to invest in long-term soil fertility improvements using biochar and biofertilizer (Zhao et al. 2024).

While the economic feasibility of biochar–biofertilizer systems is well-supported at the smallholder level, substantial scaling requires concerted investments, strategic partnerships, and comprehensive incentive mechanisms. Ensuring robust financing options, promoting targeted governmental support, and fostering collaborative public–private partnerships will be critical to realizing the full economic and sustainability benefits of biochar–biofertilizer systems in African agriculture.

6.7 Future Directions and Research Priorities

6.7.1 Toxicity as a Biochar–Biofertilizer Limitation

Although the biochar–biofertilizer synergy may appear advantageous in complementing sustainable farming, there are specific limitations within this association that require critical consideration prior to implementation. Primarily, these limitations arise from biochar, which may exhibit toxicity to microorganisms in biofertilizers. Certain biochar products contain potentially toxic elements (PTEs) that may be detrimental to soil microbes upon mixing, as well as to the soil itself, ultimately affecting plant growth. This phenomenon is predominantly influenced by the type of biochar feedstock and, to a lesser extent, pyrolysis conditions.

Biochar derived from sewage sludge and certain animal manures and manure composts are known to contain toxic heavy metals such as Pb, Zn, Cu, Cr, Ni, and As (Shen et al. 2015; Choudhary et al. 2019). In some instances, their concentration exceeds minimum requirements for stable growth of soil microbes, significantly affecting their structural and functional diversity. However, there is limited research examining how heavy metals in biochar are detrimental to microbes in biofertilizers, although numerous studies exist on how biochar generally affects microorganisms. For instance, Zhu et al. (2023) reported that sewage sludge biochar decreased bacterial and fungal populations, with the impact intensifying as pyrolysis temperature increased. Additionally, they observed that the effect was less pronounced in the fungal community compared to the bacterial community. The heavy metal content for the sludge biochar produced under different pyrolysis conditions in this study (Zhu et al. 2023) ranged between 0.029–0.032, 0.61–0.63, 0.16–0.21, 0.067–0.078, and 2.08–2.38 mg/kg for Cd, Cr, Cu, Pb, and Zn, respectively. Biochar toxicity may be directly influenced by biochar types other than sewage sludge, which are less commonly known to contain heavy metals. This phenomenon may be attributed to the presence of a prior source from which the feedstock potentially acquired these heavy metals, and the observed effects may represent the residual impact of that predecessor source. For example, plant-based biochar feedstock may have been cultivated in soil contaminated with heavy metals, or animal-based biochar may have been derived from animals fed with plants grown in heavy metal-contaminated soil conditions. Thus, before utilizing any form of biomass feedstock in this context, it is crucial to carefully monitor its origin.

In addition to the toxicity of biochar resulting from its feedstock, particularly sewage sludge, the pyrolysis temperature of biochar may generate toxic substances that are detrimental to microbes (Shen et al. 2015). Biochar contains various chemical compounds that potentially inhibit microbial activity. These include polyaromatic hydrocarbons (PAHs), volatile organic compounds, furans, phenols, ketones, methoxyphenols, and carboxylic acids (Ahmad et al. 2015; Choudhary et al. 2019; Murtaza et al. 2022). The presence of these compounds is further influenced by temperature and conditions during pyrolysis, feedstock utilized, elemental composition, and heating methods employed. These deleterious substances induce oxidative stress in microorganisms, particularly compromising the integrity of cell membranes through the generation of reactive oxygen species (ROS). Such ROS include hydroxyl radicals (·OH), superoxide radical anions (O_2), and hydrogen peroxide (H_2O_2).

6.7.2 Research Gaps and Policy Implications

The optimal use of biochar and biofertilizer in Africa has been hindered by insufficient private sector involvement and inadequate government backing for their manufacture and distribution. Despite the potential of biochar and biofertilizers to enhance agricultural practices, as demonstrated in other global regions, their adoption in Africa remains limited. This suggests a lack of substantial research interest, which is necessary to gather evidence for policy implications and strategies to encourage large-scale production. Therefore, future research actions aimed at enhancing the adoption of biochar–biofertilizer should address the following identified knowledge gaps.

Currently, numerous research efforts are concentrated on analyzing the cost-effectiveness of biochar in an isolated case. There is a limited number of studies examining the cost-effectiveness analysis of biochar–microorganism combination, to fully understand the economic advantages. Even though combining biochar and microorganisms leads to increased input costs, the resulting economic and environmental benefits remain little known.

The combination of biochar with supplementary materials such as nitrogen, phosphorus, and/or potassium fertilizers, composts, and manures may exert a more significant influence on microbial populations when combined with biochar. The addition of these supplements may enhance soil fertility through application in small quantities, ensuring improved nutrient use efficiency from biochar. However, research investigating the implications of including these supplemental materials and biochar–biofertilizer must be prioritized to observe any potential toxicity, specifically on microorganisms. Investigating these scenarios could yield insights that support sustainable practices in agriculture through integrated nutrient management and enhance plant growth.

The long-term effects of biochar–biofertilizer interaction on soil quality and plant development remain unclear. After its introduction to soil, biochar experiences an aging progression that can be affected by various factors, including the type of biochar, soil characteristics, and climatic conditions, among others. This aging may result in unexpected outcomes, such as the physical breakdown of biochar, dissolution of ash components, and oxidation (including photo-oxidation). As biochar ages, this weak interaction, for instance, electrostatic interactions, is likely to decrease because the biochar's surface becomes increasingly negative due to the development of additional oxygen-containing functional groups. Furthermore, the easily degradable carbon components may be rapidly consumed within several months of adding biochar, leaving behind stable aromatic structures that are challenging for microorganisms to utilize. These processes will ultimately impact the long-term interactions between microorganisms and biochar in the soil environment. Further field studies are necessary to assess the prolonged effectiveness of biochar with immobilized microorganisms.

Nanotechnology has garnered significant attention in recent years within scientific research and industrial applications across various fields. This aspect has also been incorporated into engineering biochar, termed nano-biochar, for soil quality enhancement and pesticide remediation (Udomkun et al. 2024). However, the application of nanotechnology remains unexplored in biochar–biofertilizer associations and holds the potential to revolutionize our understanding of these interactions. Consequently, there is a need to investigate this aspect on biochar–biofertilizer associations and its subsequent effects on soil and plants.

The toxicity of certain types of biochar is one of its limitations, particularly when combined with biofertilizers, as explained in the previous sections regarding how biochar toxicity affects microorganisms. In this section, we observed that there are limited studies that have examined how biochar directly affects biofertilizers, which is an important aspect that requires considerable attention in research. The research may primarily include how certain types of biochar also affect specific microorganisms found in major biofertilizer formulations. These investigations would be significant in maximizing the optimal potential of the biochar–biofertilizer combinations in achieving maximum yields in agriculture.

Lastly, the dissemination of these materials in Africa to the farmers remains a challenge as farmers remain unaware of such critical solutions in improving their farming ventures. It is crucial for agricultural researchers and educational institutions responsible for farmer training to promptly enhance awareness among farmers and relevant parties. This should be accomplished through various communication channels and collaborative research efforts. In addition, the current access to information has been improved through the sporulation of Internet, which can also be used as a tool for such initiatives. Key strategies for success include effective agricultural extension programs, involvement of the private sector, and government-provided subsidies. In particular, the commercialization and adoption of biochar–biofertilizers heavily depend on adequate funding. African governments should increase their financial support for biochar–biofertilizer technology and actively create policies that promote its development and make it easier for farmers to adopt.

6.7.3 Conclusion

This chapter explores biochar–biofertilizer interactions in sustainable agriculture, showing their potential as a solution to Africa's agricultural challenges. It demonstrates how these interactions improve soil fertility, increase crop yields, capture carbon, and lower greenhouse gas emissions. The mechanisms governing this synergy depend on how biochar improves soil conditions for enhancing microbes in biofertilizers. Biochar modifies soil acidity levels for microbial abundance, supplies energy through labile carbon, provides habitat for microorganisms, and improves other soil properties essential for microbial life. Additionally, factors like biochar to biofertilizer ratio, crop type, climate variables, and biochar porosity and surface area impact the synergy between biochar and biofertilizers. Nevertheless, this review also highlights the emergence of toxicity from biochar as a significant challenge, which may pose a limitation to the interactions between biochar and biofertilizers. Biochar products may contain potentially toxic elements like PTEs harmful to soil microbes and plant growth, primarily due to feedstock type and pyrolysis conditions. Such inconsistencies in this interaction require mechanistic studies to optimize its use across contexts. Looking ahead, key areas for future research include cost-effectiveness of biochar–biofertilizer interaction, combination with organic or inorganic fertilizers, nanotechnology, and long-term effects on soil quality and plant productivity.

References

Afshar M, Mofatteh S (2024) Biochar for a sustainable future: environmentally friendly production and diverse applications. Results Eng 23:102433. https://doi.org/10.1016/j.rineng.2024.102433

Ahmad MT, Asghar HN, Saleem M, Yahya Khan M, Zahir ZA (2015) Synergistic effect of rhizobia and biochar on growth and physiology of maize. Agron J 107:2327–2334

Ahmad W, Nepal J, Zou Z, Munsif F, Khan A, Ahmad I, Zaheer S, Khan MS, Jadoon SA, Tang D (2023) Biochar particle size coupled with biofertilizer enhances soil carbon-nitrogen microbial pools and CO_2 sequestration in lentil. Front Environ Sci 11:1–12

Allaire SE, Baril B, Vanasse A, Lange SF, Mackay J, Smith DL (2015) Carbon dynamics in a biochar-amended loamy soil under switchgrass. Can J Soil Sci 95:1–13

Antonangelo JA, Zhang H, Sun X, Kumar A (2019) Physicochemical properties and morphology of biochars as affected by feedstock sources and pyrolysis temperatures. Biochar 1:325–336

Arabi Z, Eghtedaey H, Gharehchmaghloo B, Faraji A (2018) Effects of biochar and bio-fertilizer on yield and qualitative properties of soybean and some chemical properties of soil. Arab J Geosci 11:1–9

Archana TS, Verma R, K KC, Haseena K, Rao G GE, Savitha MS (2025) Synergistic impact of biochar and organic amendments on field bean (*Vicia faba*) growth and soil characteristics. Int J Environ Clim Change 15(1):67–80. https://doi.org/10.9734/ijecc/2025/v15i14675

Bai SH, Omidvar N, Gallart M, Kämper W, Tahmasbian I, Farrar MB, Singh K, Zhou G, Muqadass B, Xu CY, Koech R (2022) Combined effects of biochar and fertilizer applications on yield: a review and meta-analysis. Sci Total Environ 808:152073

Balasubramanian P, Karthickumar P (2017) Biofertilizers and biopesticides: a holistic approach for sustainable agriculture. In: Sustainable utilization of natural resources. Agricultural and Environmental Biotechnology Group, Department of Biotechnology and Medical Engineering, National Institute of Technology Rourkela, Rourkela, pp 256–284

Billa SF, Evaristus T, Ajebesone A, Ngome F (2019) Agro-environmental characterization of biochar issued from crop wastes in the humid forest zone of Cameroon. Int J Recycl Org Waste Agric 8:1–13

Biswas DR, Basak BB (2014) Mobilization of potassium from waste mica by potassium-solubilizing bacteria (*Bacillus mucilaginosus*) as influenced by temperature and incubation period under in vitro laboratory conditions. Agrochimica 58:309–320

Břendová K, Száková J, Lhotka M, Krulikovská T, Punčochář M, Tlustoš P (2017) Biochar physicochemical parameters as a result of feedstock material and pyrolysis temperature: predictable for the fate of biochar in soil? Environ Geochem Health 39:1381–1395

Choudhary TK, Khan KS, Hussain Q, Ahmad M, Ashfaq M (2019) Feedstock-induced changes in composition and stability of biochar derived from different agricultural wastes. Arab J Geosci 12:617

Cotrufo MF, Soong JL, Horton AJ, Campbell EE, Haddix ML, Wall DH, Parton WJ (2015) Formation of soil organic matter via biochemical and physical pathways of litter mass loss. Nat Geosci 8:8–13

Cruz-Méndez AS, Ortega-Ramírez E, Lucho-Constantino CA, Arce-Cervantes O, Vázquez-Rodríguez GA, Coronel-Olivares C, Beltrán-Hernández RI (2021) Bamboo biochar and a nopal-based biofertilizer as improvers of alkaline soils with low buffer capacity. Appl Sci (Switz) 11:6502

Daniel AI, Fadaka AO, Gokul A, Bakare OO, Aina O, Fisher S, Burt AF, Mavumengwana V, Keyster M, Klein A (2022) Biofertilizer: the future of food security and food safety. Microorganisms 10:1220

Das SK, Ghosh GK, Avasthe RK, Sinha K (2021) Compositional heterogeneity of different biochar: effect of pyrolysis temperature and feedstocks. J Environ Manag 278:111501

Deluca TH, Profile S, Mackenzie D, Jones DL (2015) Biochar effects on soil nutrient transformations soil protease: the bottleneck of the terrestrial nitrogen cycle view project ecological context for bioavailable phosphorus in sub-boreal forest soils of northern Japan View project. In: Books.Google.Com

Dzvene AR, Chiduza C, Mnkeni PN, Peter PC (2019) Characterisation of livestock biochars and their effect on selected soil properties and maize early growth stage in soils of eastern cape province, South Africa. S Afr J Plant Soil 36(3):199–209

Fasusi OA, Cruz C, Babalola OO (2021) Agricultural sustainability: microbial biofertilizers in rhizosphere management. Agriculture 11:1–19

Fox A, Kwapinski W, Griffiths BS, Schmalenberger A (2014) The role of sulfur- and phosphorus-mobilizing bacteria in biochar-induced growth promotion of *Lolium perenne*. FEMS Microbiol Ecol 90:78–91

Fwanyanga FM, Horn LN, Sibanda T, Reinhold-Hurek B (2022) Prospects of rhizobial inoculant technology on Bambara groundnut crop production and growth. Front Agron 4:94. https://doi.org/10.3389/FAGRO.2022.1004771/BIBTEX

Ghorbani M, Asadi H, Abrishamkesh S (2019) Effects of rice husk biochar on selected soil properties and nitrate leaching in loamy sand and clay soil. Int Soil Water Conserv Res 7:258–265

Ghorbani M, Konvalina P, Neugschwandtner RW, Soja G, Bárta J, Chen W-H, Amirahmadi E (2024) How do different feedstocks and pyrolysis conditions effectively change biochar modification scenarios? A critical analysis of engineered biochars under H_2O_2 oxidation. Energy Convers Manag 300:117924

Guo J, Zheng L, Li Z, Zhou X, Cheng S, Zhang L, Zhang Q (2021) Effects of various pyrolysis conditions and feedstock compositions on the physicochemical characteristics of cow manure-derived biochar. J Clean Prod 311:127458

Huang H, Reddy NG, Huang X, Chen P, Wang P, Zhang Y, Huang Y, Lin P, Garg A (2021) Effects of pyrolysis temperature, feedstock type and compaction on water retention of biochar amended soil. Sci Rep 11:1–19

Jaafar NM, Clode PL, Abbott LK (2015) Soil microbial responses to biochars varying in particle size, surface and pore properties. Pedosphere 25:770–780

Jatuwong K, Aiduang W, Kiatsiriroat T, Kamopas W, Lumyong S (2024) Effects of biochar and arbuscular mycorrhizal fungi on soil health in Chinese Kale (*Brassica oleracea* var. *alboglabra* L.) cultivation. Microbiol Res 15:404–421

Kamran S, Shahid I, Baig DN, Rizwan M, Malik KA, Mehnaz S (2017) Contribution of zinc solubilizing bacteria in growth promotion and zinc content of wheat. Front Microbiol 8:2593

Katakula AAN, Gawanab W, Itanna F, Mupambwa HA (2020) The potential fertilizer value of Namibian beach-cast seaweed (*Laminaria pallida* and *Gracilariopsis funicularis*) biochar as a nutrient source in organic agriculture. Sci Afr 10:e00592

Katakula AAN, Handura B, Gawanab W, Itanna F, Mupambwa HA (2021) Optimized vermicomposting of a goat manure-vegetable food waste mixture for enhanced nutrient release. Sci Afr 12:e00727. https://doi.org/10.1016/j.sciaf.2021.e00727

Katakula AAN, Gawanab W, Handura B, Itanna F, Mupambwa HA (2022) Seaweed (*Gracilariopsis funicularis*) biochar incorporation into a goat manure–food waste vermicompost for optimized vermidegradation and nutrient release. Front Sustain Food Syst 6:1005740

Khan Z, Zhang K, Khan MN, Zhu K, Hu LW (2024) Effects of biochar persistence on soil physiochemical properties, enzymatic activities, nutrient utilization, and crop yield in a three-year rice-rapeseed crop rotation. Eur J Agron. https://doi.org/10.1016/j.eja.2024.127096

Laird DA, Fleming P, Davis DD, Horton R, Wang BQ, Karlen DL (2010) Impact of biochar amendments on the quality of a typical midwestern agricultural soil. Geoderma 158:443–449

Li W, Hou Y, Long M, Wen X, Han J, Liao Y (2023) Long-term effects of biochar application on rhizobacteria community and winter wheat growth on the Loess Plateau in China. Geoderma 429:116250

Liu S, Tang W, Yang F, Meng J, Chen W, Li X (2017a) Influence of biochar application on potassium-solubilizing *Bacillus mucilaginosus* as potential biofertilizer. Prep Biochem Biotechnol 47:32–37

Liu L, Wang Y, Yan X, Li J, Jiao N, Hu S (2017b) Biochar amendments increase the yield advantage of legume-based intercropping systems over monoculture. Agric Ecosyst Environ 237(2017):16–23. https://doi.org/10.1016/j.agee.2016.12.026

Liu X, Trogisch S, He JS, Niklaus PA, Bruelheide H, Tang Z, Erfmeier A, Scherer-Lorenzen M, Pietsch KA, Yang B, Kühn P, Scholten T, Huang Y, Wang C, Staab M, Leppert KN, Wirth C, Schmid B, Ma K (2018) Tree species richness increases ecosystem carbon storage in subtropical forests. Proc R Soc Lond B Biol Sci 285:20181240

Mącik M, Gryta A, Frąc M (2020) Biofertilizers in agriculture: an overview on concepts, strategies and effects on soil microorganisms. Adv Agron 162:31–87

Mashamaite CV, Motsi H, Manyevere A, Poswa SB (2024) Assessing the potential of biochar as a viable alternative to synthetic fertilizers in sub-Saharan Africa smallholder farming: a review. Agronomy 14:1215

Moosavi SA, Shokuhfar A, Lak S, Mojaddam M, Alavifazel M (2020) Integrated application of biochar and bio-fertilizer improves yield and yield components of cowpea under water-deficient stress. Ital J Agron 15:94–101

Murtaza G, Ahmed Z, Usman M (2022) Feedstock type, pyrolysis temperature and acid modification effects on physiochemical attributes of biochar and soil quality. Arab J Geosci 15:305

Namaswa T, Burslem DF, Smith J (2023) Emerging trends in appropriate kiln designs for small-scale biochar production in low to middle income countries. Bioresour Technol Rep 24:101641

Nyambo P, Zhou L, Chuma T, Sokombela A, Malobane ME, Musokwa M (2023a) Prospects of vermicompost and biochar in climate smart agriculture. In: Vermicomposting for sustainable food systems in Africa. Springer Nature Singapore, Singapore, pp 145–159

Nyambo P, Motsi H, Chiduza C, Malobane E (2023b) Biochar ageing improves soil properties, growth and yield of red radish (*Raphanus sativus*) in a Haplic Cambisol. PLoS One 18:1–14

Nyambo P, Zhou L, Mapuka FN, Slayi M, Muchaku S, Dzvene AR (2024) Nature-based solutions to climate change for enhanced nutrition, food, and income security of rural communities in sub-Saharan Africa. In: Handbook of nature-based solutions to mitigation and adaptation to climate change, pp 1–24

Pastore G, Kernchen S, Spohn M (2020) Microbial solubilization of silicon and phosphorus from bedrock in relation to abundance of phosphorus-solubilizing bacteria in temperate forest soils. Soil Biol Biochem 151:108050

Pituello C, Francioso O, Simonetti G, Pisi A, Torreggiani A, Berti A, Morari F (2015) Characterization of chemical–physical, structural and morphological properties of biochars from biowastes produced at different temperatures. J Soils Sediments 15:792–804

Rivelli AR, Libutti A (2022) Effect of biochar and inorganic or organic fertilizer co-application on soil properties, plant growth and nutrient content in Swiss chard. Agronomy 12(9):2089. https://doi.org/10.3390/agronomy12092089

Sarfaraz Q, Silva L, Drescher G, Zafar M, Severo F, Kokkonen A, Molin G, Shafi M, Shafique Q, Solaiman Z (2020) Characterization and carbon mineralization of biochars produced from different animal manures and plant residues. Sci Rep 10:2–10

Sarfraz R, Hussain A, Sabir A, Ben Fekih I, Ditta A, Xing S (2019) Role of biochar and plant growth promoting rhizobacteria to enhance soil carbon sequestration—a review. Environ Monit Assess 191:251

Schütz L, Gattinger A, Meier M, Müller A, Boller T, Mäder P, Mathimaran N, Scotti R (2018) Improving crop yield and nutrient use efficiency via biofertilization — a global meta-analysis. Front Plant Sci 8:2204

Shen X, Huang G, Yang Z, Han L (2015) Compositional characteristics and energy potential of Chinese animal manure by type and as a whole. Appl Energy 160:108–119

Shikha FS, Rahman MM, Sultana N, Mottalib MA, Yasmin M (2023) Effects of biochar and biofertilizer on groundnut production: a perspective for environmental sustainability in Bangladesh. Carbon Res 2:10. https://doi.org/10.1007/s44246-023-00043-7

Sifton MA, Smith SM, Thomas SC (2023) Biochar-biofertilizer combinations enhance growth and nutrient uptake in silver maple grown in an urban soil. PLoS One 18:e0288291

Singh D, Thapa S, Geat N, Mehriya ML, Rajawat MVS (2021) Biofertilizers: mechanisms and application. In: Biofertilizers. Woodhead Publishing, pp 151–166

Sokol NW, Kuebbing SE, Karlsen-Ayala E, Bradford MA (2019) Evidence for the primacy of living root inputs, not root or shoot litter, in forming soil organic carbon. New Phytol 221:233–246

Song W, Guo M (2012) Quality variations of poultry litter biochar generated at different pyrolysis temperatures. J Anal Appl Pyrolysis 94:138–145

Srivastava P, Gadi Y, Supriya B, Lytand W, Singh BV, Katiyar D (2023) Biochar's influence on soil microorganisms: understanding the impacts and mechanisms. Int J Plant Soil Sci 35:455–464

Study C, Lshankiti AA, Ill SG (2016) Integrated plant nutrient management for sandy soil using chemical fertilizers, compost, biochar and biofertilizers-case study in UAE-and biofertilizers. J Arid Land Stud 26:101–106

Sun H, Zhang Y, Yang Y, Chen Y, Jeyakumar P, Shao Q, Zhou Y, Ma M, Zhu R, Qian Q, Fan Y, Xiang S, Zhai N, Li Y, Zhao Q, Wang H (2021) Effect of biofertilizer and wheat straw biochar application on nitrous oxide emission and ammonia volatilization from paddy soil. Environ Pollut 275:116640

Tan Z, Zou J, Zhang L, Huang Q (2018) Morphology, pore size distribution, and nutrient characteristics in biochars under different pyrolysis temperatures and atmospheres. J Mater Cycles Waste Manag 20:1036–1049

Touray N, Tsai WT, Chen HR, Liu SC (2014) Thermochemical and pore properties of goat-manure-derived biochars prepared from different pyrolysis temperatures. J Anal Appl Pyrolysis 109:116–122

Udomkun P, Chandi K, Boonupara T, Kaewlom P (2024) Innovative approaches: exploring nano-biochar technology's impact on soil properties, alachlor retention, and microbial populations. Environ Technol Innov 35:103659

Upadhayay VK, Singh AV, Khan A, Singh J, Pareek N, Raghav A (2022) FE-SEM/EDX based zinc mobilization analysis of *Burkholderia cepacia* and *Pantoea rodasii* and their functional annotation in crop productivity, soil quality, and zinc biofortification of Paddy. Front Microbiol 13:1–20

Waqar A, Bano A, Ajmal M (2022) Effects of PGPR bioinoculants, hydrogel and biochar on growth and physiology of soybean under drought stress. Commun Soil Sci Plant Anal 53:826–847

Weiskopf SR, Rubenstein MA, Crozier LG, Gaichas S, Griffis R, Halofsky JE, Hyde KJ, Morelli TL, Morisette JT, Muñoz RC, Pershing AJ (2020) Climate change effects on biodiversity, ecosystems, ecosystem services, and natural resource management in the United States. Sci Total Environ 733:137782

Xu Y, Liao H, Zhang J, Lu H, He X, Zhang Y, Wu Z, Wang H, Lu M (2022) A novel ca-modified biochar for efficient recovery of phosphorus from aqueous solution and its application as a phosphorus biofertilizer. Nanomaterials 12:2755

Zhang L, Jing Y, Chen C, Xiang Y, Rezaei Rashti M, Li Y, Deng Q, Zhang R (2021) Effects of biochar application on soil nitrogen transformation, microbial functional genes, enzyme activity, and plant nitrogen uptake: a meta-analysis of field studies. GCB Bioenergy 13:1859–1873

Zhang K, Khan Z, Khan MN, Luo T, Luo L, Bi J, Hu LW (2024) The application of biochar improves the nutrient supply efficiency of organic fertilizer, sustains soil quality and promotes sustainable crop production. Food Energy Secur 13:e520. https://doi.org/10.1002/fes3.520

Zhao X, He L, Geng K, Zhang H, Wang J, Gan T, Mao X, Zhang X (2024) Effects of combined biochar and chemical fertilizer application on soil fertility and properties: a two-year pot experiment. Sustainability 16(20):8780. https://doi.org/10.3390/su16208780

Zhu Q, Liang Y, Zhang Q, Zhang Z, Wang C, Zhai S, Li Y, Sun H (2023) Biochar derived from hydrolysis of sewage sludge influences soil properties and heavy metals distributed in the soil. J Hazard Mater 442:130053

Chapter 7
Vermi-leachates as Sources of Bionutrients in Organic Horticulture Farming

Hupenyu Allan Mupambwa, Adornis D. Nciizah, Patrick Nyambo, Binganidzo Muchara, and Nguza Siyambango-Mulisa

Abstract The global drive towards sustainable agriculture has intensified interest in organic nutrient sources that enhance soil fertility and plant productivity without compromising environmental integrity. Vermicomposting, the bioconversion of organic waste through earthworms and microorganisms, produces not only nutrient-rich vermicompost but also vermi-leachates,liquid by-products that contain a complex mixture of soluble nutrients, phytohormones, beneficial microorganisms, and bioactive compounds. This chapter explores the production, composition, and applications of vermi-leachates, emphasizing their role as biofertilizers and plant growth promoters in horticultural systems. It highlights their ecological and agronomic benefits, including improved soil structure, increased microbial activity, enhanced nutrient cycling, and greater resistance to pests and abiotic stresses. Additionally, this chapter evaluates the economic feasibility and market potential of vermi-leachates, considering both their comparative advantages over synthetic fertilizers

H. A. Mupambwa (✉)
Sam Nujoma Marine and Coastal Resources Research Centre, Sam Nujoma Campus, University of Namibia, Henties Bay, Namibia
e-mail: hmupambwa@unam.na

A. D. Nciizah
Agricultural Research Council—Natural Resources and Engineering, Pretoria, South Africa

Department of Agriculture and Animal Health, University of South Africa, Florida Park, Roodepoort, South Africa

P. Nyambo
Agricultural Research Council—Vegetables, Industrial and Medicinal Plants, Pretoria, South Africa

B. Muchara
University of South Africa (UNISA), Graduate School of Business Leadership, Pretoria, South Africa

N. Siyambango-Mulisa
University of Namibia, Multidisciplinary Research, Centre for Research Services, Windhoek, Namibia

A. D. Nciizah et al. (eds.), *Biofertilizers for Sustainable Agriculture in Africa*, Sustainability Sciences in Asia and Africa, https://doi.org/10.1007/978-981-95-6009-7_7

and the challenges associated with organic certification, scalability, and cost. By presenting a comprehensive overview, this chapter advocates for the integration of vermi-leachates into regenerative and climate-resilient agricultural practices aimed at achieving long-term food security and environmental sustainability.

Keywords Liquid nutrients · Hydroponics · Inorganic nutrients · Growth hormones · Vegetables · Economics

7.1 Introduction

The world population is expected to reach 9.7 billion by 2050 and 11 billion by the turn of the century (United Nations Department of Economic and Social Affairs 2019). This rapid growth of the global population has exerted unprecedented pressure on agricultural systems to meet the rising demand for food. To meet the rising demand for food, agricultural production, including both crop and livestock outputs, is projected to increase by 60% between 2005/2007 and 2050 (Wongkiew and Polprasert 2023). In response to increasing food needs, the mid-twentieth century saw the emergence of the Green Revolution, characterized by the widespread adoption of high-yielding crop varieties, synthetic fertilizers, pesticides, and irrigation technologies (Mupambwa et al. 2024; Hamdan et al. 2022). This transformation significantly boosted agricultural productivity, particularly in developing countries, contributing to reduced hunger and improved food security. These Green Revolution technologies led to the widespread adoption of intensive and often unsustainable cropping practices that degrade soil health, reduce biodiversity, and compromise long-term productivity. The resulting negative effects of these intensive farming technologies have underscored the urgent need for more sustainable agricultural practices.

Conventional approaches to soil fertility management, particularly those reliant on synthetic inorganic fertilizers, have proven effective in the short term but raise significant environmental and economic concerns. Most of these conventional systems of managing soil fertility tend to focus on feeding the crop and ignoring the feeding of the soil, which is an important resource (Withers et al. 2014). The direct effects of these approaches include soil acidification, nutrient imbalances, water pollution, and declining soil organic matter, all of which threaten the sustainability of agricultural systems, especially in regions with fragile ecosystems. Considering these challenges, there is an urgent need to transition from traditional, inorganic soil fertility strategies to more ecologically sound and sustainable alternatives (Mupambwa et al. 2022). One such promising approach is vermicomposting, the bioconversion of organic waste into nutrient-rich compost through the action of mainly epigeic earthworms (Mnkeni and Mupambwa 2023). Vermicomposting not only improves soil fertility by enriching it with essential nutrients and beneficial microorganisms, but also supports waste recycling, enhances soil structure, and promotes a more balanced and resilient agroecosystem. During this process of vermicomposting, a liquid by-product called vermi-leachate or vermi-wash is also produced, which contains soluble nutrients, enzymes, and beneficial microorganisms. Vermi-leachate has gained increasing attention as a sustainable biofertilizer in horticulture due to its ability to enhance plant growth, improve soil health, and reduce the reliance on synthetic inputs. Its application in horticultural systems offers a promising avenue for promoting environmentally sound and productive agricultural practices.

This chapter explores the potential of vermicomposting and its by-products, particularly vermi-leachates, as sustainable solutions for soil fertility management. It highlights the limitations of conventional practices and delves into the biological and ecological processes that underpin vermicomposting. Special emphasis is placed on vermi-leachates, nutrient-rich liquid extracts produced during the composting process, which have shown great promise as natural biofertilizers and growth stimulants in horticultural systems. By examining how mainly the liquid outputs of vermicomposting contribute to improved soil health, crop productivity, and environmental sustainability, this chapter supports the advancement of regenerative agriculture in response to the pressing challenges of global food security and ecological degradation.

7.2 Vermi-leachates and Their Production

Vermi-leachate is a nutrient-rich aqueous solution that is generated as a natural by-product of the vermicomposting process (Singh et al. 2010). It is formed when water, either from the decomposition of organic material or from external addition, percolates through a vermicomposting system containing organic waste and earthworms (Churilova and Midmore 2019). As the liquid drains through the decomposing substrate, it extracts soluble nutrients, organic acids, enzymes, hormones, and beneficial microorganisms, resulting in a bioactive liquid often referred to as "worm tea" or "vermi-liquid," though it is distinct from intentionally brewed vermi-tea (Jiang et al. 2023).

Unlike vermicompost, which is a solid, stabilized compost product consisting of fine humus and worm castings, vermi-leachate is a liquid extract that contains both chemical and microbial components in a readily available form. Vermi-tea, on the other hand, is typically produced by steeping vermicompost in water under either aerated or nonaerated conditions, often with additives such as molasses to stimulate microbial activity. In contrast, vermi-leachate is passively produced and collected at the base of the vermicomposting unit, particularly in systems that are not fully drained or where moisture input exceeds evaporation. Due to its rich and diverse composition, vermi-leachate has shown considerable promise in sustainable agriculture and horticulture, where it can be used as a foliar spray, root drench, or fertigation input to enhance plant growth, suppress pathogens, and improve soil biological health (Oyege and Balaji Bhaskar 2023).

7.2.1 *Vermi-leachate Production Process*

The production of vermi-leachate occurs simultaneously with the decomposition of organic matter in a controlled vermicomposting environment. This process is primarily driven by epigeic earthworm species, such as *Eisenia fetida*, *Eudrilus eugeniae*, and *Perionyx excavatus*, which feed on partially decomposed organic material (Mupambwa and Mnkeni 2018). These earthworms, along with a diverse community of bacteria, fungi, and actinomycetes, break down complex organic matter into simpler, plant-available compounds. The system is typically loaded with organic feedstocks such as vegetable and fruit scraps, livestock manure, crop residues, coffee pulp, paper waste, and other biodegradable agricultural inputs. As microbial decomposition occurs and the worms digest and excrete the material, water naturally present in the waste or added externally (e.g., through irrigation or rainfall) moves through the substrate. During this process, the water picks up soluble nutrients and microbial metabolites, becoming enriched with compounds such as nitrate, ammonium, potassium, phosphorus, auxins, cytokinins, and enzymes.

The leachate collects at the bottom of the unit and can be drained through a tap or outlet in a well-designed vermicomposting bin. In some cases, the leachate may also contain anaerobic by-products if the system becomes overly saturated, underscoring the importance of good drainage and moisture control in vermicompost system design. The quality and efficacy of vermi-leachate as a biofertilizer or biostimulant are influenced by several interrelated factors as shown in Table 7.1.

Table 7.1 Key factors affecting vermi-leachate quality and their impacts

Factor	Description	Impact on leachate quality
Moisture content	Water present in the system; can come from feedstock or external additions	Optimal moisture (60%–80%) promotes aerobic microbial activity; excess moisture may cause anaerobic, phytotoxic leachate
Feedstock composition	Type and quality of organic inputs (e.g., manure, kitchen waste, crop residues)	Determines nutrient content (N, P, K), organic acids, and microbial load in leachate
Earthworm species	Species used in the system (e.g., *Eisenia fetida, Eudrilus eugeniae*)	Affects digestion rate, casting quality, and microbial interactions, influencing leachate bioactivity
System design	Physical setup (bins, pits, windrows) and drainage features	Well-drained systems reduce anaerobic risk and allow controlled leachate collection
Feeding frequency	Rate and regularity of adding organic material to the system	Consistent feeding maintains steady microbial activity and nutrient release
Temperature and pH	Environmental conditions and acidity/alkalinity of the composting environment	Optimal temperature (20–30 °C) and near-neutral pH enhance microbial and worm activity

7.3 The Role of Organic Nutrients in Plant Growth and Sustainable Agriculture

Organic farming techniques have gained significant momentum in recent years, driven by growing consumer demand for organically grown produce and the increasing commitment of farmers to maintaining or improving soil health (Gamage et al. 2023). This increase in consumer preference is largely fueled by the perceived environmental and health benefits associated with organic agriculture. As a result, organic farming practices have become central to promoting sustainability by reducing dependency on synthetic inputs and fostering an ecological balance (Deokar and Fernades 2024).

A core component of organic farming is the use of natural nutrient sources, such as compost, manure, cover crops, and vermi-leachates, to support plant growth while preserving soil fertility and structure (Gamage et al. 2023). Organic nutrient management enhances microbial diversity, boosts organic matter, and promotes efficient nutrient cycling, all of which contribute to long-term agricultural sustainability (Lori et al. 2017). Organic fertilizers not only fulfill the nutritional needs of plants, but also enhance the physical, chemical, and biological properties of the soil (Assefa and Tadesse 2019). Unlike synthetic fertilizers, they gradually release nutrients, supporting consistent plant growth and reducing the risk of leaching.

7.4 Significance of Organic Nutrients in Crop Growth

Organic farming systems rely on sustainable nutrient management practices to enhance soil fertility and promote healthy crop growth, without the use of synthetic chemicals. One of the key strategies is the utilization of organic nutrients that supply essential macronutrients such as nitrogen (N), phosphorus (P), and potassium (K), along with vital micronutrients such as iron (Fe), zinc (Zn), manganese (Mn), and copper (Cu). These nutrients are crucial for plant development and productivity, and their proper management underpins the success of organic agriculture.

Unlike synthetic fertilizers, which deliver nutrients in readily soluble forms, organic nutrient sources such as compost, farmyard manure, green manure, and vermi-leachates decompose gradually. This slow-release mechanism ensures a sustained supply of nutrients throughout the growing season, enhances nutrient use efficiency, and improves soil structure and biological activity (Badagliacca et al. 2024). The gradual decomposition also helps prevent nutrient leaching and supports long-term soil health.

Nitrogen (N) is essential for the synthesis of chlorophyll, amino acids, and proteins, making it vital for photosynthesis and vegetative growth. In organic systems, nitrogen is often derived from decomposing legume residues, compost, and vermi-leachates, which not only supply (N) but also stimulate beneficial microbial activity and improve soil aeration and water-holding capacity (Hassan et al. 2023).

Phosphorus (P), on the other hand, plays a central role in root development and the formation of flowers and seeds. Organic sources of P, such as rock phosphate and bone meal, release nutrients slowly, aligning with plant uptake rates and reducing the risks of runoff and pollution. Additionally, symbiotic relationships with mycorrhizal fungi and enhanced microbial activity in organic systems improve P solubility and availability to plants (Sande et al. 2024). Potassium (K) is involved in numerous physiological functions, including enzyme activation, water regulation, and plant immunity. Sources such as wood ash and plant residues contribute to potassium in a form that gradually becomes available. Adequate K levels improve crop tolerance to drought and other stresses, thereby enhancing resilience and yield stability (Hasanuzzama et al. 2018).

Micronutrients, although required in smaller quantities, are critical for plant metabolism. Elements, such as Fe, Zn, Mn, and Cu, act as cofactors for enzymes and play roles in processes such as respiration, photosynthesis, and hormone synthesis. Organic amendments enhance the bioavailability of these micronutrients by promoting favorable microbial interactions and maintaining optimal soil pH and organic matter levels (Essel et al. 2023). To achieve sustainable and resilient agricultural production, organic farming must continue to be explored and adopt alternative nutrient strategies that not only meet crop nutritional demands but also conserve ecological integrity. This includes integrating practices, such as crop rotation, cover cropping, compositing, and the use of biofertilizers to enhance the availability and cycling of organic nutrients (Sande et al. 2024).

7.5 Comparison of Organic Nutrients and Synthetic Fertilizers

In modern agriculture, the choice between organic and synthetic fertilizers significantly affects crop productivity, soil health, and environmental sustainability. Although both types of fertilizers aim to meet plant nutrient requirements, they differ widely in terms of nutrient release patterns, their effects on soil properties, and broader environmental impacts.

7.5.1 Nutrient-Release Patterns

Organic nutrients, such as those found in vermi-leachates, compost, and manure, are released through microbial decomposition. This process ensures a slow and consistent supply of nutrients that aligns with the plant uptake throughout the growing season. Gradual release reduces the risk of nutrient loss and enhances nutrient use efficiency (Loera-Muro et al. 2021); for instance, vermi-leachates provide a balanced mix of macronutrients and micronutrients while fostering microbial activity, which plays a critical role in nutrient mineralization. In contrast, synthetic fertilizers, such as urea and ammonium phosphate, offer nutrients in readily soluble forms, allowing for immediate uptake by crops. However, this availability often exceeds plant needs, leading to nutrient runoff, leaching, and environmental contamination. Excess nitrogen and phosphorus can leach into water bodies, causing eutrophication and disrupting aquatic ecosystems (Loera-Muro et al. 2021).

7.5.2 Soil Health Benefits

Organic fertilizers contribute substantially to the improvement of soil health. They enhance soil structure, increase organic matter content, and support the proliferation of beneficial soil microbes (Mupambwa and Mnkeni 2018). These factors collectively improve soil aeration, water retention, and nutrient cycling, thereby creating a fertile and resilient environment for plant growth. Organic amendments, such as vermi-leachates and compost, not only nourish plants but also enrich the soil's biological activity and long-term fertility. Conversely, prolonged use of synthetic fertilizers can degrade the soil structure and reduce microbial diversity. Continuous chemical inputs may lead to soil acidification, compaction, and nutrient imbalance, ultimately diminishing the natural productivity of the soil.

Biochar has emerged as a complementary organic soil amendment with promising benefits. Derived from the pyrolysis of organic biomass, biochar improves soil porosity, enhances nutrient retention, and increases cation exchange capacity. Its ability to sequester carbon and stabilize soil pH makes it a sustainable option for improving both soil quality and crop performance under stressful conditions (Siddika and Hossain 2023).

7.5.3 Environmental Impact

The environmental impact of organic fertilizers contributes to carbon sequestration and reduces pollution risks. It contributes positively to soil health by improving soil structure, increasing microbial activity, and enhancing water retention (Aytenew and Bore 2020). They gradually release nutrients, reducing the risk of nutrient leaching and groundwater contamination. Additionally, biofertilizers enhance carbon sequestration, thus mitigating climate change by reducing greenhouse gas emissions. In contrast, synthetic fertilizers, which increase yield, have serious environmental drawbacks. These include water pollution, eutrophication, soil acidification, greenhouse effects, and long-term soil degradation. Additionally, synthetic fertilizers are chemically manufactured to provide immediate nutrient availability and significantly boost crop yield. However, their excessive use has detrimental environmental effects. One major issue is nutrient leaching, particularly nitrogen and phosphorus, which contribute to water pollution and eutrophication, leading to algal blooms and aquatic ecosystem degradation (Madjar and Scăet 2024).

7.6 Plant Growth-Promoting Substances in Vermi-leachates

Plant growth-promoting substances (PGPS) play a crucial role in enhancing crop productivity by improving nutrient availability, stimulating plant metabolism, and increasing resistance to both biotic and abiotic stresses (Kumari et al. 2023). These substances encompass a range of phytohormones, beneficial microbes, and bioactive compounds that work synergistically to support plant health and improve yield (Bhattacharyya et al. 2016). Vermi-leachate, a nutrient-rich liquid by-product of the vermicomposting process, is a notable source of these bioactive agents. In addition to being a source of essential nutrients, vermi-leachate is particularly valuable because of its abundance of plant growth-promoting substances.

7.7 Vermi-leachates as a Source of Growth-Promoting Substances

Vermi-leachates are rich in essential nutrients and a diverse array of bioactive compounds that significantly enhance crop growth. These leachates contain key plant growth hormones such as auxins, gibberellins, and cytokinins that promote root elongation, stem development, and flowering. Moreover, the microbial diversity present in vermi-leachates plays a crucial role in suppressing plant diseases, offering a natural form of biological control and reducing reliance on synthetic pesticides (Koskey et al. 2022). In addition to these microbial benefits, vermi-leachates also contain essential macronutrients such as nitrogen (N), phosphorus (P), and potassium (K), which are fundamental to plant health. The natural presence of auxins and gibberellins within vermi-leachates supports vital physiological processes: auxins regulate cell elongation and root formation, whereas gibberellins drive stem elongation, seed germination, and flower development (Aremu et al. 2015). Together, these components make vermi-leachates a powerful organic input that enhances plant growth and resilience without the harmful side effects associated with synthetic growth regulators. Sustainable crop production is enhanced by the integration of vermicompost and biofertilizers. The integration of vermicomposting and biofertilizer is essential for sustainable crop production (Sande et al. 2024).

7.8 Environmental and Agricultural Benefits

The use of vermi-leachates as biofertilizers is gaining increased attention due to its ecologically friendly nature and contribution to sustainable agriculture. It significantly reduces reliance on synthetic fertilizers by providing an organic nutrient-rich alternative that improves soil fertility and enhances microbial activity (Mohite et al. 2024). Unlike synthetic fertilizers, which contribute to soil acidification, nutrient leaching, and water pollution, vermi-leachates are biodegradable and environmentally benign (Edwards et al. 2010). Their application helps maintain soil structure, reduces chemical runoff, and preserves aquatic ecosystems, particularly in regions vulnerable to eutrophication (Gajalakshmi and Abbasi 2004).

7.9 Chemistry and Biology of Vermi-leachates

7.9.1 Nutrient Composition

Vermi-leachates are rich in a variety of essential plant nutrients. These include both macronutrients and micronutrients, as well as organic acids and bioactive compounds. They typically contain substantial amounts of nitrogen (N), phosphorus (P), and potassium (K), which are usually found in ammonium and nitrate forms that are readily available in plants. Phosphorus exists in both inorganic and organic forms, playing a key role in root development and energy transfer, whereas potassium aids in stomatal regulation and overall plant vigor (Kumari et al. 2022). Micronutrients such as iron (Fe), manganese (Mn), copper (Cu), boron (B), and zinc (Zn) were also found in varying concentrations. These micronutrients contribute to various enzymatic and metabolic processes in plants. The presence of these elements depends on the source of organic waste used in the vermicomposting process (Jadhav et al. 2024).

7.9.2 Biological Activity

The biological components of vermi-leachates are one of the most valuable attributes. They contain a diverse population of beneficial microorganisms and active enzymes that support plant growth and soil health. Microbial diversity and vermi-leachates are teeming with beneficial bacteria and fungi. Among these are nitrogen-fixing bacteria (e.g., Azotobacter and Rhizobium), which fix atmospheric nitrogen into bioavailable forms. Phosphate-solubilizing bacteria (e.g., *Pseudomonas* and *Bacillus* spp.) make insoluble phosphorus available to plants (Yang et al. 2024). Actinomycetes and fungi such as *Trichoderma* are known for their disease-suppressive properties. They also contain a variety of enzymes, such as dehydrogenase, urease, and phosphatase, which facilitate the breakdown of organic matter into simpler nutrients.

7.9.3 pH, Electrical Conductivity (EC), and Stability

Vermi-leachates usually have a near-neutral pH ranging from 6.5 to 7.5, making them suitable for most crops and beneficial for maintaining the soil pH balance (Pathma and Sakthivel 2012). The EC in vermi-leachates is the moderate range of

0.5–2.5 dS/m, which makes it generally favorable, as it implies the presence of sufficient nutrients without causing salinity stress (Churilova and Midmore 2019). Stability and well-prepared vermi-leachates are microbiologically and chemically stable during short-term storage. However, owing to their biological content, it is advisable to use them fresh or refrigerate them for preservation.

7.9.4 Potential Role in Soil Microbial Health

The biological constituents of vermi-leachates significantly enhance the soil microbial biodiversity and nutrient cycling. When applied to soil, vermi-leachates stimulate native microbial populations by providing nutrients and introducing beneficial microbes. This leads to increased microbial biomass and enzymatic activity and improves soil fertility (Liu et al. 2023). Nutrient cycling and microbes introduced through vermi-leachates enhance the decomposition of organic matter and nutrient mineralization, particularly nitrogen and phosphorus cycling. The symbiotic relationships between introduced microbes and native soil flora create a dynamic, balanced ecosystem that is conducive to sustainable agriculture (Aira et al. 2011). In disease suppression, certain microbes present in vermi-leachates compete with or inhibit plant pathogens, thereby reducing the incidence of soil-borne diseases.

7.10 Vermi-leachates as Biofertilizers in Hydroponic Crop Production

Hydroponics systems have made it possible to produce crops, particularly vegetables, in a nutrient-rich solution (Rajaseger et al. 2023). Interest in the system has been growing since the last decade due to the reduced need for the already scarce arable land, and improved growth, yield, and quality of produce (Maboko et al. 2011). This is because essential nutrients are delivered directly to the roots in the required quantities and at the right time through a controlled aqueous environment. This method also minimizes pest and disease infestations, reduces production costs associated with land preparation, and allows reuse of water and mitigates environmental degradation (de Sousa et al. 2024). Hydroponic systems can either be "open" or "closed" systems (Venter 2017). According to Maboko et al. (2011) and Yuvaraj and Subramanian (2020), hydroponic systems can be categorized into nutrient film technique, deep water culture, ebb and flow, and aeroponics, with each system being best suited to specific crop types and operational condition.

Hydroponics can be used in any environment, but they are particularly suited to areas with no or limited arable land, such as urban and industrial areas (Pomoni et al. 2023) and regions that receive insufficient rainfall to support cropping systems, such as deserts. Although hydroponic systems have many advantages, their main disadvantages include the high costs associated with large-scale setup, highly trained labor, and the reliance on synthetic fertilizers (Velazquez-Gonzalez et al. 2022). Threats to the environment and potential contribution to global warming caused by the high and continuous use of synthetic fertilizers and the high energy demands of hydroponic systems have necessitated efforts to improve efficiency by adopting eco-friendly energy sources and sustainable nutrient inputs. Researchers are turning to the adoption of biofertilizers like vermi-leachates as sustainable nutrient sources in hydroponic crop production systems (Churilova and Midmore 2019; Ortega-Torres et al. 2023).

7.11 Potential of Vermi-leachates in Hydroponics

Vermi-leachate, also referred to as vermiwash, worm tea, or vermiliquer (Ortega-Torres et al. 2023), is a liquid organic fertilizer by-product containing earthworm mucus, nutrients, microorganisms, and plant growth-promoting constituents (Loera-Muro et al. 2021; Sharma et al. 2025). According to Churilova and Midmore (2019), vermi-leachate contains nutrients, microorganisms, and biologically active substances, such as fulvic acids and humic acids. Factors such as earthworm species and feedstock type determine the composition and usefulness of a vermi-leachate.

Various researchers have reported on the agronomic, environmental, and economic potential of vermi-leachates in hydroponic systems (Table 7.2) (Sharma et al. 2025; Churilova and Midmore 2019). The brown liquid fertilizer can be applied as a foliar fertilizer or mixed in hydroponic systems solely or in combination with other synthetic fertilizer sources (Rehman et al. 2024). The humic and fulvic acids found in vermi-leachates promote metabolism and micronutrient uptake (Struyk and Sposito 2001). Additionally, the microbial content in vermi-leachate promotes

Table 7.2 Potential of vermi-leachates in hydroponic systems

Benefit	Mechanism	References
Increased morphometric growth of *M. spicata* (mint), and *R. officinalis* (rosemary)	High amount of plant hormones, such as auxins, gibberellins, and cytokinins from microbial origin, enhance plant growth.	Loera-Muro et al. (2021)
Enhanced plant nutrient uptake	Increased uptake of Fe, Cu, Mn, and Zn in muralis plants grown that can be attributed to the capability of humic substances to form complexes with Fe	Rehman et al. (2024)
Increased disease resistance	Microbes in vermi-leacheate subdue soil-borne pathogens while promoting resistance in plants. Inhibit plant fungal pathogen growth	Gopalakrishnan et al. (2014)
Sustainable fertilizer	Combining with synthetic fertilizers reduces reliance on them and promotes sustainability	Mupambwa et al. (2024)

growth and usefulness of root-associated microbes like zinc-solubilizing bacteria that improve nutrient cycling, suppress pathogens, and promote plant vigor (Karnwal 2021).

7.12 Vermi-leachates as Biofertilizers in Soil-Based Crop Production

(a) *Nutrient-rich composition*
Vermi-leachates, derived from the liquid produced during the vermicomposting process, are rich in essential nutrients that significantly enhance soil fertility and microbial activity. These leachates contain a variety of beneficial components that contribute to improved agricultural practices and sustainable farming. Vermi-leachates contain essential macronutrients (nitrogen, phosphorus, potassium) and micronutrients (iron, manganese, zinc, copper, and boron) that are crucial for plant growth (Kasal et al. 2015; Singh et al. 2024; Manyanga et al. 2024; Tammam et al. 2023). They also include plant growth-regulating hormones, enzymes, and beneficial microorganisms that enhance soil fertility and plant health (Singh et al. 2024; Manyanga et al. 2024; Tammam et al. 2023; Barooah et al. 2024).

(b) *Improved soil health*
Application of vermi-leachates improves soil physical properties such as porosity, aeration, and water retention capacity, which are vital for root development and nutrient uptake (Manyanga et al. 2024; Tammam et al. 2023). They enhance soil microbial activity and enzymatic processes, leading to better nutrient cycling and availability (Manyanga et al. 2024; Dhaliwal et al. 2022; Zhang et al. 2020). Vermi-leachates promote soil aggregation through enhanced microbial activity, particularly from fungi and actinomycetes that bind soil particles together. This improves porosity, root penetration, and resistance to erosion (Gómez-Brandón and Domínguez 2014). The organic acids and humic substances in vermi-leachates enhance the soil's ability to retain moisture, which is particularly beneficial in drought-prone or sandy soils (Nagavallemma et al. 2006). Regular application of vermi-leachates increases labile and hydrolysable organic carbon pools, which feed soil microorganisms and improve nutrient cycling, contributing to more resilient and self-sustaining agroecosystems.

(c) *Enhanced crop yield and quality*
Field experiments have demonstrated that vermi-leachates can significantly increase crop yields. For instance, potato yields were notably higher when treated with vermi-leachates compared to control samples (Kasal et al. 2015). Vermi-leachates have also been shown to improve the nutritional quality of crops, such as increased levels of boron and copper in squash (Malal et al. 2024). Numerous studies have confirmed the positive impact of vermi-leachates on crop yield, nutrient content, and disease resistance in soil-based horticultural

systems. For example, Alabi et al. (2024) reported that tomato plants treated with vermi-leachate showed substantial increases in vegetative growth, fruit size, and yield compared to those treated with conventional fertilizers. Similarly, trials involving Swiss chard and lettuce showed improved germination rates and biomass accumulation under vermi-leachate treatments (Mupambwa et al. 2024). Vermi-leachates not only enhance yield but also improve crop nutritional quality. Their rich nutrient composition, combined with beneficial microbes, supports more efficient nutrient uptake and root development, leading to higher levels of key plant nutrients. In addition, the presence of biocontrol agents in the leachates helps suppress soil-borne pathogens, reducing the incidence of disease and minimizing the need for synthetic pesticides. This contributes to both higher productivity and more sustainable crop protection strategies.

(d) *Disease resistance and stress tolerance*
Vermi-leachates help in enhancing plant resistance against diseases and pests, reducing the need for chemical pesticides (Singh et al. 2024; Mohite et al. 2024). They also improve plant tolerance to abiotic stresses, such as salinity, by bolstering the antioxidative defense system of plants (Manzoor et al. 2024). The beneficial microorganisms present in vermi-leachates can suppress soil-borne pathogens, thereby promoting healthier crops (Mohite et al. 2024). Research indicates that crops treated with vermi-leachate exhibit reduced incidences of pest attacks, contributing to sustainable agricultural practices.

(e) *Applications and advancements*
Vermi-leachates can be used as foliar sprays to directly supply nutrients and growth-promoting substances to plants, leading to improved growth and yield (Kasal et al. 2015; Torres-García et al. 2024). When applied to soil, vermi-leachates enhance soil organic matter content, microbial diversity, and overall soil health, making them a sustainable alternative to chemical fertilizers (Manyanga et al. 2024; Tammam et al. 2023; Zhang et al. 2020). Due to their microbial content, vermi-leachates can act as biopesticides, providing a natural means of pest and disease control (Singh et al. 2024). They offer a sustainable and eco-friendly alternative to synthetic fertilizers in soil-based crop production. Their ability to improve soil health, enhance crop yield and quality, and provide disease resistance makes them a valuable tool in organic and sustainable agriculture practices. Continued research and field trials are essential to optimize their use and fully understand their benefits in various cropping systems.

7.13 Economic Issues of Using Organic Nutrient Sources

The economic viability of using organic nutrient sources, such as vermi-leachates, in farming is a key concern for stakeholders in the agricultural sector. These nutrient sources present significant environmental and soil health benefits, but they also pose challenges in terms of cost, market access, certification, and scalability. A general overview of the implications of adopting organic nutrient sources and the associated economic and financial opportunities and challenges is critical. The economics and financial implications can be analyzed through the lens of a generic agricultural value chain, starting from the backward activities (input production and supply) and forward linkages (production and output analysis). The backward activities of the nutrient value chains include an understanding of the processes and costs associated with producing and supplying the vermi-leachates, while the forward linkages focused on the benefits/costs of using the nutrients from utilization, application, and yields.

7.14 Initial Production Costs of Organic Fertilizers

The production of organic nutrient sources, particularly vermi-leachates, requires upfront investments in infrastructure, raw materials, and labor. For example, the construction of a small-scale smart vermicomposting system using Dal Lake aquatic weed as substrate to produce organic fertilizers was estimated at approximately US$130 for a capacity of 56 kg (Kumar et al. 2023). The first-year benefit–cost ratio was 0.48, indicating a loss. However, in the second year, the ratio improved to 1.78 due to reduced operational costs and an increase in production. These costs vary depending on the type of substrate as well as the environment in which the compost is made (country and geography). Labor-intensive maintenance, including feeding the earthworms, maintaining optimal environmental conditions, and harvesting the leachate, adds to the operational expenses.

It is important to note that the production cost of vermi-leachates is often higher than that of synthetic fertilizers. Although, unlike synthetic options, vermi-leachates contribute to long-term soil health, reducing future input costs. The economic burden of the initial investments is a significant barrier for small- and medium-scale farmers, particularly those in developing regions with limited access to credit or government subsidies (Crowder and Reganold 2015).

7.15 Market Demand and Premium Pricing

A major economic advantage of organic farming is the ability to market produce at premium prices. Organic crops typically command prices that are 20%–100% higher than their conventional counterparts due to their perceived health and environmental benefits (Reganold and Wachter 2016). Since organic products are often perceived as healthier and environmentally friendly, this creates a growing demand among consumers. In Canada, for example, organic foods make up 2.6% of total food sales, with 66% of consumers purchasing organic products weekly. In the United States, the average price premium for organic fruits and vegetables ranges from US$0.30 to $1.20 per 0.5 kg, depending on the crop.

Farmers using vermi-leachates as a biofertilizer can leverage this demand, but this requires proper certification and access to premium markets. While these premium prices often offset higher production costs, fluctuations in consumer demand can pose risks. Economic downturns or shifts in consumer preferences may lead to reduced willingness to pay for organic products, affecting farmer profitability.

7.16 Yield Considerations and Economic Returns

Organic farming systems, including those relying on vermi-leachates, often achieve lower yields compared to conventional systems. On average, organic yields are 80%–90% of conventional yields, with significant variability depending on the crop, climate, and management practices (Seufert et al. 2012). However, studies have shown that organic systems can achieve higher average profitability due to reduced reliance on synthetic inputs and the ability to market produce at premium prices. For instance, Pant et al. (2009) demonstrated that crops treated with vermi-leachates experienced improved resistance to stress and diseases, leading to better-quality produce. These benefits can partially compensate for yield reductions, especially for high-value horticultural crops such as strawberries, lettuce, and tomatoes.

7.17 Certification and Compliance Costs

Certification is a critical aspect of marketing organic produce. In the United States, organic certification through agencies such as the USDA involves an annual fee ranging from US$600 to $1500, depending on the size and scale of the farm. These costs are a major constraint for small-scale farmers who already face high production costs. Additionally, farmers must comply with rigorous record-keeping and periodic inspections to maintain certification. While these processes are essential for ensuring product authenticity, they add to the overall cost of production. Government programs such as the USDA's Organic Certification Cost Share Program provide partial reimbursement to help mitigate these expenses, but access to such programs is not universal (USDA 2023). Although the certification requirements are good for the farming business and global traceability requirements of agricultural products, this process involves compliance with stringent standards, documentation, and periodic inspections, all of which entail costs. These certification costs are prohibitive for small-scale farmers, particularly in developing countries where institutional support including subsidies are limited.

7.18 Economic Benefits, Risks, and Market Uncertainty

Adopting organic nutrient sources such as vermi-leachates introduces several economic risks such as managing market volatility and climate variability.

- *Market volatility*: Consumer demand for organic produce is sensitive to economic conditions. During financial downturns, consumers may prioritize affordability over sustainability, leading to reduced demand for organic products.
- *Climate variability*: Organic farming systems, which depend heavily on natural inputs and processes, are more susceptible to climatic variability. Unpredictable weather patterns can disrupt nutrient cycling and reduce crop yields, further impacting profitability.

While these risks are not unique to organic systems, they underscore the importance of diversification and risk management strategies for farmers. Despite the challenges, organic nutrient sources offer substantial long-term economic benefits:

- *Reduced dependency on synthetic inputs*: Vermicomposting ensures that farmers do not heavily depend on synthetic inputs. By improving soil fertility and health, vermi-leachates reduce the need for synthetic fertilizers and pesticides. This lowers input costs as well as minimizing environmental remediation expenses.
- *Enhanced soil productivity*: Vermi-leachates contribute to sustained soil fertility by increasing organic matter content, microbial activity, and nutrient availability. Over time, this leads to higher productivity and reduced input requirements.
- *Environmental cost savings*: Organic farming practices mitigate soil degradation, water pollution, and greenhouse gas emissions, reducing the environmental costs associated with conventional farming.

A study by Crowder and Reganold (2015) found that organic farming systems achieved higher profitability globally, even with yield reductions, due to premium prices and lower environmental costs. Table 7.3 compares the costs and benefits of vermi-leachates with synthetic fertilizers.

The economic issues of using organic nutrient sources such as vermi-leachates are multifaceted. While the initial costs and certification requirements pose challenges, the long-term benefits, including reduced input costs, improved soil health, and access to premium markets, make them a viable option for sustainable farming.

Table 7.3 Economic comparison of costs and benefits of vermi-leachates with synthetic fertilizers

Factor	Vermi-leachates	Synthetic fertilizers
Initial investment	High (e.g., $130 setup cost)	Moderate (varies by type)
Long-term input costs	Low	High (due to continuous use)
Environmental impact	Low	High (soil and water pollution)
Certification requirements	Yes	No
Market access	Premium prices	Standard market rates

Strategic investments in policy support, education, and research can help overcome these challenges, ensuring the economic feasibility of organic nutrient sources.

To address the economic challenges of using organic nutrient sources, the following policy measures can be considered:

- Provide financial incentives to offset the initial costs of adopting organic practices
- Establish microfinance programs to support small-scale farmers transitioning to organic systems
- Invest in research to optimize vermi-leachate production and application methods
- Promote consumer awareness about the benefits of organic products to sustain demand

7.19 Conclusions

The mounting pressures of climate change, population growth, and environmental degradation necessitate a transformative shift in how soil fertility is managed within agricultural systems. Vermi-leachates represent a compelling alternative to synthetic fertilizers. They are rich in essential nutrients, plant growth hormones, beneficial microbes, and enzymatic activity, components that synergistically support plant growth, enhance soil health, and reduce the ecological footprint of farming practices. The unique ability of vermi-leachates to improve soil microbial biodiversity, suppress plant diseases, and increase crop resilience makes them particularly valuable in horticultural systems. This chapter has demonstrated that vermi-leachates offer multifunctional benefits: agronomic, environmental, and economic. Agronomically, they enhance plant productivity and quality; environmentally, they promote sustainable nutrient cycling and reduce chemical pollution; and economically, they provide access to premium markets for organically grown produce, though with challenges in production cost and certification. While barriers such as high initial investment and limited scalability exist, policy interventions, research advancements, and farmer training can significantly bolster their adoption. Ultimately, the integration of vermi-leachates into mainstream agricultural systems aligns with global efforts to promote regenerative, low-input, and climate-smart farming. Their application should be further explored, refined, and scaled up to support more resilient food systems capable of feeding a growing population while preserving ecological integrity.

References

Aira M, Monroy F, Domínguez J (2011) Earthworms stimulate plant growth: a soil-mediated mechanism. Appl Soil Ecol 50:57–62

Aremu AO, Stirk WA, Kulkarni MG, Tarkowská D, Turečková V, Grúz J, Šubrtová M, Pěnčík A, Novák O, Doležal K, Strnad M, van Staden J (2015) Evidence of phytohormones and phenolic acids variability in garden-waste-derived vermicompost leachate, a well-known plant growth stimulant. Plant Growth Regul 75(2):483–492. https://doi.org/10.1007/s10725-014-0011-0

Assefa S, Tadesse S (2019) The principal role of organic fertilizer on soil properties and agricultural productivity—a review. Agric Res Technol Open Access J 22(2):1–5. https://doi.org/10.1980/ARTOAJ

Aytenew M, Bore G (2020) Effects of organic amendments on soil fertility and environmental quality: a review. J Plant Sci 8:112

Alabi OD, Arthur GD, Coopoosamy RM, Naidoo KK, Mbanjwa ST, Tshapha LP (2024) Vermicompost Leachate a Viable Bio-stimulant for Tomato Growth and Yield (Solanum lycopersicum). Indian Journal of Agricultural Research. 58(1):139–144. https://doi.org/10.18805/IJARe.AF-750.

Badagliacca G, Testa G, Giovanni S (2024) Organic fertilizers and bio-waste for sustainable soil management to support crops and control greenhouse gas emissions in Mediterranean agroecosystems: a review. Horticulturae 10:427. https://doi.org/10.3390/horticulturae1005042

Barooah A, Sarma I, Mahanta S (2024) Biofertilizer and vermi-compost: sources of nutrient management in organic agriculture. In: Advances in organic farming crop production and management. Apple Academic Press

Bhattacharyya P, Goswami M, Bhattacharyya L (2016) Perspective of beneficial microbes in agriculture under changing climatic scenario: a review. J Phytol 8:26

Churilova E, Midmore D (2019) Vermiliquer (vermicompost leachate) as a complete liquid fertilizer for hydroponically-grown pak choi (*Brassica chinensis* L.) in the tropics. Horticulturae 5:26

Crowder DW, Reganold JP (2015) Financial competitiveness of organic agriculture on a global scale. Proc Natl Acad Sci 112(24):7611–7616. https://doi.org/10.1073/pnas.1423674112

de Sousa R, Bragança L, da Silva MV, Oliveira RS (2024) Challenges and solutions for sustainable food systems: the potential of home hydroponics. Sustainability 16(2):817. https://doi.org/10.3390/su16020817

Deokar A, Jadhar M, Fernades A (2024) A study on understanding organic farming practices for sustainable and resilient agricultural development. IJARIIE 10(5):904–910

Dhaliwal SS, Sharma V, Sharma S, Walia SS, Singh J, Verma V, Rani N (2022) Effect of vermicompost, farm yard manure and biofertilizers on soil health and plant nutrition. In: Earthworms and their ecological significance. Nova Publishers

Edwards CA, Arancon NQ, Sherman R (2010) Vermiculture technology: earthworms, organic wastes, and environmental management. CRC Press

Essel D, Abaidoo R, Opoku A, Ewusi-Mensah N (2023) Mechanisms for nutrient interactions from organic amendments and mineral fertilizer inputs under cropping systems: review. PeerJ 11:e15135

Gajalakshmi S, Abbasi SA (2004) Earthworms and vermicomposting. Indian J Biotechnol 3(3):486–494

Gamage A, Gangahagedara R, Gamage J, Jayasinghe N, Kodikara N, Suraweera P, Merah O (2023) The role of organic farming for achieving sustainability in agriculture. Farm Syst 1:100005

Gopalakrishnan S, Vadlamudi S, Bandikinda P, Sathya A, Vijayabharathi R, Rupela O & Varshney RK (2014) Evaluation of Streptomyces strains isolated from herbal vermicompost for their plant growth-promotion traits in rice. Microbiological Research, 169(1):40–48

Gómez-Brandón M, & Domínguez J (2014) Recycling of Solid Organic Wastes Through Vermicomposting: Microbial Community Changes Throughout the Process and Use of Vermicompost as a Soil Amendment. Critical Reviews in Environmental Science and Technology, 44(12):1289–1312. https://doi.org/10.1080/10643389.2013.763588

Hasanuzzama M, Hossen S, Masud A, Moumita, Bhuyan M, Nahar K, Mahmud J (2018) Potassium: a vital regulator of plant responses and tolerance to abiotic stresses. Agronomy 8:31

Hassan A, Tabassum B, Hashim M, Khan N (2023) Role of plant growth promoting rhizobacteria (PGPR) as a plant growth enhancer for sustainable agriculture: a review. Bacteria 3:59–75. https://doi.org/10.3390/bacteria3020005

Hamdan MF, Mohd Noor SN, Abd-Aziz N, Pua T-L, & Tan BC (2022) Green Revolution to Gene Revolution: Technological Advances in Agriculture to Feed the World. Plants, 11(10):1297

Jadhav A, Gosavi A, Jadhav S, Patili A (2024) Influence of different substrates on nutrient composition of vermicompost and vermiwash. Int J Plant Soil Sci 36(6):340–352

Jiang X, Lu C, Hu R, Shi W, Zhou L, Wen P, Jiang Y, Lo YM (2023) Nutritional and microbiological effects of vermicompost tea in hydroponic cultivation of maple peas (*Pisum sativum* var. *arvense* L.). Food Sci Nutr 11(6):3184–3202. https://doi.org/10.1002/fsn3.3299. PMID: 37324884; PMCID: PMC10261758

Karnwal A (2021) Pseudomonas spp., a zinc-solubilizing vermicompost bacteria with plant growth-promoting activity moderates zinc biofortification in tomato. Int J Veg Sci 27(4):398–412

Kasal P, Červená K, Lyčková B, Kučerová R (2015) Testing of a fertilizing plant stimulator vermesfluid in potatoes. Inz Miner 16(1):57–60

Koskey G, Avio L, Turrini A, Sbrana C, Bàrberi P (2022) Biostimulatory effect of vermicompost extract enhances soil mycorrhizal activity and selectively improves crop productivity. Plant Soil 484:183

Kumar A, Muzamil M, Rashid S, Banday RUZ, Rasool S (2023) Qualitative evaluation and economic assessment of Dal-Lake aquatic weed vermicompost prepared in conventional vermi bed and smart vermicomposting bin. Waste Biomass Valoriz 15:3187–3197. https://doi.org/10.1007/s12649-023-02335-w

Kumari VV, Banerjee P, Verma VC, Sukumaran S, Chandran MAS, Gopinath KA, Venkatesh G, Yadav SK, Singh VK, Awasthi NK (2022) Plant nutrition: an effective way to alleviate abiotic stress in agricultural crops. Int J Mol Sci 23:8519. https://doi.org/10.3390/ijms23158519

Kumari E, Kumari S, Das SS, Mahapatra M, Sahoo JP (2023) Plant growth-promoting bacteria (PGPB) for sustainable agriculture: current prospective and future challenges. Agro-Environ Sustain 1(3):274–285. https://doi.org/10.59983/s2023010309

Liu W, Yang Z, Ye Q, Peng Z, Zhu S, Chen H, Liu D, Li Y, Deng L, Shu X (2023) Positive effects of organic amendments on soil microbes and their functionality in agro-ecosystems. Plants 12:3790

Loera-Muro A, Troyo-Diéguez E, Murillo-Amador B, Barraza A, Caamal-Chan G, Lucero-Vega G, Nieto-Garibay A (2021) Effects of vermicompost leachate versus inorganic fertilizer on morphology and microbial traits in the early development growth stage in mint (*Mentha spicata* L.) and rosemary (*Rosmarinus officinalis* L.) plants under closed hydroponic system. Horticulturae 7:100. https://doi.org/10.3390/horticulturae7050100

Lori M, Symnaczik S, Mäder P, De Deyn G, Gattinger A (2017) Organic farming enhances soil microbial abundance and activity—a meta-analysis and meta-regression. PLoS One 12:e0180442. https://doi.org/10.1371/journal.pone.0180442

Maboko MM, Du Plooy CP, Bertling I (2011) Comparative performance of tomato cultivars cultivated in two hydroponic production systems. S Afr J Plant Soil 28(2):97–102

Madjar RM, Vasile Scăeteanu G, Sandu MA (2024) Nutrient water pollution from unsustainable patterns of agricultural systems, effects and measures of integrated farming. Water 16:3146. https://doi.org/10.3390/w16213146

Malal H, Romero VS, Horwath WR, Dore S, Beckett P, Ait Hamza M, Lakhtar H, Lazcano C (2024) Vermifiltration and sustainable agriculture: unveiling the soil health-boosting potential of liquid waste vermicompost. Front Sustain Food Syst 8:1383715. https://doi.org/10.3389/fsufs.2024.1383715

Manyanga MA, Marumure J, Chigede N, Mubvuma M, Mudzengi CP, Nyambiya I, Muteveri M (2024) Vermicomposting for improved soil health: prospects for degraded soils. In: Marginal soils of Africa rethinking uses management and reclamation. https://doi.org/10.1007/978-3-031-55185-7_17

Manzoor A, Naveed MS, Ali RMA, Naseer MA, UL-Hussan M, Saqib M, Hussain S, Farooq M (2024) Vermicompost: a potential organic fertilizer for sustainable vegetable cultivation. Sci Hortic 336:113443. https://doi.org/10.1016/j.scienta.2024.113443

Mnkeni PNS, Mupambwa HA (2023) A decade of vermicomposting research at the University of Fort Hare: selected insights. In: Mupambwa HA, Horn LN, Mnkeni PNS (eds) Vermicomposting for sustainable food systems in Africa. Sustainability sciences in Asia and Africa. Springer, Singapore. https://doi.org/10.1007/978-981-19-8080-0_1

Mohite DD, Chavan SS, Jadhav VS, Kanase T, Kadam MA, Singh AS (2024) Vermicomposting: a holistic approach for sustainable crop production, nutrient-rich bio fertilizer, and environmental restoration. Discov Sustain 5(1):60. https://doi.org/10.1007/s43621-024-00245-y

Mupambwa HA, Mnkeni PNS (2018) Optimizing the vermicomposting of organic waste materials for production of nutrient rich organic fertilizers: a review. Environ Sci Pollut Res 25:10577–10595. https://doi.org/10.1007/s11356-018-1328-4

Mupambwa HA, Nciizah AD, Nyambo P (2022) Can organic soil fertility management sustain farming and increase food security among African smallholder farmers? In: Mupambwa HA, Nciizah AD, Nyambo P, Muchara B, Gabriel NN (eds) Food security for African smallholder farmers. Sustainability sciences in Asia and Africa. Springer, Singapore. https://doi.org/10.1007/978-981-16-6771-8_6

Mupambwa HA, Nyambe GI, Matomola B, Handura B, Howoses VA, Hausiku-Ikechukwu MK, Gawanab W (2024) Goat and sheep manure based vermicompost and vermi-leachate as a nutrient source in drip hydroponic tomatoes (*Solanum lycopersicum*) and Swiss chard (*Beta vulgaris*). J Sustain Agric Environ 3(4):e70011. https://doi.org/10.1002/sae2.70011

Nagavallemma KP, Wani SP, Stephane L, Padmaja VV, Vineela C, Babu RM and Sahrawat KL, (2006) Vermicomposting: Recycling wastes into valuable organic fertilizer. SAT eJournal. 2(1):1–16

Ortega-Torres AE, Herrera-Matallana TS, Rico-García E (2023) Vermiliquer as a biostimulant and antioxidant in hydroponic lettuce (*Lactuca sativa*) production. Hortic Sci 50(1):25

Oyege I, Balaji Bhaskar MS (2023) Effects of vermicompost on soil and plant health and promoting sustainable agriculture. Soil Syst 7(4):101. https://doi.org/10.3390/soilsystems7040101

Pant AP, Radovich TJK, Hue NV, Talcott ST (2009) Vermicompost extracts influence growth, mineral nutrients, phytonutrients, and antioxidant activity in pak choi (*Brassica rapa* Chinensis group). HortScience 44(4):1034–1039

Pathma J, Sakthivel N (2012) Microbial diversity of vermicompost bacteria that exhibit useful agricultural traits and waste management potential. Springerplus 1(1):26

Pomoni DI, Koukou MK, Vrachopoulos MG, Vasiliadis L (2023) A review of hydroponics and conventional agriculture based on energy and water consumption, environmental impact, and land use. Energies 16:1690. https://doi.org/10.3390/en16041690

Rajaseger G, Chan KL, Yee Tan K, Ramasamy S, Khin MC, Amaladoss A, Kadamb Haribhai P (2023) Hydroponics: current trends in sustainable crop production. Bioinformation 19(9):925–938. https://doi.org/10.6026/97320630019925. PMID: 37928497; PMCID: PMC10625363

Reganold JP, Wachter JM (2016) Organic agriculture in the twenty-first century. Nat Plants 2(2):1–8. https://doi.org/10.1038/nplants.2015.221

Rehman Su, Aprile A, De Castro F, Negro C, Migoni D, Benedetti M, Sabella E, Fanizzi FP (2024) Assessing the effectiveness of vermi-liquids as a sustainable alternative to inorganic nutrient solutions in hydroponic agriculture: a study on *Diplotaxis muralis*. Agronomy 14:1310. https://doi.org/10.3390/agronomy14061310

Sande TJ, Tindwa HJ, Alovisi AMT, Shitindi MJ, Semoka JM (2024) Enhancing sustainable crop production through integrated nutrient management: a focus on vermicompost, bio-enriched rock phosphate, and inorganic fertilisers—a systematic review. Front Agron 6:1422876. https://doi.org/10.3389/fagro.2024.1422876

Seufert V, Ramankutty N, Foley JA (2012) Comparing the yields of organic and conventional agriculture. Nature 485(7397):229–232. https://doi.org/10.1038/nature11069

Sharma BB, Kodandoor SC, Sridhar KR (2025) Insights into the leachates of vermicompost and millipede compost. World News Nat Sci 60:368–337

Siddika A, Rabbani A, Parveen Z, Hossain MD (2023) Effects of biochar and modified biochar on chromium contaminated soil properties. J Sci Eng (AJSE) 22:82–93

Singh R, Gupta RK, Patil RT, Sharma RR, Asrey R, Kumar A, Jangra KK (2010) Sequential foliar application of vermicompost leachates improves marketable fruit yield and quality of strawberry (*Fragaria×ananassa* Duch.). Sci Hortic 124(1):34–39. https://doi.org/10.1016/j.scienta.2009.12.002

Singh T, Kapila R, Kumar M (2024) A minireview on vermicompost and vermiwash as green pesticide for sustainable crop production: approaches, applications, and advancements. Jsfa Rep 4(1):4–10. https://doi.org/10.1002/jsf2.172

Struyk Z, Sposito G (2001) Redox properties of standard humic acids. Geoderma 102:329–346

Tammam AA, Rabei Abdel Moez Shehata M, Pessarakli M, El-Aggan WH (2023) Vermicompost and its role in alleviation of salt stress in plants—I. Impact of vermicompost on growth and nutrient uptake of salt-stressed plants. J Plant Nutr 46(7):1446–1457. https://doi.org/10.1080/01904167.2022.2072741

Torres-García A, Héctor-Ardisana EF, León-Aguilar R, Zambrano-Gavilanes FE, Fosado-Téllez OA (2024) Vermicompost leachate-based biostimulant and its effects on physiological variables and yield of different crops in Manabí, Ecuador. Cienc Tecnol Agropecu 25(1). https://doi.org/10.21930/rcta.vol25_num1_art:3388

United Nations Department of Economic and Social Affairs (2019) Growing at a slower pace, world population is expected to reach 9.7 billion in 2050 and could peak at nearly 11 billion around 2100. UN Sustainable Development. https://www.un.org/en/desa/growing-slower-pace-world-population-expected-reach-97-billion-2050-and-could. Accessed 9 July 2025

USDA (2023) Organic certification cost share programs. United States Department of Agriculture. Retrieved from https://www.usda.gov/organic

Velazquez-Gonzalez RS, Garcia-Garcia AL, Ventura-Zapata E, Barceinas-Sanchez JDO, Sosa-Savedra JC (2022) A review on hydroponics and the technologies associated for medium- and small-scale operations. Agriculture 12:646. https://doi.org/10.3390/agriculture12050646

Venter G (2017) Different types of hydroponic systems: farming for tomorrow. Farmer's Wkly 2017(17009):26–27

Vermicompost Leachate a Viable Bio-stimulant for Tomato Growth and Yield (Solanum lycopersicum) O.D. Alabi1, G.D. Arthur1, R.M. Coopoosamy1, K.K. Naidoo1 S.T.Mbanjwa1, L.P.Tshapha1

Withers PJA, Sylvester-Bradley R, Jones DL, Healey JR, Talboys PJ (2014) Feed the crop not the soil: rethinking phosphorus management in the Food Chain. Environ Sci Technol 48(12):6523–6530. https://doi.org/10.1021/es501670j

Wongkiew S, Polprasert C, Noophan P(L), Koottatep T, Kanokkantapong V, Surendra KC, Khanal SK (2023) Effects of vermicompost leachate on nitrogen, phosphorus, and microbiome in a food waste bioponic system. J Environ Manage 339:117860. https://doi.org/10.1016/j.jenvman.2023.117860

Yang P, Condrich A, Scranton S, Hebner C, Lu L, Ali MA (2024) Utilizing plant growth promoting rhizobacteria (PGPR) to advance sustainable agriculture. Bacteria 3:434–451. https://doi.org/10.3390/bacteria3040030

Yuvaraj M, Subramanian KS (2020) Different types of hydroponics system. Biotica Res Today 2(8):835–837

Zhang Z-J, Wang X-Z, Wang H, Huang E, Sheng J-L, Zhou L-Q, Jin W-Z (2020) Housefly larvae (*Musca domestica*) vermicompost on soil biochemical features for a chrysanthemum (*Chrysanthemum morifolium*) farm. Commun Soil Sci Plant Anal 51(10):1315–1330. https://doi.org/10.1080/00103624.2020.1763389

Chapter 8
Rethinking Biological Nitrogen Fixation (BNF) and Soil Fertility for Sustainable and Equitable Practices in Smallholder Farming Systems

Tirivashe Phillip Masere, Rangarirai Lucia Mhindu, Tavengwa Chitata, Cosmas Parwada, and Ronald Mandumbu

Abstract Biological nitrogen fixation (BNF) is vital for enhancing soil fertility, particularly in smallholder farming areas with limited access to synthetic fertilizers. Despite this fact, the adoption of BNF is still very low. This chapter sought to explore the role of BNF in improving soil fertility among smallholder farmers and its adoption. We argue that the current understanding is devoid of the farmers' practices for enhancing BNF, and that both field and controlled experiments may inflate the performance of BNF by negating farmer typologies and practices. In addition, current literature is focused on the economic returns from BNF and maintaining soil fertility, and this is limited in driving the sustainability of BNF. There is also limited knowledge of factors, values, and logics—like generational wealth, heritage, and care for the environment—which can spiral and/or curtail the adoption of BNF practices. We base our arguments on a review of 85 peer-reviewed journal articles from four bibliographic databases—Scopus, Web of Science, JSTOR, and Science Direct; gray literature from Zimbabwe's primary public extension agency, AGRITEX; and

T. P. Masere (✉) · R. L. Mhindu (✉)
Department of Land and Water Resources Management, Faculty of Agriculture, Environment and Natural Resources Management, Midlands State University, Gweru, Zimbabwe
e-mail: maseretp@staff.msu.ac.zw

T. Chitata
Department of Water Governance, IHE Delft, Institute for Water Education, Delft, Netherlands

C. Parwada
Department of Agronomy and Horticulture, Faculty of Agriculture, Environment and Natural Resources Management, Midlands State University, Gweru, Zimbabwe

R. Mandumbu
Department of Crop Science, Faculty of Agriculture and Environmental Science, Bindura University of Science Education, Bindura, Zimbabwe

A. D. Nciizah et al. (eds.), *Biofertilizers for Sustainable Agriculture in Africa*, Sustainability Sciences in Asia and Africa, https://doi.org/10.1007/978-981-95-6009-7_8

limited interviews with smallholder farmers. The journal articles were analyzed for trends and themes using NVIVO. The analysis showed yield, soil health, and autonomy benefits for smallholder farmers when using BNF. However, the reported benefits may be comparatively inflated because most of the research work uses the worst farmer practice of monocropping with no fertilizer as the control. Yet, there are other farmer practices to which BNF can be compared. We propose a rethinking of BNF not as a purely field-localized practice but as a cycle mediated by farmers' practices and experimentation with leguminous residues, animal manure, and everything in between for better integration of BNF into smallholder farming systems—dryland and irrigated—for sustainable and equitable soil fertility management.

Keywords Economic returns · Farmer typology · Fertility management · Legumes · Soil health · Sustainable

8.1 Introduction

Biological nitrogen fixation (BNF) is a natural process through which certain bacteria, known as diazotrophs, convert atmospheric nitrogen (N_2) into ammonia (NH_3), making it available for plant uptake. This process is critical for maintaining soil fertility, as nitrogen is an essential nutrient for plant growth, but is often limited in soils (Grandy et al. 2022). Symbiotic BNF, which occurs in leguminous plants in partnership with nitrogen-fixing bacteria like rhizobia, is particularly significant in agricultural systems (Soumare et al. 2020). By reducing the need for synthetic fertilizers, BNF offers a sustainable alternative for improving soil nitrogen levels, fostering soil health, and promoting crop productivity (Kuyah et al. 2021). In agroecosystems, the incorporation of leguminous plants that fix nitrogen enhances nutrient cycling, biodiversity, soil structure, and moisture retention, contributing to more resilient farming systems (Veste et al. 2024).

Soil fertility is a major concern in smallholder farming systems, especially in Sub-Saharan Africa, where many soils are inherently low in nitrogen and phosphorus (Falconnier et al. 2023). In these regions, smallholder farmers typically have limited access to modern technologies, farm machinery, and agricultural inputs, leading to low use of fertilizers and subsequent soil degradation (Giller et al. 2021; Masere 2022). As a result, these farmers often face challenges such as nutrient mining, which depletes the soil of essential nutrients (Nezomba et al. 2015). This situation is exacerbated by poor soil management practices and the increasing impacts of climate change, such as erratic rainfall patterns and prolonged droughts, which further reduce soil productivity (Mupambwa et al. 2022). In this context, sustainable soil fertility management practices, such as BNF, are critical for improving soil quality, boosting crop yields, and enhancing food security (Tang 2025).

Smallholder farmers are key contributors to global food production, particularly in developing regions where they represent a large portion of the agricultural workforce. Despite cultivating less than 2 hectares of land on average, smallholders play a crucial role in food security (Lowder et al. 2019; Masere 2022). However, these farmers face significant challenges in accessing financial resources, modern farming technologies, and irrigation infrastructure, which limits their ability to adopt practices that could improve soil fertility (Muoni et al. 2022). The situation is particularly dire in semi-arid and arid areas, where water scarcity further hampers agricultural productivity (OECD/FAO 2020). In Zimbabwe, for instance, smallholder farmers predominantly grow subsistence crops such as maize, groundnuts, and sorghum in areas with less than 750 mm of annual rainfall, where soil fertility is already compromised by nutrient mining and limited manure availability (Manatsa et al. 2020).

Furthermore, smallholder farming in these regions often integrates livestock, which can contribute to nutrient cycling using manure as fertilizer. However, limited livestock herds and low-quality manure restrict the effectiveness of this practice (Tavirimirwa et al. 2013; Masere 2022). Understanding and integrating practices like BNF into smallholder farming systems is essential for enhancing soil fertility and improving long-term productivity in these resource-constrained settings (Tittonell and Giller 2021).

This chapter explores the role of BNF in improving soil fertility within smallholder farming systems, with a particular focus on Sub-Saharan Africa. This chapter will discuss the potential benefits of BNF in addressing nitrogen limitations and enhancing soil health. It will also address the current understanding of BNF and its limitations including challenges to the widespread adoption of BNF among smallholder farmers, including the socioeconomic, cultural, and technical factors that influence uptake. By examining current research and the practical experiences of smallholder farmers, this chapter seeks to offer insights into how BNF can be more effectively integrated into sustainable farming practices. Additionally, this chapter will highlight the need for a more holistic understanding of BNF, one that considers the broader agricultural system, farmer practices, and sociocultural dynamics to ensure its long-term success and adoption.

8.2 Methodology

This chapter is based on a comprehensive review of existing literature, coupled with insights from gray literature and interviews with smallholder farmers. The methodology combines both qualitative and quantitative approaches to offer a robust understanding of the role of BNF in improving soil fertility within smallholder farming systems, specifically in Sub-Saharan Africa. The approach ensures that both academic perspectives and on-the-ground experiences are captured to explore the effectiveness, challenges, and potential for integrating BNF into sustainable farming practices.

A central component of this study was a thorough review of relevant literature. A systematic search was conducted across four major bibliographic databases: Scopus, Web of Science, JSTOR, and Science Direct. These databases were selected due to their extensive collection of high-quality academic research in agricultural sciences, soil fertility, and nitrogen fixation. The review included 85 peer-reviewed articles published between 2020 and 2025. The selected articles were primarily focused on the following topics: the role of BNF in soil fertility enhancement and sustainable agricultural practices; the relationship between BNF and smallholder farming productivity in Sub-Saharan Africa; and socioeconomic, technical, and environmental factors influencing the adoption of BNF in smallholder systems. The selection criteria ensured a balance between global studies and those focused on Sub-Saharan Africa, with particular attention to Zimbabwe's smallholder agricultural systems.

Gray literature from Sub-Saharan Africa's extension agencies were *used* to complement the peer-reviewed literature. This included reports, policy briefs, and unpublished studies that provided insights into local agricultural practices, extension efforts, and government initiatives aimed at promoting soil fertility and BNF adoption among smallholder farmers. These materials were particularly valuable in understanding the practical challenges faced by farmers and the gaps in knowledge that may hinder the widespread adoption of BNF practices.

To gain deeper insights into the practical realities of BNF adoption, a limited set of semi-structured interviews were conducted with 15 smallholder farmers in Zimbabwe. The 15 farmers were selected based on their involvement in smallholder farming, particularly those growing legumes or practicing crop rotations, which are associated with BNF. Participants were chosen from various agroecological zones in Zimbabwe to capture the diversity of farming systems across different environmental and socioeconomic conditions. A mix of genders, age groups, and farming practices was considered in selecting the 15 farmers. The main theme of questioning included knowledge and awareness of BNF and its benefits for soil fertility; perceptions of the effectiveness and challenges of incorporating BNF into their farming systems; socioeconomic, cultural, and technical barriers to the adoption of BNF; their experiences with leguminous crops and crop rotation practices and their interaction with agricultural extension services and access to support for BNF-based practices. The interviews were conducted in the local languages of the farmers and were transcribed verbatim for analysis.

Literature review and interview data were analyzed using thematic analysis, which involved identifying, analyzing, and reporting patterns (themes) within the data (Maguire and Delahunt 2017). NVivo software was used to assist in the coding and categorization of key themes related to the benefits of BNF, challenges to adoption, and sociocultural factors. This approach allowed for the synthesis of trends and gaps in the current understanding of BNF's role in smallholder systems.

8.3 Role of Biological Nitrogen Fixation in Improving Soil Fertility Among Smallholder Farmers of Sub-Saharan Africa

8.3.1 Biological Nitrogen Fixation: Key Processes and Influencing Factors

BNF is a natural process facilitated by diazotrophs—bacteria that convert atmospheric nitrogen into ammonia, making it available to plants. This conversion occurs via the enzyme nitrogenase, which transforms dinitrogen gas into ammonia and subsequently into nitrogen forms usable by plants and soil. BNF can occur in three primary forms (Table 8.1).

BNF plays a crucial role in nitrogen cycling within both natural and agroecosystems. Although associative and free-living nitrogen fixers contribute less inorganic nitrogen to soils, they provide essential nitrogen sources for natural vegetation, particularly in degraded or nutrient-poor soils (Guo et al. 2023). The nitrogen captured through BNF and incorporated into biomass enriches soils upon decomposition, enhancing plant nitrogen availability and promoting soil and water conservation.

The agricultural significance of BNF is immense; it lessens reliance on synthetic fertilizers, which can degrade soil quality and pollute water sources. Incorporating legumes into cropping systems not only boosts soil fertility but also enhances biodiversity, soil structure, moisture retention, and overall ecosystem health (Kuyah et al. 2021; Veste et al. 2024). For example, Xavier et al. (2023) highlight that common pulses support diverse microbial communities in their rhizosphere that promote plant growth, improve nutrient availability, and increase resilience to environmental stressors. The absence of legumes and their BNF capabilities hampers soil organic matter accumulation and sustainable agricultural practices.

Table 8.1 Forms of BNF, key organisms, and types of interactions

Form of BNF	Description	Key organisms	Type of interaction
Symbiotic	Bacteria like rhizobia and brady-rhizobia convert atmospheric nitrogen into ammonia in exchange for carbohydrates from leguminous plants (Guo et al. 2023)	Rhizobia, Brady-rhizobia, leguminous plants	Mutualistic (carbohydrates for nitrogen)
Free-living	Bacteria fix nitrogen independently, without reliance on plants (Guo et al. 2023)	*Bacillus*, *Azotobacter*, *Klebsiella*, *Clostridium*	Independent
Associative	Bacteria fix nitrogen in the rhizosphere of grasses, benefiting the plants (Rascio and La Rocca 2013; Soumare et al. 2020)	*Azospirillum*, grasses	Beneficial (in the rhizosphere)

Smallholder farmers can leverage BNF by incorporating leguminous crops into their farming systems. These crops form symbiotic relationships with nitrogen-fixing bacteria, significantly enriching the soil with nitrogen. By rotating crops with legumes or integrating them into mixed farming systems, smallholder farmers can naturally improve soil fertility, increase yields, and reduce dependence on synthetic fertilizers, which are often costly and environmentally harmful. In regions with poor soil conditions, such as in semi-arid areas where many smallholder farmers operate, promoting practices like crop rotation, reduced tillage, and organic farming can improve the effectiveness of BNF. These practices help smallholder farmers adapt to climate change and variability while ensuring that their soils remain fertile and productive.

The efficiency of BNF is influenced by various biotic, abiotic, and climatic factors, such as soil type, pH, temperature, and the availability of essential nutrients (Abdullah 2002; Mohammadi 2012; Hu et al. 2024). In intensively farmed ecosystems reliant on chemical fertilizers, the natural BNF process often declines. Practices like crop rotation with legumes, reduced tillage, and organic farming can enhance BNF, promoting sustainable agricultural systems aligned with ecological principles. Overall, BNF is a fundamental process that supports plant growth and ecological balance, making its integration into agricultural practices essential for promoting sustainable farming, improving soil health, and mitigating the environmental impacts of chemical fertilizers.

BNF promotes sustainable agricultural practices by reducing the environmental impact associated with synthetic fertilizers, such as nitrate contamination, soil acidification, and greenhouse gas emissions. By enhancing soil nitrogen levels naturally, BNF helps mitigate soil degradation and supports better moisture retention, soil structure, and organic matter build-up, which are essential for long-term soil health.

The adoption of BNF-based practices, such as intercropping legumes or using legume-based cover crops, can be an effective way to improve soil fertility without incurring high costs (Ncube et al. 2009). While some smallholder farmers may already use manure or crop residues, the integration of legumes for BNF offers a more sustainable and low-cost solution. However, for widespread adoption, it is important to address factors such as the availability of quality legume seeds, knowledge about proper crop rotations, and overcoming any socioeconomic barriers that limit access to such technologies.

8.3.2 Challenges of Smallholder Farmers and Soil Fertility in Sub-Saharan Africa

Smallholder farmers, typically cultivating less than 2 hectares of land, play a critical role in global food production, constituting approximately 84% of all farms and contributing about 35% of food needs (Lowder et al. 2019). Their significance extends to the pursuit of sustainable development goals aimed at eradicating hunger and poverty (United Nations 2015). In Sub-Saharan Africa, smallholder farmers face numerous challenges, including poverty, limited access to financial resources, modern technologies, and irrigation infrastructure. This has led to low fertilizer application rates and nutrient mining, ultimately resulting in low productivity and declining soil fertility (Giller et al. 2021; Muoni et al. 2022; Masere 2022). Consequently, smallholder agriculture often prioritizes subsistence crops over cash crops.

Inherently, the soils in Sub-Saharan Africa, including Zimbabwe, are low in phosphorus and nitrogen, further exacerbated by nutrient depletion (Falconnier et al. 2023). Smallholder farmers predominantly cultivate subsistence crops such as maize, groundnuts, and sorghum in drier areas with less than 750 mm of annual rainfall, while large-scale commercial agriculture thrives in regions with higher rainfall (OECD/FAO 2020). Livestock integration into smallholder mixed farms promotes nutrient cycling, where manure is utilized as fertilizer and crop residues serve as livestock feed. This practice not only enhances soil fertility but also acts as a safety net for smallholders, particularly during droughts (Tavirimirwa et al. 2013).

Soil fertility—the ability of soil to sustain plant growth—is influenced by various factors, including nutrient content and balance (Tang 2025). Nutrient depletion occurs due to plant removal, erosion, and leaching, whereas maintaining soil fertility involves adding organic and inorganic amendments. Healthy ecosystems typically recycle nutrients, maintaining organic matter and nutrient levels (Stewart et al. 2020). Tang (2025) emphasizes the importance of managing soil fertility and nutrient availability for sustainable agriculture, highlighting the need for soil testing and understanding nutrient budgets to avoid toxicity and ensure optimal plant growth.

Unfortunately, smallholder farmers often have limited access to fertility resources and livestock, contributing to poor soil fertility (Tittonell and Giller 2021; Masere 2022). In arid and semi-arid conditions, restricted soil moisture further limits biomass production, exacerbating these challenges. Notably, nitrogen, despite its atmospheric abundance, remains low in soil (0.1%–0.6%), often limiting plant growth (Grandy et al. 2022). Effective nitrogen management is vital, given its connection to environmental quality issues such as groundwater contamination and greenhouse gas emissions. Therefore, exploring alternative sources of nitrogen, including BNF, is crucial for enhancing soil fertility in smallholder farming systems across Sub-Saharan Africa.

Another challenge facing most smallholder farmers of Sub-Saharan Africa is the complexity of land tenure systems (Masere 2022). Historically, indigenous land tenure systems were characterized by shifting cultivation and unlimited access to land, with no formal tenure rights. However, these systems were gradually replaced under colonial rule by land rights tied to nuclear and extended families or chieftainships, as large tracts of land were occupied by parastatals and shifted to a formalized system with title deeds (Masere 2022). While land tenure among smallholder farmers was originally customary, with no formal land rights, increasing land scarcity has led to the commercialization of land that was once inherited or allocated through chieftainships. As a result, early models of land tenure are now considered too simplistic, prompting a shift toward institutional pluralism, which brings uncertainty due to the diversity of institutions involved.

8.3.3 Contribution of Biological Nitrogen Fixation to Soil Fertility Improvement in Smallholder Farms

Tables 8.2 and 8.3 provide a snippet view of the literature studied. Enhanced productivity, improved yield, organic carbon, or nitrogen content of soils following cropping arrangements associated with BNF were considered indications of soil fertility improvements.

Generally, improvements in soil organic carbon and nitrogen content as well as crop productivity were registered under legume or cereal-legume intercrops (Diop et al. 2022; Gwenambira-Mwika et al. 2021; Wamalwa 2024; Witcombe and Tiemann 2022). Namatsheve et al. (2021) observed improved crop productivity as a result of intercropping maize with cowpea and associated this with substantial amounts of nitrogen added to the soil. However, the research also reported lower cowpea biomass accumulation in the sole crop, resulting in less nitrogen added to soils. The amount of nitrogen added to soils ranged from 87 to 180 kgN/Ha for maize-legume intercrops and rotation arrangements (Gwenambira-Mwika et al. 2021).

Other studies on BNF and its benefits demonstrate the dilemma encountered in attempting to exploit and maximize economic benefits by harnessing BNF using cereal-legume rotations. For example, Gwenambira-Mwika et al. (2021) obtained the best pigeon pea biomass production when grown in rotation with maize, making it effective for N cycling. However, this biomass production was not matched with economic returns as agronomic performance was compromised, resulting in low grain yield of the pigeon pea crop. In another multisite study with a range of maize legume cropping systems, sole maize outyielded all intercropping strategies in all environments on-farm, although the maize/pigeon pea intercropping strategy outyielded sole maize in most on-station environments (Madembo et al. 2020). Intercropping systems studied resulted in depressed yields, although the productivity of these systems exceeded that of a sole maize crop.

Using two legumes (pigeon pea and groundnut) in a doubling-up arrangement in rotation or as intercrops improved maize yields relative to the sole crop in a couple of multiyear research studies in Malawi (Gwenambira-Mwika et al. 2021; John et al. 2021). John et al. (2021) observed relatively lower variation (up to 27%–34%) and hence better stability in maize yields for intercrops compared to the sole crop (62%). In their study that considered diverse soil fertility levels, gains in organic C content of soils on sandy soils (>50% sand) were insignificant. Sandy soils could not accumulate organic carbon beyond 1.5% due to the low equilibrium level of soil organic matter as dictated by prevailing climatic conditions. Despite the positives registered in a study by Muoni et al. (2022), adoption was low as smallholder farmers considered high-performing varieties to be priced beyond their reach.

Table 8.3 shows that much research has been recorded in Sub-Saharan Africa on agroforestry (Bamboriya et al. 2022; Domínguez-Núñez 2022; Kuyah et al. 2021; Sileshi et al. 2020; Xavier et al. 2023), that is, cropping systems that integrate perennial plants, annual crops, and/ or livestock keeping (Sileshi et al. 2020). Agroforestry

Table 8.2 Empirical research on BNF, soil fertility, and productivity in smallholder farms and beneficial effects to soil fertility and productivity

Researcher	BNF intervention	Crops	Country	Organic C	Tot N	Yield
Wamalwa (2024)	Intercropping		Malawi	+	+	+
Gwenambira-Mwika et al. (2021)	Intercropping crop rotation				+	+
Namatsheve et al. (2021)	Intercropping					
Rapholo (2020)	Intercropping	Maize-lablab	South Africa	n/a	+	+
John et al. (2021)	Intercropping and rotation	Maize–ground/nut and/or pigeon pea	Malawi	–	n/a	+
Madembo et al. (2020)	Intercropping			n/a	+	+
Sánchez-Navarro et al. (2024)	Intercropping	Melon and cowpea Favabean and broccoli		n/a	+	+
Nyirenda and Balaka (2021)	Intercropping CARP vs FP	Maize, groundnut	Malawi	+	+	+
Mugi (2022)	Intercropping	Pigeon pea and lablab in maize	Tanzania	n/a	n/a	+
Diatta et al. (2020)	Intercropping	Mungbean and pearl millet	n/a	n/a	+	+
Mpongwana et al. (2024)	Agroforestry- dual inoculation/forage legumes	*Mucuna pruriens* (mucuna), *Lablab purpureus* (lablab), and *Vigna unguiculata* (cowpea)	South Africa	+	+	+
Shumba et al. (2024)	Rotations/CA	Maize-cowpea		+	n/a	+
Nyamayevu et al. (2024)	Rotations/CA	Maize-legume (groundnut, soybean, or cowpeas)	Malawi	n/a	n/a	+
Traore et al. (2023)	Intercropping	Sorghum with cowpea	Mali	n/a	n/a	+
Phiri (2024)	Intercropping	Pigeon pea (*Cajanus cajan* (L.) Millsp)		n/a	n/a	+
Kuyah et al. (2021)	Intercropping	Legume–cereal	Africa Asia	+	+	+
Assogba et al. (2023)	Farming system	Legume–cereal livestock	Burkina Faso	n/a	–	n/a
Witcombe and Tiemann (2022)	Groundnuts cropping	Groundnut	Uganda	–	–	n/a

Organic C, organic carbon; total N, total nitrogen; +, that authors/the study reported on that parameter, for example, organic carbon, or total nitrogen or yield; –, the study did not report on that parameter; n/a, the parameter was not applicable to the study

Table 8.3 Review of research on BNF, soil fertility, and crop productivity in smallholder farms

Reference	BNF intervention	Crops	Location	Soil carbon and nitrogen	Yield and productivity
Falconnier et al. (2023)	Agroecology	Legumes	Multiple locations in Sub-Saharan Africa	+	+
Muoni et al. (2022)	Intercropping	Legumes	Multiple locations in Sub-Saharan Africa	+	+
Sileshi et al. (2020)	Agroforestry	Trees, shrubs, bamboos and palms	–	+	–
Veste et al. (2024)	Agroforestry	Nitrogen-fixing trees and shrubs	Southern Africa	n/a	+
Dagar et al. (2020)	Agroforestry		Sub-Saharan Africa	n/a	+
Rupak et al. (2022)	Legume-based cropping systems	Legumes		–	–
Domínguez-Núñez (2022)	Agroforestry			+	+
Bamboriya et al. (2022)	Polyculture-agroforestry Intercropping	Maize based polyculture		–	–
Kuyah et al. (2021)	Intercropping	Legume–cereal	Africa Asia	+	+
Xavier et al. (2023)		Pulses			

+, authors/the study reported on that parameter, for example, soil carbon and nitrogen or yield and productivity; –, the study did not report on that parameter; n/a, the parameter was not applicable to the study

enhances nutrient cycling, which is effective and efficient by facilitating deep nutrient uptake by tree roots and enhancing BNF, thereby encouraging biomass accumulation and soil biological activity. Kehinde et al. (2024) identified agroforestry as the presence of trees on farmland, thereby contextualizing it in farming systems, as a range of land management systems typified by cropping, livestock, and trees and satisfying specific requirements depending on their structural and functional purpose. Agroforestry improves nutrient use efficiency as deep-rooted plants (trees) capture and recover nutrients lost beyond the reach of annual crops (Sileshi et al. 2020). It is valued most in semi-arid lands, especially for its potential to revive degraded lands and avert negative environmental consequences associated with leaching and use of inorganic fertilizers in cultivated lands as described by Masso et al. (2024).

Leguminous trees and shrubs, including genera like *Leucaena*, *Acacia*, *Gliricidia*, *Clitoria*, *Sesbania*, *Inga*, and *Calliandra*, have been widely researched (Domínguez-Núñez 2022). Owing to their huge potential for N fixation and biomass production, leguminous and other trees capable of fixing N are an important part of agroforestry systems. A wide range of the agroforestry trees and their functional and structural roles in farming systems are covered in writings by Kehinde et al. (2024). Alley cropping is an agroforestry practice that involves growing annual crops between rows of trees or perennial shrubs, usually legumes spaced by than 10 m or less. Benefit is derived by pruning the perennials and using the harvested biomass on alleys as green manure (Nair et al. 2021). Annual crop production in this system thrives on deposits created by nitrogen-fixing trees growing simultaneously with them or preceding them just like in shifting cultivation.

8.4 Strategies for Enhancing Biological Nitrogen Fixation

Enhancing BNF in legume and non-legume-based cropping systems requires a multifaceted approach that considers local conditions, integrated soil management, and the use of beneficial microbial partnerships. Research has shown that tailoring legume–cereal rotations to the specific needs of individual farmers is crucial for maximizing the benefits of BNF. For instance, doubling legumes in Malawi led to different outcomes under varying ecological conditions, demonstrating that practices applied without regard to context are unlikely to yield consistent benefits (Wamalwa 2024; Giller et al. 2021). Therefore, a site-specific approach that adapts to local soil types, climate, and farmer knowledge is essential for achieving optimal nitrogen fixation.

The judicious integration of inorganic fertilizers and manure also plays a key role in improving BNF. Fertilizers ensure that crops have access to all the necessary nutrients for growth, while organic amendments like manure enhance soil fertility, improve organic matter content, and contribute to better nitrogen cycling (Giller et al. 2021; Muoni et al. 2022; Namatsheve et al. 2021). This balanced nutrient management approach ensures that nitrogen fixation does not come at the expense of other essential nutrients, ultimately supporting robust crop growth.

Co-inoculation of soil with mycorrhizal fungi and rhizobia has been identified as another promising strategy to enhance BNF. Mycorrhizal fungi improve nutrient uptake, particularly phosphorus, and contribute to better soil health by enhancing soil structure and water retention (Sileshi et al. 2020). This symbiotic partnership between fungi and legumes, when combined with nitrogen-fixing rhizobia, can significantly improve both plant growth and the efficiency of nitrogen fixation, especially in soils where phosphorus is a limiting factor.

The selection of high-nitrogen-fixing legume varieties tailored to specific soil and climatic conditions further strengthens BNF efficiency. Research suggests that the incorporation of legume species with superior nitrogen-fixing capabilities, adjusted to the particular characteristics of the farming environment, can yield better results (Giller et al. 2021). Additionally, using a diversity of legume species can maximize nitrogen fixation by taking advantage of different ecological niches and soil conditions.

Agroecological approaches, such as diversified cropping systems, offer additional pathways for improving BNF. Practices like intercropping or agroforestry enhance soil biodiversity and improve nitrogen cycling through more complex symbiotic relationships among plants, soil microbes, and soil fauna. Conservation tillage practices also play a role by maintaining soil structure, improving water retention, and fostering microbial communities that support nitrogen fixation.

Managing soil pH is another critical factor, as BNF is sensitive to pH levels. Ensuring that soils are within an optimal pH range for rhizobial activity can enhance the effectiveness of BNF. Furthermore, the use of microbial inoculants tailored to specific soil conditions can improve rhizobial growth and compete with detrimental microbes, further supporting nitrogen fixation.

Lastly, farmer education and knowledge dissemination are fundamental for the successful implementation of BNF-enhancing strategies. Educating farmers on the benefits of legumes and how to integrate them into cropping systems can increase adoption of these practices. Extension services, workshops, and demonstration plots, along with farmer-to-farmer networks, can facilitate the exchange of best practices and ensure that farmers are well-equipped to manage BNF technologies.

8.5 The Current Understanding of Biological Nitrogen Fixation

The current understanding of BNF is often disconnected from the actual practices of smallholder farmers in Zimbabwe, as revealed through limited SSIs and existing literature. While BNF is widely recognized for its potential to reduce reliance on chemical fertilizers, maintain soil fertility, and enhance crop productivity, the literature often overlooks farmer typologies and their practices (Table 8.4). This gap leads to an inflated perception of BNF's potential, especially in developing countries, where farmers play a key role in managing agroecosystems.

8.5.1 Neglect of Farmers' Practices in Enhancing Biological Nitrogen Fixation

BNF research often focuses on idealized experimental conditions that fail to reflect farmers' real-world practices. For example, studies highlight the role of legumes in enhancing BNF but often overlook farmer-driven practices like intercropping, agroforestry, or the use of organic amendments (Vanlauwe et al. 2019). In tropical regions, farmers use leguminous plants (e.g., beans, peas, groundnuts) in crop

Table 8.4 Key gaps in BNF understanding

Issue/gap	Description	Consequences
Neglect of farmers' practices	Literature often assumes idealized, standardized conditions and ignores the diversity of farmer practices.	Disconnect between scientific research and farmers' actual BNF practices, limiting adoption (Ncube et al. 2009)
Inflated performance in experiments	Field and controlled experiments focus on optimal conditions, ignoring practical farmer realities	Distorted perceptions of BNF'S effectiveness, leading to skewed policy recommendations
Narrow economic focus	Focus on economic benefits (e.g., cost savings on fertilizers) without considering broader sustainability	Risks marginalizing critical sustainability factors such as soil health, biodiversity, and resilience

rotations to improve soil nitrogen. However, these farmer practices are often underrepresented or poorly integrated into BNF research, leading to a disconnect between scientific findings and on-the-ground practices (Tittonell and Giller 2021).

8.5.2 *Inflated Performance of Biological Nitrogen Fixation in Field and Controlled Experiments*

While field and controlled experiments are important for assessing BNF, they often overlook farm-level complexities. These studies tend to assume ideal conditions, such as optimal soil management and irrigation, which are not reflective of low-input, subsistence-based systems (Vanlauwe et al. 2019). Furthermore, the heterogeneity of farmer typologies—ranging from subsistence to market-oriented farmers—greatly influences BNF outcomes, yet these factors are often ignored in experimental settings (Gatzweiler and von Braun 2016).

8.5.3 *Narrow Focus on Economic Returns and Soil Fertility Maintenance*

BNF research has primarily focused on economic returns, such as reduced fertilizer costs, and soil fertility maintenance. However, these studies often fail to consider broader socioecological factors, including soil health, biodiversity, and resilience to climate change—critical elements for long-term sustainability (Tittonell and Giller 2021).

Sustainability in the context of BNF requires more than just maintaining soil fertility or increasing immediate economic returns. It also involves factors such as soil health, biodiversity, and resilience to climate change, which are not always captured in the conventional economic assessments of BNF (Vanlauwe et al. 2019). Farmers, especially in resource-constrained settings, may not always prioritize soil fertility in isolation from broader considerations of livelihood sustainability, food security, or access to external inputs. The current literature's narrow economic focus thus risks marginalizing these essential aspects of BNF and limiting the long-term sustainability of nitrogen fixation practices in diverse farming systems.

Furthermore, the sustainability of BNF is intricately tied to soil health management, which is often addressed separately from BNF in the literature. While BNF does contribute to soil nitrogen levels, it is equally important to consider the broader ecosystem services provided by integrated soil fertility management (ISFM) practices, which involve combinations of biological, chemical, and physical soil fertility management strategies (Tittonell and Giller 2021). BNF, if considered in isolation from such comprehensive approaches, may fail to address the broader challenges of soil degradation and declining fertility in many smallholder systems.

8.6 Rethinking Biological Nitrogen Fixation: Toward a Cycle Mediated by Farmers' Practices for Sustainable and Equitable Soil Fertility Management

BNF is a cornerstone of sustainable agriculture, but its potential is often constrained by how it is conceptualized and studied. Traditional views treat BNF as a field-localized phenomenon primarily driven by leguminous crops, neglecting broader ecological processes. We propose a shift toward understanding BNF as a dynamic, integrated cycle that involves not just legumes but also the management of residues, animal manure, and other organic inputs. This more holistic perspective can enhance BNF integration into smallholder farming systems, whether dryland or irrigated, and promote sustainable and equitable soil fertility management.

8.6.1 Biological Nitrogen Fixation as an Integrated Cycle: Beyond Legumes as Isolated Contributors

Traditionally, BNF has been viewed through the lens of leguminous crops alone, disregarding the broader nutrient cycling that occurs within the farming system. BNF should be seen as part of a dynamic cycle involving legumes, animal manure, crop residues, and other organic inputs, all contributing to soil fertility (Table 8.5).

BNF is not just a practice but an ongoing process involving the management of organic matter and nitrogen-fixing bacteria, thus promoting sustainable soil fertility. Leguminous residues and manure together improve microbial environments, fostering long-term soil health (Vanlauwe et al. 2019; Giller et al. 2021).

Table 8.5 Key BNF components and their roles

BNF component	Role in BNF cycle
Legumes	Fix nitrogen and enhance soil fertility through root symbiosis with nitrogen-fixing bacteria
Animal manure	Improves soil structure, microbial activity, and enhances nitrogen-fixing potential
Crop residues	Provide organic matter that supports microbial nitrogen fixation and improves soil fertility
Organic inputs	Integrated use of organic amendments like compost enhances the nitrogen cycle

Table 8.6 Farmers' practices and their impacts on BNF

Farmers' practice	Impact on BNF
Crop rotations with legumes	Enhances soil nitrogen levels and structure by alternating legume and nonlegume crops
Agroforestry systems	Integrates legumes into tree-based systems, improving nitrogen fixation and providing ecological benefits like carbon sequestration and soil erosion control
Cover crops and intercropping	Improves soil fertility by enhancing nitrogen fixation and preventing nutrient depletion

8.6.2 Inclusion of Farmers' Practices in Biological Nitrogen Fixation in Smallholder Systems

Smallholder farmers have developed context-specific practices over generations (indigenous knowledge systems) to enhance nitrogen fixation. These practices, including crop rotations, intercropping, and the use of cover crops, optimize BNF across diverse agroecological contexts (Table 8.6).

These adaptive practices are crucial for enhancing the effectiveness of BNF in both dryland and irrigated systems. Unfortunately, such practices are often underrepresented and sometimes dismissed in formal research (Masere and Worth 2021), which typically focuses on isolated experiments. By incorporating farmer-driven approaches, BNF research can better reflect the realities of smallholder farming systems. By acknowledging the diversity of farmers' practices and promoting participatory research in BNF, the sustainability of these systems can be better understood and scaled up.

8.6.3 Context-Specific Experiments: Integrating Biological Nitrogen Fixation into Dryland and Irrigated Systems

The heterogeneity of smallholder systems requires context-specific solutions to enhance BNF. In both dryland and irrigated systems, experimenting with BNF through adaptive management—guided by farmer experimentation—is essential for maximizing nitrogen fixation and improving soil fertility management. Dryland systems, characterized by low rainfall and resource constraints, present specific challenges and opportunities for BNF. Farmers often rely on leguminous crops like cowpeas and groundnuts, which are drought-resistant and can fix nitrogen even in water-limited conditions. Additionally, mulching with leguminous residues and incorporating manure can improve soil moisture retention and enhance microbial activity, creating a more conducive environment for nitrogen fixation (Giller et al. 2021). Dryland farmers are also known to adapt leguminous systems to incorporate both crop residues and animal manure, demonstrating innovative practices that integrate BNF with local resource management strategies.

Conversely, in irrigated systems, where water availability is more stable, BNF can be intensified through strategic integration with irrigated crops. For instance, leguminous crops in irrigated systems, such as clover and vetch, can be managed through intercropping or as cover crops to enhance nitrogen fixation while improving soil structure and preventing nutrient leaching (Vanlauwe et al. 2019). Moreover, the use of animal manure in irrigated systems can complement BNF by enhancing nutrient cycling, fostering soil microbial activity, and improving overall soil fertility.

8.6.4 The Role of Organic Inputs: Leguminous Residues and Animal Manure

The cycling of nutrients through leguminous residues and animal manure is critical for maintaining sustainable soil fertility. Leguminous plants fix nitrogen, but their residues and the integration of animal manure provide long-term benefits for microbial nitrogen fixation. The leguminous residues provide organic matter that stimulates soil microbial activity and enhances nitrogen cycling. Animal manure, on the other hand, contributes essential minerals and organic matter that foster soil health and complement nitrogen fixation. Composting often combines residues, and manure improves nitrogen availability for future crops and boosts microbial activity. Research shows that integrating these organic inputs improves the microbial environment, which is crucial for optimizing nitrogen fixation and overall soil fertility (Vanlauwe et al. 2019).

8.6.5 Promoting a Holistic, Integrated Approach to Biological Nitrogen Fixation for Sustainable and Equitable Soil Fertility Management

BNF must be understood as part of a broader, integrated approach to soil fertility management. This approach requires acknowledging the diversity of farming systems, incorporating local knowledge, and ensuring equity in access to resources and technology. Such an inclusive approach to BNF will ensure that it is not only ecologically sustainable but also equitable, benefiting all farmers, especially those with limited resources.

Thus, the key steps in enhancing BNF include the following:

- Multidisciplinary research: combining soil science, agroecology, and farmer knowledge to create context-specific strategies that optimize BNF (Tittonell and Giller 2021).
- Participatory research: engaging farmers in research design and evaluation to ensure BNF strategies are grounded in local practices and needs (Giller et al. 2021).
- Equity considerations: ensuring that marginalized farmers have access to resources, markets, and extension services to fully benefit from BNF practices (Gatzweiler and von Braun 2016).

8.7 Conclusion

Improving soil fertility in smallholder farming systems requires investment in approaches that support an equitable green revolution—one that advances sustainable livelihoods and delivers both nutritional and economic benefits to enhance overall well-being. Biological nitrogen fixation (BNF) should be central to this transformation due to its capacity to increase nitrogen availability, reduce dependence on costly synthetic fertilizers, and promote environmentally sustainable agriculture.

However, the current discourse around BNF often fails to reflect the lived realities of smallholder farmers. While experimental studies highlight its potential, they frequently overlook local knowledge systems, farmer typologies, and the agroecological diversity within resource-poor contexts. Moreover, an overemphasis on economic returns and yield outcomes risks narrowing the broader sustainability agenda necessary for long-term adoption and impact.

To position BNF as a viable and lasting solution, a shift toward a more integrated, systems-based approach is essential. This includes recognizing farmers as innovators, promoting co-learning processes, and embedding BNF within holistic soil fertility strategies that incorporate leguminous crops, animal manure, and context-specific practices across both dryland and irrigated systems. Support mechanisms—such as access to inputs, farmer training, and extension services—must be designed to reflect this complexity.

Ultimately, the success of BNF will depend on how well it is embedded within the socioecological systems of smallholder agriculture. By embracing a participatory, agroecological approach, we can foster resilient farming systems that not only improve soil fertility but also enhance productivity, equity, and long-term sustainability.

References

Abdullah MKA-F (2002) Factors affecting the efficiency of symbiotic nitrogen fixation by rhizobium. Pak J Biol Sci 5(11):1277–1293. https://doi.org/10.3923/pjbs.2002.1277.1293

Assogba GGC, Berre D, Adam M, Descheemaeker K (2023) Can low-input agriculture in semi-arid Burkina Faso feed its soil, livestock and people? Eur J Agron 151:126983. https://doi.org/10.1016/j.eja.2023.126983

Bamboriya SD, Bana RS, Kuri BR, Kumar V, Bamboriya SD, Meena RP (2022) Achieving higher production from low inputs using synergistic crop interactions under maize-based polyculture systems. Environ Sustain 5(2):145–159. https://doi.org/10.1007/s42398-022-00228-7

Dagar JC, Sileshi GW, Akinnifesi FK (2020) Agroforestry to enhance livelihood security in Africa: research trends and emerging challenges. In: Agroforestry for degraded landscapes: recent advances and emerging challenges, vol 1, pp 71–134. https://doi.org/10.1007/978-981-15-4136-0_3

Diatta AA, Abaye O, Thomason WE, Lo M, Thompson TL, Vaughan LJ, Gueye F, DIagne M (2020) Evaluating pearl millet and mungbean intercropping in the semi-arid regions of Senegal. Agron J 112(5):4451–4466

Diop M, Chirinda N, Beniaich A, El Gharous M, El Mejahed K (2022) Soil and water conservation in Africa: state of play and potential role in tackling soil degradation and building soil health in agricultural lands. Sustainability 14(20):13425. https://doi.org/10.3390/SU142013425

Domínguez-Núñez JA (2022) Leguminous trees for sustainable tropical agroforestry. In: Advances in legumes for sustainable intensification. Elsevier, pp 483–504. https://doi.org/10.1016/B978-0-323-85797-0.00006-9

Falconnier GN, Cardinael R, Corbeels M, Baudron F, Chivenge P, Couëdel A, Ripoche A, Affholder F, Naudin K, Benaillon E, Rusinamhodzi L, Leroux L, Vanlauwe B, Giller KE (2023) The input reduction principle of agroecology is wrong when it comes to mineral fertilizer use in sub-Saharan Africa. Outlook Agric 52(3):311–326. https://doi.org/10.1177/00307270231199795

Gatzweiler FW, von Braun J (2016) Innovation for marginalized smallholder farmers and development: an overview and implications for policy and research. In: Gatzweiler F, von Braun J (eds) Technological and institutional innovations for marginalized smallholders in agricultural development. Springer, Cham. https://doi.org/10.1007/978-3-319-25718-1_1

Giller KE, Hijbeek R, Andersson JA, Sumberg J (2021) Regenerative Agriculture: An agronomic perspective. Outlook Agric 50(1):13–25. https://doi.org/10.1177/0030727021998063

Grandy AS, Daly AB, Bowles TM, Gaudin ACM, Jilling A, Leptin A, McDaniel MD, Wade J, Waterhouse H (2022) The nitrogen gap in soil health concepts and fertility measurements. Soil Biol Biochem 175:108856. https://doi.org/10.1016/J.SOILBIO.2022.108856

Guo K, Yang J, Yu N, Luo L, Wang E (2023) Biological nitrogen fixation in cereal crops: progress, strategies, and perspectives. Plant Commun 4(2):100499. https://doi.org/10.1016/j.xplc.2022.100499

Gwenambira-Mwika CP, Snapp SS, Chikowo R (2021) Broadening farmer options through legume rotational and intercrop diversity in maize-based cropping systems of Central Malawi. Field Crop Res 270:108225. https://doi.org/10.1016/J.FCR.2021.108225

Hu W, Wang X, Xu Y, Wang X, Guo Z, Pan X, Dai S, Luo Y, Teng Y (2024) Biological nitrogen fixation and the role of soil diazotroph niche breadth in representative terrestrial ecosystems. Soil Biol Biochem 189:109261. https://doi.org/10.1016/j.soilbio.2023.109261

John I, Snapp S, Nord A, Chimonyo V, Gwenambira C, Chikowo R (2021) Marginal more than mesic sites benefit from groundnut diversification of maize: increased yield, protein, stability, and profits. Agric Ecosyst Environ 320:107585. https://doi.org/10.1016/j.agee.2021.107585

Kehinde OC, Ujah GO, Ogunsola MO, James AB (2024) Integrating trees into agricultural landscapes: benefits and challenges of agroforestry systems. Int J Agric Environ Sci 10(7):94–121. https://doi.org/10.56201/ijaes.v10.no7.2024.pg94.121

Kuyah S, Sileshi GW, Nkurunziza L, Chirinda N, Ndayisaba PC, Dimobe K, Öborn I (2021) Innovative agronomic practices for sustainable intensification in sub-Saharan Africa. A review. Agron Sustain Dev 41(2):1–21. https://doi.org/10.1007/S13593-021-00673-4

Lowder SK, Sánchez MV, Bertini R (2019) Farms, family farms, farmland distribution and farm labour: what do we know today? FAO Agricultural Development Economics Working Paper 19-08. http://www.fao.org/3/ca7036en/ca7036en.pdf

Madembo C, Mhlanga B, Thierfelder C (2020) Productivity or stability? Exploring maize-legume intercropping strategies for smallholder conservation agriculture farmers in Zimbabwe. Agric Syst 185:102921. https://doi.org/10.1016/J.AGSY.2020.102921

Maguire M, Delahunt B (2017) Doing a thematic analysis: a practical, step-by-step guide for learning and teaching scholars. All Irel J High Educ 9(3). https://doi.org/10.62707/aishej.v9i3.335

Manatsa D, Mushore TD, Gwitira I, Wuta M, Chemura A, Shekede MD, Mugandani R, Sakala LC, Ali LH, Mupuro J (2020) Revision of Zimbabwe's agro-ecological zones. Produced by the government of Zimbabwe under Zimbabwe National Geospatial and space agency (ZINGSA) for the Ministry of Higher and Tertiary Education, innovation. Science and Technology Development. Zimbabwe Government Printers

Masere TP (2022) Evaluation of the role of small-scale farmers in soil and water conservation management in the context of climate change. In: Ondrasek G, Zhang L (eds) Resource management in agroecosystems. IntechOpen. https://doi.org/10.5772/intechopen.108889

Masere TP, Worth S (2021) Influence of public agricultural extension on technology adoption by small-scale farmers in Zimbabwe. S Afr J Agric Ext 49(2):25–42

Masso C, Gweyi-Onyango J, Luoga H, Yemefack M, Vanlauwe B (2024) A review on nitrogen flows and obstacles to sustainable nitrogen management within the Lake Victoria Basin, East Africa. Sustainability 16(11):4816. https://www.mdpi.com/2071-1050/16/11/4816

Mohammadi K (2012) Effective factors on biological nitrogen fixation. Afr J Agric Res 7(12):1782–1788. https://doi.org/10.5897/ajarx11.034

Mpongwana S, Manyevere A, Mupangwa J, Mpendulo CT, Mashamaite CV (2024) Improving soil fertility through dual inoculation with arbuscular mycorrhizal fungi and rhizobium on a eutric cambisol cultivated with forage legumes in a semi-arid region. Heliyon 10(2):e24817. https://doi.org/10.1016/J.HELIYON.2024.E24817

Mugi E (2022) Diversification and intensification of smallholder farming through intercropping with grain legumes in northern Tanzania. Wageningen University. https://doi.org/10.18174/571717

Muoni T, Jonsson M, Duncan AJ, Watson CA, Bergkvist G, Barnes AP (2022) Effects of management practices on legume productivity in smallholder farming systems in sub-Saharan Africa. Food Energy Secur 11:366. https://doi.org/10.1002/fes3.366

Mupambwa H, Nciizah A, Nyambo P (2022) Can organic soil fertility management sustain farming and increase food security among African smallholder farmers? In: Mupambwa HA, Nciizah AD, Nyambo P, Muchara B, Gabriel N (eds) Food security for African smallholder farmers. Sustainability sciences in Asia and Africa. Springer, pp 89–112. https://doi.org/10.1007/978-981-16-6771-8_6

Nair PKR, Kumar BM, Nair VD (2021) Tropical alley cropping and improved fallows. In: An introduction to agroforestry—four decades of scientific developments, 2nd edn. Springer, Cham, pp 87–111. https://doi.org/10.1007/978-3-030-75358-0_6

Namatsheve T, Chikowo R, Corbeels M, Mouquet-Rivier C, Icard-Vernière C, Cardinael R (2021) Maize-cowpea intercropping as an ecological intensification option for low input systems in sub-humid Zimbabwe: productivity, biological N2-fixation and grain mineral content. Field Crop Res 263:108052. https://doi.org/10.1016/j.fcr.2020.108052

Ncube B, Twomlow SJ, Dimes JP, van Wijk MT, Giller KE (2009) Resource flows, crops and soil fertility management in smallholder farming systems in semi-arid Zimbabwe. Soil Use Manag 25(March):78–90. https://doi.org/10.1111/j.1475-2743.2009.00193.x

Nezomba H, Mtambanengwe F, Tittonell P, Mapfumo P (2015) Geoderma point of no return? Rehabilitating degraded soils for increased crop productivity on smallholder farms in eastern Zimbabwe. Geoderma 239–240:143–155. https://doi.org/10.1016/j.geoderma.2014.10.006

Nyamayevu D, Nyagumbo I, Liang WL, Li RQ, Silva JV (2024) Grain and nutritional yield merits of sustainable intensification through maize-legume rotations in land constrained smallholder farms of Malawi. Field Crop Res 318:109565. https://doi.org/10.1016/J.FCR.2024.109565

Nyirenda H, Balaka V (2021) Conservation agriculture-related practices contribute to maize (*Zea mays* L.) yield and soil improvement in Central Malawi. Heliyon 7(3):e06636. https://doi.org/10.1016/j.heliyon.2021.e06636

OECD/FAO (2020) OECD-FAO agricultural outlook 2020–2029. FAO/OECD Publishing, Rome. https://www.oecd-ilibrary.org/agriculture-and-food/oecd-fao-agricultural-outlook-2020-2029_1112c23b-en

Phiri M (2024) Effect of biochar and crop rotation with pigeon pea on plant available phosphorus under conservation agriculture. Norwegian University of Life Sciences

Rapholo SE (2020) Modelling the effects of maize/lablab intercropping on soil water content and nitrogen dynamics using APSIM model. https://univendspace.univen.ac.za/server/api/core/bitstreams/56430140-8bf8-4e75-8638-cf97b20ee459/content

Rascio N, La Rocca N (2013) Biological nitrogen fixation. In: Rascio N, La Rocca N (eds) Encyclopedia of ecology, vol 2(10). Elsevier, pp 264–279. https://doi.org/10.1016/B978-0-444-63768-0.00685-5

Rupak G, Krishnendu R, Sarbaswarup G, Sudarshan D, Subhasis M, Nayak J, Chakraborty S, Hirak B (2022) Multiple pathways of legume-based systems towards environmental, social, and economic sustainability in smallholder systems. In: Meena RS, Kumar S (eds) Advances in legumes for sustainable intensification. Elsevier, pp 657–678. https://doi.org/10.1016/B978-0-323-85797-0.00017-3

Sánchez-Navarro V, Marcos-Pérez M, Contreras J, Zornoza R (2024) Biological nitrogen fixation and intra- and interspecific competition in two vegetable/legume intercropping systems and their relationship to crop yield. Sci Hortic 337:113502. https://doi.org/10.1016/j.scienta.2024.113502

Shumba A, Chikowo R, Thierfelder C, Corbeels M, Six J, Cardinael R (2024) Mulch application as the overarching factor explaining increase in soil organic carbon stocks under conservation agriculture in two 8-year-old experiments in Zimbabwe. Soil 10(1):151–165. https://doi.org/10.5194/SOIL-10-151-2024

Sileshi GW, Mafongoya PL, Nath AJ (2020) Agroforestry systems for improving nutrient recycling and soil fertility on degraded lands. In: Dagar JC, Gupta SR, Teketay D (eds) Agroforestry for degraded landscapes. Springer Singapore, pp 225–253. https://doi.org/10.1007/978-981-15-4136-0_8

Soumare A, Diedhiou AG, Thuita M, Hafidi M, Ouhdouch Y, Gopalakrishnan S, Kouisni L (2020) Exploiting biological nitrogen fixation: a route towards a sustainable agriculture. Plants 8(9):1011. https://doi.org/10.3390/plants9081011

Stewart ZP, Pierzynski GM, Middendorf BJ, Vara Prasad PV (2020) Approaches to improve soil fertility in sub-Saharan Africa. J Exp Bot 71(2):632–641. https://doi.org/10.1093/jxb/erz446

Tang L (2025) Soil fertility, plant nutrition and nutrient management. Plants 14(1):34. https://doi.org/10.3390/plants14010034

Tavirimirwa B, Mwembe R, Ngulube B, Banana N, Nyamushamba G, Ncube S, Nkomani D (2013) Communal cattle production in Zimbabwe: a review. Livest Res Rural Dev 25(12):16

Tittonell P, Giller KE (2013) When yield gaps are poverty traps: the paradigm of ecological intensification in African smallholder agriculture. Field Crop Res 143:76–90. https://doi.org/10.1016/j.fcr.2012.10.007

Traore A, Falconnier GN, Couëdel A, Sultan B, Chimonyo VGP, Adam M, Affholder F (2023) Sustainable intensification of sorghum-based cropping systems in semi-arid sub-Saharan Africa: the role of improved varieties, mineral fertilizer, and legume integration. Field Crop Res 304:109180. https://doi.org/10.1016/J.FCR.2023.109180

United Nations (2015) Transforming our world: the 2030 agenda for sustainable development. In: A/RES/70/1. United Nations, pp 1–41. http://www.un.org/ga/search/viewm_doc.asp?symbol=A/RES/70/1

Vanlauwe B, Hungria M, Kanampiu F, Giller KE (2019) The role of legumes in the sustainable intensification of African smallholder agriculture: lessons learnt and challenges for the future. Agric Ecosyst Environ 284:106583. https://doi.org/10.1016/j.agee.2019.106583

Veste M, Sheppard JP, Abdulai I, Ayisi KK, Borrass L, Chirwa PW, Funk R, Kapinga K, Morhart C, Mwale SE, Ndlovu NP, Nyamadzaw G, Nyoka BI, Sebola P, Seifert T, Senyolo MP, Sileshi GW, Syampungani S, Kahle H-P (2024) The need for sustainable agricultural land-use systems: benefits from integrated agroforestry systems. In: Maltitz GP et al (eds) Sustainability of southern African ecosystems under global change. Ecological studies, vol 248. Springer, Cham, pp 587–623. https://doi.org/10.1007/978-3-031-10948-5_21

Wamalwa SW (2024) Integrated soil fertility management practices effects on soil chemical properties and maize (*Zea mays* L.) yield in Machakos County, Kenya. Kenyatta University. https://ir-library.ku.ac.ke/server/api/core/bitstreams/7923fc1a-86a5-4778-ac96-a05a3a9d2b45/content

Witcombe AM, Tiemann LK (2022) Potential contribution of groundnut residues to soil N and the influence of farmer management in Western Uganda. Front Sustain Food Syst 5:691786. https://doi.org/10.3389/FSUFS.2021.691786/PDF

Xavier GR, Jesus EDC, Dias A, Coelho MRR, Molina YC, Rumjanek NG (2023) Contribution of biofertilizers to pulse crops: from single-strain inoculants to new technologies based on microbiomes strategies. Plants 12(4):954. https://doi.org/10.3390/PLANTS12040954

[illegible]

[illegible] Front [illegible]

[illegible]

[illegible] https://doi.org/[illegible]

[illegible]

Chapter 9
Vermicompost as Potential Biofertilizers in Smallholder Farming Systems

Dhitri Borah, Priti Giri, Sonal Saluja, Sudeshna Mandal, Varunendra Singh Rawat, and Ravi Kumar Goswami

Abstract The current vulnerability of agriculture to climate change emphasizes the urgent need for farmers, agricultural experts, and stakeholders to adopt sustainable solutions to prevent future food scarcities. It is imperative to promptly enhance the application of organic fertilizers to guarantee the sustained production of crops. Vermicomposting is a method of amending soil with nutrient-enriched compost by using earthworms. This approach has the potential to mitigate the adverse impacts of climate change on agricultural productivity. Vermicompost is abundant in vital macronutrients like nitrogen (N), phosphorus (P), and potassium (K), together with micronutrients such as zinc (Zn), copper (Cu), and iron (Fe), rendering it a financially viable and ecologically sustainable substitute for artificial fertilizers. By improving the soil structure, water retention, and microbial activity, vermicompost stimulates plant growth, resilience, bioavailable nutrient composition, crop production, and reduces soil-borne diseases, hence producing better crops without the need for chemical pesticides. Thus, the advancement of vermicomposting technology has the potential to enhance agricultural and land productivity in a sustainable manner. Moreover, the implementation of vermicomposting can also offer opportunities for small-scale farmers in the organic product industry, thereupon enhancing the economic worth of their means of subsistence. This chapter delves into the manifold advantages of vermicompost as a biofertilizer, encompassing its capacity to enhance soil health, augment crop productivity, and mitigate input expenses for local farmers.

D. Borah
Department of Zoology, Biswanath College, Gauhati University, Biswanath, Assam, India

P. Giri
Department of Botany, University of Delhi, Delhi, India

S. Saluja
Department of Zoology, Dyal Singh College, Karnal, Kurukshetra University, Haryana, India

S. Mandal
Department of Medicine, Indiana University School of Medicine, Indianapolis, IN, USA

V. S. Rawat · R. K. Goswami (✉)
Department of Zoology, Hindu College, University of Delhi, Delhi, India

A. D. Nciizah et al. (eds.), *Biofertilizers for Sustainable Agriculture in Africa*,
Sustainability Sciences in Asia and Africa, https://doi.org/10.1007/978-981-95-6009-7_9

Keywords Vermicompost · Earthworms · Smallholder farmers · Biofertilizers · Soil health · Crop productivity

Abbreviations

AMF	Arbuscular Mycorrhizal Fungi
CASP	Comprehensive Agriculture Support Programme
COVID-19	Coronavirus Disease of 2019
CSD	Commission on Sustainable Development
EU	European Union
FAO	Food and Agriculture Organization
HLPE	High Level Panel of Experts
IAASTD	International Assessment of Agricultural Science and Technology for Development
INM	Integrated Nutrient Management
IPCC	Intergovernmental Panel on Climate Change
MLAFWRD	Ministry of Lands, Agriculture, Fisheries, Water and Rural Development
Non-AMF	Non-Arbuscular Mycorrhizal Fungi
OFBF	Organic Fertilizers and Biofertilizers
PGPR	Plant Growth-Promoting Rhizobacteria
ROAM	Rwanda Organic Agriculture Movement
SSA	Sub-Saharan Africa
UN	United Nations

9.1 Introduction

Global food security is a major problem, especially for smallholder farmers who face many obstacles in raising agricultural yields. The importance of finding a balance between crop absorption, soil nutrient availability, yield quality, and soil health is highlighted by the adoption of integrated nutrient management (INM) strategies in modern agricultural approaches to address these issues (Leoni et al. 2019). It is concerning that with the advancement of agricultural technology, natural supplements are consistently being neglected in favor of synthetic substitutes. This reduces soil fertility and lowers agricultural yields and quality (Mondal et al. 2017). Sustainable agriculture has become a major global priority due to concerns about food security, climate change, and environmental degradation (Ratnasari et al. 2023).

Vermicomposting is regarded as one of the most promising approaches to environmental restoration, producing nutrient-rich biofertilizers, and cultivating sustainable crops. It has gained lots of attention due to its potential to improve agricultural sustainability and productivity for a variety of field crops, such as fruit trees, floral plants, and vegetables (Ali et al. 2015; Lim et al. 2015). Vermicompost is an eco-friendly and nutrient-rich biological fertilizer for small-scale farmers, generated through the natural abilities of earthworms to breakdown organic waste. In the African context, vermicompost provides a sustainable and economical substitute to chemical fertilizers, where a lack of resources limits their use. It contains vital nutrients, boosts microbial activity, and improves soil structure and water retention (Edwards 2004). Promoting its use can improve rural life, increase food security, and strengthen resilient farming systems by increasing soil fertility and managing waste effectively. Although in the last few decades, food security in Africa has improved significantly, yet, agricultural productivity across the continent particularly in Sub-Saharan Africa (SSA) remains among the lowest in the world (Aslam et al. 2019). Land degradation and the depletion of natural resources are widespread challenges in many African countries (Jones et al. 2013; Nkonya et al. 2016). Following the COVID-19 pandemic and the Russia–Ukraine conflict, food and nutrition insecurity has become a more serious concern across the region (FAO 2020). These global events disrupted value chains, energy markets, and led to export restrictions. Consequently, access to food imports, agricultural goods, and essential production inputs such as inorganic fertilizers has decreased significantly. This has caused a sharp increase in prices and a greater market volatility (Abu Hatab 2022; FAO 2021, 2022; UN 2022). These disruptions are expected to have a long-term negative impacts on agricultural productivity and food availability.

African smallholder farmers are particularly affected by the rising cost of fertilizer import (UN 2022), and these smallholders are important to African agricultural production, accounting for 70%–80% of the continent's agricultural output (IAASTD 2009; Lowder et al. 2014). Although projections for 2050 and beyond are cautiously optimistic, sustainably feeding future generations will depend on addressing a range of interconnected challenges, including rapid population growth, urbanization, climate change, shifting dietary preferences, and the unequal distribution of natural resources (Keats and Wiggins 2017). Urbanization, in particular, is increasing the demand for non-locally produced food commodities, such as premium rice, which could worsen Africa's trade imbalance if policy reforms are not implemented. Nevertheless, the continent possesses substantial potential to achieve long-term food security, owing to its rich endowment of natural resources such as arable land, a sizable labor force, freshwater availability, and energy reserves (FAO 2015). However, realizing this potential requires substantial financial investment and the adoption of improved agricultural practices (Williams 2015). In this context, integrating vermicomposting into smallholder farming presents a promising solution that not only supports sustainable agricultural intensification but also aligns with circular economy principles by recycling organic waste and improving soil fertility. It helps to improve soil fertility and crop yields while having no negative impact on the environment. Its adoption is an effective way to improve the livelihoods of smallholder farmers while conserving the health of agroecosystems. This chapter delves into the potential of vermicompost as a sustainable biofertilizer in smallholder farming practices.

9.2 Biofertilizers

The terms "biofertilizers" and "biostimulants" are often used interchangeably in the scientific literature primarily due to the lack of a standardized and universally accepted definition that clearly differentiates between them. However, the term "plant biostimulants" has been defined by the EU as (2019, Article 47a): a "plant biostimulant means a product stimulating plant nutrition processes independently of the product's nutrient content with the sole aim of improving one or more of the following characteristics of the plant or the plant rhizosphere: (a) nutrient use efficiency; (b) tolerance to abiotic stress; (c) quality traits (d) availability of confined nutrients in soil or rhizosphere." Furthermore, EU (2019) and du Jardin (2015) differentiate between microbial plant biostimulants and nonmicrobial plant biostimulants. Various authors use the term "biofertilizer" for microbial plant biostimulants such as arbuscular mycorrhizal fungi (AMF) and non-arbuscular mycorrhizal fungi (non-AMF), as well as beneficial bacteria, especially plant growth-promoting rhizobacteria (PGPR) such as phosphate (P) and potassium (K)-solubilizing bacteria and nitrogen-fixing bacteria (e.g., *Azotobacter*, *Azospirillum*, *Cyanobacteria/blue-green algae*), and nitrogen-fixing mutualistic endosymbionts of the type *Rhizobium* (Table 9.1) (Bamdad et al. 2022; du Jardin 2015; Schütz et al. 2018). According to du Jardin (2015), the word "biofertilizers" should be referred to as a subclass of biostimulants. The current production and use of biofertilizers in Africa are still minimal, despite the fact that the initial applications of biofertilizers in Africa, such as arbuscular mycorrhizal fungi (AMF) and especially nitrogen-fixing mutualistic endosymbionts such as *Rhizobium*, date back decades (Raimi et al. 2021). The largest market for biofertilizers is North America, which accounts for almost half of global sales. Europe and Asia-Pacific follow, with Africa accounting for approximately 5% share (BioFit 2015; Raimi et al. 2021). An overview of the current inoculant production and connected (research) institutions, together with associated problems and possible remedies, is given by Raimi et al. (2021). The most developed market for both manufacturers and consumers of biofertilizers is South Africa. While Central and West Africa have the lowest levels of production and utilization (mostly rhizobia-based goods), North Africa, including Egypt, Kenya, Malawi, and Zimbabwe, seems to have some "advanced" levels.

9.3 Vermicompost: A Good Solution for Conventional Farming

Vermicompost farming has the potential to replace inorganic farming as the primary way of producing green agricultural products (Dominguez and Edwards 2004). Organic waste can be recycled from a variety of sources, including the agricultural industry, municipal solid waste, animal manure, and agriculture (Aira and Dominguez 2009; Gómez-Brandón et al. 2011a). Animal excreta such as

Table 9.1 Functional role of microbial communities in enhancing vermicomposting processes

Types of biofertilizers	Name of bacteria/ fungi	Advantages	References
Beneficial bacteria-based			
Nitrogen fixer	*Rhizobium*, *Azotobacter*, *Azospirillum*, *Cyanobacteria* (blue-green algae)	Improve soil fertility, including N supply, soil aggregation Induce plant defense against bacterial, fungal, and viral pathogens Improve crop growth, yield, soil formation, soil surface stabilization, crust formation, enhance soil organic carbon, foster soil microbiota	Ahemad and Kibret (2014), Schütz et al. (2018), Thomas and Singh (2019), Vacheron et al. (2013) and Ammar et al. (2022)
Phosphate and potassium-solubilizing bacteria	*Enterobacter* sp.; *Bacillus* sp.	Solubilize and make P available, significant improvements in plant growth, yield, and P/K uptake (limited yield effect when soil P is low)	Bamdad et al. (2022), Schütz et al. (2018)
N fixers and P solubilizers	*Staphylococcus sarprophytes*, *Anabaena azollae*	Positive effects are higher when N-fixers and P-solubilizers are combined	Schütz et al. (2018)
Beneficial fungal-based			
Arbuscular mycorrhizal fungi (AMF)	*Glomus* sp.	Improve nutrition efficiency for macro- and micronutrients, water balance, and abiotic and biotic stress protection Yield improvements (specifically in dry and tropical climates)	du Jardin (2015), Schütz et al. (2018)
Non-arbuscular mycorrhizal fungi (non-AMF)	*Aspergillus niger; Trichoderma* sp.	Improve nutrient uptake by crops (even better compared to bacteria) Biocontrol function, increase nutrient use efficiency, abiotic stress tolerance, organ growth, and morphogenesis	Bamdad et al. (2022), du Jardin (2015)

vermicompost are a valuable resource for the production of organic fertilizer, which gives a high amount of macronutrients and micronutrients (Gómez-Brandón et al. 2011b). It has been observed that the gradual release of nutrients from vermicompost allows for easy absorption during plant growth while additionally increasing soil moisture content. This encourages the development of a more sustainable replacement for synthetic mineral fertilizers as an alternative to existing farming methods. Since 1960, agrochemicals have been widely utilized to increase agricultural productivity at the expense of societal and environmental health (Gómez-Brandón et al. 2021). It impaired the crops' ability to tolerate biological pressures by diminishing natural fertility and removing beneficial soil organisms. Vermicomposting techniques may help in plant development, yield, and maintenance recovery. It has been highlighted as a beneficial pathway where organic matter improves soil fertility and overall soil structure. Organic farming is more advantageous and economically viable than inorganic farming (Wonnapinij et al. 2022), as it effectively controls pests and diseases while protecting the environment (Fig. 9.1). Furthermore, it reduces pollution and improves soil nutrient availability. Crops cultivated in these conditions have excellent nutritional content and can be purchased at a competitive price (Atteya et al. 2021). It also benefits the ecosystem by lowering greenhouse gas emissions, which allows the decomposition of residual matter as a basis for increasing earth abundance (Vyas et al. 2022). Green gardening plays an important role in modern food production. Using vermicompost could result in a big and valuable yield for the green garden (Robatjazi 2023). This study may assist farmers in better understanding how vermicomposting can raise crop value, improve soil fertility, and lower the expense of buying man-made manure for sustainable development.

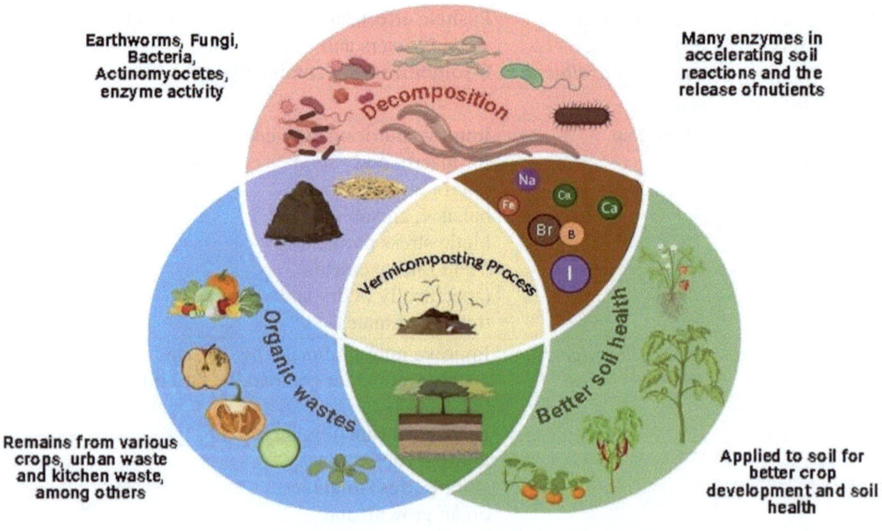

Fig. 9.1 Vermicomposting and its impact on soil health, agricultural growth, and the environment. (Generated with BioRender)

9.4 Roles of Smallholder Farming in Africa

The United Nations Commission on Sustainable Development (CSD) and other significant global bodies increasingly recognize the significance of smallholder farmers for maintaining food security and an adequate supply of food around the world. Smallholder farm production strengthens the local food chain by providing an achievable way to improve household nutrition and food security. Moreover, smallholder farming helps the food security condition of low-income households in developing countries such as South Africa. Less than two hectares of land are used by more than 80% (475 million) of farms around the world. Despite accounting for only 12% of global farmland, these farms supply an estimated 80% of food produced in Asia and Sub-Saharan Africa (Bekele et al. 2022). In order to enhance food availability, generate income, provide livelihoods, support dietary diversity, and act as a buffer against market-related shocks, smallholder farmers must play a central role in food production and in ensuring national food security (HLPE 2013). Consequently, smallholder farms offer multiple advantages and make substantial contributions to improving food security for a large portion of the population (Hazell 2020). Even though smallholder farms are essential to global nutrition and food security, they are a vulnerable group accounting for the majority of the world's hungry and underprivileged that development policies typically overlook (Fan and Rue 2020). Smallholders are now more susceptible to a variety of new financial, health, pricing, and climatic risks and challenges. Family workers, especially women, children, and the elderly, are regularly employed to manage smallholder farms (Sharma and Garg 2019). Smallholder farms may influence family members' nutrition by direct consumption or by giving revenue that allows them to purchase locally farmed food (Bekele et al. 2022). Many families face daily hunger and poverty as a consequence of rural unemployment, which has a detrimental effect on the early years of children (Kaur 2020). These families lack long-term employment and food security, while receiving government grants and assistance through school feeding programs (Aslam et al. 2019). Additionally, malnutrition adversely affects all aspects of an individual's life, with inadequate nutrition during early childhood leading to long-term consequences and lasting changes within households. To mitigate the impact of food price shocks, there is an increasing global interest in promoting and expanding local food production initiatives (Oyege and Balaji Bhaskar 2023).

9.4.1 Challenges of Small-Scale Farmers

Smallholder farms are facing a number of challenges, both natural and man-made, that impact their capacity to increase productivity and migrate to profitable farming methods. These obstacles force farmers to adopt lower-yielding, lower-risk farming strategies, such as those that produce little to no return, which perpetuates the cycle of poverty. Small-farm women, who account for 43% of the agricultural labor force in developing nations, confront special challenges in accessing productive resources such as markets, innovations, farm animals, land, farming commodities, extension, and funding (FAO 2020). However, women are important for enhancing agricultural productivity, improving nutritional status and security of food, and advancing overall growth. In many developing countries, young people, who have the potential to enter the agricultural labor force with enthusiasm, joy, and innovation, find agriculture uninteresting due to high productivity constraints (Harvey et al. 2014). In recent decades, several Asian and African countries have experienced high population densities and shrinking of the sizes of farms as a result of inheritance-based land fragmentation and rapid population growth (Hernández and Reardon 2012). In addition to reducing the farm size and earnings, an analysis of the relationship between Kenya's smallholder farming systems and growing rural population density shows that, above a certain threshold, increasing rural population density is associated to declining agricultural labor productivity (Muyanga and Jayne 2014). The vast majority of smallholders live in rural areas with little access to formal financial services (Demirgüç-Kunt and Klapper 2012). By 2020, certain African countries may see a 50% reduction in rainfed agricultural yields, with small-scale farmers being the most affected (IPCC 2014). According to Harvey et al. (2014), smallholder farmers in Malawi suffered bigger financial losses during droughts than large landowners, partly because they grew more drought-sensitive crops.

Smallholder farms are particularly vulnerable to extreme weather more often due to a number of factors, including high reliance on climate-dependent agriculture, chronic food insecurity, a lack of formal safety nets, and an absence of resources and capacity to mitigate and adjust to the impacts of climate change (Harvey et al. 2014). Smallholders' income rises as they join in modern market channels; nonetheless, engagement is influenced by a range of nonland assets, and there is a conflicting evidence on how much a farm's size affects involvement. Nonland resources include rural infrastructure (such as irrigation and road access), cooperative membership, education, participation in modern agricultural markets, and non-farm jobs in rural regions (Hernández and Reardon 2012; Rao and Qaim 2011). Due to a lack of knowledge (about price, supply and demand, and quality requirements), smallholder farmers experience lower access to markets and higher pricing from opportunistic middlemen and businessmen (Omiti et al. 2009). Food supply chains are now geographically longer but have fewer participants owing to rapid urbanization, globalization, and economic growth (Reardon et al. 2009).

9.5 Vermicompost as Biofertilizer for Improving Productivity in African Smallholder Agroecosystems: Opportunities and Challenges

Smallholder farmers in Zimbabwe rely on integrated crop-livestock farming to achieve food security, nutrition, and income. However, low productivity in both sectors poses challenges. Crop yields are limited by poor soil fertility and reliance on unpredictable rain-fed agriculture, while livestock production suffers from inadequate and poor-quality feed, particularly during dry seasons (MLAFWRD 2022). Free-range livestock depends heavily on grass and agricultural residues, which are both climate-sensitive and low in nutrients (Habermann et al. 2019). Soils in much of Zimbabwe are nitrogen-deficient and infertile (Soropa et al. 2018). Smallholders attempt to improve soil fertility using both organic and inorganic fertilizers. Organic options include livestock manure, crop residues, nitrogen-fixing crops, leaf litter, and termitaria (Giller et al. 2011). Due to the high cost of synthetic fertilizers, smallholders often rely on local organic inputs. However, overuse of manure and chemicals leads to environmental issues such as nitrogen volatilization, soil acidification, and water eutrophication (Joensuu et al. 2021). In South Africa, although fertilizer production meets domestic demand, rising global prices—driven by crises such as the Ukraine–Russia war—have raised costs for farmers (Rajkhowa et al. 2017; Rehman et al. 2023). While subsidies exist under the Comprehensive Agriculture Support Programme (CASP), their impact is diminishing due to inflation. Organic fertilizers are emerging as alternatives but face economic and competitive limitations (Sharma and Garg 2019). Moreover, many smallholders lack access to adequate extension services and reliable agronomic advice (Bekele et al. 2022). On-farm composting, as practiced in biodynamic farming, offers a promising solution, reducing dependence on external inputs and enhancing sustainability (Anbazhagan et al. 2024; Sande et al. 2024). To improve productivity and resilience, integrating biofertilizers like vermicompost into smallholder systems is vital for soil health, climate mitigation, and long-term food security. According to Ali et al. (2015), vermicompost application has been shown to improve crop yields in vegetables, cereals, and fruit trees under smallholder settings. Furthermore, it enhances soil organic matter, promotes carbon sequestration, and supports long-term soil health—an essential requirement for sustainable agriculture. Vermicomposting presents numerous opportunities for smallholder farmers across Africa, offering a sustainable, low-cost solution to persistent challenges such as declining soil fertility, limited access to chemical fertilizers, and the impacts of climate change (Fig. 9.2).

- *Soil fertility enhancement*: Vermicompost is rich in macro- and micronutrients, humus, and beneficial microbes, which improve soil structure, increase water retention, and enhance nutrient availability (Gómez-Brandón et al. 2011b). This is especially beneficial in African soils that are often nutrient-depleted (Nyamangara et al. 2014).

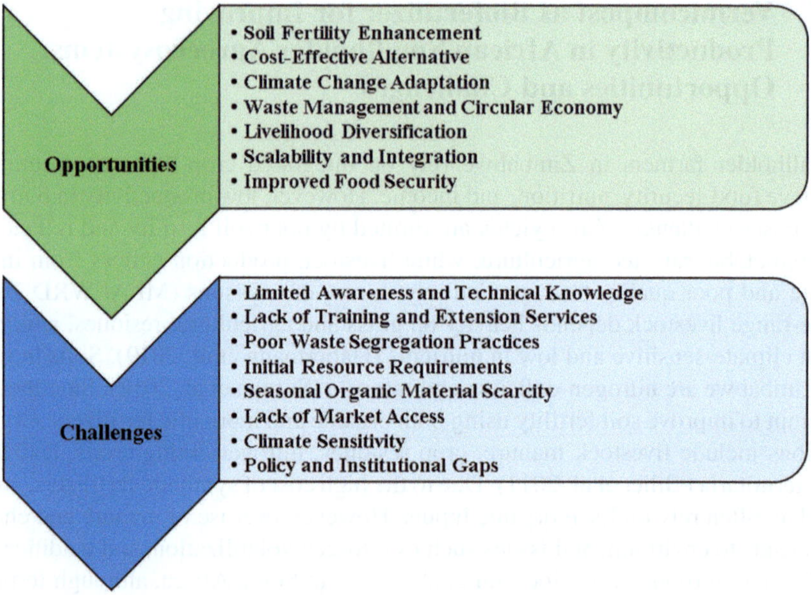

Fig. 9.2 Opportunities and challenges faced by smallholder farmers in Africa

- *Cost-effective alternative*: With rising costs and limited availability of synthetic fertilizers due to global crises (FAO 2022), vermicomposting offers a practical and affordable alternative using on-farm organic waste like crop residues and livestock manure resources readily available to most smallholders (Giller et al. 2011).
- *Climate change adaptation*: Vermicomposting contributes to climate-smart agriculture by reducing greenhouse gas emissions from unmanaged manure and by sequestering carbon in soils (Lim et al. 2015). It also helps farmers build more resilient agroecosystems in the face of erratic weather patterns.
- *Waste management and circular economy*: This practice supports the circular economy by recycling farm and household organic waste into valuable fertilizer, thus reducing environmental pollution and enhancing farm sustainability (Ali et al. 2015).
- *Livelihood diversification*: Surplus vermicompost can be packaged and sold locally, offering an additional income stream for rural households. It also promotes agroecological entrepreneurship, particularly among youth and women.
- *Scalability and integration*: Vermicomposting can be scaled from household to cooperative levels and easily integrated into conservation agriculture, agroforestry, or mixed farming systems already practiced by smallholders (Sharma and Garg 2019).
- *Improved food security*: By enhancing crop yields and soil health, vermicomposting can directly contribute to increased food production and improved household nutrition among smallholder communities (Dixon et al. 2020)

Despite its promises, vermicomposting also presents several challenges for smallholder farmers in Africa (Fig. 9.2), limiting its widespread adoption and effectiveness:

- *Limited awareness and technical knowledge*: Many smallholder farmers lack proper awareness of vermicomposting and its benefits. Knowledge gaps about species selection (e.g., *Eisenia fetida*), composting ratios, and optimal environmental conditions can hinder successful implementation (Anbazhagan et al. 2024).
- *Lack of training and extension services*: Government extension services are often underfunded or focused on promoting synthetic inputs, with limited support for biofertilizer education or demonstration projects (Bekele et al. 2022). This prevents smallholders from accessing the necessary skills to begin and maintain vermicomposting units.
- *Initial resource requirements*: While low-cost compared to synthetic fertilizers, vermicomposting still requires some initial inputs such as suitable earthworms, bins, and organic waste materials, which may not be readily available or affordable in remote rural areas (Sharma and Garg 2019).
- *Seasonal organic material scarcity*: During dry seasons, agricultural residues and animal manure essential for composting are often redirected for livestock feeding, limiting their availability for vermicomposting (Mapfumo and Giller 2001).
- *Poor waste segregation practices*: In many rural areas, organic and inorganic waste are mixed, which complicates the collection of clean feedstocks for vermicomposting and may pose risks to earthworm health (Lim et al. 2015).
- *Lack of market access*: While vermicompost can be sold for income, weak rural market infrastructure, low consumer awareness, and limited certification or quality control can prevent smallholders from monetizing their product effectively (Rajkhowa et al. 2017).
- *Climate sensitivity*: Vermicomposting requires specific temperature and moisture ranges for optimal performance. Extreme weather conditions, such as prolonged drought or floods are common due to climate change, and can reduce the efficiency and worm survival (Bennett et al. 2019).
- *Policy and institutional gaps*: Most agricultural policies in Africa continue to favor chemical fertilizers over organic inputs. A lack of supportive policies, incentives, or subsidies for composting technologies hampers large-scale adoption among smallholders (FAO 2020; MLAFWRD 2022).

9.6 Future Perspective

Organic fertilizers and biofertilizers (OFBFs), including vermicompost, represent a promising path toward sustainable agricultural intensification in African smallholder systems. However, realizing their full potential requires substantial investment in research and development. Much of the current knowledge on OFBF, including vermicomposting techniques, originates from studies conducted outside Africa and often dates back several decades. There is a pressing need for updated, locally relevant research that explores efficient, low-cost, and scalable production technologies. Innovations that simplify vermicomposting practice such as optimizing feedstock mixtures or developing locally adaptable earthworm species could greatly enhance productivity while reducing input cost for farmers. Future studies may also establish how best to integrate all forms of fertilizers, organic, mineral, and internally sourced inputs like vermicompost to achieve optimal soil fertility and yield outcomes under varying agroecological conditions. Vermicomposting fits well within this integrated approach due to its dual role in nutrient recycling and waste management, aligning closely with circular economy principles. The global push for a circular economy, particularly strategies aimed at minimizing carbon emissions by 2050, is expected to positively influence the uptake and institutional support for vermicomposting and related OFBF technologies. Policies such as Rwanda's Green Growth and Climate Resilience Strategy offer valuable models, emphasizing low-emission agricultural practices and sustainable soil fertility management. If governments scale up production, improve distribution, and implement subsidies for vermicompost and other organic inputs, these products could become far more accessible and affordable. In the wake of mineral fertilizer shortages and price surges exacerbated by geopolitical conflicts like the Russia–Ukraine war, recognizing OFBF as viable alternatives is increasingly urgent. Empowering farmers through targeted training in composting and vermicomposting techniques is essential. Extension services should emphasize best practices, such as moisture control and optimal C:N ratios, while promoting vermicomposting through community outreach, radio, and social media. Training should be coupled with the provision of key inputs, such as lime to combat soil acidity and starter cultures of earthworms. Moreover, the expansion of OFBF production hinges on supportive government policies. The Rwanda Organic Agriculture Movement (ROAM) and other stakeholders advocate for including vermicompost and related inputs in subsidy programs. Financial support in the form of grants can also stimulate private-sector investment in vermicompost production. Although initial funding remains a challenge, the growing demand for organic fertilizers due to mineral fertilizer constraints presents an opportunity. With improved access to organic waste and increased farmer awareness, vermicomposting can play a transformative role in closing the nutrient loop and enhancing resilience in African smallholder agroecosystems.

9.7 Conclusions

- Vermicompost has enormous potential as a nutrient-rich biofertilizer in agricultural systems. Its high nutrient content improves soil health, plant growth, and crop yields, making it an invaluable resource for sustainable agriculture.

- Vermicompost is a cost-effective, eco-friendly, and locally adaptable substitute to chemical fertilizer, making it beneficial for small-scale farmers.
- Integrating vermicompost into agricultural practices fosters circular resource utilization, waste reduction, and resilience to environment.
- Overall, vermicompost is a realistic, efficient, and sustainable solution for improving global food security and livelihoods in smallholder farming systems.

References

Abu Hatab A (2022) Africa's food security under the shadow of the Russia-Ukraine conflict. Strateg Rev South Afr 44(1):37–46

Ahemad M, Kibret M (2014) Mechanisms and applications of plant growth promoting rhizobacteria: current perspective. J King Saud Univ Sci 26(1):1–20

Aira M, Dominguez J (2009) Microbial and nutrient stabilization of two animal manures after the transit through the gut of the earthworm *Eisenia fetida* (Savigny, 1826). J Hazard Mater 161:1234–1238

Ali U, Sajid N, Khalid A, Riaz L, Rabbani MM, Syed JH, Malik RN (2015) A review on vermicomposting of organic wastes. Environ Prog Sustain Energy 34(4):1050–1062

Ammar EE, Aioub AA, Elesawy AE, Karkour AM, Mouhamed MS, Amer AA, El-Shershaby NA (2022) Algae as bio-fertilizers: between current situation and future prospective. Saudi J Biol Sci 29:3083–3096

Anbazhagan M, Meenambal M, Banu GS (2024) From waste to wonder: Temple waste transformed into vermicompost. World J Pharm Pharm Sci 13:1115–1122. https://doi.org/10.20959/wjpps20244-26997

Aslam Z, Bashir S, Hassan W, Bellitürk K, Ahmad N, Niazi NK, Maitah M (2019) Unveiling the efficiency of vermicompost derived from different biowastes on wheat (*Triticum aestivum* L.) plant growth and soil health. Agronomy 9(12):791

Atteya AK, Albalawi AN, El-Serafy RS, Albalawi KN, Bayomy HM, Genaidy EA (2021) Response of *Moringa oleifera* seeds and fixed oil production to vermicompost and NPK fertilizers under calcareous soil conditions. Plants 10(10):1998

Bamdad H, Papari S, Lazarovits G, Berruti F (2022) Soil amendments for sustainable agriculture: microbial organic fertilizers. Soil Use Manag 38(1):94–120

Bekele G, Dechassa N, Tana T (2022) Effect of inorganic and organic fertilizers on productivity of groundnut (*Arachis hypogaea* L.) varieties in East Hararghe, Eastern Ethiopia. Oil Crop Sci 7(3):112–121

Bennett NJ, Cisneros-Montemayor AM, Blythe J, Silver JJ, Singh G, Andrews N, Sumaila UR (2019) Towards a sustainable and equitable blue economy. Nat Sustain, 2(11):991–993

BioFit (2015) Biofertilizers towards sustainable agricultural development. BioFit Project Report—2015 1 BG01 KA202 014258—Funded by Erasmus of the European Union. Retrieved from https://www.bio-fit.eu/bio-fit-book

Demirgüç-Kunt A, Klapper LF (2012) Measuring financial inclusion: the global findex database. In: World Bank Policy Research Working Paper, vol 6025

Dixon J, Garrity DP, Boffa J, Williams TO, Amede T, Auricht C, Lott R, Mburathi G (2020) Farming systems and food security in Africa: priorities for science and policy under global change, Earthscan food and agriculture series. Routledge - Earthscan, Oxon, p 638. https://hdl.handle.net/10568/109061

Dominguez J, Edwards CA (2004) Vermicomposting organic wastes: a review. In: Soil zoology for sustainable development in the 21st century. Cairo, pp 369–395

du Jardin P (2015) Plant biostimulants: definition, concept, main categories and regulation. Sci Hortic 196:3–14
Edwards CA (2004) The importance of earthworms as key representatives of the soil fauna. Earthworm Ecol 2:3–11
EU (2019) Regulation (EU) 2019/1009 of the European Parliament and of the Council of 5 June 2019 laying down rules on the making available on the market of EU fertilising products and amending Regulations (EC) No 1069/2009 and (EC) No 1107/2009 and repealing Regulation (EC) No 2003/2003. Off J Eur Union (EU), 25 June 2019
Fan S, Rue C (2020) The role of smallholder farms in a changing world. In: The role of smallholder farms in food and nutrition security. Springer, pp 13–28
FAO (2015) Regional overview of food insecurity: African food security prospects brighter than ever. Food and Agriculture Organization of the United Nations, Rome. https://openknowledge.fao.org/server/api/core/bitstreams/7333f34a-b229-409f-adfd-baa16ab465c1/content
FAO (2020) Africa regional overview of food security and nutrition 2019. Food and Agriculture Organization of the United Nations (FAO), Rome. https://doi.org/10.4060/CA7343EN
FAO (2021) Assessing the impact of the COVID-19 pandemic on agriculture, food security and nutrition in Africa. Food and Agriculture Organization of the United Nations (FAO), Rome. Retrieved from https://www.fao.org/3/cb5911en/cb5911en.pdf
FAO (2022) The importance of Ukraine and Russian Federation for global agriculture markets and the risks associated with the war in Ukraine. Food and Agriculture Organization of the United Nations (FAO), Rome. Retrieved from https://www.fao.org/3/cb9013en/cb9013en.pdf
Giller KE, Corbeels M, Nyamangara J, Triomphe B, Affholder F, Scopel E, Tittonell P (2011) A research agenda to explore the role of conservation agriculture in African smallholder farming systems. Field Crop Res 124(3):468–472
Gómez-Brandón M, Aira M, Lores M, Dominguez J (2011a) Epigeic earthworms exert a bottleneck effect on microbial communities through gut associated processes. PLoS One 6(9):e24786
Gómez-Brandón M, Aira M, Lores M, Dominguez J (2011b) Changes in microbial community structure and function during vermicomposting of pig slurry. Bioresour Technol 102:4171–4178
Gómez-Brandón M, Martínez-Cordeiro H, Dominguez J (2021) Changes in the nutrient dynamics and microbiological properties of grape marc in a continuous-feeding vermicomposting system. Waste Manag 135:1–10
Habermann E, Cohn AS, Herrero M (2019) Climate change effects on livestock feed resources: a review. J Fish Livest Prod 7(2):259
Harvey CA, Rakotobe ZL, Rao NS, Dave R, Razafimahatratra H, Rabarijohn RH et al (2014) Extreme vulnerability of smallholder farmers to agricultural risks and climate change in Madagascar. Philos Trans R Soc B Biol Sci 369(1639):20130089
Hazell P (2020) Importance of smallholder farms as a relevant strategy to increase food security. In: The role of smallholder farms in food and nutrition security. Springer, pp 29–43
Hernández R, Reardon T (2012) Duration analysis of farmer adoption of new technologies and markets in Nicaragua: quality-niche versus commodity vegetables. UC Davis
HLPE (2013) Investing in smallholder agriculture for food security. A report by the High-Level Panel of Experts on Food Security and Nutrition of the Committee on World Food Security. Committees on World Security, p 111. Available at: http://www.fao.org/
IAASTD (2009) Agriculture at a crossroads: sub-Saharan Africa (SSA) report, International Assessment of Agricultural Knowledge, Science and Technology for Development (IAASTD), vol V. Island Press, Washington, DC
Intergovernmental Panel on Climate Change (2014) Climate change 2014: impacts, adaptation, and vulnerability. In: Contribution of working group II to the fifth assessment report of the Intergovernmental Panel on Climate Change. Cambridge University Press. https://www.ipcc.ch/report/ar5/wg2/
Joensuu K, Rimhanen K, Heusala H, Saarinen M, Usva K, Leinonen I, Palosuo T (2021) Challenges in using soil carbon modelling in LCA of agricultural products the devil is in the detail. Int J Life Cycle Assess 26(9):1764–1778

Jones A, Breuning-Madsen H, Brossard M, Dampha A, Deckers J, Dewitte O et al (2013) Soil atlas of Africa. European Commission
Kaur T (2020) Vermicomposting: an effective option for recycling organic wastes. In: Organic agriculture. IntechOpen, pp 1–17
Keats S, Wiggins S (2017) Future diets: implications for agriculture and food prices. Res Rep Stud. https://policycommons.net/artifacts/4847832/future-diets/5684546/
Leoni B, Loconsole D, Cristiano G, De Lucia B (2019) Comparison between chemical fertilization and integrated nutrient management: yield, quality, N, and P contents in Dendranthema grandiflorum (Ramat.) Kitam. cultivars. Agronomy, 9(4):202
Lim SL, Wu TY, Lim PN, Shak KPY (2015) The use of vermicompost in organic farming: overview, effects on soil and economics. J Sci Food Agric 95(6):1143–1156
Lowder SK, Skoet J, Singh S (2014) What do we really know about the number and distribution of farms and family farms in the world? Background paper for The State of Food and Agriculture 2014. https://www.fao.org/4/i3729e/i3729e.pdf
Mapfumo P, Giller KE (2001) Soil fertility management strategies and practices by smallholder farmers in semi-arid areas of Zimbabwe. ICRISAT. https://research.wur.nl/en/publications/soil-fertility-management-strategies-and-practices-by-smallholder
MLAFWRD (2022) Zimbabwe First Round Crop and Livestock Assessment Report 2021/2022 Season March 2022 (Issue March). MIN, p 97
Muyanga M, Jayne TS (2014) Effects of rising rural population density on smallholder agriculture in Kenya. Food Policy, 48:98–113
Mondal T, Datta JK, Mondal NK (2017) Chemical fertilizer in conjunction with biofertilizer and vermicompost induced changes in morpho-physiological and bio-chemical traits of mustard crop. J Saudi Soc Agric Sci 16(2):135–144
Nkonya E, Johnson T, Kwon HY, Kato E (2016) Economics of land degradation in sub-Saharan Africa. In: Economics of land degradation and improvement–a global assessment for sustainable development. Springer, Cham, pp 215–259
Nyamangara J, Marondedze A, Masvaya EN, Mawodza T, Nyawasha R, Nyengerai K et al (2014) Influence of basin-based conservation agriculture on selected soil quality parameters under smallholder farming in Zimbabwe. Soil Use Manag 30(4):550–559
Omiti JM, Otieno DJ, Nyanamba TO, McCullough EB (2009) Factors influencing the intensity of market participation by smallholder farmers: a case study of rural and peri-urban areas of Kenya. Afr J Agric Resour Econ 3(1):57–82
Oyege I, Balaji Bhaskar MS (2023) Effects of vermicompost on soil and plant health and promoting sustainable agriculture. Soil Syst 7(4):101
Raimi A, Roopnarain A, Adeleke R (2021) Biofertilizer production in Africa: current status, factors impeding adoption and strategies for success. Sci Afr 11:e00694
Rajkhowa DJ, Sarma AK, Mahanta K, Saikia US, Krishnappa R (2017) Effect of vermicompost on green gram productivity and soil health under hilly ecosystem of North East India. J Environ Biol 38(1):15
Rao EJ, Qaim M (2011) Supermarkets, farm household income, and poverty: insights from Kenya. World Dev 39(5):784–796
Ratnasari A, Syafiuddin A, Mehmood MA, Boopathy R (2023) A review of the vermicomposting process of organic and inorganic waste in soils: additives effects, bioconversion process, and recommendations. Bioresour Technol Rep 21:101332
Reardon T, Barrett CB, Berdegué JA, Swinnen JFM (2009) Agrifood industry transformation and small farmers in developing countries. World Dev 37(11):1717–1727
Rehman SU, De Castro F, Aprile A, Benedetti M, Fanizzi FP (2023) Vermicompost: enhancing plant growth and combating abiotic and biotic stress. Agronomy 13(4):1134
Robatjazi J (2023) An overview on vermicompost environmental impacts (from past to future). J Agric Sci Food Res 14(4). https://doi.org/10.35248/2469-9837.23.14.161

Sande TJ, Tindwa HJ, Alovisi AMT, Shitindi MJ, Semoka JM (2024) Enhancing sustainable crop production through integrated nutrient management: a focus on vermicompost, bio-enriched rock phosphate, and inorganic fertilizers—a systematic review. Front Agron 6:1422876

Schütz L, Gattinger A, Meier M, Müller A, Boller T, Mäder P, Mathimaran N (2018) Improving crop yield and nutrient use efficiency via biofertilization—a global meta-analysis. Front Plant Sci 8:2204

Sharma K, Garg VK (2019) Vermicomposting of waste: a zero-waste approach for waste management. In: Taherzadeh MJ, Bolton K, Wong J, Pandey A (eds) Sustainable resource recovery and zero waste approaches. Elsevier, Missouri, pp 133–164

Soropa G, Nyamangara J, Nyakatawa EZ (2018) Nutrient status of sandy soils in smallholder areas of Zimbabwe and the need to develop site-specific fertiliser recommendations for sustainable crop intensification. S Afr J Plant Soil 35(3):199–202. https://doi.org/10.1080/02571862.2018.151790

Thomas L, Singh I (2019) Microbial biofertilizers: types and applications. In: Biofertilizers for sustainable agriculture and environment. Springer, pp 1–19

UN (2022) Global impact of the war in Ukraine: billions of people face the greatest cost-of-living crisis in a generation. UN Global Crisis Response Group on Food, Energy and Finance. Policy Brief No. 2, 8 June 2022. https://unsdg.un.org/sites/default/files/2022-06/GCRG_2nd-Brief_Jun8_2022_FINAL.pdf

Vacheron J, Desbrosses G, Bouffaud ML, Touraine B, Moënne-Loccoz Y, Muller D, Prigent-Combaret C (2013) Plant growth-promoting rhizobacteria and root system functioning. Front Plant Sci 4:356

Vyas P, Sharma S, Gupta J (2022) Vermicomposting with microbial amendment: implications for bioremediation of industrial and agricultural waste. Biotechnologia 103(2):203

Williams TO (2015) Reconciling food and water security objectives of MENA and sub-Saharan Africa: is there a role for large-scale agricultural investments? Food Secur 7:1199–1209

Wonnapinij P, Sriboonlert A, Surat W (2022) Exploration of microbial communities in the guts and casts of *Eudriluseugeniae*, *Perionyx excavatus*, and *Eisenia fetida*. Folia Microbiol 67:329–337

Chapter 10
Microorganisms in Sustainable Soil Fertility Management

D. Dlamini, V. Makuvaro, B. T. Makaure, N. Munyuki, T. T. Haripo, and R. M Zishiri

Abstract Soil is a complex product of nature's forces that provides a home for countless organisms and serves as the foundation for agriculture. Soil microorganisms play a pivotal role in enhancing the function and structure of soil ecosystems. In the past century, science has revealed a crucial link between soil productivity and the activity of microorganisms. This chapter explores key groups of these microorganisms—bacteria, fungi, actinomycetes, and nematodes—and highlights their diverse roles in improving soil fertility, with a particular focus on agriculture in Africa. In this chapter, we provide an in-depth overview of the roles of soil microorganisms in soil health and soil nutrient management. We reviewed factors influencing the efficacy of soil microorganisms and explored ways to improve their effectiveness in sustainable soil fertility management. The importance of maintaining a diverse and healthy microbial community for sustainable agriculture and ecosystem management is emphasized throughout, particularly in addressing the unique challenges faced by African farmers. Potential applications of microbial-based approaches and microbe-centered best practices for soil fertility enhancement, specifically tailored to the needs of African farmers, are also explored. This chapter concludes by highlighting the need for further research and development of biofertilizers specifically adapted to climate change and other challenges of African agriculture. It advocates for policies and initiatives that promote the widespread adoption of biofertilizers, recognizing their potential to revolutionize sustainable farming practices and contribute to food security and environmental well-being across the continent.

Keywords Food security · Biofertilizers · Microorganisms · Microbial associations · Sustainable farming practices

---

D. Dlamini (✉) · V. Makuvaro · B. T. Makaure · N. Munyuki · T. T. Haripo · R. M. Zishiri
Department of Agronomy & Horticulture, Faculty of Agriculture, Environment & Natural Resources Management, Midlands State University, Gweru, Zimbabwe
e-mail: dlaminid@staff.msu.ac.zw

A. D. Nciizah et al. (eds.), *Biofertilizers for Sustainable Agriculture in Africa*,
Sustainability Sciences in Asia and Africa, https://doi.org/10.1007/978-981-95-6009-7_10

10.1 Introduction

Africa possesses a vast agricultural capacity capable of feeding not only its own rapidly growing population but also contributing to the world's food supply. Agriculture is a livelihood source for approximately 60% of the population in low-income African countries (Africa Development Bank 2021). This population comprises smallholder farmers who mostly practice both crop and animal husbandry (mixed farming). Declining soil fertility and limited capacity of these farmers to acquire fertilizers, which are largely the non-ecofriendly chemical fertilizers, hinder agricultural production. The situation is further exacerbated by increasing climate variability and change, as changes in precipitation and temperature affect nutrient availability and use efficiency. A combination of these unpleasant scenarios calls for effective and sustainable soil fertility management practices.

Soil fertility management influences the yield and quality of crops through the maintenance of optimum nutrient levels and improving nutrient use efficiency. It also offers environmental benefits such as erosion control and improvement in water infiltration and retention through the use of suitable organic matter. The use of organic sources rather than inorganic fertilizer sources reduces environmental pollution. The addition of organic matter to the soil and crop rotations can encourage beneficial microbes involved in nutrient cycling and soil structure improvement. Efficient soil fertility management is critical for sustainable agriculture and the choice of method to use is becoming even more important in the face of climate variability and change. The use of biofertilizers is one sustainable way of maintaining an optimum supply of nutrients to crops as well as enhancing nutrient use efficiency (Ortiz and Sansinenea 2022; Fadiji et al. 2024). The biofertilizers can be applied to seed surfaces, soil, or plant surfaces such as roots (Abbasniayzare et al. 2012). Despite the numerous benefits and research done on biofertilizers, their global market share, compared to chemical fertilizers, is only approximately 1% (Sansinenea 2021; Joshi and Gauraha 2022). For Africa, the biofertilizer market size was estimated at USD 226.59 million in 2024 and is expected to reach USD 354.28 million by 2029, growing at a Compound Annual Growth Rate (CAGR) of 9.35% during the period 2024–2029 (Africa Biofertilizer Market Report 2024). Rhizobium biofertilizer has the largest market share in Africa, accounting for approximately 46% (Africa Biofertilizer Market Report 2024). There is potential to increase the adoption of biofertilizers in Africa given their ability to support plant growth, eco-friendly nature, and the increasing demand for organically produced food.

The objective of this chapter is to elucidate the nature, significance, and potential use of microbes in soil fertility management with the ultimate goal of promoting the adoption of biofertilizers. This chapter comprises four broad subject areas: (i) major groups of microorganisms and their role in soil fertility management; (ii) soil factors influencing the effectiveness of microorganisms in soil fertility management; (iii) ways of improving the efficacy of microorganisms for sustainable soil management; and (iv) opportunities and challenges in the adoption of biofertilizers.

This chapter outlines the fundamental characteristics and roles of microbes beneficial in soil fertility management and provides insights into current opportunities and challenges in the use of biofertilizers. It also offers recommendations for improved effectiveness of microorganisms in soil fertility management and increased adoption of biofertilizers.

10.2 Microorganisms Beneficial in Soil Fertility Management

Soil fertility is crucial for sustainable agriculture, as it significantly impacts crop yields, food security, and overall ecosystem health. Fungi, bacteria, nematodes, and actinomycetes are examples of beneficial microorganisms that are essential to preserving soil fertility. This section describes various microbes and their roles in soil fertility management.

10.2.1 Roles of Fungi in Soil Fertility Management

Fungi are especially important among other microorganisms because of their many roles in disease prevention, soil structure enhancement, and nutrient cycling (Zhang et al. 2023). They play a vital role in maintaining and enhancing soil fertility through symbiotic relationships, organic matter decomposition, and microbial interactions. Mycorrhizal fungi, for example, arbuscular mycorrhizae, form mutualistic associations with plant roots, thereby increasing phosphorus and nitrogen uptake while improving drought resistance and pathogen defense (Smith and Read 2010; Liu et al. 2018). Saprophytic fungi, such as *Trichoderma* and *Aspergillus*, break down lignin and cellulose in organic matter, releasing bioavailable nutrients and stabilizing soil carbon (De Boer et al. 2015; Baldrian 2017). Fungi increase soil fertility by fixing nitrogen, breaking down organic matter, solubilizing phosphorus, and enhancing soil structure. Table 10.1 shows different types of fungi and their roles in soil fertility management.

10.2.2 Roles of Bacteria in Soil Fertility Management

Bacteria are the largest subgroup of microorganisms and are essential in maintaining ecological balance because of their presence across varying conditions of temperature, pH, and low water potential (Shaughnessy 2023). The most important functions of bacteria in crop production are in soil fertility enhancement and include biological nitrogen fixation (BNF), phosphorus solubilization, potassium mobilization, and nutrient cycling, which are the sought-after properties in biofertilizer formulation (Shaughnessy 2023). They are classified based on various criteria, including morphology, metabolism, and genetic makeup. Bacteria exhibit diverse

Table 10.1 Soil fertility-enhancing fungi: functions and references

Fungi family/genus	Functions	Mechanism	References
Mycorrhizal fungi (e.g., arbuscular mycorrhiza, ectomycorrhizal fungi, and ericoid mycorrhizal fungi)	Fix atmospheric nitrogen, involved in nitrogen metabolism in legumes, enhance nutrient uptake by plants, improve plant drought tolerance, soil structure, and water retention	Form tripartite symbioses with legumes and nitrogen-fixing bacteria (e.g., *Rhizobium*), form hyphal networks for nutrient exchange and produce glomalin for soil aggregation	Mohamed et al. (2021) and Wang et al. (2023)
Saprohytic fungi (e.g., *Aspergillus* and *Trichoderma*)	Decompose organic materials to release nutrients, recycle carbon and nutrients, improve phosphorus solubility, enhance phosphorus uptake, and promote bioremediation	Secrete enzymes such as cellulases and ligninases, and mineralize nitrogen and phosphorus	Yu et al. (2024) and Sachdev and Singh (2020)
Endophytic fungi (e.g., *Epichloe*)	Suppress disease, promote plant growth, and induce stress resistance	Produce phytohormones such as auxins, synthesize stress-related metabolites (osmolyte accumulation)	Baron and Rigobelo (2022) and Rodriguez and Sanders (2015)

shapes, and the most common are spherical (cocci), rod-shaped (bacilli), spiral (spirilla), and comma-shaped (vibrios) (Mohamad et al. 2014). In soil ecosystems, several bacterial families contribute significantly to nutrient cycling, plant growth promotion, and disease suppression. Some key families are summarized in Table 10.2.

In BNF, specialized bacteria (diazotrophs) convert atmospheric nitrogen (N_2) into plant-available ammonia (NH_3) using the nitrogenase enzyme complex (Dunfield and Germida 2021). While N_2 constitutes about 80% of the atmosphere, plants cannot utilize it directly. Certain bacteria, including both free-living species and those in symbiotic relationships with legumes, transform this inert nitrogen into usable forms, enriching soil fertility (Khosro and Yousef 2012; Dunfield and Germida 2021). Phosphorus, though essential for plant growth, is frequently locked in insoluble soil forms inaccessible to plants. Phosphorus-solubilizing bacteria (PSB) transform these unavailable forms into soluble phosphates through three key mechanisms summarized below (Rodŕiguez and Fraga 1999; Bargaz et al. 2021):

- Organic acid secretion (citric, malic, oxalic acids) that chelates metal ions, releasing bound phosphates.
- Enzymatic action (phosphatases) to mineralize organic phosphorus.
- Proton (H^+) release that acidifies the rhizosphere, enhancing phosphate solubility.

These microbial processes improve phosphorus bioavailability, reducing dependence on synthetic fertilizers.

Table 10.2 Soil fertility-enhancing bacteria: functions and references

Bacterial family/genus	Role in soil fertility enhancement	References
Rhizobiaceae (e.g., *Rhizobium*)	Biological nitrogen fixation (BNF) in legumes, converting atmospheric N_2 into plant-available NH_3	Dunfield and Germida (2021) and Khosro and Yousef (2012)
Azotobacteraceae (e.g., *Azotobacter*)	Phosphorus solubilization, potassium mobilization, and production of phytohormones (e.g., cytokinins)	Bargaz et al. (2021), Nawaz et al. (2023), and Blake et al. (2021)
Pseudomonadaceae (e.g., *Pseudomonas fluorescens*)	Phosphorus solubilization, siderophore production (iron chelation), and disease suppression	Rodríguez and Fraga (1999); Rigobelo et al. (2022), and Bonaterra et al. (2022)
Actinomycetaceae (e.g., *Streptomyces*)	Decomposition of organic matter, nutrient cycling, and production of antimicrobial compounds	Rigobelo et al. (2022) and Dutilloy et al. (2022)
Thiobacillus spp.	Sulfur oxidation, and converting sulfide/elemental sulfur into plant-available sulfate	Friedrich et al. (2001); Narayan et al. (2022)
Burkholderia spp. (e.g., B. phytofirmans)	Stress tolerance enhancement (drought, salinity, and heat) via ACC deaminase activity	Muhammad et al. (2024) and Gamalero and Glick (2015)
Azospirillum spp.	Phytohormone production (auxins), promoting root growth, and nutrient uptake	Patten and Glick (2002)
Desulfovibrio spp.	Sulfate reduction in anaerobic conditions, and influencing sulfur cycling in waterlogged soils	Friedrich et al. (2001)

Potassium (K) is a vital macronutrient for plant functions like photosynthesis and enzyme activation, yet much soil K exists in insoluble mineral forms (Olaniyan et al. 2022). Potassium-solubilizing bacteria (KSB) enhance K availability through two primary mechanisms (Nawaz et al. 2023):

- Organic acid secretion (citric, malic, oxalic) to dissolve K-bearing minerals (e.g., feldspar, mica).
- Chelation of metal ions that bind K, liberating K^+ ions into soil solution.

These processes mirror phosphorus solubilization strategies, underscoring KSB's role in sustainable agriculture by reducing reliance on chemical K fertilizers.

Bacteria play a vital role in influencing the availability of sulfur and iron for plants. *Thiobacillus* spp. oxidize sulfur compounds (e.g., sulfide) into plant-absorbable sulfate, while *Desulfovibrio* reduces sulfate under anaerobic conditions (Friedrich et al. 2001). Additionally, siderophore-producing bacteria, such as *Pseudomonas* and *Bacillus*, solubilize iron by binding Fe^{3+}, thereby enhancing its uptake (Rigobelo et al. 2022). These processes are critical for plant growth, amino acid synthesis (e.g., cysteine), and metabolic functions (Narayan et al. 2022).

Plant-growth-promoting bacteria enhance development through phytohormones: auxins (root growth), gibberellins (stem elongation, germination), and cytokinins (cell division, delayed senescence). They also enhance nutrient availability, produce antimicrobials, and induce systemic resistance (Blake et al. 2021). In addition, stress-related hormones such as abscisic acid (ABA) and ethylene regulate growth by coordinating cellular processes and metabolic signaling (Brookbank et al. 2021).

Many bacteria exhibit antagonistic activity against plant pathogens, offering a natural form of disease control. Through the production of antibiotics, bacteria produce a range of antimicrobial compounds that inhibit the growth of fungal and bacterial pathogens (Dutilloy et al. 2022). Beneficial bacteria compete with pathogens for nutrients and colonization sites, effectively excluding them from the plant. Induced Systemic Resistance (ISR) is a plant defense mechanism activated by bacteria, enhancing resistance against pathogens (Blake et al. 2021). This involves signaling molecules such as jasmonic and salicylic acid, as well as lytic enzymes (e.g., chitinase, glucanase) that degrade fungal cell walls (Bonaterra et al. 2022).

Beneficial bacteria play a crucial role in sustainable African agriculture by enhancing plant growth and protecting crops (Ghimirey et al. 2024). Utilizing these microbes as biofertilizers can boost soil fertility, increase yields, and reduce dependency on synthetic inputs, fostering a more resilient farming system.

10.2.3 Roles of Nematodes in Soil Fertility Management

Soil nematodes play pivotal roles in maintaining soil ecosystem functions, including microbial biomass turnover, nutrient cycling, and organic matter decomposition, all of which directly influence plant nutrient availability (Huang et al. 2024). They accelerate the breakdown of crop residues and organic matter decomposition, thereby enhancing soil fertility (Mekonen et al. 2017). According to Maina et al. (2021), there are three mechanisms used by both direct and nondirect feeders to increase microbial activity:

- *Bioturbation*, which improves the diffusion of oxygen and nutrients.
- *Mucus secretion*, which enriches nutrient availability.
- *Grazing*, which keeps bacterial communities active and excretes ammonia to encourage nitrogen mineralization.

Some nematode species, for example, *Diplolaimelloides meyli*, *Diplolaimelloides oschei*, *Panagrolaimus paetzoldi*, and *Diplolaimella dievengatensis*, have been particularly recognized for their stimulatory effects of these processes (Mekonen et al. 2017). Nematodes drive energy flow and nutrient cycling through their broad interactions with other soil species, which extend beyond direct soil processes (Huang et al. 2024). According to Mekonen et al. (2017), they also function as bioindicators, reflecting the structure of soil communities and the operation of ecosystems.

10.2.4 Roles of Actinomycetes in Soil Fertility Management

The rhizosphere is home to the Gram-positive filamentous bacteria known as actinomycetes, which colonize plant tissues as endophytes and persist through dormant sporulation. Medicinal herbs, wheat, maize, cereal legumes, and other plants all contain them. Actinomycetes are essential for preserving the ecology of the rhizosphere and recycling nutrients (Mohamed et al. 2021). Notably, *Streptomyces* and other species are essential for soil health because they generate enzymes, antibiotics, and plant growth-promoting (PGP) chemicals. Actinomycetes play an essential role in soil fertility management, significantly contributing to sustainable agricultural practices (Zhou et al. 2024). Among their various functions, phosphate solubilization is prominently exhibited by *Streptomyces* and *Arthrobacter*, which enhance phosphorus availability in rock phosphate (RP) and tricalcium phosphate (TCP) systems, similar to *Pseudomonas* and *Bacillus* strains (Jog et al. 2016). Additionally, nitrogen fixation, which is also a critical process for plant growth, is facilitated by *Frankia*, an actinobacterium that forms symbiotic relationships with actinorhizal plants, contributing to 15% of global nitrogen fixation. Notably, *Frankia* produces specialized vesicles (*Streptomyces, Micromonospora, Actinoplanes*, and *Nocardia*) containing hopanoid lipids to protect nitrogenase from oxygen, enabling nitrogen fixation even in free-living conditions (Mohamed et al. 2021). Given their multifaceted roles in nutrient cycling, actinomycetes are indispensable for maintaining soil fertility and advancing sustainable agriculture.

10.3 Factors Influencing the Efficacy of Microorganisms in Soil Fertility Management

The efficacy of microorganisms in soil nutrient management is governed by a complex interaction of environmental factors and management practices. The optimum ranges of these factors vary among the different kinds and species of microbes. These factors and their effects are summarized in Table 10.3.

10.4 Ways of Improving the Efficacy of Microorganisms in Sustainable Soil Fertility Management

Recent advances in microbial research and biotechnology present exciting opportunities to improve soil fertility management. Several promising directions should be prioritized to maximize the efficacy of microorganisms in sustainable agriculture.

Table 10.3 Factors affecting the efficacy of microorganisms in soil nutrient management

Factor	Description	References
Soil pH	Affects microbial population composition and activity. Most bacteria prefer pH 6–7, while fungi tolerate wider ranges. Changes in pH alter bacterial growth rates and favor pH-adapted communities. Extreme pH reduces nutrient availability. Free-living nitrogen fixers show genus-specific responses to pH gradients.	Fernández-Calviño and Bååth (2010), Mayerhofer et al. (2021), Wang et al. (2017), and Abdul Rahman et al. (2021)
Temperature	Influences microbial growth rates, enzyme activity, and community composition. Metabolic rate doubles for every 10 $°C$ temperature rise to optimum temperature (Q_{10} effect). Low temperature slows metabolic activities while high temperature causes heat stress. Affects decomposition, nutrient cycling, and nitrogenase efficiency. Arbuscular mycorrhizal fungi (AMF) colonization shows temperature-dependent responses.	Briones et al. (2014), Domeignoz-Horta et al. (2020), Abdul Rahman et al. (2021), and Goicoechea et al. (2020)
Soil moisture	Directly controls microbial respiration and activities. Excess soil moisture results in the inhibition of microbial growth and activity, and there is a reduction in the mineralization of N and C. Optimal moisture content varies by microbial group; for example, fungi favor 60% moisture content. Low moisture content induces dormancy, while waterlogging favors anaerobes. Controls oxygen and nutrient diffusion, which affects microbial respiration and growth.	Borowik and Wyszkowska (2016), Domeignoz-Horta et al. (2020), Van Horn et al. (2014), and Nisa et al. (2021)
Soil aeration	Beneficial soil microorganisms primarily require oxygen for aerobic respiration and to perform various activities. However, strict anaerobes experience oxidative stress and growth inhibition at oxygen concentrations greater than 0.5%. Facultative anaerobes preferentially utilize aerobic respiration but can switch to fermentation and anaerobic respiration. Oxygen levels also affect microbial metabolism. Aerobic conditions inhibit nitrogenase; diazotrophs use biofilms/spatial separation to protect nitrogenase.	Smercina et al. (2019), Großkopf and LaRoche (2012), and Inomura et al. (2017)
Soil organic matter (SOM)	Primary energy source and nutrient reservoir that provides nutrients and energy for microbes. High SOM increases microbial biomass and diversity. Enhances soil structure, and nutrient cycling. The quality of SOM (C:N ratio) affects decomposition speed and influences plant-microbe competition for N.	Abdul Rahman et al. (2021), Mayerhofer et al. (2021), Tang et al. (2024), and Nisa et al. (2021)
Soil texture	Determines pore space, aeration, and moisture retention. Clay protects microbes but may reduce O_2 availability; sand improves nematode mobility.	Rabot et al. (2018), Hartmann and Six (2023), Papatheodorou (2004), and Nisa et al. (2021)

(continued)

Table 10.3 (continued)

Factor	Description	References
Soil structure	Affects microbial habitat and protection. Good aggregation enhances microbial activity by providing habitat and aeration. Aggregate stability enhances root-microbe interactions.	Rabot et al. (2018), Hartmann and Six (2023), Papatheodorou (2004), and Nisa et al. (2021)
Management practices	Several management practices can influence microbial activities for example tillage (disrupts hyphae and alters soil structure, organic amendments (boosts biomass), and use of agrochemicals (alter microbial community composition).	Lupwayi et al. (2017) and Dong et al. (2022)

10.4.1 Advances in Microbial Biotechnology and Soil Health Optimization

Genetic engineering offers the potential to develop microorganisms with stress tolerance, pathogen resistance, and improved nutrient solubilization (Singh and Pathak 2024a, b). Current research focuses on exploring the potential use of genetically modified microorganisms in improving nutrient cycling efficiency, for instance, strains of nitrogen-fixing bacteria in non-leguminous crops. This approach may involve adding organic amendments or specific microbes to shape beneficial microbial communities. According to Dar et al. (2013), understanding plant–microbe interactions allows agronomists to improve soil health, crop productivity, and stress resilience. Additionally, microbes can be engineered to thrive in extreme conditions such as varying temperatures, rainfall, and soil pH, thereby broadening their agricultural applicability.

10.4.2 Precision Agriculture

Advanced technologies such as transcriptomics, soil sensing, metagenomics, and remote sensing enable real-time soil condition monitoring and precision management of soil fertility (Lumbini et al. 2023). These tools can facilitate accurate tracking of microbial populations, nutrient levels, and overall soil health. According to Dar et al. (2013), farmers can identify the best microbial products for specific crops or soil types by analyzing soil microbiomes. Precision agriculture optimizes the timing and placement of microbial inoculants, thereby reducing wastage and enhancing effectiveness (Singh and Pathak 2024a, b).

10.4.3 Microbial Consortia and Synergistic Interaction

Research efforts should be focused on the development of microbial associations between a mixture of complementary species of microorganisms. It should extend deeper into the complex interactions among different soil microorganisms and their roles in nutrient cycling, disease suppression, and soil health (Lumbini et al. 2023). Singh and Pathak (2024a, b) stated that these consortia can work synergistically to effectively enhance soil fertility compared to single strains. A consortium of nitrogen-fixing bacteria, phosphate-solubilizing bacteria, and mycorrhizal fungi has the potential to support plant growth and improve overall soil health, comprehensively. Understanding and harnessing these interactions of microorganisms can lead to more robust and versatile microbial products (Dinesh et al. 2010) that improve their efficacy in sustainable soil fertility management.

10.4.4 Formulations and Deliverable Innovations

Research into novel formulations, such as encapsulations and the combination of biochar, organic inputs, and microbe-rich compost, has the potential to provide more effective and versatile soil fertility solutions (Lumbini et al. 2023; Singh and Pathak 2024a, b). This can be achieved through improved shelf life and viability of microbial products that ensure that they remain active for a longer period of time and perform better once applied to the soil. More gradual and sustained activity of microbes in the soil can be possible through controlled-release formulations and slow-release carriers. This innovation helps microorganisms to establish more effectively, even in less ideal soil conditions (Dar et al. 2013), and reduces the need for frequent reapplications (Singh and Pathak 2024a, b).

10.4.5 Enhanced Knowledge Dissemination and Farmer Training

It is essential to bridge the gap between research and farmers. This can be achieved through putting priority on the development of effective methods of disseminating research and educating farmers about the benefits of biological methods (Lumbini et al. 2023). These can include demonstration programs, demonstration farms, and farmer field schools, which can be instrumental in promoting microbial management practices (Singh and Pathak 2024a, b). The involvement of farmers in research design and implementation can lead to more practical and context-specific solutions to enhance the adoption of biological soil fertility management (Lumbini et al. 2023). Moreover, farmers that are using microbial solutions can be incentivized for adopting sustainable technology (Pahalvi et al. 2021) and allowed to conduct peer education of other farmers. Governments and agricultural organizations can support this initiative through playing a pivotal role in making microbial products more accessible and affordable, particularly to smallholder farmers in developing regions.

10.4.6 Crop-Specific Approaches

Crop nutrition and balanced fertilization, both from organic and inorganic sources, were suggested as some of the primary considerations for satisfactory crop growth and production priorities (Noulas et al. 2023). Balanced fertilization refers to the application of plant nutrients in optimum quantities, in the right proportions through appropriate methods and at the right time for a specific crop's needs and agro-climatic conditions (IFA 2009; IPNI 2012; Johnston and Bruulsema 2014). Therefore, tailoring biological soil fertility management to specific crop requirements can help to optimize results (Lumbini et al. 2023). Research should focus on the development of practices that are crop-specific, at the same time aligning with different growth stages and nutritional needs of the crop.

10.4.7 Integrated Nutrient Management (INM)

Microbial inoculant cultures are now considered a key component of plant nutrient management and fertility restoration that enhance microbial health, maintain soil fertility, and improve crop growth and productivity, thus making a contribution to a sustainable agro-ecosystem (Harish et al. 2022; Kumar et al. 2022). Nutrient availability for crops, combined with maintenance of soil health and sustainability, can be optimized through detailed and vast research in INM, a holistic approach that combines various sources of nutrients, including both organic and inorganic sources.

10.4.8 Scaling Up for Large-Scale Adoption

Biological soil fertility management is seemingly limited to small-scale production. Research should address the scalability of biological methods to suit large-scale agricultural operations, ensuring their feasibility and impact on a broader scale (Lumbini et al. 2023). According to Lumbini et al. (2023) and (Singh and Pathak 2024a), policies should be put in place to incentivize and promote the adoption of sustainable agricultural practices at large-scale production. Certification systems that validate microbial products can build farmer confidence among large-scale farmers and ensure that effective, viable strains are made available in the market (Dinesh et al. 2010).

10.5 Opportunities in the Adoption of Microorganisms in Soil Fertility Management

There is great potential in the use of beneficial microbes as biofertilizers in soil fertility management. Several opportunities presented to farmers impose the adoption of strategies that increase crop yield and reduce environmental impact (Altomare and Tringovska 2011). Improving soil health is important and critical with reference to the changing climate. Some of the opportunities are presented below.

10.5.1 Enhancement of Soil Health

Microorganisms play a crucial role in enhancing soil health. The use of microorganisms, particularly plant growth-promoting microorganisms (PGPMs), can significantly improve soil health by enhancing nutrient availability and cycling (AbdElgawad et al. 2021). These microorganisms contribute to the breakdown of organic matter, making essential nutrients more accessible to plants. This process not only boosts crop yields but also promotes a more resilient soil ecosystem. Over time, the use of biofertilizers derived from microorganisms enhances the fertility and structure of the soil (Mihoub et al. 2023). Fungi play an important role in improving soil structure by means of their mycelium networks (Ammar et al. 2022). As they bind soil particles together, they create stable aggregates that enhance soil water retention and aeration. Better root penetration and access to moisture and nutrients are made possible by well-structured soil. Additionally, by stabilizing the soil matrix, microbes help stop soil erosion. Furthermore, through bioremediation processes, some microorganisms can break down contaminants in disturbed soils (Abatenh et al. 2017). By converting dangerous compounds like pesticides or heavy metals into less toxic forms, they can enhance the general health of the soil and make it safer for plant growth.

10.5.2 Improved Nutrient Cycling and Availability

Generally, beneficial rhizosphere microorganisms can boost plant growth via multiple regulatory biochemical pathways that include manipulating the plant hormonal signaling, preventing pathogenic microbial strains, and increasing the bioavailability of soil nutrients through processes discussed in Table 10.1 (Bharti et al. 2017; Singh 2019). P-solubilizing microorganisms and mineral fertilizers like rock phosphate have frequently produced positive outcomes when used together in field settings (Singh 2019; Beleri 2023). Microorganisms may lessen the negative effects of an overabundance of micronutrients, which can happen in acidic or contaminated soils. Using PGPM as biofertilizers is a practical way to obtain safer food while lowering the financial and environmental costs associated with the widespread use of synthetic fertilizers (Beleri 2023).

10.5.3 Improved Crop Resilience

Microorganisms can help crops tolerate biotic (pests and diseases) and abiotic (drought, salinity) stresses. For instance, certain PGPMs can enhance plant resistance to pathogens or improve water uptake efficiency during periods of drought, thereby ensuring better crop performance under challenging conditions. When it comes to providing resistance against different pathogen attacks, PGPMs have proved to be a viable alternative to chemical agents in the form of biological agents (Bhardwaj et al. 2014). Microbial biofertilizers contribute to disease resistance by promoting a healthy rhizosphere where beneficial microorganisms outcompete pathogenic ones for resources (Mihoub et al. 2023). The presence of these beneficial microbes creates a more balanced ecosystem in the soil, reducing the incidence of diseases that could otherwise compromise crop yield. Antibiotics and other antimicrobial substances that prevent the growth of diseases in the soil are produced by some bacteria and fungi.

10.5.4 Economic Benefits for Smallholder Farmers

Utilizing beneficial microorganisms can lead to cost savings for smallholder farmers who often face financial constraints due to expensive agrochemical inputs. Microbial-based biofertilizers might lessen the demand for conventional fertilizers, which can be expensive and worsen soil quality (Nabyonga et al. 2022). According to some research, farmers may be able to cut their fertilizer expenses by as much as 50% by utilizing biofertilizers (Mihoub et al. 2023). Using biofertilizers can also enhance soil health, lowering the demand for additional agricultural inputs and limiting farming's negative environmental effects. Similarly, demand for organic produce that is produced using environmentally friendly methods such as microbial-based biofertilizers is rising as customers' awareness of food safety and environmental sustainability increases (Domínguez and Gómez-Brandón 2013). Their income can be greatly increased by having access to premium markets. Furthermore, biofertilizers derived from microorganisms can aid smallholder farmers in adapting to climate change by enhancing plant resistance to stressors like disease and drought (Singh and Pathak 2024a). These biofertilizers lower the risk of crop failure by improving nutrient uptake efficiency and fostering general plant health in response to unfavorable weather conditions (Acee-Eke and Mac-Kingsley 2020; Nabyonga et al. 2022).

10.6 Challenges Faced in the Adoption of Microorganisms in Soil Fertility Management

Whilst there are significant opportunities associated with the adoption of microorganisms in soil fertility management, there are also notable challenges that must be addressed through education, research, regulatory frameworks, and farmer engagement strategies (AI-Youbi et al. 2021). There are several obstacles that may prevent the successful application of microorganisms for managing soil fertility. These limitations include lack of scientific understanding, practical economic considerations, regulatory bottlenecks, viability and stability of microorganisms, potential risks of invasive species, practical application, and cultural considerations.

10.6.1 Lack of Scientific Understanding

The lack of scientific knowledge on the intricate relationships between soil ecosystems and microorganisms is one of the main obstacles. Although studies have revealed several helpful microorganisms, including nitrogen-fixing bacteria and mycorrhizal fungi, much remains unknown about their precise functions, modes of action, and ideal environments (Morais et al. 2022). Furthermore, defined procedures for separating and using these microbes are frequently lacking. Farmers may find it difficult to trust these solutions due to this inconsistency, which may also cause inconsistent outcomes in field trials. To advance the use of microorganisms in soil fertility management, more thorough research that fills in these gaps is essential.

10.6.2 Economic Considerations

The adoption of microbiological remedies is significantly influenced by economic issues as well. Concerns over the cost-effectiveness of microbial inoculants in comparison to conventional fertilizers may make farmers reluctant to invest in biofertilizers. Even though biofertilizers are frequently promoted as more affordable options to chemical fertilizers, smallholder farmers in developing nations may still face financial obstacles due to the upfront expenditures of purchasing and applying them. Farmers may not be able to effectively increase their production if they are unable to pay for these inputs or do not have access to them because of distribution problems or market limitations. This puts overall food security at risk and exacerbates already-existing disparities in food production capability (Dar et al. 2013). Furthermore, there can be doubts about the return on investment (ROI) of utilizing these solutions. Biofertilizers frequently take longer to create beneficial microbial communities in the soil than chemical fertilizers, which provide plants instant access to nutrients. During crucial growing seasons, when farmers must find rapid solutions to meet the demand for food, this delay may be harmful. Therefore, if biofertilizers are used without proper planning, there may be short-term food shortages (Hamzaoui-Essoussi and Zahaf 2012).

10.6.3 Regulatory Hurdles and Quality Issues

Regulatory obstacles can affect the use of microorganisms in farming. The use of biological products in farming methods is subject to strict laws in several areas. Although these rules are frequently intended to guarantee efficacy and safety, they may unintentionally make it more difficult for novel microbial products to enter the market. The approval and commercialization of microbial products for agricultural application are fraught with regulatory issues in several areas of Africa. The adoption of advantageous microbes in conventional farming methods may be slowed by these restrictions (Singh and Pathak 2024a) Additionally, the microbial product market can be inconsistent, with variations in the quality and viability of products due to lack of standardization. Quality control measures are essential to ensure that farmers receive products containing effective, viable strains (Singh and Pathak 2024a).

10.6.4 Practical Application and Cultural Considerations

There are other difficulties associated with using microorganisms in practice. Application methods are also critical. Inoculants need to be applied at the right time and under the right conditions to be effective, which can require specialized knowledge or equipment that may not be readily available, especially for small-scale farmers in developing regions (Singh and Pathak 2024a). For example, to maintain their life, microorganisms in biofertilizers should be kept at the ideal temperature of 4–30 °C (Guo et al. 2022). Temperatures above 30 °C can cause heat stress-induced cell death or decreased viability, whereas temperatures below 4 °C can slow down metabolic activities. Farmers need to know how to integrate these microbes into their present farming methods without interfering with workflows or jeopardizing crop yields. Programs for education and training that provide farmers with the know-how to properly use microbial products at the right time are also necessary. It can be challenging for novel microorganism-based farming techniques to get accepted in some areas due to the strong cultural ties to old farming systems. Changes in management strategies are necessary to include microbial solutions into conventional farming processes; however, some farmers may be reluctant to implement them because of habit or mistrust of novel techniques. Over time, opinions may change with the aid of educational initiatives that highlight the advantages of employing beneficial bacteria.

10.6.5 Potential Risks of Invasive Species

The introduction of non-native or genetically engineered microbes presents threats to biodiversity and ecological balance. Ecological balance and biodiversity loss may be threatened if non-native or genetically modified microorganisms outcompete native species or disrupt established ecosystems. Anomalies also frequently arise when these microbes create defense mechanisms against other soil species, a phenomenon known as antibiosis (Csorba et al. 2023). Established native microorganisms compete with microbial inoculants when they are introduced into the soil. The capacity of the imported strains to colonize and carry out their intended tasks may be impacted by this competition, which may lower their survival rate Dar et al. 2013. It is also important to take into account how well the introduced microbes get along with the natural soil microbiota. It could be difficult for certain introduced species to adapt.

10.6.6 Viability and Stability of Microorganisms

Numerous helpful microorganisms are susceptible to changes in pH, humidity, and temperature. The microbial populations may drastically decrease before they reach the soil or plant roots if these conditions are not ideal, which would lessen their potency as biofertilizers. Furthermore, microbial inoculants must remain in the soil in order to provide long-term advantages. However, the long-term viability of these microbes can be impacted by a number of variables, including crop rotation, soil type, and tillage techniques. Maintaining microbial activity may need routine treatments, which could add to farmers' expenses and labor (Dinesh et al. 2010).

10.7 Conclusion

- Beneficial soil microorganisms, which include fungi, actinomycetes, bacteria, and nematodes, have diverse and unique roles that enhance nutrient availability and utilization through processes such as biological nitrogen fixation, decomposition of organic matter, solubilization of phosphorus and potassium, aggregation of soil, and suppression of pathogen growth and activity.
- Soil pH, temperature, moisture content, aeration, organic matter content, and texture may alter soil microbial activity and composition, and reduce or inhibit growth and colonization rates if these factors are at suboptimal levels.
- Improving the effectiveness of microorganisms in sustainable soil fertility management necessitates combined approaches, including microbial biotechnology, precision agriculture, novel formulations, farmer education, crop-specific strate-

gies, integrated nutrient management, and large-scale adoption backed by policy and certification.

- Microorganisms present important opportunities for improving soil health, increasing nutrient availability and crop resilience, as well as promoting economic sustainability, particularly for smallholder farmers, by decreasing reliance on inorganic fertilizers and supporting climate-resilient, eco-friendly farming practices.
- Despite the identified opportunities, the broad use of microbes as biofertilizers faces obstacles such as insufficient scientific understanding, economic and regulatory challenges, difficulties in practical application, and potential ecological dangers that need to be addressed through research, education, and supportive policy measures.
- By promoting viable and stable microbial communities in soils, nutrient availability, and use efficiency can be enhanced, plant growth improved, and ecological balance supported, thereby leading to more resilient agricultural systems.

References

Abatenh E, Gizaw B, Tsegaye Z, Wassie M (2017) The role of microorganisms in bioremediation—a review. Open J Environ Biol 2(1):38–46. https://doi.org/10.17352/ojeb

Abbasniayzare MR, Yousefi AR, Mohammadi G (2012) Biofertilizers and their role in soil fertility management. J Agric Sci 4(3):45–53

AbdElgawad H, Abuelsoud W, Madany MM, Selim S, Zinta G, Mousa AS, Hozzein WN (2021) Actinomycetes enrich soil rhizosphere and improve seed quality as well as productivity of legumes by boosting nitrogen availability and metabolism. Biomolecules 11(3):380. https://doi.org/10.3390/biom11030380

Abdul Rahman NSN, Abdul Hamid NW, Nadarajah K (2021) Effects of abiotic stress on soil microbiome. Int J Mol Sci 22(16):9036. https://doi.org/10.3390/ijms22169036

Acee-Eke G, Mac-Kingsley O (2020) Climate-smart agriculture and smallholder farmers' resilience: a review. Sustainability 12(20):8621. https://doi.org/10.3390/su12208621

Africa Biofertilizer Market Report (2024) Market analysis and future trends. Publisher

Africa Development Bank. (2021). Agriculture in Africa: a vital sector for development. Retrieved from https://www.afdb.org

Al-Youbi A, Al-Zahrani H, Al-Mutairi K (2021) Challenges in adopting microbial biofertilizers in modern agriculture. J Agric Sci 13(4):112–125. https://doi.org/10.5539/jas.v13n4p112

Altomare C, Tringovska I (2011) Beneficial soil microorganisms, an ecological alternative for soil fertility management. Sustain Agric Rev 7:161–214. https://doi.org/10.1007/978-94-007-1521-9_6

Ammar EE, Aioub AAA, Elesawy AE, Karkour AM, Mouhamed MS, Amer AA, El-Shershaby NA (2022) Algae as bio-fertilizers: between current situation and future prospective. Saudi J Biol Sci 29(5):3083–3096. https://doi.org/10.1016/j.sjbs.2022.03.020

Baldrian P (2017) Forest microbiome: diversity, complexity and dynamics. FEMS Microbiol Rev 41(2):109–130

Bargaz A, Elhaissoufi W, Khourchi S, Benmrid B, Borden KA, Rchiad Z (2021) Benefits of phosphate solubilizing bacteria on belowground crop performance for improved crop acquisition of phosphorus. Microbiol Res 252:126842. https://doi.org/10.1016/j.micres.2021.126842

Baron NC, Rigobelo EC (2022) Endophytic fungi: a tool for plant growth promotion and sustainable agriculture. Mycology 13(1):39–55

Beleri T (2023) Microbial phosphate solubilization: a sustainable approach for phosphorus management in agriculture. Front Microbiol 14:1123456. https://doi.org/10.3389/fmicb.2023.1123456

Bhardwaj D, Ansari MW, Sahoo RK, Tuteja N (2014) Biofertilizers function as key player in sustainable agriculture by improving soil fertility, plant tolerance and crop productivity. Microb Cell Factories 13(1):66. https://doi.org/10.1186/1475-2859-13-66

Bharti N, Pandey SS, Barnawal D, Patel VK, Kalra A (2017) Plant growth promoting rhizobacteria (PGPR): a sustainable agriculture to rescue the vegetation from the effect of biotic stress. J Plant Growth Regul 36(4):1006–1022. https://doi.org/10.1007/s00344-017-9693-7

Blake C, Christensen MN, Kovács ÁT (2021) Molecular aspects of plant growth promotion and protection by Bacillus subtilis. Mol Plant-Microbe Interact 34(1):15–25. https://doi.org/10.1094/MPMI-08-20-0225-CR

Bonaterra A, Badosa E, Daranas N, Francés J, Roselló G, Montesinos E (2022) Bacteria as biological control agents of plant diseases. Microorganisms 10(9):1759. https://doi.org/10.3390/microorganisms10091759

Borowik A, Wyszkowska J (2016) Soil moisture as a factor affecting the microbiological and biochemical activity of soil. Plant Soil Environ 62(6):250–255. https://doi.org/10.17221/158/2016-PSE

Briones MJI, McNamara NP, Poskitt J, Crow SE, Ostle NJ (2014) Interactive biotic and abiotic regulators of soil carbon cycling: evidence from controlled climate experiments on peatland and boreal soils. Glob Chang Biol 20(9):2971–2982. https://doi.org/10.1111/gcb.12585

Brookbank BP, Patel J, Gazzarrini S, Nambara E (2021) Role of basal ABA in plant growth and development. Genes 12(12):1936. https://doi.org/10.3390/genes12121936

Csorba C, Farkas E, Biró B (2023) Ecological risks of microbial inoculants in agriculture: a review. Appl Soil Ecol 182:104712. https://doi.org/10.1016/j.apsoil.2022.104712

Dar MH, de Souza R, Zaidi NW (2013) Biofertilizers: a sustainable approach for pulse production. Legum Res 36(4):331–339

De Boer W, Hundscheid MP, Klein Gunnewiek PJ, de Ridder-Duine AS, Thion C, Van Veen JA, Van Der Wal A (2015) Antifungal rhizosphere bacteria can increase as response to the presence of saprotrophic fungi. PLoS One 10(9):e0137988

Dinesh R, Srinivasan V, Hamza S, Manjusha A (2010) Short-term incorporation of organic manures and biofertilizers influences biochemical and microbial characteristics of soils under an annual crop [turmeric (Curcuma longa L.)]. Bioresour Technol 101(12):4697–4702

Domeignoz-Horta LA, Pold G, Liu X-JA, Frey SD, Melillo JM, DeAngelis KM (2020) Microbial diversity drives carbon use efficiency in a model soil. Nat Commun 11:3684. https://doi.org/10.1038/s41467-020-17502-z

Domínguez J, Gómez-Brandón M (2013) The influence of earthworms on nutrient dynamics during the process of vermicomposting. Waste Manag Res 31(8):859–868. https://doi.org/10.1177/0734242X13497079

Dong W, Liu E, Yan C, Tian J, Zhang H, Zhang Y (2022) Impact of no-tillage and conventional tillage systems on soil microbial communities. Appl Soil Ecol 169:104208. https://doi.org/10.1016/j.apsoil.2021.104208

Dunfield KE, Germida JJ (2021) Rethinking crop nutrition in times of modern microbiology: innovative biofertilizer technologies. Front. Sustain. Food Syst. 5:606815

Dutilloy E, Oni FE, Esmaeel Q, Clément C, Barka EA (2022) Plant beneficial bacteria as bioprotectants against wheat and barley diseases. J Fungi 8(6):632. https://doi.org/10.3390/jof8060632

Fadiji AE, Afolabi RO, Olabode OS (2024) Sustainable soil fertility management using biofertilizers. J Sustain Agric 12(2):78–95

Fernández-Calviño D, Bååth E (2010) Growth response of the bacterial community to pH in soils differing in pH. FEMS Microbiol Ecol 73(1):149–156. https://doi.org/10.1111/j.1574-6941.2010.00873.x

Friedrich CG, Rother D, Bardischewsky F, Quentmeier A, Fischer J (2001) Oxidation of reduced inorganic sulfur compounds by bacteria: emergence of a common mechanism? Appl Environ Microbiol 67(7):2873–2882. https://doi.org/10.1128/AEM.67.7.2873-2882.2001

Gamalero E, Glick BR (2015) Bacterial modulation of plant ethylene levels. Plant Physiol 169(1):13–22. https://doi.org/10.1104/pp.15.00284

Ghimirey V, Chaurasia J, Acharya N, Dhungana R, Chaurasiya S (2024) Biofertilizers: a sustainable strategy for enhancing physical, chemical, and biological properties of soil. Innov Agric 7:1–11. https://doi.org/10.3897/ia.2024.128697

Goicoechea N, Bettoni MM, Fuertes-Mendizábal T, González-Murua C, Aranjuelo I (2020) Durum wheat yield and N uptake as influenced by AM fungi under different N rates in a Mediterranean climate. Agric Ecosyst Environ 295:106918. https://doi.org/10.1016/j.agee.2020.106918

Großkopf T, LaRoche J (2012) Direct and indirect costs of dinitrogen fixation in Crocosphaera watsonii WH8501 and possible implications for the nitrogen cycle. Front Microbiol 3:236. https://doi.org/10.3389/fmicb.2012.00236

Guo Y, Chen X, Wu Y, Zhang L, Cheng J, Wei G (2022) Challenges in the application of microbial inoculants in agriculture. Appl Microbiol Biotechnol 106(1):69–82. https://doi.org/10.1007/s00253-021-11729-6

Hamzaoui-Essoussi L, Zahaf M (2012) Adoption of organic farming in the context of sustainable development. J Int Food Agribus Mark 24(2):121–140. https://doi.org/10.1080/08974438.2012.652759

Harish MN, Choudhary P, Viswanathan R (2022) Contributions of beneficial microorganisms in soil remediation and quality improvement of medicinal plants. Plants 11(23):3200. https://doi.org/10.3390/plants1123320015

Hartmann M, Six J (2023) Soil structure and microbiome functions in agroecosystems. Nat Rev Earth Environ 4(1):4–18. https://doi.org/10.1038/s43017-022-00379-5

Huang C, Yao Z, Wang T, Wang X, Zhang Y, Zhu B (2024) Application of synthetic fertilizers with crop straw facilitates optimization of soil nematode community and supports crop yields. Appl Soil Ecol 197:105340

IFA (2009) The global “4R” nutrient stewardship framework for developing and delivering fertilizer best management practices. International Fertilizer Industry Association, Paris

IPNI (2012) 4R plant nutrition: a manual for improving the management of plant nutrition. International Plant Nutrition Institute, Norcross, GA

Inomura K, Bragg J, Follows MJ (2017) A quantitative analysis of the direct and indirect costs of nitrogen fixation: a model based on Azotobacter vinelandii. ISME J 11(1):166–175. https://doi.org/10.1038/ismej.2016.97

Johnston AM, Bruulsema TW (2014) 4R nutrient stewardship for improved nutrient use efficiency. Procedia Eng 83:365–370. https://doi.org/10.1016/j.proeng.2014.09.029

Jog R, Nareshkumar G, Rajkumar S (2016) Enhancing soil health and plant growth promotion by actinomycetes. In: Plant growth promoting actinobacteria: a new avenue for enhancing the productivity and soil fertility of grain legumes. Springer, Singapore, pp 33–45

Joshi HC, Gauraha AK (2022) Global market trends in biofertilizers. J Agric Econ 8(1):112–125

Khosro M, Yousef S (2012) Bacterial biofertilizers for sustainable crop production: A review. ARPN J Agri Biol Sci 7(5):307–316

Kumar R, Kumari S, Singh P (2022) Biofertilizers function as key players in sustainable agriculture by improving soil fertility, plant tolerance, and crop productivity. Microb Cell Fact 13:66. https://doi.org/10.1186/1475-2859-13-668

Liu J, Zhang X, Wang H, Hui X, Wang Z, Qiu W (2018) Long-term nitrogen fertilization impacts soil fungal and bacterial community structures in a dryland soil of loess plateau in China. J Soils Sediments 18:1632–1640

Lumbini Development Trust (2023) Second Buddhist International Travel Mart 2023 concluded in Lumbini. https://lumbinidevtrust.gov.np/post/2_66811b414af44

Lupwayi NZ, Larney FJ, Blackshaw RE, Kanashiro DA, Pearson DC (2017) Phospholipid fatty acid biomarkers show positive soil microbial community responses to conservation soil management of irrigated crop rotations. Soil Biol Biochem 113:273–283. https://doi.org/10.1016/j.soilbio.2017.06.018

Maina S, Karuri H, Ng'endo RN (2021) Free-living nematode assemblages associated with maize residues and their ecological significance. J Nematol 53(1):1–12

Mayerhofer J, Dietrich M, Schintlmeister A, Woebken D (2021) Revisiting the role of microbial diversity in nitrogen cycling in soils. Nat Commun 12:4647. https://doi.org/10.1038/s41467-021-24881-4

Mekonen S, Petros I, Hailemariam M (2017) The role of nematodes in the processes of soil ecology and their use as bioindicators. Agric Biol JN Am 8(4):132–140

Mihoub A, Ahmad M, Radicetti E (2023) Biofertilizers in agroecology: a review on the role of soil microorganisms in plant nutrition. Agronomy 13(3):831. https://doi.org/10.3390/agronomy13030831

Mohamad NA, Jusoh NA, Zaw Htike Z, Lei Win S (2014) Bacteria identification from microscopic morphology: a survey. Int J Soft Comput Artif Intell Appl 3(2):1–12. https://doi.org/10.5121/ijscai.2014.3201

Mohamed HI, Sofy MR, Almoneafy AA, Abdelhamid MT, Basit A, Sofy AR, Lone R, Abou-El-Enain MM (2021) Role of microorganisms in managing soil fertility and plant nutrition in sustainable agriculture. In: Plant growth-promoting microbes for sustainable biotic and abiotic stress management. Springer, Cham, pp 93–114

Morais TP, Brito CH, de Souza JT (2022) Microbial inoculants in agriculture: challenges and limitations. Appl Microbiol Biotechnol 106(1):1–15. https://doi.org/10.1007/s00253-021-11720-1

Muhammad A, Kong X, Zheng S, Bai N, Li L, Khan MHU, Fiaz S, Zhang Z (2024) Exploring plant-microbe interactions in adapting to abiotic stress under climate change: a review. Front Plant Sci 15:1482739. https://doi.org/10.3389/fpls.2024.1482739

Nabyonga L, Basamba TA, Nyakoojo C (2022) Economic viability of biofertilizers for smallholder farmers in sub-Saharan Africa. Sustainability 14(5):2876. https://doi.org/10.3390/su14052876

Narayan OP, Kumar P, Yadav B, Dua M, Johri AK (2022) Sulfur nutrition and its role in plant growth and development. Plant Signal Behav 18(1):2030082. https://doi.org/10.1080/15592324.2022.2030082

Nawaz A, Qamar ZU, Marghoob MU, Imtiaz M, Imran A, Mubeen F (2023) Contribution of potassium solubilizing bacteria in improved potassium assimilation and cytosolic K+/Na+ ratio in rice (Oryza sativa L.) under saline-sodic conditions. Front Microbiol 14. https://doi.org/10.3389/fmicb.2023.1196024

Nisa RU, Tantray AY, Kouser N, Allie KA, Wani SM, Alamri SA, Alyemeni MN, Wijaya L (2021) Influence of ecological and edaphic factors on biodiversity of soil nematodes. Saudi J Biol Sci 28(5):3049–3059. https://doi.org/10.1016/j.sjbs.2021.02.046

Noulas C, Torabian S, Qin R (2023) Crop nutrient requirements and advanced fertilizer management strategies. Agronomy 13(8):2017. https://doi.org/10.3390/agronomy13082017

Olaniyan FT, Alori ET, Adekiya AO, Ayorinde BB, Daramola FY, Osemwegie OO, Babalola OO (2022) The use of soil microbial potassium solubilizers in potassium nutrient availability in soil and its dynamics. Ann Microbiol 72(1):45. https://doi.org/10.1186/s13213-022-01701-8

Ortiz MA, Sansinenea E (2022) Biofertilizers: A sustainable approach to soil fertility. Environ Sci Technol 10(4):567–580

Pahalvi HN, Shukla R, Singh S (2021) Challenges in the commercialization of microbial biofertilizers. In: Microbial technology for sustainable environment, vol 1, pp 345–362. https://doi.org/10.1007/978-981-16-3840-4_21

Papatheodorou EM (2004) The effects of large soil fauna on microbial activity in decomposing leaf litter. Appl Soil Ecol 26(3):211–218. https://doi.org/10.1016/j.apsoil.2003.12.009

Patten CL, Glick BR (2002) Role of pseudomonas putida Indoleacetic acid in development of the host plant root system. Appl Environ Microbiol 68(8):3795–3801. https://doi.org/10.1128/AEM.68.8.3795-3801.2002

Rabot E, Wiesmeier M, Schlüter S, Vogel H-J (2018) Soil structure as an indicator of soil functions: a review. Geoderma 314:122–137. https://doi.org/10.1016/j.geoderma.2017.11.009

Rigobelo EC, Kandasamy S, Saravanakumar D (eds) (2022) Plant growth-promoting microorganisms for sustainable agricultural production. Frontiers Media SA. https://doi.org/10.3389/978-2-88974-833-4

Rodríguez H, Fraga R (1999) Phosphate solubilizing bacteria and their role in plant growth promotion. Biotechnol Adv 17(4):319–339. https://doi.org/10.1016/S0734-9750(99)00014-2

Rodriguez A, Sanders IR (2015) The role of community and population ecology in applying mycorrhizal fungi for improved food security. ISME J 9(5):1053–1061

Sachdev S, Singh RP (2020) Trichoderma: a multifaceted fungus for sustainable agriculture. In: Ecological and practical applications for sustainable agriculture. Springer, Singapore, pp 261–304

Sansinenea E (2021) Market dynamics of biofertilizers: challenges and opportunities. J Environ Sci 9(3):345–357

Shaughnessy H (2023) An overview on bacterial ecology and its importance. J Bacteriol Parasitol 0(0):1–2. https://doi.org/10.35248/2155-9597.23.S21.048

Singh JS (2019) Microbes play major roles in ecosystem services. Clim Change Environ Sustain 7(1):1–8. https://doi.org/10.5958/2320-642X.2019.00001.5

Singh R, Pathak D (2024a) Microbial biofertilizers: opportunities and challenges in sustainable agriculture. J Appl Microbiol 136(2):1–15. https://doi.org/10.1093/jambio/lxad305

Singh CP, Pathak GS (2024b) Role of microorganisms in soil fertility: agronomic perspectives. Int J Res Agron 7(10):409–414. https://doi.org/10.33545/2618060X.2024.v7.i10f.1778

Smercina DN, Evans SE, Friesen ML, Tiemann LK (2019) To fix or not to fix: controls on free-living nitrogen fixation in the rhizosphere. Appl Environ Microbiol 85(6):e02546–e02518. https://doi.org/10.1128/AEM.02546-18

Smith SE, Read DJ (2010) Mycorrhizal symbiosis. Academic

Tang Q, Xia Y, Ti C, Li Y, Zhou W (2024) Soil organic matter stability and microbial community in relation to different plant cover: a case study in Yunnan, China. Sci Total Environ 806:150568. https://doi.org/10.1016/j.scitotenv.2021.150568

Van Horn DJ, Okie JG, Buelow HN, Gooseff MN, Barrett JE, Takacs-Vesbach CD (2014) Soil microbial responses to increased moisture and organic resources along a salinity gradient in a polar desert. Appl Environ Microbiol 80(10):3034–3043. https://doi.org/10.1128/AEM.03414-13

Wang C, Liu D, Bai E (2017) Decreasing soil microbial diversity is associated with decreasing microbial biomass under nitrogen addition. Soil Biol Biochem 120:126–133. https://doi.org/10.1016/j.soilbio.2018.02.003

Wang Y, Liu Z, Hao X, Wang Z, Wang Z, Liu S et al (2023) Biodiversity of the beneficial soil-borne fungi steered by Trichoderma-amended biofertilizers stimulates plant production. npj Biofilms Microbiomes 9(1):46

Yu Y, Chen L, Duan W (2024) Responses of bacterial and fungal communities to short-term nitrogen and phosphorus additions in temperate forest soil aggregates in northeastern China. Appl Soil Ecol 197:105341

Zhang H, Dong L, Yao X, Wang W (2023) Soil fertility shifts the relative importance of saprotrophic and mycorrhizal fungi for maintaining ecosystem stability. Glob Chang Biol 29(4):1206–1216

Zhou Z, Shi X, Bhople P, Jiang J, Chater CC, Yang S et al (2024) Enhancing C and N turnover, functional bacteria abundance, and the efficiency of biowaste conversion using Streptomyces-Bacillus inoculation. J Environ Manage 358:120895

Chapter 11
Harnessing Biofertilizers for Sustainable Cereal Production

Sucheta Dahiya, Sachin Kumar, Akanksha, and Ankita Chatterjee

Abstract Global food security is seriously threatened by several issues facing modern intensive agriculture methods. To boost agricultural productivity and fulfill the nutritional demands of the world's increasing population, there is extensive reliance on chemical fertilizers and pesticides. However, the unconcerned application of agrochemicals has led to environmental contamination that poses hazards to human health. Furthermore, the quality and physical characteristics of agricultural soils, together with their chemical (nutrient imbalance) and biological health, are progressively declining. Today, biofertilization is one of the environmentally friendly methods used in sustainable agriculture. Utilizing biofertilizers boosts plant productivity through raising nitrogen and fertility levels of the soil, as they are biodegradable. Biofertilizers consist of living microorganisms that enhance soil fertility naturally, without causing environmental pollution. The organisms include microbes such as bacteria, microfungi, and microalgae, as well as macroorganisms, macrofungi, and macroalgae. The biofertilizers are biodegradable in nature; they did not build up any residues in soil like chemical or synthetic fertilizers. The use of biofertilizers helps maintain soil health and supports the pursuit of sustainable agriculture. Because of their unique properties, biofertilizers contribute to healthy plant development, resulting in improved produce quality and increased plant biomass. They promote better plant growth by enhancing nutrient availability, regulating phytohormones, controlling phytopathogens, and reducing both biotic and abiotic stresses. The formulation of biofertilizers involves preparing inoculums with cell protectants such as glycerol, lactose, and starch, which serve as suitable carrier materials. Proper packaging and efficient distribution methods are also essential components of the preparation process. Recent advancements in biofertilizer technology include techniques like entrapment or microencapsulation, nano-immobilization of microbial inoculants, and biofilm-based formulations. This chapter provides a comprehensive overview of the current progress in utilizing

S. Dahiya · S. Kumar (✉) · Akanksha · A. Chatterjee
Department of Agronomy, Faculty of Agricultural Sciences, SGT University Gurugram, Gurugram, Haryana, India

A. D. Nciizah et al. (eds.), *Biofertilizers for Sustainable Agriculture in Africa*, Sustainability Sciences in Asia and Africa, https://doi.org/10.1007/978-981-95-6009-7_11

microbial strains as biofertilizers and highlights their significant role in improving soil fertility and crop productivity.

Keywords Biofertilizers · Beneficial microorganisms · Rhizosphere · Soil fertility · Sustainable agriculture

11.1 Introduction

In recent decades, there has been a rapid rise in the global human population, accompanied by a substantial increase in food production. This surge in both food demand and supply has largely been achieved through enhanced plant nutrition. However, the heavy reliance on chemical fertilizers to sustain this growth has raised significant environmental and public health concerns (Kawalekar et al. 2013). The agricultural sector accounts for nearly one-third of the global gross domestic product (GDP). However, with the continuous rise in the global population, it is projected that by 2050 the world's population will reach around 9.5 billion, leading to an enormous increase in food demand (Green et al. 2005; Gerland and Dakora 2014). The main obstacles to the production of various crops include abiotic and biotic stressors, urbanization, the inadequacy of fertile land, and unpredictable weather linked to climate. In addition, crucial factors for increasing crop production per unit land and reaching the desired objective of food security include soil quality, nutrient status, environmental factors, and the biological activity of the soil (Tilman et al. 2011). Excessive use of chemical fertilizers (which include N, P, or K) in modern high-input farming systems and various technologies are employed to fulfill plant nutritional needs and enhance global agricultural productivity. However, due to the low efficiency of fertilizer utilization, only about 30–40% of the applied nutrients are absorbed by plants. The remaining portion is lost through processes such as leaching, runoff, infiltration, and seepage, which ultimately contribute to environmental pollution.

Furthermore, chemical-based fertilizers have radio nuclides such as uranium (particularly U-238), radium (Ra-226), and thorium, and heavy metals, including cadmium, lead, arsenic, magnesium, copper, nickel, and zinc, primarily originating from the phosphate rock used in fertilizer production, which is difficult to break down, making them persistent contaminants in the environment. Eutrophication of water sources is a significant problem associated with the overuse of chemical fertilizers. Herbicides, insecticides, and inorganic fertilizers have seriously damaged the soil environment. Moreover, the desertification of agricultural land, decreased crop output, and a rise in severe human diseases were all due to the improper use of pesticides and inorganic fertilizers (Bedair et al. 2022). The environmental pollution threatening public health has necessitated the development of eco-friendly and sustainable technologies aimed at minimizing the use of synthetic fertilizers. Therefore, there is an urgent need to adopt eco-friendly approaches to enhance soil fertility, such as Integrated Nutrient Management (INM) strategies that incorporate biofertilizers. Currently, the use of natural plant biostimulants is advocated as an innovative approach to tackle the challenges of sustainable agriculture, enhancing crop yield, nutrient uptake, and tolerance to abiotic stresses. As synthetic fertilizers are chemical based, overuse of these damages soil and plants by eutrophication (Das et al. 2023). Artificial fertilizers infiltrate the nearby streams and rivers, where they damage fish and other aquatic life by contaminating the water. Conversely, plants are safe and gradually add nutrients to the earth. In order to stop soil erosion, a healthy soil with ideal plant cover is necessary. Hence, to improve soil fertility researchers are actively looking for novel ways to create environmentally benign fertilizers using living microbes, particularly microorganisms, which may biodegrade without leaving behind residues.

Biostimulants are natural or organic-based substances or microorganisms applied to plants to enhance nutrient use efficiency, tolerance to abiotic stresses, and overall plant performance, regardless of the crop's inherent nutrient content. Biofertilizers, a subset of biostimulants, consist of microbial inoculants containing both active and inactive beneficial microbes that improve plant nutrition and fertilizer efficiency. These substances may include living microorganisms, plant residues, and organic matter, and are applied as seed treatments, foliar sprays, or soil drenches. Once introduced, they colonize the root zone, enhancing plant growth by increasing the availability of essential primary nutrients (N, P, and K) and micronutrients (Fe, Mn, and Zn). Microorganisms in biofertilizers also secrete growth-promoting hormones such as auxins, cytokinins, biotin, and vitamins, which are crucial for plant development. Additionally, they contribute to higher protein content, essential amino acids, and nitrogen fixation, potentially boosting crop yield by 10–40%. Biofertilizers were first recognized by a Dutch scientist in 1888, and in 1895, Nobbe and Hiltner produced the first commercial preparation under the trade name "Nitragin" in the United States. Given their significant role in enhancing plant growth and productivity, biofertilizers are a key component of organic and sustainable agriculture. They provide a cost-effective alternative to conventional nitrogenous fertilizers. For instance, *Azotobacter* can supply approximately 20–25 kg N ha^{-1} in crops such as maize, wheat, and cotton under favorable conditions, while phosphorus-solubilizing bacteria (PSB) can solubilize 20–30% of insoluble phosphates, leading to yield increases of up to 20%. *Azospirillum* sp. increases yield of cereals and grasses by improving root development and increasing the water use efficiency and nutrient uptake from rhizosphere and by biological nitrogen fixation (Chatterjee et al. 2014).

Diverse, intricate, and interactive communities of microorganisms occupy all terrestrial plants. Microbes can colonize plants in one or two ways: rhizosphere, entophytic, or epiphytic. Over the past 20 years, microbial communities that live in various environments have been the interest of study and how they all work together or in a synergistic way to support the plant growth, development, and protection (Afzal et al. 2019). The plant rhizosphere hosts a diverse range of microorganisms, including prokaryotic bacteria, unicellular protozoa, multicellular eukaryotic nematodes, fungi, and some viruses. Many of these bacteria are integral to the plant microbiome, playing a vital role in supporting the growth and development of agriculturally important crops. To develop effective biofertilizers, beneficial microbial strains must be carefully identified, capable of colonizing the rhizosphere, and rigorously tested for their compatibility and performance in this environment. Growth-promoting microbes have several advantages for use as biofertilizer, such as improving nutrient availability, producing phytohormones, breaking down organic matter, and helping to reduce biotic and abiotic stressors. Comprising live or dormant microbial cells, biofertilizers are applied to soil, seeds, or seedlings to improve nutrient availability and facilitate plant uptake. Nowadays, using biofertilizers has become a more economical and environmentally responsible option than chemical-based fertilizers.

Biofertilizers are substances containing living microorganisms, plant residues, and organic matter. They can be applied to seeds, soil, or plant surfaces, where they colonize the rhizosphere and enhance plant growth by increasing the availability of primary nutrients to the plant (Yilmaz and Sönmez 2017). Additionally, biofertilizers improve soil quality by enriching the rhizosphere with essential nutrients. The microorganisms in biofertilizers play a critical role in supplying nitrogen, phosphorus, potassium, and other nutrients necessary for plant development. They also contribute to higher grain yield and enrich soil micronutrients such as Fe, Mn, and Zn (Malik et al. 2009). Furthermore, these microorganisms secrete growth-promoting hormones like auxins, cytokinins, biotin, and vitamins, which are essential for plant growth (Sahoo et al. 2012). Biofertilizers are cost-effective, safe, and offer promising applications in organic farming and sustainable agriculture. Their pivotal role in enhancing plant productivity makes them indispensable in eco-friendly farming practices. This chapter focuses on commonly used biofertilizers, including *Azotobacter*, phosphorus-solubilizing bacteria, *Azospirillum*, blue–green algae, *Rhizobium*, and nutrient-mobilizing microorganisms (Yilmaz and Sönmez 2017; Malik et al. 2009; Sahoo et al. 2012).

11.2 Features of Biofertilizers

- Biofertilizers are cost-effective and eco-friendly fertilizers containing microbial inoculants of algae, fungi, and bacteria, either individually or in combination. It helps in improving soil health and fertility.
- They are easy to apply and required in less amount in comparison to inorganic fertilizers.
- Biofertilizers also increases the microfloras build up in soil.

Classification and methods of biofertilizers application are shown in Figs. 11.1 and 11.2.

- *Nitrogen-fixing bacteria*: Although nitrogen is the most abundant element in the atmosphere, it often becomes a limiting nutrient for plants due to the challenges associated with its fixation and uptake. Certain microorganisms, however, can establish associations with plants and are capable of significant nitrogen fixation.

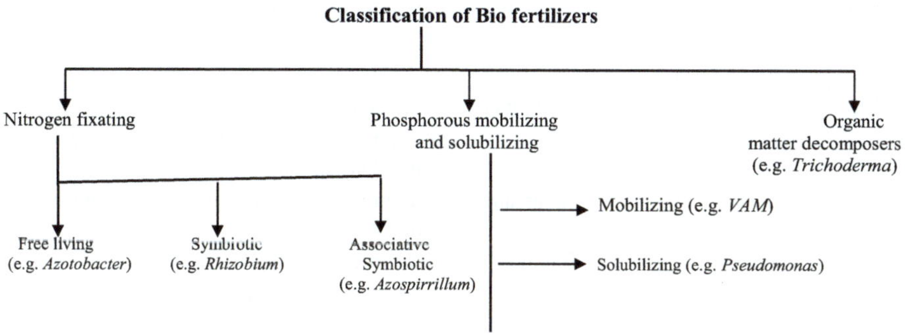

Fig. 11.1 Classification of N fixing, P mobilizing, and solubilizing and organic matter decomposers

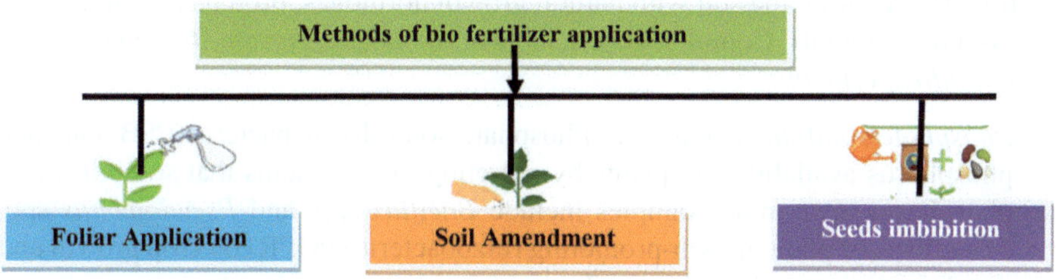

Fig. 11.2 Application methods of biofertilizer

Table 11.1 N-fixing bacteria as biofertilizer and their effects

Bacteria	Example	Crop plant	Effect
Symbiotic	*Anabaena azollae*	Rice	By increasing microbial population, it enhances soil fertility and growth
Free living	*Azotobacter*	Rice	Stimulate plant growth
Associative symbiotic	*Azospirillum*	Wheat, maize	Mitigation of abiotic stress

The role of nitrogen-fixing bacteria and their effects on various crop plants are summarized in Table 11.1.

- Free living: Free-living bacteria in the soil helps in nitrogen fixation, for example, *Azotobacter.*
- Symbiotic: Having mutual and other endophyte associations (a mutual relationship between a plant and a microorganism that lives inside the plant's tissues without affecting the plant) with plants, for example, *Rhizobia* and *Frankia.*
- Associative: Living in a rhizosphere with or without endophytes symbiosis, for example, *Azospirillum.*
- *Blue–Green Algae (Cyanobacteria)*: Blue–green algae (BGA) contribute to improving soil fertility, enhancing crop yield, and reducing the dependence on chemical fertilizers. They are especially effective in increasing the fertility of rice fields and are widely used in rice cultivation across various regions. In addi-

Table 11.2 Plant growth-promoting rhizobacteria (PGPR) and their effects on crop growth and development

Growth-promoting rhizobacteria (PGPR)	Host plant	Effect on crops
Azotobacter chroococcum	Wheat	Showed improved growth with phosphate-solubilizing mutants
Bacillus megaterium	Sugarcane	Promotion of yield and yield-attributing characters
Pantoea agglomerans	Tomato	Helps in growth improvement
Pseudomonas fluorescens	Sweet potato	Helps in yield enhancement
Rhizobium leguminosarum	Faba bean	Helps in yield enhancement

tion, BGA enrich the soil with natural growth hormones, proteins, vitamins, and essential minerals. Common genera include *Anabaena*, *Nostoc*, *Tolypothrix*, and *Cylindrospermum.*

1. *Phosphate-solubilizing bacteria*: Phosphate-solubilizing bacteria (PSB) enhance phosphorus availability to plants by secreting organic acids that solubilize soil phosphorus. Common examples include *Bacillus* spp. and *Pseudomonas* spp. The effects of plant growth-promoting rhizobacteria (PGPR) on crop growth and development are summarized in Table 11.2.
2. *Phosphate mobilizing:* They are the beneficial microorganisms that help plants access phosphorus that is otherwise unavailable in the soil (*Laccaria* sp., *Boletus* sp., *Amanita* sp., *Glomus* sp., *Gigaspora*sp., *Acaulospora* sp.).
3. *Organic matter decomposers*: Organic matter is broken down into compost with the help of various decomposing microorganisms, including *Aspergillus niger*, *A. terreus*, *Trichoderma viride*, *Bacillus* spp., and bacteria such as *Pseudomonas*, *Serratia*, *Klebsiella*, and *Enterobacter*. These microorganisms possess cellulolytic and lignolytic activities that degrade plant cell walls, facilitating nutrient recycling (Table 11.3).

11.3 Influence of Various Biofertilizers on the Growth, Yield, and Quality of Cereal Crops and Their Mechanisms of Action

1. *Azotobacter*

Azotobacter is a type of gram-negative, free-living, nitrogen-fixing aerobic bacteria inhabiting in the soil (Sumbul et al. 2020). They are known to fix atmospheric nitrogen for their cellular protein synthesis, which is mineralized in the soil, imparting the crop plants a considerable part of nitrogen available from the soil source. They show advantageous impacts on the crop growth and yield through the biosynthesis of biologically active substances, instigation of rhizospheric microbes, production

Table 11.3 Mechanism of different biofertilizers

Biofertilizers	Mechanism	Groups	Examples
Phosphorus solubilizing	The fixing of atmospheric nitrogen increases the content of nitrogen in soil and fixes the plants to get it	Free living	*Nostoc, Stigonema, Klebsiella, Desulfovibrio, Rhodospirillum*
		Symbiotic	*Frankia, Rhizobium, Anabaena Azollae and Trichodesmium*
		Associative symbiotic	*Azospirillum* sp., *Herbaspirillum* sp.,*Enterobacter, Azoarcus* sp. *Alcaligenes, Acetobacter diazotrophicus*
Nitrogen fixing	By taking nitrogen from the atmosphere and fixing and supplying it to plants and soil, the amount of nitrogen will rise	Bacteria	*Bacillus circulans, B. subtilis, B.polymyxa, Pseudomonas striata, Penicillium* sp., *Microccocus, Aereobacter,* and *Agrobacterium, Flavobacterium*
		Fungi	*Penicillium* sp., *Aspergillus awamori,* and *Trichoderma*
Potassium solubilizing	Organic acids production helps in the breakdown of silicates and the elimination of metal ions Potassium (silicates) can be dissolved and made accessible to plants	Bacteria	*Bacillus mucilaginosus, B. circulanscan, B. edaphicus,* and *Arthrobacter* sp.
		Fungi	*Aspergillus niger*
Plant growth promoting	Produce hormones that improve root health, crop yield, and nutrient availability expansion	Plant growth-promoting rhizobacteria	*Pseudomonas* sp., *Arthrobacter, Agrobacterium, Pseudomonas fluorescens, Rhizobium, Erwinia, Bacillus, Enterobacter, Streptomyces,* and *Xanthomonas*
	Oxidation of sulfur creates sulfates that can be used by plants	Sulfur oxidizing	*Thiobacillus* sp.
Micronutrient	It is necessary to make zinc soluble via acidification, oxidoreductive systems, protonation and chelated ligands	Zinc solubilizing	*Mycorrhiza, Pseudomonas* sp., *Bacillus* sp.

of phytopathogenic inhibitors, alteration of nutrient uptake, and eventually magnifying the biological nitrogen fixation (Lenart 2012). The *Azotobacter chroococcum* in research experiments applied as microbial inoculants releases growth-promoting substances and hence improves the growth of the plant in agriculture (Gothandapani et al. 2017). *Azotobacter* is able to transform atmospheric nitrogen to ammonia, which can be easily utilized by the plants (Prajapati et al. 2008). *Azotobacter* plays an important role in the growth of plants, especially if it improves the yield of wheat. Similarly, the combined application of *Azospirrillum* and *Azotobacter* significantly increases the spikes, number of tillers, grain weight, grain size, spikelets per plant, and spike length in wheat (Chauhan et al. 2011). However, Katiyar et al.

(2011) reported that seed inoculation with *Azotobacter* increases the yield by up to 1.92–2.0% compared to noninoculated seed. The inoculation with *Azotobacter* significantly increased plant height to a maximum, that is, 79.7 cm, and gave the highest leaf area index to 4.40 than the control at all the stages of crop growth (Ram et al. 2005). Nitrogen uptake by the crop and the amount of available nitrogen in post-harvest soil increased significantly with *Azotobacter* inoculation over no inoculation. Furthermore, application of *Azotobacter* increased grain yield from 16.5% to 19.42% over control. Hence, *Azotobacter* combined with farmyard manure and inorganic fertilizer (NPK) results in the highest yield (Mahato and Kafle 2018).

2. *Rhizobium*

Rhizobium forms nitrogen-fixing nodules that establish a symbiotic association with leguminous crops, capable of fixing 50–100 kg N/ha (Barman et al. 2019). Well-nodulated legumes also leave substantial residual nitrogen in the soil, which can partially fulfill the nitrogen requirements of subsequent crops in rotation. Intercropping cereals with legumes has been shown to improve grain and nitrogen yield per unit area compared with sole cropping (Rodriguez et al. 2020). *Rhizobium* can also associate with the roots of non-leguminous plants without forming nodules, promoting growth through direct and indirect mechanisms. These include the production of phytohormones, chemical signals such as lipo-chito-oligosaccharides (LCOs) and lumichrome, solubilization of precipitated phosphorus, mineralization of organic phosphorus, modification of root morphology, production of siderophores, and reduction of ethylene levels (Mehboob et al. 2009). Furthermore, rhizobia enhance non-legume growth via biocontrol of pathogens through antibiosis, parasitism, or competition for nutrients and space, and by improving soil adhesion to roots through exopolysaccharide secretion, which regulates water movement and facilitates root development (Mehboob et al. 2009).

Rhizobia can function as endophytes in gramineous crops like maize, wheat, millets, barley, rice, and other cereals, without forming nodules or causing disease symptoms (Mia and Shamsuddin 2010). In wheat, *Rhizobium* inoculation has been reported to increase tiller number per plant by 42%, plant height by 13%, 100-grain weight by 10%, grains per spike by 16%, and overall yield by 10% compared with uninoculated controls (Adnan et al. 2014). Similarly, field inoculation of rice with *Rhizobium* significantly improved grain yield, agronomic nitrogen-use efficiency, harvest index, and total protein content per hectare (Yanni et al. 2001). The mechanisms of plant growth-promoting rhizobacteria are illustrated in Fig. 11.3.

3. *Azospirillum*

Azspirillum sp., a free-living and associative nitrogen fixer in the rhizosphere, promotes root development in cereals and pasture grasses, increasing water and nutrient absorption, as well as atmospheric nitrogen fixation, resulting in higher production (Kaur et al. 2022). *Azospirillum* inoculation increased yield by 43% in wheat, 60% in barley, and 44% in finger millet, while combined application with PSB in wheat resulted in 23.2% and 11.9% higher grain yield, 21.6% and 9.9% higher straw yield, and 32.3% and 15.7% higher net returns over the control and *Azospirillum* alone, respectively (Kaushik et al. 2012). In addition, minor increases

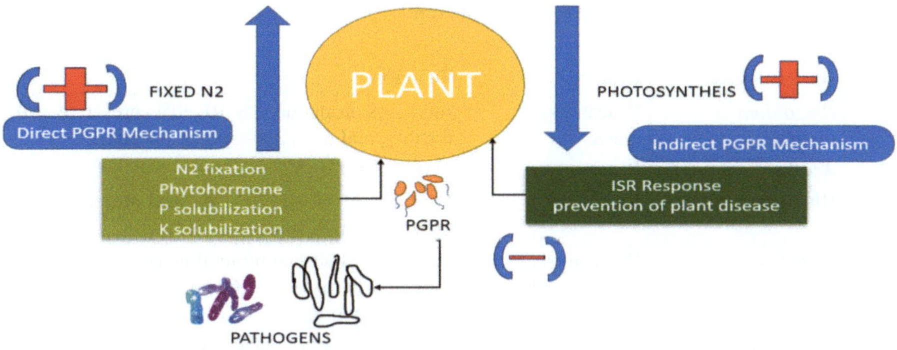

Fig. 11.3 Mechanisms of plant growth-promoting rhizobacteria

in N, P, and K concentrations in maize and wheat leaves and grains were observed following *Azospirillum* seed inoculation.

4. *Blue–Green Algae (BGA)*
BGA biofertilizer use in rice, often known as "algalization," promotes an ecologically sound agroecosystem and increases the monetary potential of paddy agriculture by minimizing the demand for energy-intensive inputs (Pabbi 2015). These microbes have significance in agriculture as they have their ability to metabolize molecular nitrogen, release a portion of fixed nitrogen and growth-promoting compounds as extra metabolites, solubilize insoluble phosphates, add organic matter, and strengthen soil physical and chemical properties. Feyisa et al. (2013) found that inoculating rice with *Azolla* once into the soil improved yield by 911 kg/ha on *Azolla filiculoides* plots and 721 kg ha^{-1} on *Azolla microphylla* plots. However, foliar application of BGA (6.0 g/L) gives the maximum growth, yield, and quality parameters of maize and reduces the pollution caused due to high levels of mineral nitrogen fertilizer. Likewise, inoculating rice with *Anabaena* sp. promoted plant growth, chlorophyll, leaf area, percent nitrogen, and protein by 70%, 22%, 60%, and 25%, respectively, over the control (Priya et al. 2022). Similarly, Davari and Sharma (2010) concluded that the combinations of FYM + wheat residue + biofertilizers (BGA and PSB) and vermicompost + wheat residue + biofertilizers (BGA and PSB) resulted in the highest grain yield of basmati rice over control by 51–58%, and net returns by 37–47%.

5. *Phosphorous-Solubilizing Bacteria (PSB)*
The majority of phosphorus in soil exists in insoluble, fixed forms that plants cannot absorb, hence the application of PSB is critical for mobilization as well as accessibility to crops. Phosphorus-solubilizing bacteria (PSB) mobilize fixed phosphorus in the soil, which makes it accessible to uptake by a wide range of crops—including wheat, rice, maize, cowpea, soybean, gram, and potato—under field conditions across India. Kumar (2018) found that combining vermicompost, zinc, and PSB with inorganic fertilizers (100% NPK) resulted in the highest net return and B:C ratio. Saxena et al. (2013) reported a significant increase in growth parameters, yield

Table 11.4 Biofertilizers required for different crops

S. no.	Biofertilizers	Crops	Functions
1.	*Rhizobium* (symbiotic)	Wheat, bajra, maize	Increases production by 10–30%, preserves soil fertility, and fixes 200–300 kg N/ha/year
2.	*Azolla*	Rice	Fixes 30–60 kg N/ha/year
3.	Blue–green algae	Rice	Provides 20–40 kg N/ha/year and promotes growth through substances (IAA)
4.	*Azospirillum*	Rice, sugarcane, wheat, sorghum	Improves water and mineral intake, promotes root growth, fixes 20–160 kg N/ha/year, and boosts crop output
5.	Arbuscular Mycorrhizal fungi (symbiotic)	Wheat, corn	Improve nutrient uptake, expand nutrient availability, solubilize phosphate, and boost crop productivity

attributes, and nutrient uptake of wheat plants was noticed in PSB + VAM fungi inoculated plants and increase in grain yield was recorded notably by 92.8%. However, Khan and Zaidi (2007) stated that triple inoculation of *Azotobacter chroococcum*, *Bacillus*, and VAM (*Glomus fasciculatum*) significantly improved wheat growth, increasing dry matter by 2.6-fold, doubling grain yield, and improving nitrogen, phosphorus, and grain quality in comparison to noninoculated plants (Table 11.4).

11.4 Commercial Biofertilizers: Preparation and Application

A product will work on crops if there is perfect formulation and the right application of the product. In the current scenario, effective microbial inoculants need to be characterized so that they can be utilized as inoculants for broader soil conditions. But the bioinoculants commercialized are of poor quality or are contaminated with unwanted microbes (Herrmann and Lesueur 2013). Consequently, the performance of bioinoculants when applied in the field is inconsistent. Moreover, some of these inoculants may not yield the same stimulant under field conditions due to inappropriate formulation. Inoculant formulation is the key point for preparation of PGPR that needs to be carried out in such a manner that ensures high storage viability from storage period to application. Bioinoculant formulations comprise inoculum preparation, additive inclusion, carrier selection and sterilization, upscaling, quality control, and packaging employing efficient delivery systems (Fasusi et al. 2021). Formulations can be solid or liquid, with solid kinds classified as dry or wet (Kumar et al. 2022).

1. Liquid Formulation

Inoculants formed by blending one or more microbial cultures with various liquids such as water, oil, and polymers called liquid inoculants and oil or polymer are incorporated into the culture to enhance its stability, adhesion, and dispersion efficiency (Vassilev et al. 2020). Since liquid inoculants are easy to prepare and relatively cheap compared to solid formulation, they are gaining more market share. The additional polymer has been described to form a microenvironment, with high water activity limiting transfer of heat, thus further maintaining conditions needed for bacterial survival. Nevertheless, most of these microbes are sensitive to some classical abiotic stressors (e.g., nutrient limitation, thermal shock) as they cause a dramatic decrease in the number of viable cells (Vassilev et al. 2020; Trimurtulu et al. 2014). However, Li et al. (2021) reported the impact of liquid organic fertilizer on sunflower was investigated, with an emphasis on the structure of the rhizosphere microbial population and the patterns of microbial co-occurrence. However, the application of organic fertilizer led to an increase in soil nutrient concentrations ranging from 13.8% to 137.1% compared to untreated soils. Furthermore, the treatment resulted in a drop of soil pH and salinity of 5.6% and 54.7%, respectively. Furthermore, the organic fertilizer treatment contributed to improvements in sunflower yield, the number of plants, and plant height, with increases observed between 28.6% and 67.3%. Following the application, the relative abundance of specific halotolerant microorganisms and phytopathogenic fungi declined in soils treated with organic fertilizer. In contrast to this, the relative abundance of favorable plant growth-promoting microorganisms and organic matter-decomposing organisms, such as *Nocardioides*, *Rhizophagus*, and *Stachybotrys*, increased substantially. A greater number of keystone taxa, such as *Amycolatopsis*, *Variovorax,* and *Gemmatimonas*, exhibited positive correlations with soil nutrient concentrations as well as crop yield-related characteristics in soils treated with organic fertilizer (Li et al. 2021). To summarize, the utilization of liquid organic fertilizer minimized the negative impacts of salt and alkalinity stress on sunflower output by enhancing soil quality and optimizing the structure and co-occurrence patterns of the rhizosphere microbial population.

2. Solid Formulations

Solid bioinoculant formulations, prepared using organic or inorganic carrier materials, are typically available in granule or powder form and can be classified based on particle size and intended application (Khan et al. 2023). Formulations that incorporate protectants, additives, and carriers may present as either wet or dry; however, their overall composition remains relatively consistent. In the case of wet formulations, the absence of a drying process results in a consistently high water content during both storage and application (Lobo et al. 2019). In solid wet formulations, PGPR is immobilized within the carrier matrix to lessen the impact of unfavorable circumstances. Either biofilm formation or microbial attachment to a support structure, like alginate beads, can accomplish this procedure. Biofilm development was used to encapsulate *Azospirillum brasilense* Ab-V5 in biodegradable foam (Lobo et al. 2019). Conventional components were used to create the foam, and these were then mixed with industrial waste products such glycerol, sugarcane bagasse, and cassava starch. For improved survivability, certain adhesive or protective materials are applied in addition to the carrier material (Zheng et al. 2024). These materials include disaccharides like lactose and sucrose, arabic gum carboxymethyl cellulose, and maltodextrin. The microbial species involved determines the effectiveness of cell protectants, and the former concentration must be tailored to the species. Glutamate of sodium enhances the bacterial resistance to drying and stabilizes the head groups, which makes the cell membrane more fluid and affects membrane conservation; additionally, milk proteins' amino groups interact with those of carrier molecules to maintain the protein structure. Whey is an inexpensive substance that functions as an extra nutrient and a protective agent, much like glycerol and maize bran residue.

11.5 Biofertilizer Formulations: Approaches and Application

Any biofertilizer strain needs to be competitive to survive, persist, and establish itself in the given environmental circumstances. Following establishment, the PGPR strain's population begins to increase and exhibits the intended growth-promoting characteristics. The impact of PGPR activity on plant development and crop improvement is shown in Fig. 11.4. Compared to chemical fertilizers, the effects of biofertilizer inoculation manifest more slowly. However, they have a long-lasting impact that raises soil fertility. Applying biofertilizers may be done in several ways, including dipping roots, applying soil, and inoculating seeds using a liquid or dry formulation (Kumar et al. 2022). However, before applying biofertilizer, more care must be taken. For example, the produced microbe solution should not be kept overnight and direct exposure to sunlight. Furthermore, the ideal temperature range for storing biofertilizers is between 0 and 35 °C. The process of rehydrating dry bioinoculant creates a suspension that may be applied to seeds, roots, or soil (Kumar et al. 2022). Conversely, liquid formulations do not need to be rehydrated and are utilized as is. Sterilized seeds are put to a slurry made by diluting carrier biofertilizer with water or jaggery to create a homogeneous coating before the seeds are mixed in. The covered seeds are then allowed to air dry before being planted in the ground (Reddy et al. 2019). For transplanted crops, the root dipping method is used,

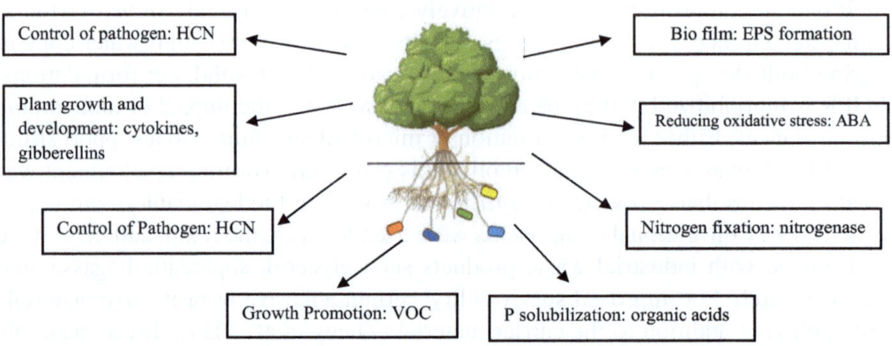

Fig. 11.4 PGPR effects on crop growth and development

in which water and biofertilizer are combined. Origins are dipped in the suspension of diluted biofertilizer, held for a period, and then transplanted. However, biofertilizer is applied to the soil before planting, either by spraying or distribution, when the farmer is ready to sow seeds (Reddy et al. 2019).

11.6 Limitations in Biofertilizer Production

Large-scale commercial production of biofertilizers is hampered by several factors, including poor soil quality, competition from inoculation strains, abiotic stress, contamination, and a lack of efficient stress (Mahmud et al. 2021), abilities, storage, awareness, carrier material, and even a lack of rules and security requirements. The product's label must include the microorganisms' name, activity, and expiration date for the biofertilizers to be considered credible. The sturdy exotic species must be chosen by the manufacturers and subsequently tested in various environmental settings, such as with various crops and soil types. Every stage of the manufacture of biofertilizer should involve quality control. The product must have a shelf life of 2–3 months, which can be increased by adding nutrients and additives in addition to a carrier (Kumar et al. 2022). Additionally, attention needs to be placed on amending with organic matter such as vermicompost and farmyard manure to reduce the use of chemical fertilizer and to preserve PGPR effectiveness and soil quality without compromising crop growth and production.

A microbial consortium or co-inoculation may improve plant development and boost nutrient absorption as well, but appropriate experimentation is required to examine the synergistic effect of bacteria on various crops (Cirillo et al. 2023) Regulations must be established for the creation, upkeep, and advertising of biofertilizers. Additionally, the government must provide farmers with incentives and subsidies to encourage the use of PGPR as biofertilizers to support organic farming. Promoting the usage of biofertilizers among farmers may aid in restoring soil fertility and provide the added benefit of sustainable crop production.

Biofertilizers: Their Role in Agriculture

- Biofertilizers complement synthetic fertilizers to fulfill crop requirements for nutrition.
- Biofertilizers contribute approximately 20–180 kg N/ha to the soil, improving crop output and nutrient efficiency in favorable conditions.
- Biofertilizers significantly reduce the usage of synthetic fertilizers and produce chemical-free yields.
- Biofertilizers stimulate plant growth by releasing growth-promoting compounds, leading to improved nutrient and water uptake, enhanced soil quality, and better rhizosphere development.
- Biofertilizers consist of diverse microbial communities capable of inhibiting pathogenic organisms, thereby minimizing the occurrence and spread of multiple plant diseases.

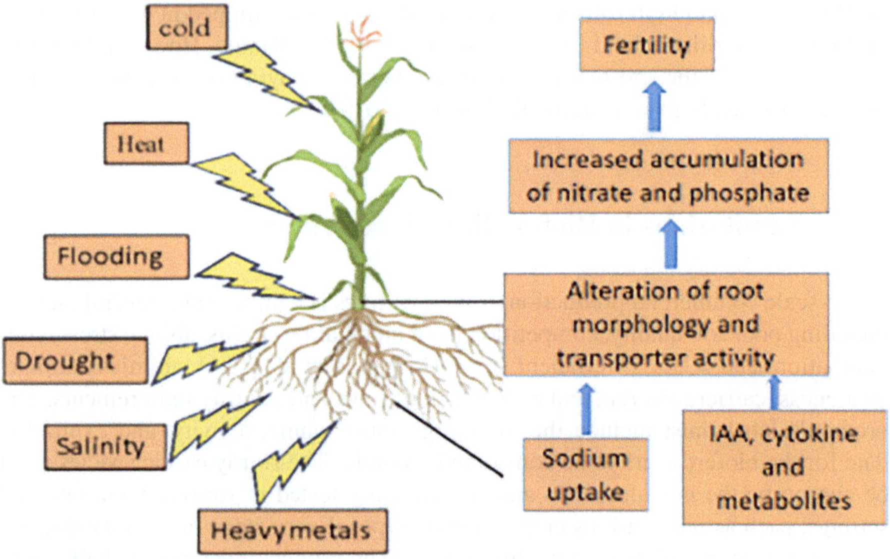

Fig. 11.5 Biofertilizers promoting abiotic stress management in plants

- Biofertilizers promote abiotic stress management in plants (Fig. 11.5).
- Biofertilizers play a vital role in improving soil quality, enhancing its structure and fertility, and promoting better crop performance.

Key Factors Affecting Biofertilizer Performance

- The effectiveness of an inoculant or microorganism largely depends on the host plant and its genotype.
- The effectiveness of inoculants in nitrogen fixation and nutrient solubilization is strongly influenced by their quality.
- The outcome of biofertilizer application depends on the crop's management practices and cultural operations.
- The soil characteristics, including its physical and chemical attributes, play a major role in determining the impact of microbial inoculants.

Limitations in Biofertilizer Application

- There is a lack of acceptable, high-quality strains that are capable of supplying nutrients to the soil.
- Inadequate storage facilities limit the effective utilization of biofertilizers.
- Inoculant efficacy and the benefits they provide are constrained by field factors such as extreme pH levels, temperature stress, and nutrient limitations.
- There is a lack of awareness among farmers.
- There is a lack of skilled personnel and insufficiently trained staff.
- Biofertilizers have a low nutritional concentration compared to inorganic fertilizers, which may cause plant deficiencies.

11.7 Potential of Biofertilizer in the Agricultural Market

Nitragin, a *Rhizobium* inoculation for legume, had its registration about 120 years ago, representing the start of the history of biofertilizers (Kumar et al. 2022). There have been rhizobia strains on the market for almost a century as bioinoculants. It is estimated that biofertilizers make up around 5% of all fertilizers sold on the market. About 150 of the biofertilizers are microbial strains that are approved for use in agriculture, out of which a few are listed in Table 11.5. Consumers are now more inclined to choose chemical-free products due to growing concerns about food quality and safety, the environment, and pesticide residues in food (Nitzko et al. 2024). The following nations have had recent increases in organic retail sales: the United States, Germany, India, China, Switzerland, and Denmark. However, in the biofertilizer industry, rhizobia comprise the majority share at 79%, with PSB contributing roughly 15% of the global market. Many manufacturing firms control the worldwide market. Some of them areMadras Fertilizers Limited (India), Novozymes (Denmark), International Panaacea Limited (India), SOM Phytopharma Ltd. (India), Symborg (Spain), Kan Biosys (India), Kiwa Biotech (China), and Rashtriya Chemicals & Fertilizers Ltd. (India), National Fertilizers Limited (India), Mapleton Agribiotech (Australia), Lallemand Inc. (Canada), Fertilizers and Chemicals Ltd. (India), and Rizobacter Argentina S.A. (Argentina). A small business can readily develop biofertilizers for usage in small fields. For example, in America, a strain of *Azospirillum* is chosen from the field following thorough testing, and an appropriate

Table 11.5 Major bacterial strains and their product name

Categories of biofertilizer	Bacterial strain	Commercial name	Crop usages
Nitrogen-fixing biofertilizer	*Azotobacter chroococcum*	Bioazoto, Bhoomi Rakshak, Azonik	Wheat, sorghum, maize, cotton, vegetables, horticulture crops, flowers
Phosphate-solubilizing/-mobilizing biofertilizer	*Pseudomonas striata*, *B. polymyxa* and *B. megaterium*	P sol B	For all crops
Potassium-solubilizing biofertilizer	*Bacillus mucilaginous* and *B. subtilis*	CBF	For all crops
Zinc-solubilizing biofertilizer	*Thiobacillus thiooxidans*	Zn sol B	Wheat, paddy, pulses, citrus, ginger, etc.
Sulfur-oxidizing biofertilizer	*Thiobacillus thiooxidans*	S sol B	For cereals, oilseeds, fiber crops, medicinal crops, vegetables, flowers, orchards, and forage crops
Plant growth-promoting biofertilizer	*Azotobacter chroococcum* and *Pseudomonas fluorescens*	Bio gold	For all crops

formulation is then created and released into the market. There are currently more than 100 commercially available *Azospirillum* strains that are meant to increase crop output in South America, mostly in wheat, maize, and soybeans (Aizawa 2014).

11.8 Future Prospects

Biofertilizers are undoubtedly the way of the future in agriculture, and it is expected that they will replace chemical fertilizers. They facilitate and safeguard the biodegradation of soil carried out by living organisms, which eventually leads to a safe method of increasing soil fertility without the use of chemical leftovers. Furthermore, we believe that incorporating nanoparticles into biofertilizers may provide effective and environmentally beneficial alternatives. Thus, to lessen environmental stress and improve plant quality and production, plant diseases might be managed and plant resistance raised. Using cloud computing, big data, IoT apps, and artificial intelligence approaches, as well as simulation systems for every crop in smart agriculture, they are utilized to identify the biofertilizers required to promote a species' development and which organisms are most effective at extracting this biofertilizer, as well as investigate plant variety across different locations. They encourage farmers to adopt the practice of using biofertilizers by means of voluntary awareness campaigns and surveys to evaluate their knowledge and elucidate the advantages of biofertilizers over chemical fertilizers, particularly in agricultural regions like North Africa, East Asia, the Caribbean, and some parts of Latin America that are still unspoiled in their natural habitat. Nations with assistance and microinnovative ventures, such as microbial biostimulants, which are founded on the concept of entrepreneurship and use their surroundings to generate essential local fertilizers and support them at a cheaper cost for farmers commercially and biofertilizers based on biofilm.

11.9 Conclusion

Biofertilizers will be used as part of sustainable agriculture and are one example of new ecologically friendly technology. Biofertilizers have a higher nutritional content, raise plant production, and improve soil fertility. Biofertilizers may be made from a range of living things such as bacteria (*Azotobacter*, *Cyanobacter*, and *Nitrobacter*), micro fungi (*Glomus* sp.), and microalgae (*Chlorella vulgaris*, *Spirulina platensis*, *Haematococcus pluvialis*, and *Acutodesmus dimorphus*). Biofertilizers can also be made from the leftovers of macroorganisms, such as plants, fungus, or algae. However, residues from microbes and macroorganisms improve soil fertility in an environmentally safe manner, being fully biodegradable and pollutant-free. This approach safeguards ecosystem integrity and promotes the ideals of sustainable development. Biofertilizer application offers a practical means to boost crop productivity and reduce chemical fertilizer usage, promoting eco-friendly and sustainable farming. Numerous microbes that promote plant growth have been identified for their advantageous characteristics, which are important in enhancing the availability of nutrients like N, P, K, Zn, and S, regulating phytohormones, preventing plant diseases, and reducing abiotic stressors. It has been discovered that, in both greenhouse and field settings, the inoculation of a single or group of advantageous microbes increases plant biomass and agricultural output. Nonetheless, several limitations have been shown to occasionally restrict crop development in field settings across various agricultural environments, and the inoculation with viable PGPR strains has no effect on crop output, plant growth, or agri-product quality. The survival and functionality of injected microorganisms under field circumstances are impacted by the typical differences in the growing environment of bacteria between laboratory and greenhouse settings. Moreover, analyzing multiple microbial genomes and pinpointing specific genes enable enhanced production, as beneficial microbes release PGP metabolites that offer both rapid and enduring solutions to increase crop productivity for sustainable global food security.

References

Adnan M, Shah Z, Khan A, Khan GA, Ali A, Khan NA, Zaib K (2014) Integrated effects of rhizobial inoculum and inorganic fertilizers on wheat yield and yield components. Am J Plant Sci 5:2066–2073. https://doi.org/10.4236/ajps.2014.513222

Afzal I, Shinwari ZK, Sikandar S, Shahzad S (2019) Plant beneficial endophytic bacteria: mechanisms, diversity, host range and genetic determinants. Microbiol Res 221:36–49

Aizawa S (2014) *Azospirillum brasilense* — a bushy hook of the polar flagellum. In: The flagellar world electron microscopic images of bacterial flagella and related surface structures. https://doi.org/10.1016/B978-0-12-417234-0.00003-7

Barman S, Das S, Bhattacharya SS (2019) The prospects of bio-fertilizer technology for productive and sustainable agricultural growth. In: New and future developments in microbial biotechnology and bioengineering. Elsevier, pp 233–253

Bedair H, Rady HA, Hussien AM et al (2022) Pesticide detection in vegetable crops using enzyme inhibition methods: a comprehensive review. Food Anal Methods 15(7):1979–2000. https://doi.org/10.1007/S12161-022-02254-X

Chatterjee A, Singh S, Agrawal C et al (2014) Role of algae as a biofertilizer. Algal Green Chem:189–200. https://doi.org/10.1016/B978-0-444-63784-0.00010-2

Chauhan DS, Sharma RK, Tripathi SC, Kharub AS, Chhokar RS (2011) News paradigm in tillage technology for wheat production. Research Bulletin No. 8, DWR, Karnal, 16

Cirillo V, Romano I, Woo SL, Di Stasio E, Lombardi N, Comite E, Pepe O, Ventorino V, Maggio A (2023) Inoculation with a microbial consortium increases soil microbial diversity and improves agronomic traits of tomato under water and nitrogen deficiency. Front Plant Sci 14:1304627. https://doi.org/10.3389/fpls.2023.1304627

Das H, Devi N, Venu N, Borah A (2023) Chemical fertilizer and its effects on the soil environment. Res Rev Agric Sci 7:31–51

Davari MR, Sharma SN (2010) Effect of different combinations of organic materials and biofertilizers on productivity, grain quality and economics in organic farming of basmati rice (*Oryza sativa*). Indian J Agron 55(4):290–294

Fasusi OA, Cruz C, Babalola OO (2021) Agricultural sustainability: microbial biofertilizers in rhizosphere management. Agriculture 11(2):163

Feyisa T, Amare T, Adgo E, Selassie YG (2013) Symbiotic blue green algae (Azolla): A potential biofertilizer for paddy rice production in Fogera Plain, northwestern Ethiopia. Ethiop J Sci Technol 6(1):1–11

Gerland VN, Dakora FD (2014) Potential use of rhizobial bacteria as promoters of plant growth for increased yield in landraces of African cereal crops. Afr J Biotech 3(1):1–7

Gothandapani S, Sekar S, Padaria JC (2017) *Azotobacter* chroococcum: utilization and potential use for agricultural crop production: an overview. Int J Adv Res Biol Sci 3:35–42

Green N, Goltapeh EM, Amini J, Mehnaz S, Zarea MJ (2005) Effect of *Azospirillum zeae* and seed priming with zinc, manganese and auxin on growth and yield parameters of wheat, under dryland farming. Agric Res 10:44–55

Herrmann L, Lesueur D (2013) Challenges of formulation and quality of biofertilizers for successful inoculation. Appl Microbiol Biotechnol 97:8859–8873

Katiyar NK, Ranawat S, Pathak RK, Kumar A (2011) Effect of *Azotobacter* and nitrogen levels on yield and quality of wheat. Ann Plant Soil Res 13(2):152–155

Kaushik MK, Bishnoi NR, Sumeriya HK (2012) Productivity and economics of wheat as influenced by inorganic and organic sources of nutrients. Ann Plant Soil Res 14(1):61–64

Kawalekar KR, Sabir M, Ozturk M, Akhtar MS, Ibrahim FH (2013) Nitrate and nitrogen oxides: sources, health effects and their remediation. Rev Environ Contam Toxicol 242:183–217

Kaur T, Devi R, Kumar S, Sheikh I, Kour D, Yadav AN (2022) Microbial consortium with nitrogen fixing and mineral solubilizing attributes for growth of barley (Hordeum vulgare L.). Heliyon 8:e09326. https://doi.org/10.1016/j.heliyon.2022.e09326

Khan A, Singh AV, Gautam SS, Agarwal A, Punetha A, Upadhayay VK, Goel R (2023) Microbial bioformulation: a microbial assisted biostimulating fertilization technique for sustainable agriculture. Front Plant Sci 14:1270039

Khan MS, Zaidi A (2007) Synergistic effects of the inoculation with plant growth-promoting rhizobacteria and an arbuscular mycorrhizal fungus on the performance of wheat. Turkish J Agri For 31(6):355–362

Kumar AU (2018) Impact of biofertilizers in enhancing growth and productivity of wheat: a review. Int J Chem Stud 6(4):360–362

Kumar S, Sindhu SS, Kumar R (2022) Biofertilizers: an ecofriendly technology for nutrient recycling and environmental sustainability. Curr Res Microb Sci 3:100094

Lenart A (2012) Occurrence, characteristics, and genetic diversity of *Azotobacter* chroococcum in various soils of southern Poland. Pol J Environ Stud 21(2):415–424

Li H, Luo N, Ji C, Li J, Zhang L, Xiao L, Lai H (2021) Liquid organic fertilizer amendment alters rhizosphere microbial community structure and co-occurrence patterns and improves sunflower yield under salinity-alkalinity stress. Microb Ecol 84(2):423–438

Lobo CB, Tomás MSJ, Viruel E, Ferrero MA, Lucca ME (2019) Development of low-cost formulations of plant growth-promoting bacteria to be used as inoculants in beneficial agricultural technologies. Microbiol Res 219:12–25

Mahato S, Kafle A (2018) Comparative study of *Azotobacter* with or without other fertilizers on growth and yield of wheat in Western hills of Nepal. Ann Agrar Sci 16(3):250–256

Mahmud AM, Upadhyay SK, Srivastava AK, Bhojiya AA (2021) Bio fertilizers: a nexus between soil fertility and crop productivity under abiotic stress. Curr Res Environ Sustain 3:100063. https://doi.org/10.1016/j.crsust.2021.100063

Malik B, Mandal B, Bandopandhyay PK, Gangopadhyay A, Mani PK, Kundu AL, Majumdar D (2009) Organic amendment influence on soil organic pools and crop productivity in a nineteen years old rice-wheat agro-ecosystem. Soil Sci Am J 72:775–785

Mehboob I, Naveed M, Zahir ZA (2009) Rhizobial association with non-legumes: mechanisms and applications. Crit Rev Plant Sci 28(6):432–456

Mia MB, Shamsuddin ZH (2010) *Rhizobium* as a crop enhancer and biofertilizer for increased cereal production. Afr J Biotech 9(37):6001–6009

Nitzko S, Bahrs E, Spiller A (2024) Consumer willingness to pay for pesticide-free food products with different processing degrees: does additional information on cultivation have an influence? Farming Syst 2:100059. https://doi.org/10.1016/j.farsys.2023.100059
Pabbi S (2015) Blue green algae: a potential biofertilizer for rice. In: The algae world, pp 449–465
Prajapati K, Yami KD, Singh A (2008) Plant growth promotional effect of *Azotobacter* chroococcum, *Piriformospora indica* and vermicompost on rice plant. Nepal J Sci Tech 9:85–90
Priya H, Dhar DW, Singh R et al (2022) Co-cultivation approach to decipher the influence of nitrogen-fixing cyanobacterium on growth and N uptake in rice crop. Curr Microbiol 79:53. https://doi.org/10.1007/s00284-021-02732-1
Ram T, Yadav SK, Sheoran RS (2005) Growth analysis of wheat (*Triticum aestivum* L.) under varying fertility levels and *Azotobacter* strains. Indian J Agric Res 39(4):295–298
Reddy KK, Babu VS, Gavali RS (2019) Sustainable livelihoods and adaptation to climate change (SLACC), Training Manual. National Institute of Rural Development and Panchayati Raj, Ministry of Rural Development, Government of India. http://mksp.gov.in/portal/images/SLACC_training_manual_English.pdf
Rodriguez C, Carlsson G, Englund JE, Flöhr A, Pelzer E, Jeuffroy MH, Jensen ES (2020) Grain legume-cereal intercropping enhances the use of soil-derived and biologically fixed nitrogen in temperate agroecosystems. A meta-analysis. Eur J Agron 118:126077
Sahoo RK, Bhardwaj D, Tuteja N (2012) Biofertilizers: a sustainable eco-friendly agricultural approach to crop improvement. In: Plant acclimation to environmental stress. Springer New York, New York, NY, pp 403–432. https://link.springer.com/chapter/10.1007/978-1-4614-5001-6_15
Saxena J, Chandra S, Nain L (2013) Synergistic effect of phosphate solubilizing rhizobacteria and arbuscular mycorrhiza on growth and yield of wheat plants. J Soil Sci Plant Nutr 13(2):511–525
Sumbul A, Ansari RA, Rizvi R, Mahmood I (2020) *Azotobacter*: a potential bio-fertilizer for soil and plant health management. Saudi J Bio Sci 27(12):3634–3640
Tilman FI, El-Seoud IA, Badr EA (2011) Response of two rice cultivars to blue-green algae, a-mycorrhizae inoculation, and mineral nitrogen fertilizer. Middle E Russ J Plant Sci Biotechnol 2(1):29–34
Trimurtulu N, Rao DLN, Trimurtulu N, Amaravathi G (2014) Liquid microbial inoculants and their efficacy on field crops, ANGRAU. Agricultural Research Station, Amaravathi, p 54
Vassilev N, Vassileva M, Martos V, Garcia del Moral LF, Kowalska J, Tylkowski B, Malusá E (2020) Formulation of microbial inoculants by encapsulation in natural polysaccharides: focus on beneficial properties of carrier additives and derivatives. Front Plant Sci 11:270
Yanni YG, Rizk RY, Abd El-Fattah FK, Squartini A, Corich V, Giacomini A, Dazzo FB (2001) The beneficial plant growth-promoting association of *Rhizobium* leguminosarumbv. Trifolii with rice roots. Funct Plant Biol 28(9):845–870
Yilmaz E, Sönmez M (2017) The role of organic/bio–fertilizer amendment on aggregate stability and organic carbon content in different aggregate scales. Soil Till Res 168:118–124
Zheng Y, Liao J, Fang Y, Gui R, Hou Y, Zhang M, Chen X (2024) Innovative fabrication of eco-friendly bio-based foam from sugarcane bagasse and sodium alginate with enhanced properties and sustainable applications for plastic replacement. Int J Biol Macromol 282:137464

Chapter 12
Mitigating Abiotic Stress in *Sorghum bicolor* Through Biofertilizer Application

Tessia Rakgotho, Beverly Mampholo, Adornis D. Nciizah, and Thembeka Mabiya

Abstract *Sorghum bicolor* is one of the world's five major cereal crops and is an important source of food, raw material for brewing, and a promising bioenergy crop. Due to unavoidable abiotic stressors, maintaining an adequate food supply is currently challenging. Abiotic stress is the primary factor affecting the growth and yield of many crops globally. Abiotic stresses, such as salinity and drought, are the major factors that potentially affect plant growth, development, and productivity, resulting in deduced crop yield by imposing ion toxicity, hyperosmotic, and oxidative stresses. These stresses cause alterations in various plant characteristics, including physiological, morphological, biochemical, and molecular aspects. It has been difficult for farmers to produce high-quality food because of the effects of these stresses. To overcome these problems, farmers use expensive tools and chemical fertilizers to produce quality crops, which in turn negatively impact food security. Studies have revealed that biofertilizers play important roles in plants by improving their growth and development. Due to their advantages in terms of cost-effectiveness, eco-friendliness, and ease of use in agricultural environments, biofertilizers have become a popular alternative to these in the modern era. Many microorganisms

T. Rakgotho
Life Sciences Building, Department of Biotechnology, University of the Western Cape, Bellville, South Africa

Agricultural Research Council – Roodeplaat, Vegetable, Industrial and Medicinal Plants, Pretoria, South Africa

B. Mampholo (✉)
Agricultural Research Council – Roodeplaat, Vegetable, Industrial and Medicinal Plants, Pretoria, South Africa
e-mail: MampholoM@arc.agric.za

A. D. Nciizah
Agricultural Research Council, Natural Resources and Engineering, Pretoria, South Africa

T. Mabiya
Life Sciences Building, Department of Biotechnology, University of the Western Cape, Bellville, South Africa

A. D. Nciizah et al. (eds.), *Biofertilizers for Sustainable Agriculture in Africa*, Sustainability Sciences in Asia and Africa, https://doi.org/10.1007/978-981-95-6009-7_12

referred to as biofertilizers might result in plant activities that, even in the presence of abiotic stress, promote soil development. Among the beneficial microbial consortia that support plant growth are rhizobacteria, arbuscular mycorrhizal fungus, azotobacter phosphate solubilizers, and nematophagous and insect frass of other biofertilizers. Biofertilizers can be used as an alternative to chemical fertilizers because they are eco-friendly, cost-effective, and easy to apply in the agricultural field. These biofertilizers help plants directly or indirectly by fixing nitrogen (N), solubilizing phosphorus (P) and potassium (K), secreting siderophores, antibiotics, and enzymes, and releasing plant growth-promoting hormones to enhance productivity and adaptability to overcome biotic and abiotic stresses. The performance of plant development can be maintained by biofertilizers even under challenging circumstances. Thus, the present review demonstrates the effectiveness of different types of biofertilizers on *Sorghum bicolor* under abiotic conditions.

Keywords Biofertilizers · Chemical fertilizers · Drought · Salinity *and Sorghum bicolor*

12.1 Introduction

The world population is growing rapidly, and this increases the demand for the production of nutritional food to feed the overgrowing population. The major challenge for global food security is to feed the increasing world population with affordable and easily accessible nutritious food. Therefore, increasing the production of high-quality food is crucial; however, it remains a challenge due to the impact of climate change, which generates abiotic and biotic stress pressures that have a detrimental impact on plant development and food availability, particularly given limited resources. Abiotic stress is the primary factor affecting the growth and yield of many crops globally. Among the stressors, both salinity and drought are key abiotic limitations that potentially impede plant growth and development, causing a decline in crop yield, alterations in most physiological, morphological, biochemical, and molecular characteristics, and agricultural productivity (Gul et al 2022). To overcome these problems, farmers use expensive tools, chemical fertilizers, and pesticides to produce quality crops. However, the excessive and long-term usage of chemical fertilizers in agriculture leads to reduced soil fertility and soil acidification, ultimately affecting crop (Bahrulolum et al. 2021) and altering the chemical ecology of the soil. Moreover, chemical fertilizers contain various nutrients such as nitrogen (N_2), phosphorus (P), potassium (K), and sulfur, which, when used in excess amounts, result in weakening of plant roots, high rate of disease incidence, and eutrophication of groundwater. Nutrients such as nitrates have neither taste nor smell but can be detected using laboratory tests and are subject to regulation (Santamaria 2006). This poses a serious public health risk such as blue baby syndrome. Thus, it is imperative to come up with new solutions to overcome environmental stressors that affect the development and growth of plants. To alleviate the impact of abiotic stress on plants and improve plant productivity, various management strategies have been implemented, ranging from breeding approaches, including conventional, molecular, and genetic engineering, and most recently, speed breeding techniques, to agronomical approaches such as nutrient management and use of biostimulants and biofertilizers. Biofertilizers can directly and indirectly improve plant growth and protect plants from abiotic and biotic stress. Compared to other chemical fertilizers, biofertilizer manufacturing emits 23.2 times less carbon emissions than nitrogen fertilizer manufacturing. In order to reduce emissions, it is important to identify and prioritize any interventions that make fertilizers less harmful to the environment.

12.2 Biofertilizers

Biofertilizers are a relatively new approach based on plant–microbe interactions, which has received a lot of interest in the agriculture industry since at least the 1980s (Aloo et al. 2022). Biofertilizer has been considered a better option than conventional fertilizers because it contains a typical population of beneficial microbes that offer nutrients to the soil required by the plant without harming the environment. They can be defined as substances containing naturally occurring microorganisms that have beneficial properties to plant growth and development, improving soil health and reducing the negative environmental effects such as pollution and soil degradation. (Kumar et al. 2017). They are microbial inoculants that contain cultures of dormant or live cells of the effective strains of N-fixation, P-solubilization/mobilizing, and K-solubilization (Mahmud et al. 2021). They are environmentally friendly, cost-effective, and guarantee sustainable agricultural production (Mącik et al. 2020; Agarwal et al. 2018; Lim and Matu 2015; Mahanty et al. 2017). Biofertilizers are classified based on their function and mechanisms of actions (Mącik et al. 2020; Daniel et al. 2022). The most used biofertilizers are nitrogen-fixers, potassium and phosphorus solubilizers, and plant growth-promoting rhizobacteria (PGPR) (Daniel et al. 2022; Nosheen et al. 2021). They can directly and indirectly improve plants' growth and development by enhancing nutrient uptake and increase crop yields as shown in Fig. 12.1 (Mącik et al. 2020). Direct mechanisms that act upon plants directly include nitrogen fixation, phosphate solubilization, micronutrient solubilization, and the production of phytohormones (Chaudhary et al. 2022). The indirect mechanism generally protects the plant from the deleterious effect of the pathogens by releasing lytic enzymes, antibiotics, siderophores, and cyanide production (Mahmud et al. 2021). Biofertilizer can be directly applied to seeds, roots, soil, or foliar spray to enhance the activity of microbial multiplication, which then mobilizes nutrients, improves soil fertility, and increases plant health and productivity (Chaudhary et al. 2022). Few studies have demonstrated that biofertilizer can improve plant growth under abiotic stress by producing growth

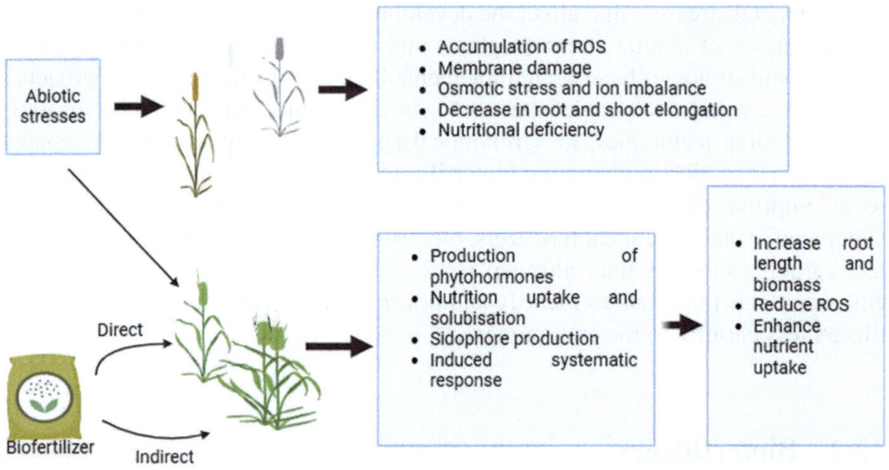

Fig. 12.1 Schematic representation of the effectiveness of biofertilizer on stresses *Sorghum bicolor.* (Bioreader software: author design)

hormones and improving nutrient availability. There is significant evidence to support the use of biofertilizers to improve sorghum response to abiotic stresses, making them ideal in a changing climate.

12.3 Sorghum Growth and Development

Sorghum [*Sorghum bicolor* L. *Moench*] is one of the most important crops in the world, ranked fifth in the world and second most important cereal after maize in Africa (Swamia et al. 2011; Mofokeng et al. 2017; Zidenga 2004). Sorghum is a versatile cereal crop with significant potential for improving food security and nutrition. For many of the world's most vulnerable populations, sorghum has long been a vital staple crop (Pereira and Hawkers 2022). Known for its resilience, the crop thrives in arid and drought-prone areas where other staple crops may fail, making it a crucial food source in regions of Africa, Asia, and certain parts of the Americas, where reliable food production is often hindered by challenging environmental conditions and competition (Khoddamia et al. 2023). Its cultivation is important because it serves as a source of food for both humans and animals, bioethanol production, and various industrial applications. Another important area is its use in the production of bioenergy to solve the global energy crisis (Monteiro et al. 2012; Calvino and Messing 2012; Rooney et al. 2007). Although sorghum is known to be a drought- and moderate salt-tolerant cereal crop, it can be impacted by other various abiotic stresses. Thus, this review aims to highlight the advantages of biofertilizer use in sorghum production systems under salinity and drought.

12.3.1 Main Uses of Sorghum as a Staple Food

Sorghum is packed with essential nutrients, making it a highly nutritious option for a wide range of people. Sorghum serves as an important source of fiber, energy, vitamins, carbohydrates, proteins, and lipids (Tanwar et al. 2023). While sorghum is commonly cooked and consumed as a dish, it can also be processed into ingredients for other recipes. The crop can be used to make bread, porridge, flour, and various baked goods (Rao et al. 2019). One of the key benefits of sorghum is its high iron content, which plays a vital role in preventing anemia (Istrati et al. 2019). Additionally, its significant fiber content supports digestive health and helps lower blood cholesterol levels. As a naturally gluten-free grain, sorghum is an excellent option for individuals with celiac disease or gluten sensitivity (Tanwar et al. 2023). It also stands out among major cereals for its unique starch composition, which is digested more slowly compared to other grains. This slower digestion helps regulate obesity by promoting the gradual release of glucose, aiding in appetite control and better blood sugar management. Furthermore, its low-calorie content makes it beneficial for those dealing with obesity, diabetes, and cardiovascular diseases and the grain's rich antioxidant properties also help reduce oxidative stress (Rezaee et al. 2021). Sorghum is not only a valuable food crop for humans but also plays a significant role in livestock feed. It is a highly nutritious and versatile crop for animal nutrition, particularly in regions where it is abundantly grown. Sorghum provides an excellent source of energy, fiber, and essential nutrients for various types of livestock, including cattle, poultry, pigs, and sheep (Ronda et al. 2019). It is utilized as a feed ingredient to increase animal health and productivity (Khalid et al. 2022). The plant's stalks, leaves, and grains can be used as forage, silage, or grain-based feed, making it a versatile resource for farmers. The ability of Sorghum to thrive in challenging growing conditions helps ensure a steady feed supply in regions with limited water resources.

12.3.2 Biofuel Production from Sorghum

In addition to its role in food and agriculture, sweet sorghum is increasingly recognized for its potential in biofuel production in various forms, including bioethanol, biogas, and biodiesel as a renewable energy source (Velmurugan et al. 2020). The plant's high sugar content in the grains and its efficient growth in regions with limited water availability make it an ideal candidate for biofuel production, helping reduce dependence on fossil fuels and contribute to sustainable energy practices (Stamenkovic et al. 2020). The use of sorghum for biofuel production is especially useful in regions where water is scarce and other crops may not grow efficiently. Obesity, by promoting the gradual release of glucose, aids in appetite control and better blood sugar management. Furthermore, its low-calorie content makes it beneficial for those dealing with obesity, diabetes, and cardiovascular diseases, and the grain's rich antioxidant properties also help reduce oxidative stress (Rezaee et al. 2021).

12.3.3 Effects of Abiotic Stress on Sorghum bicolor

Although *Sorghum bicolor* is widely recognized for its natural drought tolerance and moderate resistance to salt stress (Wagaw 2019), it can still be impacted by various abiotic stress factors. These stresses limit crop growth and productivity to varying degrees, depending on the timing, duration, and intensity of exposure. Such environmental conditions can negatively affect the growth, yield, and quality of crops, with potential yield reductions ranging from 50% to 70% (Kumar 2019). Some of these abiotic stresses that affect sorghum include drought, salinity, waterlogging, temperature, and heavy metal accumulation.

12.3.3.1 Drought Stress on *Sorghum bicolor*

Drought stress is one of the most significant environmental challenges for Sorghum plants, especially in regions with limited water availability. While sorghum is generally considered drought-tolerant compared to many other crops, prolonged or severe drought conditions can still adversely affect its growth, development, and yield. The crop's ability to withstand drought depends on factors such as the severity and duration of the drought (Assefa et al. 2010; Wagaw 2019). Furthermore, each crop has specific water-requirement critical stages; if water is insufficient during these times, yield and quality may be significantly reduced. Reactive oxygen species (ROS) are important in maintaining normal plant growth and improving stress tolerance. ROS are produced in plant cells in both ionic and molecular forms, primarily in chloroplasts, endoplasmic reticulum, mitochondria, peroxisomes, apoplasts, and plasma membranes (Sharma et al. 2012). Hydroxyl radicals (OH) and superoxide anions (O_2O_2-) are examples of ionic states, whereas hydrogen peroxide (H_2O_2) and singlet oxygen (O_2) are examples of molecular states. Additionally, every type of ROS possesses a distinct oxidative potential and impacts various physiological and biochemical processes that are controlled by distinct genes in plants (Fig. 12.1). Under normal growth conditions, plants use ROS in small concentrations as a signaling molecule (Baxter et al. 2014), thereby providing a shield against oxidative stress. All ROS are extremely harmful to organisms at high concentrations, under the extreme environmental conditions either individually or simultaneously trigger excessive production of ROS in plant cell that pose a major challenge for optimal plant growth and development, causing peroxidation of lipids, oxidation of proteins, damage to nucleic acids, enzyme inhibition, leading to activation of programmed cell death (Sharma et al. 2012) and stomatal behavior (Mittler 2017), as well as decreasing plant productivity (Raja et al. 2017). One of the first physiological responses of sorghum plants to drought is stomatal closure (Zheng et al. 2024; Agurla et al. 2018a, b). This mechanism helps conserve water, but it also reduces photosynthesis, limiting growth and biomass accumulation during extended drought conditions. Additionally, sorghum plants may adjust their root systems to better cope with water scarcity. Sorghum plants are known to develop deeper roots that can access water from lower soil layers, helping the plant maintain growth during dry periods (Chen et al. 2020). Sorghum plants also exhibit adaptive behaviors like leaf rolling or folding during drought (Kadioglu et al. 2011). This reduces the surface area exposed to the sun, minimizing water loss through transpiration. Furthermore, the leaves are adapted with waxy coatings that help conserve water by reducing evaporation. Some sorghum varieties possess physiological mechanisms to tolerate drought, through drought escape, avoidance, and drought tolerance to complete their life cycle (Begna et al. 2022).

12.3.3.2 Salinity Stress

Salt stress is a significant abiotic factor that causes considerable damage to agricultural production, particularly in arid and semi-arid regions. Albeit sorghum is more tolerant of saline conditions than many other crops, excessive salt levels can still adversely affect its growth, development, and yield (Mansour et al. 2021).

Salinity occurs due to the accumulation of soluble salts in the soil, such as potassium (K+), magnesium ($Mg^{2+)}$, calcium (Ca^{2+}), chloride (Cl^-), sulfate (SO^-), carbonate (CO^-), and sodium (Na^+) (Behera et al. 2022). High concentrations of these soluble salts can cause severe damage and reduce crop yield. Excessive salinity leads to ionic stress, osmotic stress, and ultimately oxidative stress in plants (Zhou et al. 2024). Oxidative stress results in membrane peroxidation, ion leakage, and damage to nucleic acids, cell membranes, and other cellular structures (Noreen et al. 2009). Salt stress also negatively impacts various physiological processes in sorghum, such as seed germination, potassium uptake, photosynthesis, chlorophyll synthesis, and electrolyte leakage.

To cope with salt stress, sorghum has developed several adaptive mechanisms. One of these is the production of low molecular weight organic solutes called osmoprotectants, which help maintain cellular structure and osmotic balance. Osmoprotectants help prevent osmotic stress by stabilizing cellular structures in response to high salt concentrations. They play a crucial role in maintaining cellular function under saline conditions. They serve as a defensive mechanism in plants against stress by contributing to osmotic adjustment, detoxifying reactive oxygen species (ROS), stabilizing membranes, and preserving the native structures of enzymes and proteins (Mehta and Vyas 2023).

12.3.3.3 Waterlogging stress

Waterlogging occurs when the soil becomes saturated with water to the point where air spaces in the soil are filled. This results in excess water around the plant roots, reducing oxygen availability and limiting respiration (Behera et al. 2022). Waterlogging typically occurs due to excessive irrigation, heavy rainfall, or poor drainage that prevents water from draining from the soil. Sorghum is less resistant to excessive moisture, especially when water persists for extended periods. Prolonged exposure of sorghum plants to these conditions can result in oxygen deprivation due to waterlogged soil. When the soil pores fill with water, oxygen, which is essential for root respiration, is displaced. This can damage the roots, and result in root rot, and limit the plant's ability to take up nutrients, resulting in plant death (Neira et al. 2015; Sharma and Kumar 2023).

During waterlogging, leaf stomata close, and chlorophyll degradation, along with leaf senescence and yellowing, reduces the leaves' ability to capture light, ultimately causing a decline in the photosynthetic rate (Yan et al. 2018). Plants have evolved various mechanisms to cope with the stress of waterlogging. A key adaptation is the formation of specialized root structures, such as adventitious roots (ARs), aerenchyma, and hypertrophic lenticels (Manghwar et al. 2024). Furthermore, plants increase the production of antioxidant enzymes to mitigate oxidative stress caused by oxygen deficiency and the accumulation of toxic substances. Additionally, plants adjust their root morphology, anatomy, and architecture to ensure the proper functioning of the root system (Pedersen et al. 2020).

12.3.3.4 High- and Low-Temperature Stress

Sorghum plants require minimum soil temperatures of around 18–20 °C for optimal germination (Khalifa and Eltahir 2023). When temperatures fall below this threshold, they can cause cellular damage by freezing the water inside the plant cells. These disrupt the plants' structure, damage cell membranes, which can lead to cell death (Vera Hernández et al. 2023). Low temperatures can also affect the plants' ability to take up water and nutrients, leading to stunt growth, reduced photosynthesis, and poor development of the plant. Furthermore, cold temperatures can delay maturation of sorghum plants, leading to prolonged growing seasons, which may make the crop more susceptible to other stresses like disease or late-season frost (Tari et al. 2013). As the stomata stays closed for a longer period to prevent water loss, the closing of the stomata limits the uptake of CO_2 for photosynthesis and in return slows down the plant's metabolism, affecting the use of water and nutrients of the crop.

High-temperature stress of above 35 °C can have a significant negative impact on sorghum plants, especially as global temperatures rise due to climate change. Sorghum is generally a heat-tolerant crop, but it has its limits when exposed to extreme temperatures, particularly during sensitive stages of development, such as flowering and grain filling. The ideal air temperature range for sorghum during the vegetative period is 26–34 °C, while during the reproductive period, it is 25–28 °C (Prasad et al. 2015). High temperatures can directly affect the plants' ability to perform photosynthesis, through thermal inactivation of photosynthetic enzymes, particularly ribulose-1,5-bisphosphate carboxylase/oxygenase (RuBisCO), which is crucial for carbon fixation (Mathur et al. 2014). Photosynthesis is particularly sensitive to high-temperature stress and is usually inhibited before other cellular functions are affected.

There is a reduction in carbon fixation, lower photosynthetic efficiency, and stunted plant growth. High temperatures, especially during flowering stage, can reduce the fertility of sorghum plants by causing pollen sterility (Prasad et al. 2006). This will result in reduced seed germination, lower grain yield, and poor reproductive success.

Sorghum plants have developed molecular responses for adaptation to stressful conditions such as transcription factors that can either be up- or downregulated in response to stress. Transcription factors function by regulating gene expression through binding to the *cis*-regulatory elements in the promoter region of different stress-related genes by regulating their transcription and, thus, enhance plant tolerance to harsh environmental conditions (Franco-Zorrila et al. 2014). These include the expression of transcription factors such as heat shock proteins (HSPs), which facilitate the restoration of normal function of proteins by assisting in the refolding of denatured proteins and preventing protein aggregation (Mulaudzi et al. 2020).

12.3.3.5 Heavy Metal Stress

Heavy metals are released into the environment because of rising human activity, such as mining activities. Eventually, these contaminants build up in the soil. This heavy metal has a significant negative impact on plants at multiple levels, from germination, seedling growth to photosynthesis, root development, and reproductive processes (Angon B. et al. 2024). While sorghum is relatively more tolerant to certain environmental stresses, it is still vulnerable to the toxicity of heavy metals like many plants, particularly in contaminated soils. Sorghum plants may accumulate heavy metals in various tissues, including roots. Once these metals are absorbed by the roots, they can be translocated through the plant, leading to the accumulation of toxic levels in different tissues, which results in the deterioration of plant health. The high concentrations of heavy metals in the soil of sorghum plants can reduce the ability of seeds to sprout and inhibit the metabolism of enzymes involved in germination, leading to poor seedling emergence and establishment (Soudek et al. 2014).

12.4 The Role of Biofertilizers in Managing Abiotic Stresses in *Sorghum bicolor*

Sorghum bicolor is known to have the ability to tolerant some abiotic stresses like drought, salt, and heavy stresses (Wagaw 2019; Mishra et al. 2021; Kazemi et al. 2023). Although it is able to tolerate some abiotic stresses, it is still imperative to develop new methods to improve its growth, yield, and quality under these abiotic stresses. Abiotic stresses affect all major processes of growth and development of *Sorghum bicolor*, affecting stages such as germination, growth, photosynthesis, water relation, nutrient imbalance, and oxidation (Van Zelm et al. 2020; Abreha et al. 2022). Drought is known as the most predominant abiotic stress affecting all major sorghum-producing areas (Abreha et al. 2022). Sorghum plants utilize various mechanisms to survive, abiotic stress, including ion homeostasis and its compartmentalization, osmotic adjustments, induction of antioxidant enzyme machinery, hormonal regulation, and some morphological and anatomical adaptations (Badigannavar et al. 2018; Behera et al. 2022; Zheng et al. 2024). Nanoparticles and molecules such as carbon monoxide and jasmonic acids have been previously used as biostimulants and phytohormones to alleviate abiotic and biotic stresses (Rakgotho et al. 2022; Ndou et al. 2023; Mulaudzi et al. 2020). Zinc oxide nanoparticles, iron oxide nanoparticles, and cerium oxide nanoparticles were previously used to alleviate drought and salinity stress on sorghum plants (Dimkpa et al. 2019; Djanaguiraman et al. 2018). Molecular engineering and genetic breeding are some of the techniques that were employed to improve *Sorghum bicolor* growth and development under stress (Jin et al. 2021; Zheng et al. 2024). However, some of these techniques and biostimulants are expensive, time-consuming, and not easily accessible. Biofertilizers have gained a lot of interest in the agricultural sector because they are known to be cheaper, environmentally friendly, and can directly and indirectly alleviate abiotic and biotic stresses.

Biofertilizers are derived from microorganisms and can improve plant growth under stressful environment by producing hormones, micronutrients, and other substances that play vital roles, helping plants cope with abiotic stresses (Bharti and Suryavanshi 2021). Raza et al. (2019) have reported that different abiotic stresses interfere with normal plant growth; however, the use of biofertilizers can help alleviate these effects. Few studies have been conducted to show the effectiveness of biofertilizer on different plants under different abiotic stresses (Anli et al. 2020; Al-Amri 2021; Najafi et al. 2021). Nitroxin, mycorrhiza, and *Thiobacillus* sp. were used to enhance drought stress tolerance and alter the antioxidant enzymes in medicinal pumpkin (Najafi et al. 2021). Few studies have been conducted on the effectiveness of biofertilizers on *Sorghum bicolor*, especially under abiotic stresses, and most studies have shown that biofertilizers like plant growth-promoting rhizobacteria can alleviate drought stress on sorghum. The most common and popular used biofertilizers to alleviate abiotic stress in plants are the arbuscular mycorrhizal fungus and plant growth-promoting rhizobacteria.

12.4.1 Plant Growth-Promoting Rhizobacteria (PGPR)

Plant growth-promoting rhizobacteria (PGPR) are free-living bacteria that colonize plant roots and play a crucial role in the maintenance of plant growth, nutrient management, and hormonal balance (Vejan et al. 2016; Oleńska et al. 2020; AbuQamar et al. 2024). These rhizosphere-colonizing bacteria have gained a lot of interest in the agricultural sector because of their ability to improve plant growth, withstand abiotic stress, and manage plant diseases (Benmrid et al. 2024). Species such as *Pseudomonas fluorescens, Pseudomonas putida, Pseudomonas aeruginosa, Bacillus subtilis*, and other *Bacillus* sp. are widely used commercially. PGPR are used to improve plant growth and development (Ahmad et al. 2022). PGPR are known to promote plant growth and protect plants using several direct and indirect mechanisms to overcome abiotic and biotic stress (Bhanse et al. 2022; Vocciante et al. 2022; Kumar et al. 2018). Plant growth-promoting rhizobacteria directly promote plant growth by solubilizing phosphates, producing hormones, and fixing nitrogen. They also influence plant metabolism by increasing water and minerals uptake, enhancing root development, and increasing the enzymatic activity (Shameer and Prasad 2018; Prasad et al. 2019; Kong and Liu 2022). PGPR also protect plants indirectly by competing with pathogens for scarce nutrients, biocontrolling pathogens, and inducing systemic responses in host plants (Hasan et al. 2024). Under abiotic stress, PGPR help plants cope with stress impacts through various mechanisms, including production of phytohormones, deaminase, osmolytes, antioxidants, and exopolysaccharides (EPS), and by inducing systemic tolerance (IST) (Bessai et al. 2021). A study conducted by Umapathi et al. (2022) indicated that the use of *P. intermedius, Bacillus* sp., and *A. pittii* as a biofertilizer mitigated the drought effects in sorghum plants and improved plant growth activities such as germination, root, shoot length, and dry matter production by producing plant growth-promoting substances.

Phytohormone production by PGPR is believed to increase root length and surface area and the number of root tips, leading to higher water and nutrient uptake under salinity stress (Bessai et al. 2021). It has been reported that the most common PGPR produce high amounts of indole-3-acetic acid (IAA), and the IAA was reported to alleviate salt stress in plants. PGPR produce IAA and transfer it into the rhizosphere of salt-stressed plants, where it acts in combination with the endogenous IAA plant pool to stimulate plant growth (Bessai et al. 2021).

12.4.2 Arbuscular Mycorrhizal Fungi

Arbuscular mycorrhizal fungi (AMF) is an important soil microorganism that coexists with the roots of approximately 72% of plants (Kamali and Mehraban 2020). This arbuscular mycorrhizal fungus is known as a type of biofertilizer that receives sugars, amino acids, vitamins, and other organic substances from the host plant

(Begum et al. 2019). AMF is known to improve plant development by improving nutrient acquisition, such as phosphorus, water, and mineral uptake, and it has been reported that AMF improves plant tolerance and resilience to abiotic stressors such as drought, salt, and heavy metal toxicity. These benefits come from the arbuscular mycorrhizal interface, which lets fungal and plant partners exchange nutrients, signaling molecules, and protective chemical compounds. Plants' antioxidant defense systems, osmotic adjustment, and hormone regulation are shown (Kamali and Mehraban 2020).

Mycorrhizal colonization (*Funneliformis mosseae*) is one of the AMF that was reported to enhance the tolerance of sorghum to water-deficit conditions by maintaining proline level, enhancing nutrient uptake, regulating water balance, and detoxifying reactive oxygen species (ROS) by promoting enzymatic and nonenzymatic antioxidant activity (Putri et al. 2023). Chandra et al. (2022) showed that symbiotic interaction between AMF and sorghum plants reduced water loss, enhanced photosynthesis, and improved root development under drought stress. AMF was also reported to protect plants against salt stress, and this was done by improving mineral nutrition, maintaining a high level of potassium/sodium ion (K^+/ $Na^{+)}$ ratio and a low shoot/root Na^+ ratio, increasing antioxidant capacity by enhanced antioxidant enzymes, and modifying osmotic adjustment by the accumulation of soluble sugar (Wang et al. 2019). Kamali and Mehraban 2020, have demonstrated that inoculation of plants with rhizobacteria and AMF is an effective method to alleviate drought stress in *Sorghum bicolor*. Shi et al. (2019) also showed that inoculation of sorghum plants with AMF improved the nutrient uptakes of P, N, and S, and the photosynthesis efficiency of the plant in Mo-contaminated soil.

12.4.3 Limitations to the Use of Biofertilizers in Crop Production

Biofertilizers are known to mitigate abiotic and biotic stress and improve plant growth. However, there are several limitations that hinder crop production and yield quality. These limitations include lower nutrient and distribution of nutrient because biofertilizers first need to decompose organic material and fix atmospheric nitrogen to nutrient before being readily available to the plants. Environmental factors such as temperature, moisture, and pH may reduce the effectiveness of biofertilizer depending on the type of microbes and plants (Viscardi et al. 2016). Other challenge includes short shelf life and storage issue because biofertilizers contain living organisms that have a limited shelf life and require specific storage conditions. Lack of awareness and technical support is another challenge because this can lead to misuse or underuse of biofertilizers, which may lead to the effectiveness of biofertilizers (Rai et al. 2023).

12.4.4 Conclusion

Biofertilizers are an effective and environmentally friendly alternative to chemical fertilizers, which boost crop quality and profitability. In contemporary crop agriculture, biofertilizers are used for two main reasons: the first is to increase the use of biofertilizers, which increases crop yields per unit area, and the second is that prolonged use of synthetic fertilizers damages soil and presents several risks to the environment and human health. Excellent comprehension and a great deal of hands-on expertise with a range of soil types can enhance the efficiency of biofertilizers. Even though biofertilizers are inexpensive and healthy for the environment, our farmers are still not familiar with their application in crop production.

References

Abreha, K.B, Enyew, M., Carlsson, A.S., Vetukuri, R.R., Feyissa, T., Motlhaodi, T., Ng'uni, D., Geleta, M (2022) Sorghum in dryland: morphological, physiological, and molecular responses of sorghum under drought stress. Planta.255, pp.1–23

AbuQamar SF, El-Saadony MT, Saad AM, Desoky ESM, Elrys AS, Abd El-Mageed TA, Semida WM, Abdelkhalik A, Mosa WF, Al Kafaas SS, Naser S (2024) Halotolerant plant growth-promoting rhizobacteria improve soil fertility and plant salinity tolerance for sustainable agriculture—A review. Plant Stress 12:100482

Agarwal P, Gupta R, Gill IK (2018) Importance of biofertilizers in agriculture biotechnology. Ann Biol Res. 9(3), pp.1–3

Agurla S, Gahir S, Munemasa S, Murata Y, Raghavendra AS (2018a) Mechanism of stomatal closure in plants exposed to drought and cold stress. Adv Exp Med Biol 1081:215–232

Agurla S, Gahir S, Munemasa S, Murata Y, Raghavendra AS (2018b) Mechanism of stomatal closure in plants exposed to drought and cold stress. In: Survival strategies in extreme cold and desiccation: adaptation mechanisms and their applications, pp 215–232

Ahmad HM, Fiaz S, Hafeez S, Zahra S, Shah AN, Gul B, Aziz O, Fakhar A, Rafique M, Chen Y, Yang SH (2022) Plant growth-promoting rhizobacteria eliminate the effect of drought stress in plants: a review. Front Plant Sci 13:875774

Al-Amri SM (2021) Application of bio-fertilizers for enhancing growth and yield of common bean plants grown under water stress conditions. Saudi Journal of Biological Sciences 28(7):3901–3908

Aloo BN, Tripathi V, Makumba BA, Mbega ER (2022) Plant growth-promoting rhizobacterial biofertilizers for crop production: the past, present, and future. Front Plant Sci 13:1002448

Angon PB, Islam MS, Das A, Anjum N, Poudel A, Suchi SA (2024) Sources, effects and present perspectives of heavy metals contamination: soil, plants and human food chain. Heliyon 10(7):e28357

Anli M, Baslam M, Tahiri A, Raklami A, Symanczik S, Boutasknit A, Ait-El-Mokhtar M, Ben-Laouane R, Toubali S, Ait Rahou Y, Ait Chitt M (2020) Biofertilizers as strategies to improve photosynthetic apparatus, growth, and drought stress tolerance in the date palm. Front Plant Sci 11:516818

Assefa Y, Staggenborg SA, Prasad VP (2010) Grain sorghum water requirement and responses to drought stress: A review. Crop management 9(1):1–11

Badigannavar A, Teme N, de Oliveira AC, Li G, Vaksmann M, Viana VE, Ganapathi TR, Sarsu F (2018) Physiological, genetic and molecular basis of drought resilience in sorghum [Sorghum bicolor (L.) Moench]. Indian J Plant Physiol 23:670–688

Baxter A, Mittler R, Suzuki N (2014) ROS as key players in plant stress signalling. J Exp Bot 65(5):1229–1240

Begna T, Yali W, Gichile H (2022) Impact of drought stress in sorghum (Sorghum Bicolor (L.) Moench) production and drought genetic resistance mechanisms. Eur J Agric For Res 10(2):28–49

Begum N, Qin C, Ahanger MA, Raza S, Khan MI, Ashraf M, Ahmed N, Zhang (2019) Role of arbuscular mycorrhizal fungi in plant growth regulation: implications in abiotic stress tolerance. Front Plant Sci 10:1068

Behera PP, Saharia N, Borah N, Devi SH, Sarma RN (2022) Sorghum physiology and adaptation to abiotic stresses. Int J Environ Clim Change 12(10):1005–1022

Benmrid B, Ghoulam C, Ammar I, Nkir D, Saidi R, Staropoli A, Iacomino G, ELhajjami E, Cheto S, Geist linger J, Idbella M (2024) Drought-tolerant rhizobacteria with predicted functional traits enhanced wheat growth and P uptake under moderate drought and low P-availability. Microbiol Res 285:127795

Bessai SA, Corrêa A, Cruz C, Yadav AN, Nabti E (2021) Plant growth promoting microbes as biofertilizers: promising solutions for sustainable agriculture under climate change associated abiotic stresses. Plant Sci Today 8:60–76

Bhanse P, Kumar M, Singh L, Awasthi MK, Qureshi A (2022) Role of plant growth-promoting rhizobacteria in boosting the phytoremediation of stressed soils: opportunities, challenges, and prospects. Chemosphere 303:134954

Bharti N, Suryavanshi M (2021) Quality control and regulations of biofertilizers: current scenario and future prospects. In: Biofertilizers, pp 133–141

Calvino M, Messing J (2012) Sweet Sorghum as a model system for bioenergy crop. Cur Opinion Biotechnol 23(3):323–329

Chandra P, Singh A, Prajapat K, Rai AK, Yadav RK (2022) Native arbuscular mycorrhizal fungi improve growth, biomass yield, and phosphorus nutrition of sorghum in saline and sodic soils of the semi–arid region. Environmental and Experimental Botany 201:104982

Chaudhary P, Singh S, Chaudhary A, Sharma A, Kumar G (2022) Overview of biofertilizers in crop production and stress management for sustainable agriculture. Front Plant Sci 13:930340

Chen X, Wu Q, Gao Y, Zhang J, Wang Y, Zhang R, Zhou Y, Xiao M, Xu W, Huang R (2020) The role of deep roots in sorghum yield production under drought conditions. Agronomy 10(4):611

Daniel AI, Fadaka AO, Gokul A, Bakare OO, Aina O, Fisher S, Burt AF, Mavumengwana V, Keyster M, Klein A (2022) Biofertilizer: the future of food security and food safety. Microorganisms. 10(6):1220

Dimkpa CO, Singh U, Bindraban PS, Elmer WH, Gardea-Torresdey JL, White JC (2019) Zinc oxide nanoparticles alleviate drought-induced alterations in sorghum performance, nutrient acquisition, and grain fortification. Sci Total Environ 688:926–934

Djanaguiraman M, Nair R, Giraldo JP, Prasad PVV (2018) Cerium oxide nanoparticles decrease drought-induced oxidative damage in sorghum leading to higher photosynthesis and grain yield. ACS Omega 3(10):14406–14416

Gul Z, Tang ZH, Arif M, Ye Z (2022) An insight into abiotic stress and influx tolerance mechanisms in plants to cope in saline environments. Biology 11(4):597

Franco-Zorrilla JM, López-Vidriero I, Carrasco JL, Godoy M, Vera P, Solano R (2014) DNA-binding specificities of plant transcription factors and their potential to define target genes. Proceedings of the National Academy of Sciences 111(6):2367–2372

Hasan A, Tabassum B, Hashim M, Khan N (2024) Role of plant growth promoting rhizobacteria (PGPR) as a plant growth enhancer for sustainable agriculture: a review. Bacteria 3(2):59–75

Istrati DI, Constantin OE, Vizireanu C, Dinică R, Furdui B (2019) Sorghum as source of functional compounds and their importance in human nutrition. The annals of the university Dunarea de Jos of Galati. Fasc VI-Food Technol 43(2):189–205

Jin X, Long Y, Xiong S, Yang Z, Chen W, Hawar A, Chi X, Chen Y, Luo H, Qi J, Lu G (2021) SbNAC2 enhances abiotic stress tolerance by upregulating ROS scavenging activities and inducing stress-response genes in sorghum. Environ Exp Bot 192:104664

Kadioglu A, Terzi R, Saruhan N, Saglam A (2011) Current advances in the investigation of leaf rolling caused by biotic and abiotic stress factors. Plant Sci 182:42–48
Kamali S, Mehraban A (2020) Nitroxin and arbuscular mycorrhizal fungi alleviate negative effects of drought stress on Sorghum bicolor yield through improving physiological and biochemical characteristics. Plant Signal Behav 15(11):1813998
Kazemi H, Sabouri A, Aalami A, Abedi A (2023) A. Comprehensive meta-analysis to identify the responsive genes in sorghum under salinity and drought stresses (Sorghum bicolor). J Plant Growth Regul 42(11):7096–7115
Khalifa M, Eltahir EA (2023) Assessment of global sorghum production, tolerance, and climate risk. Front Sustain Food Syst 7:1184373
Khalid W, Ali A, Arshad MS, Afzal F, Akram R, Siddeeg A, Kousar S, Rahim MA, Aziz A, Maqbool Z, Saeed A (2022) Nutrients and bioactive compounds of Sorghum bicolor L. used to prepare functional foods: a review on the efficacy against different chronic disorders. International Journal of Food Properties 25(1):1045–1062
Khoddamia A, Messinaa VM, Vadabalija Venkatab K, Farahnakyc A, Blanchardd CL, Robertsa TH (2023) Sorghum in foods: functionality and potential in innovative products. Crit Rev Food Sci Nutr 63(9):1170–1186
Kong Z, Liu H (2022) Modification of rhizosphere microbial communities: a possible mechanism of plant growth promoting rhizobacteria enhancing plant growth and fitness. Front Plant Sci 13:920813
Kumar R, Kumawat N, Sahu YK (2017) Role of biofertilizers in agriculture. Popular:63–66
Kumar A, Singh VK, Tripathi V, Singh PP, Singh AK (2018) Plant growth-promoting rhizobacteria (PGPR): perspective in agriculture under biotic and abiotic stress. In: Crop improvement through microbial biotechnology. Elsevier, pp 333–342
Kumar RP, Gnansounou E, Raman JK, Gurunathan B (2019) Refining biomass residues for sustainable energy and bioproducts: technology, advances, life cycle assessment, and economics. Academic Press.
Lim SF, Matu SU (2015) Utilization of agro-wastes to produce biofertilizer. Int J Energy Environ Eng 6:31–35
Mącik M, Gryta A, Frąc M (2020) Biofertilizers in agriculture: an overview on concepts, strategies and effects on soil microorganisms. Adv Agron 162:31–87
Mahanty T, Bhattacharjee S, Goswami M, Bhattacharyya P, Das B, Ghosh A, Tribedi P (2017) Biofertilizers: a potential approach for sustainable agriculture development. Environ Sci Pollut Res 24:3315–3335
Mahmud AA, Upadhyay SK, Srivastava AK, Bhojiya AA (2021) Biofertilizers: a nexus between soil fertility and crop productivity under abiotic stress. Curr Res Environ Sustain 3:100063
Manghwar H, Hussain A, Alam MA, Ali Q, Liu F (2024) Waterlogging stress in plants: unraveling the mechanisms and impacts on growth, development, and productivity. Environ Exp Bot 224:105824
Mansour MMF, Emam MM, Salama KHA, Morsy AA (2021) Sorghum under saline conditions: responses, tolerance mechanisms, and management strategies. Planta 254:1–38
Mathur S, Agrawal D, Jajoo A (2014) Photosynthesis: response to high temperature stress. J Photochem Photobiol B Biol 137:116–126
Mehta D, Vyas S (2023) Comparative bioaccumulation of osmoprotectants in saline stress tolerating plants. A review. Plant Stress 9:100177
Mishra D, Kumar S, Mishra BN (2021) An overview of morpho-physiological, biochemical, and molecular responses of sorghum towards heavy metal stress. Rev Environ Contam Toxicol 256:155–177
Mittler R (2017) ROS are good. Trends Plant Sci 22(1):11–19
Mofokeng MA, Shimelis H, Laing M, Shargie N (2017) Sorghum ['Sorghum bicolor' (L.) Moench] breeding for resistance to leaf and stalk anthracnose, Colletotrichum sublineolum, and improved yield: Progress and prospects. Aust J Crop Sci 11:1078–1085

Monteiro JST, Havarland M, Ivanova T (2012) Sweet Sorghum (Sorghum bicolor (L) Moench) Bioenergy Value-importance for Portugal. Agri. tropica tropicasubtropica 45(1):12–19
Mulaudzi T, Hendricks K, Mabiya T, Muthevhuli M, Ajayi RF, Mayedwa N, Gehring C, Iwuoha E (2020) Calcium improves germination and growth of Sorghum bicolor seedlings under salt stress. Plants 9(6):730
Najafi S, Nazari Nasi H, Tuncturk R, Tuncturk M, Sayyed RZ, Amirnia R (2021) Biofertilizer application enhances drought stress tolerance and alters the antioxidant enzymes in medicinal pumpkin (Cucurbita pepo convar. pepo var. Styriaca). Horticulturae 7(12):588
Ndou N, Rakgotho T, Nkuna M, Doumbia IZ, Mulaudzi T, Ajayi RF (2023) Green synthesis of iron oxide (hematite) nanoparticles and their influence on Sorghum bicolor growth under drought stress. Plants. 12(7):1425
Neira J, Ortiz M, Morales L, Acevedo E (2015) Oxygen diffusion in soils: understanding the factors and processes needed for modelling. Chil J Agric Res 75:34–44
Noreen S, Ashraf M, Hussain M, Jamil A (2009) Exogenous application of salicylic acid enhances antioxidative capacity in salt stressed sunflower (Helianthus annuus L.) plants. Pak J Bot 41(1):473–479
Nosheen S, Ajmal I, Song Y (2021) Microbes as biofertilizers, a potential approach for sustainable crop production. Sustainability 13(4):1868
Oleńska E, Małek W, Wójcik M, Swiecicka I, Thijs S, Vangronsveld J (2020) Beneficial features of plant growth-promoting rhizobacteria for improving plant growth and health in challenging conditions: a methodical review. Sci Total Environ 743:140682
Pereira LM, Hawkes C (2022) Leveraging the potential of sorghum as a healthy food and resilient crop in theSouth African food system. Frontiers in sustainable food systems 6:786151
Pedersen O, Sauter M, Colmer TD, Nakazono M (2020) Regulation of root adaptive anatomical and morphological traits during low soil oxygen. New Phytologist 229(1):42–49
Prasad PV, Boote KJJ, Allen LH Jr (2006) Adverse high temperature effects on pollen viability, seed-set, seed yield and harvest index of grain-sorghum [Sorghum bicolor (L.) Moench] are more severe at elevated carbon dioxide due to higher tissue temperatures. Agric For Meteorol 139(3–4):237–251
Prasad PVV, Djanaguiraman M, Perumal R, Amity IA (2015) Impact of high temperature stress on floret fertility and individual grain weight of grain sorghum: sensitive stages and thresholds for temperature and duration. Front Plant Sci 6:820
Prasad M, Srinivasan R, Chaudhary M, Choudhary M, Jat LK (2019) Plant growth promoting rhizobacteria (PGPR) for sustainable agriculture: perspectives and challenges. In: PGPR amelioration in sustainable agriculture, pp 129–157
Putri DALP, Widyastuti R, Idris I, Ikhwani AZN, Nugroho S, Sudiana IM, Kanti A, Purnaningsih I, Ochiai K, Kobayashi M, Wijaya L (2023) Unraveling the mechanisms of drought tolerance enhancement in Sorghum bicolor through glomus mosseae inoculation: insights from comparative analysis of super 2 and Konawe Selatan accessions. S Afr J Bot 161:293–304
Rai PK, Rai A, Sharma NK, Singh T, Kumar Y (2023) Limitations of biofertilizers and their revitalization through nanotechnology. J Clean Prod 418:138194
Rakgotho T, Ndou N, Mulaudzi T, Iwuoha E, Mayedwa N, Ajayi RF (2022) Green-synthesized zinc oxide nanoparticles mitigate salt stress in Sorghum bicolor. Agriculture-Basel 12(5):597
Rao BD, Devi S, Kiranmai E, Tonapi VA (2019) Sorghum. In: Whole grains. CRC Press, pp 197–234
Raja V, Majeed U, Kang H, Andrabi KI, John R (2017) Abiotic stress: Interplay between ROS, hormones and MAPKs. Environmental and Experimental Botany 137:142–157
Raza A, Razzaq A, Mehmood SS, Zou X, Zhang X, Lv Y, Xu, J (2019) Impact of climate change on crops adaptation and strategies to tackle its outcome: A review. Plants 8(2):34.
Rezaee N, Fernando WB, Hone E, Sohrabi HR, Johnson SK, Gunzburg S, Martins RN (2021) Potential of Sorghum polyphenols to prevent and treat Alzheimer's disease: a review article. Frontiers in Aging Neuroscience 13:729949

Ronda V, Aruna C, Visarada KBRS, Bhat BV (2019) Sorghum for animal feed. In Breeding Sorghum for diverse end uses. pp. 229–238. Woodhead Publishing.
Rooney W, Blumnthal J, Bean B, Mullet, JE (2007) Designing Sorghum as a dedicated bioenergy feedstock. Biofuels, Bioproducts and Biorefining 1(2):147–157
Santamaria P (2006) Nitrate in vegetables: toxicity, content, intake and EC regulation. Journal of the Science of Food and Agriculture 86(1):10–17
Shameer S, Prasad TNVKV (2018) Plant growth promoting rhizobacteria for sustainable agricultural practices with special reference to biotic and abiotic stresses. Plant Growth Regul 84:603–615
Sharma P, K., Kumar S (2023) Soil air and plant growth. In: Soil physical environment and plant growth. Springer, Cham, p 6
Sharma P, Jha AB, Dubey RS, Pessarakli M (2012) Reactive oxygen species, oxidative damage, and antioxidative defense mechanism in plants under stressful conditions. J Bot 1:217037
Shi Z, Zhang J, Lu S, Li Y, Wang F (2019) Arbuscular mycorrhizal fungi improve the performance of sweet sorghum grown in a Mo-contaminated soil. Journal of Fungi 6(2):44
Soudek P, Petrová Š, Vaňková R, Song J, Vaněk T (2014) Accumulation of heavy metals using Sorghum sp. Chemosphere. 104:15–24
Stamenković OS, Siliveru K, Veljković VB, Banković-Ilić IB, Tasić MB, Ciampitti IA, Đalović IG, Mitrović PM, Sikora VŠ, Prasad PV (2020) Production of biofuels from sorghum. Renewable and Sustainable Energy Reviews 124:109769
Swami AK, Alam SI, Sengupta N, Sarin R (2011) Differential proteomic analysis of salt stress response in Sorghum bicolor leaves. Environ. Experim Bot 71(2):321–328
Tari I, Laskay G, Takács Z, Poór P (2013) Response of sorghum to abiotic stresses: A review. Journal of agronomy and crop science 199(4):264–274
Tanwar R, Panghal A, Chaudhary G, Kumari A, Chhikara N (2023) Nutritional, phytochemical and functional potential of sorghum: a review. Food Chem Adv 3:100501
Umapathi M, Chandrasekhar CN, Senthil A, Kalaiselvi T, Santhi R, Ravikesavan R (2022) Isolation, characterization and plant growth-promoting effects of sorghum [*Sorghum bicolor* (L.) moench] root-associated rhizobacteria and their potential role in drought mitigation. Arch Microbiol 204(6):354
Van Zelm E, Zhang Y, Testerink C (2020) Salt tolerance mechanisms of plants. Annu Rev Plant Biol 71(1):403–433
Vejan P, Abdullah R, Khadiran T, Ismail S, Nasrulhaq Boy. A (2016) Role of plant growth promoting rhizobacteria in agricultural sustainability—a review. Molecules 21(5):573
Velmurugan B, Narra M, Rudakiya DM, Madamwar D (2020) Sweet sorghum: a potential resource for bioenergy production. In: Refining biomass residues for sustainable energy and bioproducts. Academic, pp 215–242
Vera Hernández PF, Mendoza Onofre LE, Rosas Cárdenas FDF (2023) Responses of sorghum to cold stress: a review focused on molecular breeding. Front Plant Sci 2023(14):1124335
Viscardi S, Ventorino V, Duran P, Maggio A, De Pascale S, Mora ML, Pepe O (2016) Assessment of plant growth promoting activities and abiotic stress tolerance of Azotobacter chroococcum strains for a potential use in sustainable agriculture. J Soil Sci Plant Nutr 16(3):848–863
Vocciante M, Grifoni M, Fusini D, Petruzzelli G, Franchi E (2022) The role of plant growth-promoting rhizobacteria (PGPR) in mitigating plant's environmental stresses. Appl Sci 12(3):1231
Wagaw K (2019) Review on mechanisms of drought tolerance in sorghum (Sorghum bicolor (L.) Moench) basis and breeding methods. Academic Research Journal of Agricultural Science and Research 7(2):87–99
Wang F, Sun Y, Shi Z (2019) Arbuscular mycorrhiza enhances biomass production and salt tolerance of sweet sorghum. Microorganisms 7(9):289
Yan K, Zhao S, Cui M, Han G, Wen P (2018) Vulnerability of photosynthesis and photosystem I in Jerusalem artichoke (Helianthus tuberosus L.) exposed to waterlogging. Plant Physiology and Biochemistry. 125:239–246

Zheng H, Dang Y, Diao X, Sui N (2024) Molecular mechanisms of stress resistance in sorghum: implications for crop improvement strategies. J Integr Agric 23(3):741–768
Zhou H, Shi H, Yang Y, Feng X, Chen X, Xiao F, Lin H, Guo Y (2024) Insights into plant salt stress signaling and tolerance. Journal of Genetics and Genomics 51(1):16–34
Zidenga (2004) Towards switchable crops: beyond the green revolution. ISB News Report. pp.5

Chapter 13
The Potential of Arbuscular Mycorrhizae Fungi and Biochar Biofertiliser in Managing the Parasitic Weed *Striga* spp.

Ronald Mandumbu, Cosmas Parwada, Edith Matema, C. T. Gadzirayi, and Never Mafuse

Abstract Striga is an obligate parasite that has partially or wholly lost its ability to photosynthesise. Successful parasitism of cereals by Striga is modulated by the release of strigolactones into the rhizosphere by the host. The major driver of strigolactones production is low soil fertility due to nutrient mining of nitrogen and phosphorus in smallholder farmer soils. Management of the parasite has been difficult, if not impossible, under smallholder input-constrained farming systems. Here, we propose and explore the potential use of a biofertiliser made from mixing biochar and arbuscular mycorrhiza fungi (AMF) to effect changes in the rhizosphere. AMF promotes the growth of plants by bringing morpho-physiological and biochemical changes in host plants by serving as biofertilisers and bioprotectors, and providing water and nutrients to the plants. AMF boosts the health of subsequent crops through soil aggregation improvement, enhanced abiotic stress tolerance, and protection against pathogens. The extracellular hyphal network gives a high absorbing volume for increased uptake and translocation of water and essential nutrients into the associated plants. Biochar is characterised by high specific surface area, high content of functional groups, pH regulation, and improved porosity. Biochar addition to the soil improves the soil water holding capacity, nutrient supply, cation exchange

R. Mandumbu (✉)
Department of Crop Production, Faculty of Agriculture, University of Eswatini, Luyengo, Eswatini

C. Parwada
Department of Agronomy, Midlands State University, Gweru, Zimbabwe

E. Matema
Department of Agribusiness, Women's University in Africa, Manresa, Harare, Zimbabwe

C. T. Gadzirayi
CIRAD Zimbabwe, Belgravia, Harare, Zimbabwe

N. Mafuse
Department of Agricultural Economics, Education and Extension, Bindura University of Science Education, Bindura, Zimbabwe

A. D. Nciizah et al. (eds.), *Biofertilizers for Sustainable Agriculture in Africa*, Sustainability Sciences in Asia and Africa, https://doi.org/10.1007/978-981-95-6009-7_13

capacity, and soil porosity. Higher pH destabilises the structural integrity of strigolactones, which are more stable at acidic pH. It is anticipated that the biochar/AMF biofertiliser brings a modification of the rhizosphere, which leads to the stoppage or limited release of strigolactones. If released, the strigolactones encounter an alkaline and watery environment and hence may fail to elicit germination of the *Striga* parasite. Biochar may help to provide 'safe sites' for AMF hyphae or spores during dry periods. This, therefore, gives a high possibility for sustainable management of *Striga* through affordable biofertilisers. The biofertiliser holds a huge potential in providing one of the most affordable and sustainable options for parasitic weeds management in both cereals and legumes.

13.1 Introduction

Parasitic plants such as *Striga* spp. have lost their autotrophic nature and depend on other plants to provide them with water and nutrients to reach maturity (Rich 2024). Over 50% of arable land under cereal and legume crops in most countries of sub-Saharan Africa is infested by the parasites (Rodenburg et al. 2016). Whilst other weeds compete with plants for water, nutrients, and sunlight, parasitic weeds have evolved the capacity to form a haustorium that extracts already synthesised photosynthates and diverts them into the parasite (Riopel and Timko 1995; Joel 2013; Fernández-Aparicio et al. 2020). According to Yacoubou et al. (2021), the *Striga* parasite is a major driver of food insecurity for millions of subsistence farmers in sub-Saharan Africa.

According to Rich (2024), losses to *Striga* in sub-Saharan Africa have been estimated at US$7 billion annually. In fact, yield losses due to *Striga* spp. are seriously compounded by the limited use of agricultural inputs such as fertilisers and the absence of a suppressive cropping system that prevents the proliferation of the weed (Mandumbu et al. 2019). Whilst many *Striga* control measures have been developed and they include biological, chemical, ecological, and cultural methods, the seed traits of the parasite, wide host range, genetic variability, and the subterranean development of the parasite make it difficult, if not impossible, to control.

The parasite germinates in response to the release of strigolactones following a chemotropic gradient until it reaches the host root. The radicle of the parasite secretes cell wall-digesting enzymes such as pectinolytic enzymes, polygalacturonase, and peroxidases (Kim et al. 1998; Losner-Goshen et al. 1998). The combined effects of mechanical forces and enzymatic degradation pave the way for the radicle to grow towards the host's xylem vessels and form a bridge. Eventually xylem-to-xylem connections occur, and resources are diverted to the parasite.

It is now known that strigolactones are responsible for AMF recruitment. Previously, the study of *Striga* was shrouded in 'mysteries' of the nature of the molecules that stimulate the parasite and their use. The knowledge of how *Striga* interacts with its hosts at the molecular level and the nature of the strigolactones was unveiled by several researchers in the last decade (Umehara et al. 2008; Yoneyama et al. 2007a, b; Yoneyama et al. 2010; Matusova et al. 2004; Akiyama and Hayashi 2006; Akiyama 2007; Xie et al. 2010; Lopez-Raez et al. 2008a, b). It stimulates fungal spores within the soil to germinate, develop, and grow their mycelia in the direction of the host plant. Upon contact, the mycelia form an appressorium. Intracellular arbuscules produce a large surface area for nutrient exchange. According to Kuila and Ghosh (2022), AMF hyphal diameter is 10–20 times less than root hairs. These enable the absorption and transportation of water and nutrients even under drought conditions. After the establishment of the host AMF, soil nutrients can be provided to the host plant in exchange for photosynthates. Thiombiano et al. (2024) reported that the strigolactones play a crucial role in establishing AM fungi symbiosis as they stimulate hyphal branching in AM fungi, enabling them to extend towards and colonise host roots. Gutjahr (2014) also cited that the benefits of AM fungi symbiosis explain why plants, particularly under phosphorus deficiency, release strigolactones into the soil despite the danger of parasites.

According to, the understanding of biological processes underlying symbiosis, where fungi act as biofertilisers and bioprotectors, provides tools to design biotechnological applications addressing agricultural challenges such as the *Striga* challenge. Jamil et al. (2021) reported that due to the persistence and severity of the *Striga* challenge, extensive efforts have been invested to try and develop easy and effective control methods usable by smallholder farmers. Several authors have reported decreased incidence of various *Striga* species, such as *Striga hermonthica* and *Striga asiatica,* after inoculation of AMF (Samejina and Sujimoto 2018; Oluchi et al. 2022; Odelade and Babadola 2019).

Biochar is an insoluble, stable, highly aromatised, carbon-rich solid produced by high temperature pyrolysis of biomass in the absence of oxygen. According to Kolton et al. (2017), it has a long-term sequestration of carbon, reduces nitrous oxide emissions, increases soil water and nutrient retention, raises soil pH, and increases cation exchange capacity of soils (Yuan et al. 2024). There is a growing consensus from research that biochar additions augment rhizosphere microbial diversity and activity, which contributes significantly to the positive impacts of biochar on plant health and productivity (Kolton et al. 2017; Jaiswal et al. 2018). One of the biochar effects that has been elaborated by several researchers is the alteration of the microbial community in the rhizosphere and the facilitation of beneficial microorganisms that stimulate growth (Elad et al. 2011; Kolton et al. 2011). Whilst plants take up the role of recruiting organisms in the rhizosphere through strigolactones secretions, biochar may facilitate the recruitment process through the provision of 'safe sites' for the rhizospheric microorganisms, such as AMF (Corral-Lugo et al. 2016; Shi et al. 2020; Liu et al. 2021).

Therefore, the objectives of this chapter are to explore the possibility of using a biochar/AM fungi biofertiliser for the management of the *Striga* parasite through altering the rhizosphere, making it unsuitable for strigolactones exudation and molecular stability.

13.2 Yield Losses by *Striga* spp.

The parasitic *Striga* spp. does not have the capacity to meet its carbon demands due to a deformed photosynthetic apparatus. It is therefore mandatory that the parasite locates and establishes itself on a suitable host in order to survive. According to Gutjahr (2014), after germination, the *Striga* seedlings develop a haustorium to connect to the host root and siphon off water, minerals, and assimilates, resulting in enormous yield losses.

If the parasite is not timely managed, it can induce serious yield losses of up to 100% (Dechassa and Abate 2021; Kamara et al. 2020; Oluchi et al. 2022; Mandumbu et al. 2019). Leandre et al. (2018) also reported yield losses of up to 70% by *Striga gesneroides* in cowpeas. Therefore, literature is awash with huge yield losses due to the *Striga* parasite, raising the need for affordable management of the parasite, especially under small-scale agriculture.

13.3 Strigolactones in Plants

Strigolactones are the molecules that are central to the cereal/*Striga* parasitism fiasco and the cereal/AMF symbiotic association. It is the common elicitor in the establishment of both relationships. Gutjahr (2014) reported that the *Striga* species evolved much later after the AMF symbiosis. The hormone strigolactones is a branching hormone that assists in the branching of fungal mycelium to better water and nutrients (Akiyama et al. 2005; Akiyama and Hayashi 2006; Akiyama 2007). The symbiotic relationship between plants and fungi is mutually beneficial.

Kuila and Ghosh (2022) reported that the hyphal diameter is 100 times less finer than the finest root hairs. They assist the plants in water absorption under drought conditions when the soil hydraulic potential is significantly low. The efficiency of AMF fungi in water absorption increases when they are branched. The successful establishment of the symbiotic relationship between the fungus and plants is mutually beneficial to both organisms. Lopez-Raez et al. (2008b) reported that fungal mycelium assists in amassing edaphic resources such as water, nitrogen, and phosphorus. The plant supplies the fungus with carbon mainly in the form of carbohydrates. Therefore, the more infertile the soil is, the more strigolactones are formed by the plant.

13.4 Establishment of Symbiosis

The AMF symbiosis starts with the germination of fungal spores which have been stimulated by the presence of the hormone strigolactones released by the host plant. The establishment of symbiosis includes hyphal branching, appressorium development, colonisation, formation of intracellular arbuscules and concomitant production of extraradical mycelium (Akiyama et al. 2005).

Wilkes (2021) found that AMF contribute to soil structural stability and fungi through increased growth and mass from a host's root system. AMF forms a glycoprotein and glomalin with soil adhesive properties between microaggregates (Dastogeer et al. 2020; Wright and Upadhyaya 1996).

Smith and Smith (2011) and Khaliq et al. (2022) summarised the possible mechanisms for increased uptake of phosphorus by arbuscular mycorrhizal plants as:

1. The extensive physical exploration of the soil by fungal hyphae thus decreases the distance that P ions must diffuse before uptake, taking into cognisance that the ion is highly immobile.
2. Faster P uptake from the soil solution.
3. Production of exudates by host plants and change of the rhizosphere pH.

13.5 AMF Plant Symbiosis

According to Matinez et al. (2024), agricultural expansion and intensification threaten soil resources, and plant associations with AMF have a significant impact on productivity through increased tolerance to biotic and abiotic stresses. Arbuscular mycorrhizal fungi are beneficial symbionts of most terrestrial plants that facilitate plant nutrient uptake and increase host resistance to diverse environmental stresses (Xie et al. 2022). The AMF provides novel contact surfaces through AMF hyphae into cortical cells. Nitrogen losses through leaching, runoff, as well as gaseous nitrogen use efficiency in plants are very low and rarely exceed 50% in most agricultural systems.

Chen et al. (2018) reported that the dense hyphal network of the highly ramified AHF mycelium creates a three-dimensional matrix that enmeshes and cross-links soil particles without compacting the soil. Many authors cite improved stress tolerance to a variety of stresses, including drought, salinity, herbivory, high temperatures, metals and diseases due to fungal symbiosis (Begum et al. 2019). The hyphal network by AMF with plant roots enhances increases in root surface area with subsequent improvements in plant growth.

The AMF produces glomalin, a glycoprotein that is adhesive to soil particles and helps in micro-aggregate formation (Wilkes 2021). Therefore, AMF contribute to soil structure, stability, and function through increased growth and mass from a host's root system. According to Li et al. (2012) and Blanke et al. (2011), AMF led to better aggregate stability of the soils and maintenance of a neutral pH. Meena and Dawar (2023) noted that AMF has demonstrated their ability to induce morphological, biochemical and physiological alterations in plants, thereby enhancing overall productivity. In the symbiosis relationship, energy primarily flows from the plant into the fungus, whilst inorganic resources move from the fungus to the plant.

13.6 Biochar and *Striga* Management

Biochar is a rich thermal degradation by-product from the treatment of organic materials under oxygen-limited conditions, and has been recognised as a valuable means for increasing soil carbon. According to Shi et al. (2020), biochar can improve soil nitrogen retention and thus increase nitrogen use efficiency. Zhang et al. (2022) showed that wheat straw biochar-derived biochar application increased nitrogen use efficiency by 20–55% compared to nitrogen fertilisation alone. Under N-fertiliser biochar amendment could increase nutrient use efficiency of plants by extending the retention of soil nutrients and slowly releasing them as plants require them. He et al. (2022) also reported the same and confirmed that biochar amendment could increase the plant's photosynthetic rate by increasing leaf stomatal conductance, transpiration, and chlorophyll content. These parameters, when they are enhanced in the plant, will increase the tolerance of the plants to parasitic weeds by maintaining a better carbon balance from a more efficient photosynthetic process.

According to Kolton et al. (2017), microbial diversity was higher in the rhizosphere of plants grown in biochar-amended soils, suggesting that the increased bacterial diversity is facilitated by the carbon backbone of the biochar. The improved biodiversity of microorganisms may be stemming from niche partitioning stimulated by biochar-associated modification of the soil physical properties, such as gaseous exchange and nutrient and water holding capacity. This has been referred by Zhang et al. (2020) as the biochar effect, and it helps facilitate the propagation of beneficial microorganisms.

Under this environment, strigolactones production will be limited as its production and exudation are facilitated by the existence of a poor soil environment where the plants seek to make associations with microorganisms in the rhizosphere for better nutrient and water acquisition. Strigolactones are hijacked by the parasitic weed and used as a cue for parasitism despite the noble reason for their production. Therefore, altering the rhizosphere environment is an effective way of reducing the parasitic weeds under smallholder agriculture.

13.7 AMF–Biochar Biofertiliser

Biofertilisers are a mixture of naturally occurring substances that are used to improve soil fertility (Begum et al. 2019). The mixture of biochar and AMF has the potential to be an environmentally friendly fertiliser, whose combined characteristics may help to manage parasitic weeds such as *Striga*. Whilst most studies have focused on the use of AMF to manage salinity, metal stress, high temperatures, and drought (Gebisa 2024; Kuila and Ghosh 2022), there is a great potential for the biofertiliser to lessen the parasitic weed scourge. Both AMF and *Striga* respond to the same stimuli: strigolactones. The two, however, have contrasting effects on plant growth. AMF promote plant growth and development, leading to better utilisation

Table 13.1 Traits of biochar and AMF

Trait	Biochar	AMF	References
pH	Increases soil pH	–	Sahoo et al. (2021)
Soil bulk density	Reduces	Reduces	Schimdt et al. (2021)
Cation exchange capacity	High		Yadav et al. (2023)
Water holding capacity	Increases	Increases	
Plant nutrient availability	Increases	Increases	Smith and Smith (2011) and Tomczyk et al. (2020)
Content of surface functional groups	High		Tomczyk et al. (2020)
Soil porosity	High	High	Singh et al. (2022)
Drought tolerance	Increases	Increases	Singh et al. (2022)
Nutrient use efficiency	Increases	Increases	Gebisa (2024) and Singh et al. (2022)
Soil structure	Improves	Improves	

of resources, better nutrient use efficiency and resilience. The latter uses strigolactones cues for their own advantage and to the detriment of the host plant through parasitic siphoning of nutrients, water, and photosynthates.

Both biochar and AMF have properties that if combined in an organic fertiliser have the capacity to reduce the *Striga* pandemic in small-scale agriculture (Table 13.1).

The combined effects of biochar and AMF made as an organic fertiliser can result in a sustainable cropping system and can alter the rhizosphere so that fewer *Striga* seeds are stimulated to germinate (Fig. 13.1). The acidic pH of most soils in small-scale agriculture, the leaching rates, drought stress, and poor nutrient supply are major drivers of the *Striga* parasite and pandemic. The inoculation of AMF on maize cultivars decreased the incidence of *S. hermonthica* and increased the availability of nitrogen and phosphorus (Samejina and Sugimoto 2018; Oluchi et al. 2022; Odelade and Babadola 2019; Kamara et al. 2020). There are no studies that have documented the effects of biochar on the *Striga* pandemic.

It becomes pertinent to alter the rhizospheric environment of the host plant to prevent excessive production of strigolactones. The high CEC, higher pH, nutrient availability, and moisture conservation may all work corporately to alter the rhizosphere (Fig. 13.1) to the demise of the parasite. Assuming that the elements brought by biochar/AMF fertiliser become an additive, then *Striga* management may be achieved. The mechanism will be through reduced strigolactone production as a result of alteration of the rhizosphere.

Better water economy of the plant and the regulated pH may result in better parasite management. The soil pH determines the structural stability of the strigolactones. According to Zwaneburg and Pospiosil (2013), the half-life of strigolactones is higher at acidic pH, and the stability rapidly decreases at alkaline pH. In acidic soils, an adequate persistence of 6–8 days was observed, whereas in alkaline soils, a persistence of 1–3 days was observed (Zwaneburg and Pospiosil 2013). The structural alterations of strigolactones under high pH may be inadequate to elicit

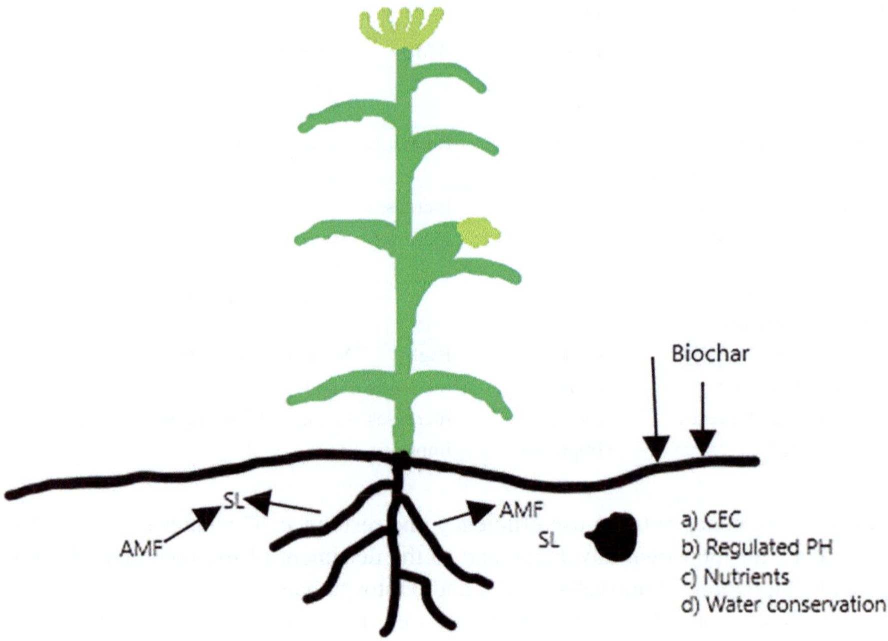

Fig. 13.1 Underground interactions between the host plant, AMF, *Striga* seed, and biochar. SL strigolactones, AMF arbuscular mycorrhiza fungi, CEC cation exchange capacity

germination of the *Striga* seed. Increasing soil pH is one of the tenets of biochar organic fertilisers hence may result in less *Striga* seed germination.

Research still needs to be done to determine the compatibility of the two components of the biofertiliser. Biochar may also help the life of AMF outside the normal growing seasons or specific sites for spores or hyphal overwintering.

Biochar-related properties of the fertilisers may help the expression of tolerance traits, such as tiller development, as they also depend on reduced stress of the plant. A better-fed, a better watered crop has the capacity to withstand the negative effects of the parasite. Better nutrient retention and the ability to resist the leaching forces also assist the crop to achieve better crop profitability even under the parasite pressure. So, the biochar/AMF fertiliser may be combining all the tenets of the constituent components (Fig. 13.2).

The AMF biofertiliser has the potential to enhance survival of both the fungal spores and mycelium under dry conditions. Several studies have shown that the addition of biochar leads to better moisture conservation. Under AMF biochar biofertiliser, the soils have been reported to conserve moisture, and they can pass through the mid-season droughts. This ensures that strigolactones production is greatly reduced. Higher cation exchange (CEC) and microbial population that comes from biochar addition may also help interfere with strigolactones molecules. Both biochar and AMF have neutralising effects on the soil pH. Strigolactones stability is enhanced under acidic conditions.

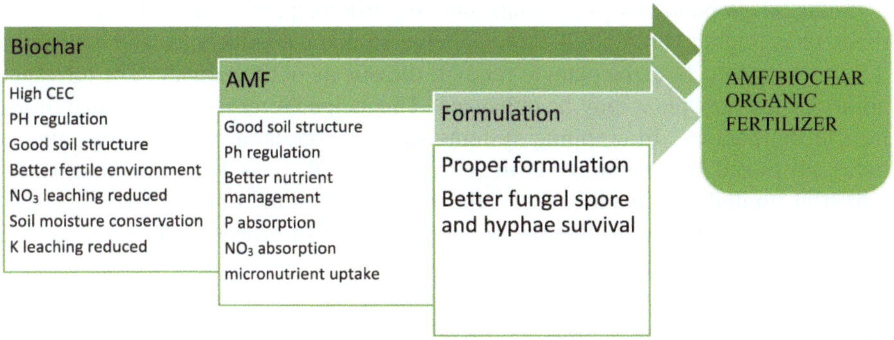

Fig. 13.2 Properties of the AMF biochar biofertiliser

The rhizospheric environment is modified by the biofertiliser such that strigolactones are either not stable enough to elicit their physiological activities or they are not produced at all. The higher microbial populations that are facilitated by the addition of biochar may also help to break down strigolactones molecules. Therefore, the biofertiliser addition makes the rhizospheric environment antagonistic to the physiological effects of the strigolactones. According to Sun et al. (2008) and Jamil et al. (2011a, b), germination of *Striga* seed was dependent upon the quantity and quality of the strigolactones that reach the parasite seed. Major research efforts were directed towards finding cereal genotypes that produced reduced strigolactone quantities (Jamil et al. 2011a, b) as a means of managing the parasite. Limited research effort was directed towards altering the rhizospheric environment, making the produced strigolactones ineffective.

Whilst increased soil fertility has long been recognised as a method of managing the parasite (Raju et al. 1990; Jamil et al. 2011b; Lopez-Raez et al. 2008a, b; Yoneyama et al. 2007a), the main strategy was the addition of inorganic fertilisers, which smallholder farmers could not afford. Few studies (Chikwari et al. 2019) focused on the use of organic fertilisers. The main challenge with organic fertilisers was probably reduced mineralization, which was not adequate to reduce strigolactones production.

AMF use strigolactones to establish a symbiotic relationship with the hosts. Once the association is established, the plant may not continue to produce large quantities of strigolactones as its mineral needs are met with AMF. One distinct advantage of the biofertiliser will be its affordability to small-scale farmers. It then becomes a sustainable option for resolving the impasse that has always existed between smallholder farmers and the *Striga* parasite. Biofertilzers may also help resolve the existence of physiological speciation and differential virulence (Bozkurt et al. 2015) that has challenged most control strategies proposed by researchers. *Striga* speciation is quite rampant in the *Striga* genus, as reported by Botanga et al. (2002), who while studying in Benin, discovered a high degree of speciation within the 14 analysed *Striga asiatica* populations. The use of the biofertiliser provides a multipronged approach to deal with strigolactones before and after production.

Some *Striga* plants may germinate and establish the parasitism. Better photosynthetic rates, higher chlorophyll content, better stomatal conductance and better root development may help the plant to fend off the effects of the parasite. The tolerance ability of the host plant may be enhanced. This helps increase the tolerance of the host to the parasite. In cropping systems such as conservation agriculture that is being promoted by governments in Africa, where there is less soil disturbance, which is limited to a planting hole, the rhizospheric environment is modified to reduce its susceptibility to the *Striga* colonisation. The crop is more fit to survive when grown in a better environment.

13.8 Conclusion

The AMF Biochar biofertiliser provides a potential option that alters the rhizospheric environment to reduce strigolactones production, provides an environment with diverse microorganisms population that may assist in interfering with strigolactones before the *Striga* seed is elicited to germinate. The PH regulation function of the AMF–biochar biofertiliser alters pH from acidic where strigolactones have better stability to neutrality where they are not stable. Better moisture conservation due to the two components of the biofertiliser provides a watery environment that decreases the stability of the strigolactones, at the same time increasing the fitness of the host plant. More moisture also helps to reduce drought effects on the host plant, which exacerbates the effects of the parasite on the host. Therefore, the fertiliser provides a multifaceted approach to negating the effects of strigolactones on *Striga* seed germination. The biofertiliser holds a vast potential to help smallholder farmers end one of the longest battles in smallholder crop production.

References

Akiyama K (2007) Chemical identification and functional analysis of apocarotenoids involved in the development of arbscular mycorhizal symbiosis. Biosci Biotechnol Biochem 71:1405–1414

Akiyama K, Hayashi H (2006) Strigolactones: chemical signals for fungal antisymbionts and parasitic weeds in plant roots. Ann Bot 97:925–931

Akiyama K, Matsuzaki K, Hayashi H (2005) Plant sesquiterpenes induce hyphal branching in arbsucular mycorhizal fungi. Nature 435:824–827

Begum N, Qin C, Ahanger MA, Raza S, Khan MI, Ashraf M, Ahmed N, Zhang L (2019) Role of arbuscular mycorrhizal fungi in plant growth regulation: implications in abiotic stress tolerance. Front Plant Sci 10:1068. https://doi.org/10.3389/fpls.2019.01068

Botanga C, Kling J, Berner D, Timko M (2002) Genetic variability of Striga asiatica based on AFLD analysis and host parasite interactions. Euphytica 128:375–388

Bozkurt ML, Muth P, Parzies HK, Hausmann BIG (2015) Genetic diversity of east and West African Striga hermonthica populations and virulence effects on a contrasting set of sorghum cultivars. Weed Research 55:71–81

Blanke V, Wagner M, Renker C, Lippert H, Michulitz M, Kuhn AJ, Buscot F (2011) Arbuscular mycorrhizas in phosphate-polluted soil: interrelations between root colonization and nitrogen. Plant Soil 343:379–392

Chen M, Arato M, Borghi L, Nouri E, Reinhardt D (2018) Beneficial services of arbuscular mycorrhizal fungi – from ecology to application. Front Plant Sci 9:1270. https://doi.org/10.3389/fpls.2018.01270

Chikwari A, Muzemu S, Mandumbu R, Mabasa S, Rugare JT (2019) Striga asiatica management in Sorghum bicolor using organic and inorganic nutrient sources in Zimbabwe. Asian Journal of Biological sciences. https://doi.org/10.3923/ajbs.2019

Corral-Lugo A, Daddaoua A, Ortega A, Espinosa-Urgel M, Krell T (2016) Rosmarinic acid is a homoserine lactone mimic produced by plants that activates a bacterial quorum-sensing regulator. Sci Signal 9:ra1

Dastogeer KMG, Zahan MI, Tahjib-Ul-Arif M, Akter MA, Okazaki S (2020) Plant salinity tolerance conferred by arbuscular mycorrhizal fungi and associated mechanisms: a meta-analysis. Front Plant Sci 11:1927

Dechassa N, Abate B (2021) Striga threat to cereal crops production and its management: a review. Adv Life Sci Technol 88:11–18

Elad Y, Cytryn E, Harel YM, Lew B, Graber ER (2011) The biochar effect: plant resistance to biotic stresses. Phytopathol Mediterr 50:335–349

Ejeta G (2007) The Striga scourge in Africa: a growing pandemic. Integrating new technologies for Striga control: towards ending the witch hunt. (Eds Ejeta G and Gressel J), World Scientific publishers.

Fernández-Aparicio M, Delavault P, Timko MP (2020) Management of infection by parasitic weeds: a review. Plants 9:1184. https://doi.org/10.3390/plants9091184

Gebisa LA (2024) Associations of arbuscular mycorrhizal fungi (AMF) for enhancement in soil fertility and promotion of plant growth: a review. Adv Biosci Eng 12:72–80

Gutjahr C (2014) Phytohormone signaling in arbuscular mycorrhizae development. Curr Opin Plant Biol 20:26–34

Gurney AL, Grimanelli D, Kanampiu F, Hoisington D, Scholes DJ, Press MC (2003) Novel sources resistance to Striga hermonthica in Tripsacum dactyloides, a wild relative of maize. New Phytologist 160:557–568

He J, Xiao Y, Yimingniyazi A (2022) Effect of Parasitic Native Plant Cuscuta australis on Growth and Competitive Ability of Two Invasive Xanthium Plants. Biology 13(1):23. https://doi.org/10.3390/biology13010023

Jaiswal AK, Elad Y, Cytryn E, Graber ER, Frenkel O (2018) Activating biochar by manipulating the bacterial and fungal microbiome through preconditioning. New Phytol 219:363–377

Jamil M, Rodenburg J, Charnikhova T, Bouwmeester HJ (2011a) Pre-attachment *Striga hermonthica* resistance of New Rice for Africa (NERICA) cultivars based on low strigolactone production. New Phytol 192:964–975

Jamil M, Charnikhova T, Cardoso C, Jamil T, Ueno K, Verstappen F, Asami T, Bouwmeester HJ (2011b) Quantification of the relationship between strigolactones and *Striga hermonthica* infection in rice under varying levels of nitrogen and phosphorus. Weed Res 51:373–385

Jamil M, Kountche BA, Al-Babili S (2021) Current progress in Striga management. Plany Physiology 185:1339–1352

Jamil M, Wang JY, Yonli D, Patil RH, Riyazaddin M, Gangashetty P, Berqdar L, Chen GTE, Traore H, Margueritte O (2022) A new formulation for strigolactone suicidal germination agents, towards successful Striga management. Plants 11:808

Joel DM (2013) Functional structure of the mature haustorium. In: Joel DM, Gressel J, Musselman LJ (eds) Parasitic orobanchaceae. Springer, Berlin/Heidelberg, pp 25–60

Kamara AY, Menkir A, Chikoye D, Tofa AI, Fagge AA, Dahini R, Solomon R, Ademulegun T, Omoigui L, Aliyu KT, Kamai N (2020) Mitigating *Striga hermonthica* parasitism and damage in maize using soyabean rotation, nitrogen application and *Striga* resistant varieties in the Nigerian Savanna. Exp Agric 56:620–632

Khaliq A, Perveen S, Alamer KH, Zia Ul Haq M, Rafique Z, Alsudays IM, Althobaiti AT, Saleh MA, Hussain S, Attia H (2022) Arbuscular mycorrhizal fungi symbiosis to enhance plant–soil interaction. Sustainability 14:7840. https://doi.org/10.3390/su14137840

Kim D, Kocz R, Boone L, Keyes WJ, Lynn DG (1998) On becoming a parasite: evaluating the role of wall oxidases in parasitic plant development. Chem Biol 5:103–117

Kolton M, Harel YM, Pasternak Z, Graber ER, Elad Y, Cytryn E (2011) Impact of biochar application to soil on the root-associated bacterial community structure of fully developed greenhouse pepper plants. Appl Environ Microbiol 77:4924–4930

Kolton M, Graber ER, Tsehensky L, Elad Y, Cytryn E (2017) Biochar stimulated plant performance is strongtly linked to microbial diversity and metabolic potential in the rhizosphere. New Phytol 213:1393–1404

Kuila D, Ghosh S (2022) Aspects, problems and utilization of Arbuscular Mycorrhizal (AM) application as bio-fertilizer in sustainable agriculture. Curr Res Microb Sci 3:100107

Leandre SP, Francis K, Richard A, Joseph B, Jean-Baptiste T, Jeremy T, Quedraogo PA, Close TJ, Roberbs PA (2018) Screening for resistance to *Striga gesneroides* and estimation if yield loss among cowpea progenies in the Upper East Region of Ghana. Afr J Agric Res 13(28):1430–1442

Li AR, Smith SE, Smith AF, Guan K (2012) Innoculation with AMF suppressesinitiation of haustoria in the root hemi-parasite *Pedicularis tricolor*. Ann Bot 109:1075–1080

Liu Z, Wu X, Li S, Liu W, Bian R, Zhang X, Zhang J, Drossos M, Li L, Pan G (2021) Quantitative assessment of the effects of biochar amendment on photosynthetic carbon assimilation dynamics in a rice-soil system. New Phytol 232:1250–1258

Lopez-Raez JA, Charnikhova T, Gomez Riolden V, Matusova R, Kohlen W, De Vos R (2008a) Tomato strigolactones are derived from carotenoids and their biosynthesis is promoted by phosphate starvation. New Phytol 178:863–874

Lopez-Raez JA, Matusova R, Cardoso C, Jamil M, Charnikhova T, Kohlen W, Ruyter-spira C, Verstappen F, Bouwmeester H (2008b) Strigolactones: ecological significance and use as a target for parasitic plant control. Pest Manag Sci 64:471–477

Losner-Goshen D, Portnoy VH, Mayer AM, Joel DM (1998) Pectolytic activity by the haustorium of the parasitic plant Orobanche L. (Orobanchaceae) in host roots. Ann Bot 81:319–326

Mandumbu R, Mutengwa C, Mabasa S, Mwenje E (2019) Challenges to the exploitation of host plant resistance for Striga management in cereals and legumes by farmers in sub-Saharan Africa: a review. Acta Agric Scand Sect B Soil Plant Sci 69(1):82–88. https://doi.org/10.1080/09064710.2018.1494302

Martinez L, Pouvreau JB, Jestin C, Montiel G, Gravot A, Berardocco S, Marnet N, Boucherau A, Delage E, Simia P, Poulin L (2024) Comparative amnalysis of two neighbouring conducive and suppressive soils towards plant parasitism caused by Phelipanche ramosa in Brassicae napus. Phytobiomes Journal 8:425–445

Matusova R, van Mourik T, Bouwmeester HJ (2004) Changes in the sensitivity of parasitic weed seeds to germination stimulants. Seed Science Research 14:335–344

Meena DK, Dawar R (2023) Arbuscular mycorrhizal fungi: role in sustainable agriculture. In: Mangaraj et al (eds) Frontiers in agriculture sustainability. Integrated Publications, New Delhi

Mohemed N, Charnikhova T, Bakker EJ, van Aast A, Babiker AGT, Boumeester HJ (2016) Evaluation of field resistance to *Striga hermonthica* in sorghum bicolor: the relationship with strigolactones. Pest Manag Sci 72:2082–2090

Odelade KA, Babadola OO (2019) Bacteria, fungi and Archarea domains in rhizosphere soil and their effects in enhancing agricultural productivity. Int J Environ Res Public Health 16:3873

Oluchi GD, Antangbenro AS, Odhiambo JJO, babadola OO (2022) Striga hermonthica: a highly destructive pathogen in maize production. Environ Challenges 8:100590

Raju, PS, Osman MA, Soman P, Peacock J.M (1990) Effects of N, P and K on Striga asiatica (L) Kunzte seed germination and infestation in sorghum. Weed Science 30:139–144

Rich PJ (2024) An agar based method for determining method for determining mechanisms of drought resistance in sorghum. In Ghanim AMA et al (eds). Mutation breeding and efficiency enhancing technologies for resistance to Striga in cereals. https://doi.org/10.1007/978-3-662-68-181-7_6

Riopel JL, Timko MP (1995) Haustorial initiation and differentiation. In: Press MC, Graves JD (eds) Parasitic plants. Chapman & Hall, London, pp 39–79
Rodenburg J, Demont M, Zwart SJ, Batiaans 1 (2016) Parasitic weed incidence and related economic losses to rice in Africa. Agric Ecosyst Environ 235:306–317
Sahoo SS, Vijay VK, Chandra R, Kumar H (2021) Production and characterization of biochar produced from slow pyrolysis of Pigeon pea stalk and bamboo. Cleaner Eng Technol 3:100101
Samejima H, Sugimoto Y (2018) Recent research progress in combatting root parasitic weeds. Biotechnol Biotechnol Equip 32: 221–240
Schimdt HP, Kamann C, Hageman N, Leifeld J, Bucheli TD, Monedero MAS, Cayuela ML (2021) Biochar in agriculture: a systematic review of 26 global meta-analysis. GCB Bioenergy 13:1708–1730
Shi W, Ju Y, Bian R, Li L, Joseph S, Mitchell DRG, Munroe P, Taherymoosavi S, Pan G (2020) Biochar bound urea boosts plant growth and reduces nitrogen leaching. Sci Total Environ 701:134424
Singh H, Northup BK, Rice CW, VaraPrasad PV (2022) Biochar applications influence soil physical and chemical properties, microbial diversity and crop productivity. A meta-analysis. Biochar 4:8
Smith FA, Smith SE (2011) Roles of arbuscular mycorrhiza in plant nutrition and growth: new paradigms from cellular to ecosystem scales. Annu Rev Plant Biol 62:227–250
Sun, Z, Hans J, Walter MH, Matusova R, Beekwilder R, Vestapen FWA, Ming Z, van Echtelt E, Strack D, Bisseling T, Boumeester HJ (2008) Cloning and Characterisation of a maize characterisation of a maize carotenoid cleavage dioxygenase (ZmCCDI) and its involvement in the biosynthesis of apocarotenoids with various roles in mutualistic and parasitic interactions. Planta 228:789–801
Thiombiano B, Flokova K, Walmsley A, Boumeester HJ (2024) Striga germination stimulant analysis. In Ghanim AMA et al (eds). Mutation breeding and efficiency enhancing technologies for resistance to Striga in cereals. https://doi.org/10.1007/978-3-662-68-181-7_86
Tomczyk A, Sokolowska Z, Boguta P (2020) Biochar physicochemical properties: pyrolysis temperature and feed stock kind effects. Rev Environ Sci Biotechnol 19:191–215
Umehara M, Hanada A, Yoshida S, Akiyama K, Arite T (2008) Inhibition of shoot branching by new terpernoid plant hormones. Nature 455:195–200
Wilkes TI (2021) Arbuscular mycorrhizal fungi in agriculture. Encyclopedia 1:1132–1154. https://doi.org/10.3390/encyclopedia1040085
Wright SF, Upadhyaya A (1996) A survey of soils for aggregate stability and glomalin, a glycoprotein produced by hyphae of arbuscular mycorrhizal fungi. Plant Soil 198:97–107
Xie X, Yoneyama K, Yoneyama K (2010) The strigolactones story. Annu Rev Phytopathol 48:93–117
Xie LC, Lin-Hong Jin, Yan-Hui Lu, Hong-Xing Xu, Lian-Sheng Zang, Jun-Ce Tian, and Zhong-Xian Lu (2022) Resistance of Lepidopteran Egg Parasitoids, Trichogramma japonicum and Trichogramma chilonis, to Insecticides Used for Control of Rice Planthoppers. Journal of Economic Entomology 115(2):446–454. https://doi.org/10.1093/jee/toab254
Yacoubou AM, Zoumarou Wallis N, Menkir A, Zinsou VA, Onzo A, Garcia-Oliveira AL, Meseka S, Wende M, Gedil M, Agre P (2021) Breeding maize (Zea mays) for Striga resistance: past, current and prospects in sub-Saharan Africa. Plant Breed 140:195–210
Yadav SPS, Bhendari S, Bhetta D, Poundel A, Bhattarai S, Yadav P, Ghimire N, Paudel P, Shrestha J, Oli B (2023) Biochar application: a sustainable approach to improve soil health. J Agric Food Res 11:100498
Yoneyama K, Yoneyama K, Takeuchi Y, Sekimoto H (2007a) Phosphorus deficiency in red clover promotes exudation of orobanchol, the signal for mycorhizal symbionts and germination stimulant for root parasites. Planta 225:1031–1038
Yoneyama K, Xie X, Kusumoto D, Sekimoto H, Sugimoto Y, Takeuchi Y, Yoneyama K (2007b) Nitrogen deficiency as well as phosphorus deficiency in sorghum promotes the production and exudation of 5-de-oxystrigol, the host recognition signal for arbscular mycorhizzal fungi and root parasites. Planta 227:125–132

Yoneyama K, Awad AA, Xie X, Yoneyama Kaori, Takeuchi Y (2010) Strigolactones as germination stimulants for root parasitic plants. Plant and Cell Physiology 51:1095–1103
Yuan HM, Xue N, Roiloa S, Yao J, Yu FH (2024) Increasing biochar diversity promotes the impacts of plant diversity on remediating cadmiuim in the soil. J Plant Ecol 17:rtae068
ZhangY, HanX, Zhao D, Wei K, Yuan Y, Li Y, Liu M, Zhang C-S (2020) Exploring biocontrol agents from microbial keystone taxa associated to suppressive soil: A new attempt for a biocontrol strategy. Front. Plant Sci 12:655673
Zhang L, Cao X, Yao Z, Dong X, Chen M, Xiao L et al (2022) Identification of risk areas for Orobanche cumana and Phelipanche aegyptiaca in China, based on the major host plant and CMIP6 climate scenarios. Ecology and Evolution, 12:e8824
Zwaneburg B, Pospiosil T (2013) Structure and activity of strigolactones: new plant hormone with a rich future. Mol Plant 6:38–62

Chapter 14
The Role of Omics in the Future Direction of Biofertiliser Innovation

Simin Emamzadeh Yazdi, Keletso C. Mohale, and Noluyolo Nogemane

Abstract Food security is increasingly at risk due to the expansion of the global population as well as concurrent threats to the ecosystem, including soil degradation and erosion, climate change, and biodiversity loss. To prevent further threats to the ecosystem, utilising plant and soil-associated microorganisms, biofertilisers offer a promising method to eliminate the application of agrochemical fertilisers. Biofertilisers are microbial substances such as bacteria, fungi, and algae that promote plant nutrition and growth by enhancing soil nutrient availability and mobility. Exploring biofertilisers, especially in conjunction with advanced multi-omics technologies such as genomics, transcriptomics, proteomics, metabolomics, and phenomics, represents a significant opportunity in agriculture. However, these technologies have not been effectively utilised, and there are limitations in the advancement of innovative and improved biofertiliser technologies. The application of multi-omics technologies in the study of biofertilisers can be a noteworthy step forward in improving the scientists' comprehension of microbiome intricacies, allowing for in-depth exploration and characterization of the microbial community structure, functionality, and mechanism of action in plant microbiomes. The aforementioned discoveries will significantly enhance our understanding of soil microbial communities and their relationships with host plants, as well as their impact on plant nutrition and other plant growth-promoting characteristics. For example, metagenomic sequences have valuable insight into the potential functions of the rhizosphere community and assessments of interactions between microbes in plants, while metabolomics analysis is being used to diagnose plant diseases and their etiological agents. However, the application to microbiome science remains limited. Despite the recent progress in omics investigation on biofertilisers, it is just the tip of the iceberg of this emerging knowledge and the related techniques. A comprehensive understanding of these traits will be achieved once we gain further insights into

S. E. Yazdi · K. C. Mohale · N. Nogemane (✉)
Department of Agriculture and Animal Health, Science Campus, University of South Africa (UNISA), Johannesburg, Gauteng, South Africa
e-mail: nogemn@unisa.ac.za

A. D. Nciizah et al. (eds.), *Biofertilizers for Sustainable Agriculture in Africa*, Sustainability Sciences in Asia and Africa, https://doi.org/10.1007/978-981-95-6009-7_14

their expression in natural environments through effective utilisation of complementary high-throughput techniques.

Keywords Omics · Genomics · Transcriptomics · Proteomics · Metabolomics · Meta-omics · Biofertiliser

14.1 Introduction

The future of agriculture is undergoing a remarkable transformation driven by technology, innovation, and sustainability. Embracing these advancements will boost productivity and profitability and create a more environmentally friendly farming system (Khan et al. 2021). In this regard, biofertilisers provide environmentally friendly alternatives for promoting sustainable agriculture. Although biofertilisers have shown plenty of benefits at low cost, they are still not able to compete with conventional chemical fertilisers in industrial agriculture due to various challenges, such as regulatory obstacles, technical complexities, adaptation of standard production processes, storage, and application (Mitter et al. 2021; Rai et al. 2023; Yadav and Yadav 2024). The use of chemical fertilisers in crop production leads to significant ecological challenges, including eutrophication, soil degradation, and disruption of microbial balance. These practices contribute to greenhouse gas emissions and a decline in biodiversity. Rising input costs and nutrient imbalances endanger farmers' economies and threaten food security. Overapplication of chemical fertilisers can further harm natural ecosystems. Addressing these challenges is crucial for sustainable agricultural practices in the future (Zhao et al. 2024). Therefore, biofertilisers that contain living microorganisms can promote plant growth through environmentally friendly systems as they are composed of plant growth-promoting microbes (PGPMs) essential for improving agricultural product growth and resistance (Daniel et al. 2022). The study of plant–microbe interactions has been profoundly enhanced by omics and meta-omics technologies, including genomics, epigenomics, transcriptomics, proteomics, and metabolomics. These technologies offer a comprehensive understanding of the various interactions between plants and microbes. Genomics and epigenomics identify important genes; transcriptomics reveal the dynamics of gene expression; proteomics identify essential proteins; and metabolomics profiles small chemical compounds. Collectively, these technologies and studies have contributed to the emergence of phenomics, an omics field/discipline that investigates an organism's observable characteristics, or phenotypes, throughout its genome.

In the studies of complex microbial communities, meta-omics aids in investigating the potential and activity of the entire functional capacity of multiple organisms. It incorporates several 'omics' analyses to investigate how organisms interact, which metabolic pathways they occupy, and how they utilise nutrients. However, there is little information and progress on the development of applications and the mode of action of omics and biofertilisers on plant growth and microecology of omics in biofertilisers (Jain et al. 2024).

This chapter discusses an overview of methodologies employed and their mode of application and action in diverse omics approaches aimed at elucidating the complex interactions that shape phenotypic traits, metabolic pathways, and biosynthetic processes in plants subjected to biofertilisers treatment.

14.2 Global Threat to Food Security

Over the next 30 years, it is predicted that the world population will exceed 10 billion people (Population-The United Nations 2025). Accordingly, a 60% increase in food supply needs to take place by 2050 to reach the required food demand (Ammar et al. 2023). Currently, over 280 million people experience acute hunger daily, according to the United Nations Office for the Coordination of Humanitarian Affairs (OCHA) (The World Bank 2025). This statistic highlights a serious issue concerning global food security. In recent decades, significant disruptions in food production and limited access to essential markets have created a dynamic environment where food insecurity both reflects and aggravates larger humanitarian challenges (The World Bank 2025). Several factors, including economic instability, violence, displacement, and climate change, have contributed to the worsening of this food crisis (The World Bank 2025).

The inadequate humanitarian response has endangered significant portions of the global population, leading to food shortages. An overview of the challenges associated with food security in countries such as Afghanistan, Ethiopia, Malawi, Mozambique, Somalia, and Zimbabwe reveals that, despite ongoing aid, the situation remains vulnerable. This vulnerability is largely due to the severe underfunding of the food security sector, which has led to a significant increase in malnutrition (The World Bank 2025). These countries face challenges related to food insecurity, exacerbated by climate-induced shocks such as drought, salinity, and flooding. As climate incidents become more frequent and severe, the vulnerability of these populations to food insecurity will only increase unless strong measures are taken to address the situation (The World Bank 2025).

In light of these challenges and limitations threatening food security, biofertilisers may be used as alternative strategies to fortify the soil and enhance agricultural production. In the face of growing food demand, using these natural solutions is a crucial step towards ensuring food security for the future.

14.3 Plant and Soil-Associated Microorganisms Biofertilisers

Recently, biofertilisers have become well-recognised as one of the eco-friendly techniques in sustainable agriculture (Ammar et al. 2023). The utilisation of biofertilisers enhances the nutrient content and fertility of the soil and promotes plant growth while simultaneously reducing soil pollution (Ammar et al. 2023).

Biofertilisers encompass a variety of organisms, including microorganisms such as bacteria, micro-algae, and micro-fungi, as well as macroorganisms including macro-fungi, macro-algae, and higher plants (Ammar et al. 2023). The identified biofertiliser's mechanism of action includes nitrogen fixation, growth hormone biosynthesis, siderophore secretion, hydrogen cyanide (HCN) production, a variety of hydrolytic enzymes, and the solubilisation of vital nutrients like potassium, zinc, and phosphorus (Prisa et al. 2023). Furthermore, recent developments in sustainable agriculture emphasised increasing interest in two emerging areas: plant biostimulants and nanomaterials. These fields are promising in improving agricultural outcomes (Sharma et al. 2023). The integration of nanotechnology with plant biostimulants highlights their potential to enhance effectiveness, ultimately supporting the sustainable production of high-quality nutrition. Furthermore, combining nanoparticles and biofertilisers into nano-biofertilisers may enhance plant growth and nutrient uptake (Magnabosco et al. 2023). Over the past decade, advancements in multi-omics technologies have significantly enhanced scientists' knowledge about the roles, processes, structures, and mechanisms of action of biofertilisers, which can lead to a better understanding of the relationship between the different microbial colonies with soil, their effect on the plant nutrient contents, as well as their interaction with other technologies such as nanotechnology. It is for this reason that the development of biofertiliser technologies has the potential to foster optimism and advance sustainable agricultural practices (Mitter et al. 2021).

14.4 The Role of Omics Technologies in Agricultural Research

Between 450,000 and 500,000 kinds of green plants have been estimated to exist. Only a portion of plant species have had their genomes sequenced thus far; the number of green plant species having sequenced genomes (1482) represents around 0.26–0.29% of all species of plants, and these estimates have expanded significantly in microbiology (Rappuoli et al. 2023; Bernal-Gallardo and de Folter 2024). Omics technologies are highly throughput analytical methods for quantitatively evaluating distinct biomolecular components found in biological samples. The omics, including genomics, epigenomics, transcriptomics, proteomics, metabolomics, and phenomics, contribute significantly to the comprehensive analysis of an organism's phenotype as they relate to its genomic architecture. Collectively, these disciplines facilitate the exploration of phenomics, allowing for a deeper understanding of the phenotypic manifestations that arise from genetic and molecular processes across the genome (Fig. 14.1) (McCoy et al. 2023; Mmbando 2024).

Genomics is the study of an individual's entire genetic code or genome (DNA), including how those genes interact with one another and with their environment. In agricultural research, scientists can identify specific genes associated with key traits such as yield, disease resistance, stress tolerance, and caloric content (Kumar et al.

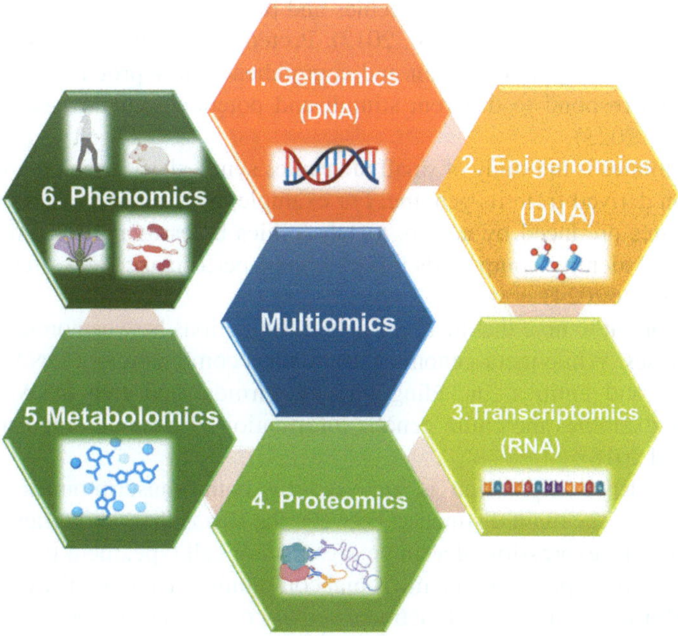

Fig. 14.1 Flow diagram of the types of omics that contribute to integrated multi-omics. The first three omics are nucleic acid–based omics methods

2023). Moreover, they can elucidate the responsible genes underlying the interactions between plants and biofertilisers, providing insights into how these relationships enhance plant performance and productivity. By using quantitative trait loci (QTLs), scientists can identify and characterise the specific gene or genomic region(s) that underpin complex properties in plants, such as quality, yield, or disease resistance. Within the QTLs, researchers can prioritise particular genes in order of importance and propose one or more putative causative genes that can be responsible for the observed characteristic (Kumar et al. 2023). Studying alterations to the genome that produce changes in the expression of genes without altering their DNA base sequence is known as epigenomics analysis. To comprehend how these alterations impact gene expression, it entails examining epigenetic markers throughout the genome (Fan et al. 2025). The two types of epigenetic changes that may be assessed at high throughput biochemically are methylation of DNA cytosine residues (at CpG) and modifications of specific histone proteins in the chromosomes (histone marks). Another pathway for epigenetic modifications in gene expression is RNA editing, which is mostly detected by transcriptomic techniques (Gogolev et al. 2021).

According to Morozova et al. (2009), transcriptomics investigates the transcriptomes or RNA molecules in cells, which include the RNAs produced by biological cells and tissues within an organism, encompassing both messenger RNAs (mRNAs) and noncoding RNAs (ncRNAs). It gives a better understanding of how DNA is exhibited as proteins and other molecules and turned on in different types of cells (Morozova et al. 2009; Lowe et al. 2017). Proteomics studies the complete set of proteins produced by an organism and comprehends how proteins are expressed, function, and respond to different stimuli and potential genetic targets in plants (Kumar et al. 2023).

Metabolomics is the study of small molecules synthesised by an organism, using information derived from its genome. For example, by examining the complete set of metabolites produced by a crop, metabolomics helps us understand metabolic pathways and identify compounds linked to crop performance and stress tolerance (Rani and Sagar 2024).

It is important to note that there are subtle distinctions between omics and meta-omics analyses. While meta-genomics examines a combination of DNA from various species and entities, including viruses, viroids, and free DNA, genomics exclusively examines the entire genetic information of a single organism (Sharon and Banfield 2013).

Although RNA expression is analysed by both transcriptomics and meta-transcriptomics, the main distinction between the two is that while transcriptomics looks at the gene expression of a single organism or cell type, meta-transcriptomics looks at the gene expression of a microbial community in a particular environment, offering information about the functional activity of various microorganisms within that community (Dadlani 2024). Whereas both meta-proteomics and meta-metabolomics refer to the analysis of the proteins and all metabolites across multiple organisms or samples, respectively; proteomics and metabolomics study all the proteins and metabolites generated by a single organism or cell type (Aguiar-Pulido et al. 2016; Starr et al. 2018).

14.5 Advancing Multi-Omics Technologies in Exploring Biofertilisers

The emergence of omics techniques has aided the study of microbial and plant–microbe interactions throughout ecological systems. Omics science utilises integrative analysis to explore the intricate interactions between plants and microbes at the molecular level. By operating multi-omics combinations, it has been possible to demonstrate the structure of microbiomes by amplicon sequencing and identify their function through meta-genomics, meta-transcriptomics, and meta-proteomics, as well as their system of interactions through metabolomics. These pathways elucidate the mechanisms by which microbes and microbial communities influence plant–microbiome interactions, thereby enhancing plant growth and development and resilience to environmental stressors (Kumar et al. 2023). Comprehensive investigations utilising various omics data sets—specifically genomics, transcriptomics, proteomics, and metabolomics—are essential to deepen our understanding of plant–microbe interactions and their underlying mechanisms. The integration of different omics methodologies facilitates the identification of significant biomarker molecules and associated metabolic pathways, enabling the monitoring of networks involved in plant–microbe interactions. Genomics, the most used omics technique, investigates environmental samples, and transcriptomics and proteomics analyse the expressed genes and the proteins in the biomass. Furthermore, metabolomics enables comprehensive analysis of small metabolites produced by microbial organisms, leveraging insights derived from their genomic data (Trivedi et al. 2021; Kumar et al. 2023). To demonstrate how multi-omics has been utilised to identify beneficial bacteria and comprehend their method of application and activity as biofertilisers, a few studies will be reviewed below. According to Kumar et al. (2023), multi-omics approaches enable the identification of microbiome composition, thus facilitating the elucidation of the roles of various genes, metabolites, proteins, and nucleic acids in plant–microbial interactions. The interplay between plants and their associated microorganisms acting as biofertilisers can be comprehensively analysed through genome sequencing techniques. This approach enables the characterisation of advantageous plant traits alongside the functional diversity of the associated soil microbiome, revealing critical insights into their symbiotic relationships. Conversely, alterations in the secondary metabolite composition within the metabolic profile can be effectively assessed through metabolomics analysis (Kumar et al. 2023).

Metabolomics can reveal the mechanism by which PGPRs assist plants in acclimating to environmental stressors. In the acclimatisation of crops to environmental stresses, Mashabela et al. (2022) clarified the rhizosphere and leaf of wheat (*Triticum aestivum* L.) metabolome changes treated with two plant growth-promoting rhizobacteria (PGPR) (*Paenibacillus alvei* (T22) and *Bacillus subtilis* strains) using untargeted metabolomics analysis. In this study, mass spectrometry (MS) data were analysed and interpreted by chemometrics and molecular networking. This analysis demonstrated a varying accumulation of diverse compounds spanning multiple classes of metabolites, such as phenylpropanoids, organic acids, lipids, organoheterocyclic compounds, and benzenoids. Several of the identified compounds were found to be integral to plant–microbe interactions, serving important functions in chemotaxonomy, biological control mechanisms, and the promotion of plant growth. The findings indicate that seed bio-priming using PGPRs can elicit specific metabolic alterations in plants, which induce systemic responses. This response enhances the plants' capacity to adapt to both biotic and abiotic stressors (Mashabela et al. 2022).

Use of metabolomics can assist in revealing metabolites playing a role/contributing to the infection-defensive mechanism of PGPRs. Another recent investigation conducted by Mashabela et al. (2023) focused on the metabolomic profiling of the defensive strategies employed by PGPRs [*P. alvei* (T22) and *B. subtilis*] on wheat cultivar infection by *Puccinia striiformis* f. sp. *tritici*. The data collected from the untargeted ultra-high performance liquid chromatography–high-definition mass spectrometry (UHPLC–HDMS) revealed that the treatment of plants with PGPRs led to significant alterations in both primary and secondary metabolite profiles when compared to the untreated wheat cultivar. Specifically, the PGPR-treated plants exhibited an upregulation of phenolic compounds, including flavonoids, hydroxycinnamic acids (HCAs), and HCA amides. Furthermore, the study demonstrated a time-dependent metabolic reprogramming in the PGPR-treated plants in response to bacterial infection, contrasting with those treated with PGPRs but not subjected to bacterial pathogens. This indicates a dynamic metabolic adaptation of the plants to *P. striiformis* infection over time (Mashabela et al. 2023).

Use of multi-omics can assist in revealing the mode of PGPR application in the field that can best enhance soil health and growth. In regard to the mode of application, the response mechanism of potato (*Solanum tuberosum* L.) to four different methods of treatments of PGPM biofertilisers containing *Bacillus amyloliquefaciens*, namely: (i) scattered application of biofertiliser, (ii) soaking seed tubers in a dissolved biofertiliser, (iii) a combination of scattered biofertiliser and sucrose, and (iv) scattered attapulgite free of any bacteria as a control, was examined through multi-omics analysis of the transcriptome and metabolomics (Tian et al. 2024). The transcriptomics investigation revealed that among the four application methods, distinct sets of differentially expressed genes play crucial roles in phenylpropanoid biosynthesis, sugar metabolism, and photosynthesis. Remarkably, the scattered application of the biofertiliser treatment emerged as the most effective, resulting in the upregulation of 828 genes, more than any other treatment. This finding underscores the potential of biofertilisers to significantly enhance plant health and productivity. When a gas chromatography–time-of-flight mass spectrometry (GC–TOF-MS) untargeted metabolomics analysis was conducted, the principal component analysis (PCA) results revealed no clear separation in both positive electrospray ionisation - (ESI+) modes and negative (ESI–) modes, except for those seed tubers which were soaked in a dissolved biofertiliser. Furthermore, the partial least squares-discriminant analysis (PLS-DA) illustrated a clear separation in the ESI+ mode of different models. However, the PLS-DA analysis could not separate the scattered attapulgite free of any bacteria and the scattered application of biofertiliser groups in ESI– mode. The metabolomics analysis annotated 952 (617 ESI+ and 335 ESI–) from a total of 2498 mass (1435 ESI+ and 1063 ESI–) and found that carboxylic acids and derivatives, glycerophospholipids, indoles and derivatives, and fatty acyls were the four most major metabolites (Tian et al. 2024).

The role and function of bacterial endophytes within plants remain inadequately understood, primarily due to the challenges associated with studying their specific functions. These challenges stem from the difficulties encountered in cultivating the majority of prokaryotic organisms. A meta-genomic analysis of the endophyte community that exists in rice (*Oryza sativa* L.) roots has shed light on the functional characteristics of these bacterial endophytes (Sessitsch et al. 2012). These findings enabled the prediction of characteristics and metabolic processes essential for an endophytic lifestyle, indicating that the endo-rhizosphere is a unique microhabitat necessitating various adaptations. Genome sequencing of the endophytic bacteria has revealed key beneficial traits, such as the presence of flagella, plant-polymer-degrading enzymes, robust protein secretion systems, mechanisms for iron acquisition and storage, quorum-sensing capabilities, and detoxification pathways for reactive oxygen species. These findings suggest that these endophytes play a crucial role in the nitrogen cycle, with identified protein domains associated with nitrogen fixation, denitrification, and nitrification processes, supported by the expression of specific genes. Genomics demonstrates that endophyte culture has significant potential for promoting plant growth, enhancing stress resistance, providing biocontrol against pathogens, and facilitating bioremediation (Sessitsch et al. 2012).

Metabolomics analysis demonstrates how microorganisms and plants interact and how their metabolic profiles alter in this relationship. It can identify variations in primary and secondary metabolites, such as sugars, amino acids, phytoalexins, and antimicrobial components of plants. Moreover, the elucidation of metabolite exchange between plant roots and mycorrhizal fungi enhances our comprehension of the mechanisms underlying nutrient transport and assimilation. (Trivedi et al. 2021; Kumar et al. 2023). Metabolomics is instrumental in identifying a diverse array of compounds, such as terpenes, alkaloids, and phytohormones, which are integral to the biosynthesis of signalling molecules. These metabolites significantly contribute to plant–microbe interactions, enhancing plant resilience against environmental stressors (Kaya 2024).

Through metabolomics, scientists can detect and quantify the small molecules or metabolites produced by a biological system. Faddetta et al. (2023) characterised three actinobacterial strains: *Streptomyces violaceoruber*, *S. coelicolor*, and *Kocuria rhizophila*. These strains exhibit multiple plant growth-promoting (PGP) traits, including indole acetic acid production, phosphate solubilisation, nitrogen fixation, and tolerance to drought and salinity. The metabolomics analysis of these strains revealed the presence of bioactive components such as antibiotics and siderophores in *S. violaceoruber*, and thus the possibility of these strains as biofertilisers that can be used under adverse conditions to promote plant growth and increase stress tolerance. The 33 annotated compounds belong to various classes of metabolites, including those related to tryptophan metabolism, polyketides associated with actinorhodin biogenesis, four alkaloids, two siderophores, and one inosine.

There is reported progress in the application of omics technologies to study biomining of microorganisms from extreme environments (Li and Wen 2021). Omics technologies can help establish a comprehensive metabolic model of biomining microorganisms and also provide insights into the evolutionary relationships among different strains, leading to the discovery of new substances, and facilitate the study of interactions between genes, proteins, and metabolites. Moreover, comprehensive metabolic models can also be established by using metagenomics, metatranscriptomes, and meta-proteomics to describe how members of a community interact and how the outside world affects their activities. Biomining microorganisms can transform toxic solid waste into a nontoxic form and are effective in catalysing mineral dissolution. This process can serve as a comprehensive solution for mineral beneficiation and help address environmental problems. Additionally, these microorganisms offer a powerful approach to soil waste management and a reduction of environmental pollution. They can also be utilised as biofertilisers to enhance soil fertility and promote plant growth (Mažylytė et al. 2022; Srivastava 2023). So, the integration of omics technologies offers a promising approach to enhance our understanding and application of these biofertilisers in agricultural systems.

The use of metabolomics in studying the plant–endophyte relationship has been advocated for by Braga et al. 2016. Endophytes are very important and beneficial biological resources that can serve various functions. The metabolites produced by endophytic compounds can be utilised in various industries such as agriculture, medicine, bioremediation, etc. Metagenomic analyses facilitated by next-generation sequencing indicate that merely 5% of bacterial species can be successfully cultured using existing methodologies. There are no exclusive databases available for endophytic microorganisms and their metabolites (Braga et al. (2016)). This indicates that many microbes, particularly bacteria, remain unknown, and their functions have not yet been identified. Braga et al. (2016) suggest that, to get a better understanding of this plant endophytic relationship, the best approach would be to combine metabolomics with other omics methodologies by connecting genotype and phenotype. Therefore, this shows that the advancement of omics and meta-omics technologies has the potential to significantly enhance our ability to identify and characterise a diverse array of microbial communities, particularly those involved in biofertiliser applications. This progress can contribute to environmental safety, promote sustainable agricultural practices, and bolster food security by leveraging the functional capacities of these microbial populations (Braga et al. 2016; Kumar et al. 2023).

Recent meta-omics studies have not only assigned functional traits to microbial populations but have also identified several nonpathogenic viruses associated with plant hosts, an area that remains underexplored. Techniques such as meta-transcriptomics and meta-proteomics hold significant potential for elucidating the microbial communities linked to specific plant species, along with their functional roles and activities. These methodologies have demonstrated their effectiveness in investigating microbial bioremediation processes as well. While the effects and economic benefits of biostimulants are well established, the physiological mechanisms and molecular pathways remain poorly understood. To effectively demonstrate this complexity, the use of multi-omics analysis is highly recommended.

14.6 Conclusion

Current research reveals that our understanding of plant–microbiome interactions is still evolving, with several gaps remaining. To achieve sustainable agriculture through agroecological practices, we must embrace the vast potential of biofertilisers in empowering staple foods. However, there has been a recent surge in studies focusing on various properties of the microbiome, including its activity and functional capabilities. Considering the enormous number of genes and metabolites in microorganisms and plants that have not yet been identified, as well as the significance of their interactions for achieving food security and sustainable agriculture, multi-omics research has a lot of promise for discovering hitherto unexplored subjects.

While the effects and economic benefits of biostimulants are well established, the physiological mechanisms and molecular pathways remain poorly understood. To effectively demonstrate this complexity, the use of multi-omics analysis is highly recommended. Recent advancements in multi-omics technologies are progressively illuminating the complexities of the plant microbiome. These technologies facilitate comprehensive analyses of PGPRs' interactions with host plants, shedding light on their environmental adaptability and informing strategies for optimising sustainable agricultural practices. These studies often employ omics strategies to gain deeper insights into these interactions. Investigating the interactions between biofertilisers and plants can be effectively accomplished through metabolomic analysis. To establish a robust validation of the study by correlating genotype with phenotype, metabolomics is commonly employed alongside complementary omics techniques. Therefore, by providing detailed insights into the genetics and metabolomics foundations of plant performance, these innovative methods could enhance agricultural practices and boost crop yield.

References

Aguiar-Pulido V, Huang W, Suarez-Ulloa V, Cickovski T, Mathee K, Narasimhan G (2016) Metagenomics, metatranscriptomics, and metabolomics approaches for microbiome analysis: supplementary issue: bioinformatics methods and applications for big metagenomics data. Evol Bioinforma 12:EBO-S36436. https://doi.org/10.4137/EBO.S36436

Ammar EE, Rady HA, Khattab AM, Amer MH, Mohamed SA, Elodamy NI, Al-Farga A, Aioub AA (2023) A comprehensive overview of eco-friendly bio-fertilizers extracted from living organisms. Environ Sci Pollut Res 30(53):113119–113137. https://doi.org/10.1007/s11356-023-30260-x

Bernal-Gallardo JJ, de Folter S (2024) Plant genome information facilitates plant functional genomics. Planta 259(5):117. https://doi.org/10.1007/s00425-024-04397-z

Braga RM, Dourado MN, Araújo WL (2016) Microbial interactions: ecology in a molecular perspective. Braz J Microbiol 47:86–98

Dadlani M (2024) Metatranscriptomics vs metagenomics: exploring the difference. Available at: https://www.cosmosid.com/blog/metatranscriptomics-vs-metagenomics-exploring-the-difference/# [09/02/2025]

Daniel AI, Fadaka AO, Gokul A, Bakare OO, Aina O, Fisher S, Burt AF, Mavumengwana V, Keyster M, Klein A (2022) Biofertilizer: the future of food security and food safety. Microorganisms 10(6):1220. https://doi.org/10.3390/microorganisms10061220

Faddetta T, Polito G, Abbate L, Alibrandi P, Zerbo M, Caldiero C, Reina C, Puccio G, Vaccaro E, Abenavoli MR, Cavalieri V (2023) Bioactive metabolite survey of actinobacteria showing plant growth promoting traits to develop novel biofertilizers. Meta 13(3):374. https://doi.org/10.3390/metabo13030374

Fan BL, Chen LH, Chen LL, Guo H (2025) Integrative multi-omics approaches for identifying and characterizing biological elements in crop traits: current Progress and future prospects. Int J Mol Sci 26(4):1466. https://doi.org/10.3390/ijms26041466

Food Security Update (2025) The world bank. Available: https://thedocs.worldbank.org/en/doc/40ebbf38f5a6b68bfc11e5273e1405d4–0090012022/related/Food-Security-Update-112-January-17-2025.pdf. [January-05-2025]

Gogolev YV, Ahmar S, Akpinar BA, Budak H, Kiryushkin AS, Gorshkov VY, Hensel G, Demchenko KN, Kovalchuk I, Mora-Poblete F, Muslu T (2021) OMICs, epigenetics, and

genome editing techniques for food and nutritional security. Plants 10(7):1423. https://doi.org/10.3390/plants10071423
Jain A, Sarsaiya S, Singh R, Gong Q, Wu Q, Shi J (2024) Omics approaches in understanding the benefits of plant-microbe interactions. Front Microbiol 15:1391059. https://doi.org/10.3389/fmicb.2024.1391059
Kaya C (2024) Microbial modulation of hormone signalling, proteomic dynamics, and metabolomics in plant drought adaptation. Food Energy Secur 13(1):513. https://doi.org/10.1002/fes3.513
Khan N, Ray RL, Sargani GR, Ihtisham M, Khayyam M, Ismail S (2021) Current progress and future prospects of agriculture technology: gateway to sustainable agriculture. Sustainability 13(9):4883. https://doi.org/10.3390/su13094883
Kumar U, Raj S, Sreenikethanam A, Maddheshiya R, Kumari S, Han S, Kapoor KK, Bhaskar R, Bajhaiya AK, Gahlot DK (2023) Multi-omics approaches in plant–microbe interactions hold enormous promise for sustainable agriculture. Agronomy 13(7):1804. https://doi.org/10.3390/agronomy13071804
Li M, Wen J (2021) Recent progress in the application of omics technologies in the study of biomining microorganisms from extreme environments. Microb Cell Factories 20(1):178. https://doi.org/10.1186/s12934-021-01671-7
Lowe R, Shirley N, Bleackley M, Dolan S, Shafee T (2017) Transcriptomics technologies. PLoS Comput Biol 13(5):e1005457. https://doi.org/10.1371/journal.pcbi.1005457
Magnabosco P, Masi A, Shukla R, Bansal V, Carletti P (2023) Advancing the impact of plant biostimulants to sustainable agriculture through nanotechnologies. Chem Biol Technol Agric 10(1):117. https://doi.org/10.1186/s40538-023-00491-8
Mashabela MD, Tugizimana F, Steenkamp PA, Piater LA, Dubery IA, Mhlongo MI (2022) Untargeted metabolite profiling to elucidate rhizosphere and leaf metabolome changes of wheat cultivars (*Triticum aestivum* L.) treated with the plant growth-promoting rhizobacteria *Paenibacillus alvei* (T22) and *Bacillus subtilis*. Front Microbiol 13:971836. https://doi.org/10.3389/fmicb.2022.971836
Mashabela MD, Tugizimana F, Steenkamp PA, Piater LA, Dubery IA, Terefe T, Mhlongo MI (2023) Metabolomic evaluation of PGPR defence priming in wheat (*Triticum aestivum* L.) cultivars infected with *Puccinia striiformis* f. sp. *tritici* (stripe rust). Front Plant Sci 14:1103413. https://doi.org/10.3389/fpls.2023.1103413
Mažylytė R, Kaziūnienė J, Orola L, Valkovska V, Lastauskienė E, Gegeckas A (2022) Phosphate solubilizing microorganism Bacillus sp. MVY-004 and its significance for biomineral fertilizers' development in agrobiotechnology. Biology 11(2):254. https://doi.org/10.3390/biology11020254
McCoy JC, Spicer JI, Ibbini Z, Tills O (2023) Phenomics as an approach to comparative developmental physiology. Front Physiol 14:1229500. https://doi.org/10.3389/fphys.2023.1229500
Mitter EK, Tosi M, Obreġon D, Dunfield KE, Germid JJ (2021) Rethinking crop nutrition in times of modern microbiology: innovative biofertilizer technologies. Front Sustain Food Syst 5:1–23. https://doi.org/10.3389/fsufs.2021.606815
Mmbando GS (2024) Omics: a new, promising technologies for boosting crop yield and stress resilience in African agriculture. Plant Stress 11(100):366. https://doi.org/10.1016/j.stress.2024.100366
Morozova O, Hirst M, Marra MA (2009) Applications of new sequencing technologies for transcriptome analysis. Annu Rev Genomics Hum Genet 10:135–151. https://doi.org/10.1146/annurev-genom-082908-145957
Population-The United Nations (2025) United Nations: peace, dignity and equality on a healthy planet. https://www.un.org/en/global-issues/population
Prisa D, Fresco R, Spagnuolo D (2023) Microbial biofertilisers in plant production and resistance: a review. Agriculture 13(9):1666. https://doi.org/10.3390/agriculture13091666
Rai PK, Rai A, Sharma NK, Singh T, Kumar Y (2023) Limitations of biofertilizers and their revitalization through nanotechnology. J Clean Prod:138194. https://doi.org/10.1016/j.jclepro.2023.138194

Rani N, Sagar NA (2024) Metabolomics: a paradigm shift in understanding biofertilizers dynamics. In: Metabolomics, proteomics and gene editing approaches in biofertilizer industry, vol II. Springer Nature Singapore, Singapore, pp 35–51

Rappuoli R, Young P, Ron E, Pecetta S, Pizza M (2023) Save the microbes to save the planet. A call to action of the International Union of the Microbiological Societies (IUMS). One Health Outlook 5(1):5. https://doi.org/10.1186/s42522-023-00077-2

Sessitsch A, Hardoim P, Döring J, Weilharter A, Krause A, Woyke T, Mitter B, Hauberg-Lotte L, Friedrich F, Rahalkar M, Hurek T, Sarkar A, Bodrossy L, van Overbeek L, Brar D, van Elsas JD, Reinhold Hurek B (2012) Functional characteristics of an endophyte community colonizing rice roots as revealed by metagenomic analysis. Mol Plant-Microbe Interact 25(1):28–36. https://doi.org/10.1094/MPMI-08-11-0204

Sharma B, Tiwari S, Kumawat KC, Cardinale M (2023) Nano-biofertilizers as bio-emerging strategies for sustainable agriculture development: potentiality and their limitations. Sci Total Environ 860:160476. https://doi.org/10.1016/j.scitotenv.2022.160476

Sharon I, Banfield JF (2013) Genomes from metagenomics. Science 342(6162):1057–1058. https://doi.org/10.1126/science.1247023

Srivastava P (2023) Recent trends in biomining microorganisms for solid waste management. In: Microbial technology for sustainable E-waste management. Springer. https://doi.org/10.1007/978-3-031-25678-3_17

Starr AE, Deeke SA, Li L, Zhang X, Daoud R, Ryan J, Ning Z, Cheng K, Nguyen LV, Abou-Samra E, Lavallée-Adam M (2018) Proteomic and metaproteomic approaches to understand host–microbe interactions. Anal Chem 90(1):86–109. https://doi.org/10.1021/acs.analchem.7b04340

Tian Y, Liu Y, Yue L, Zhao X, Zhou Q, Uwaremwe C, Wang Y, Chen G, Sha Y, Zhang Y, Wang R (2024) Multi-omics analysis reveals the effects of three application modes of plant growth promoting microbes biofertilizer on potato (Solanum tuberosum L.) growth under alkaline loess conditions. Microbiol Res 287:127855. https://doi.org/10.1016/j.micres.2024.127855

Trivedi P, Mattupalli C, Eversole K, Leach JE (2021) Enabling sustainable agriculture through understanding and enhancement of microbiomes. New Phytol 230(6):2129–2147. https://doi.org/10.1111/nph.17319

Yadav A, Yadav K (2024) Challenges and opportunities in biofertilizer commercialization. SVOA Microbiol 5(1):01–14. 10.58624/SVOAMB.2024.05.037

Zhao G, Zhu X, Zheng G, Meng G, Dong Z, Baek JH, Jeon CO, Yao Y, Xuan YH, Zhang J, Jia B (2024) Development of biofertilizers for sustainable agriculture over four decades (1980–2022). Geogr Sustainability 5(1):19–28. https://doi.org/10.1016/j.geosus.2023.09.006

Chapter 15
The Synergy of Nanotechnology and Biofertilisers: A Pathway to Sustainable Growth

Shalini Singh, Aastha Dubey, Prasun Kumar Singh, Chinmaya Sahoo, and Ankit Saini

Abstract The agricultural sector faces great challenges, nutrient deficiencies, environmental change, and a vast range of biotic and abiotic stresses as a result of climate change. As a result of this, traditional farming practices, which depend more on synthetic agrochemicals for plant protection, are becoming less effective, resulting in undesirable outputs to the crops and the environment. Nanobiofertilisers have emerged as an effective solution for improving agricultural sustainability. Nanotechnology, particularly through the development of nanomaterials, offers new approaches for enhancing agricultural production practices and farming systems. Nanotechnology, especially through development of nanomaterials, offers new approaches to augment farming practices in agriculture. Nanofertilisers and nanopesticides are developed to reduce the usage of conventional plant protection chemicals, reduce nutrient waste, and increase crop yields. These nanomaterials are designed to be target specific, reducing the risk of harming nontarget plants and the release of contaminants into the environment and hence, benefiting eco-friendly agriculture. Another alternative is the use of biofertilisers which is gaining momentum as one of the sustainable agricultural practices. These live microorganisms enhance soil fertility by directly expanding the root surface area, making access to nutrients easier and transport within the plant, promoting growth and plant disease suppression. However, the easy and commonplace use of biofertilisers is impacted by the irregular supply and other methods for their quality control. Integrating

S. Singh · A. Dubey · P. K. Singh
School of Agriculture, Galgotias University, Greater Noida, UP, India

C. Sahoo (✉)
Dr. Rajendra Prasad Central Agricultural University, Pusa, Bihar, India

A. Saini
Dr. KSG Akal College of Agriculture, Eternal University, Baru Sahib, Sirmaur, Himachal Pradesh, India

Chaudhary Sarwan Kumar Himachal Pradesh Krishi Vishvavidyalaya, Palampur, Himachal Pradesh, India

A. D. Nciizah et al. (eds.), *Biofertilizers for Sustainable Agriculture in Africa*, Sustainability Sciences in Asia and Africa, https://doi.org/10.1007/978-981-95-6009-7_15

nanotechnology with biofertilisers can be considered to be an advanced cutting-edge in agriculture. Bionanofertilisers, which include green synthesis, nanocoating, or encapsulation of beneficial microbes, increase the efficiency of nutrient utilisation for crops and reduce wastage of nutrients. To be able to introduce and market bionanofertilisers, the usage of nanofertilisers for sustainable agricultural practices requires a wide range of field testing, campaigns for sensitisation, training, and adherence to standardised methods of production.

Keywords Bio-nanofertilisers · Nanotechnology · Nanocoating · Encapsulation · Nanomaterials

15.1 Introduction

In developing countries, more than 60% of the population depends on agriculture for a living, making agriculture a vital sector of the economy and a significant contributor to gross domestic product (GDP), which denotes the total monetary value of all goods and services produced within a country. However, there exists an urgent need to enhance agricultural production through sustainable systems, considering the expanding global population. Numerous techniques have been employed to enhance soil fertility and ensure sustainable agricultural production and productivity (Sagar et al. 2021). For instance, the Green Revolution-driven modern agriculture has significantly increased production and contributed to feeding the world's expanding population. But this achievement has come at a cost. Heavy chemical fertiliser application has been essential to sustaining sufficient food supplies and improving crop production worldwide during the last 60 years (Meena et al. 2016; Chaudhary et al. 2017).

However, overreliance on chemical fertilisers has unintentionally led to environmental contamination, imbalance in nutrients and soil degradation (Bisht and Chauhan 2020). Low fertiliser use efficiency leads to leaching of nutrients into groundwater, emissions of greenhouse gases into the environment, soil acidification, eutrophication, declining soil fertility and productivity, and the loss of biological diversity (Drechsel et al. 2015; Kourgialas et al. 2017; Tomer et al. 2016; Mahanty et al. 2017). Consequently, significant efforts have been made over the past 10 years to switch from chemical synthetic fertilisers to environmentally friendly biofertilisers and nanofertilisers (Liu and Lal 2015; Mikhak et al. 2017; Davarpanah et al. 2016).

Biological fertilisers, which include biofertilisers and biosynthesised nanofertilisers, are the most recent and technologically advanced approach to supplying mineral nutrients to crops. This innovation marks a significant advancement towards sustainable agriculture. A key development lies in the biological mechanisms that enable precise control over nanoparticle morphology. The major advantage of microbial extracellular secretion is the ease of downstream processing, which yields enormous amounts of nanoparticles in a comparatively pure state, devoid of additional cellular proteins linked to the microorganism. The use of fungal enzymes in nanoparticle synthesis also shows considerable promise compared to chemical fertilisers, biological fertilisers offer superior nutrient use efficiency by reducing leaching and delivering nutrients in alignment with plant demands (Pandey and Chandra 2016; Subbarao et al. 2013; Malusà et al. 2016). Advancing the production of microbial inoculants and nanoparticles through safe, economically feasible, and eco-friendly processes is essential for promoting sustainable agricultural practices. Additionally, employing organisms for biosynthesis aligns with the principles of green chemistry, which emphasise safety and the dual role of the organism ecologically as both a reducing and capping agent (Elkhatib et al. 2015; Bagherzade et al. 2017; Hussain et al. 2016).

Microorganisms play a role in the production of biological fertilisers (Malusà et al. 2016; Liu and Lal 2015; Pandey and Chandra 2016; Tomer et al. 2016; Mikhak et al. 2017; Mahanty et al. 2017). Because of their metabolic capabilities, these microbes can directly synthesise microbial inoculants or biosynthesised nanoparticles. This biogenic production aligns with green chemistry principles while providing eco-friendly, high-purity, and cost-effective alternatives to conventional inputs. Biological fertilisers may also be generated by microbial activity that produces metabolic energy using redox processes (Patil and Solanki 2016). Therefore, bacterial, fungal, actinomycetic and plant extract-based biological systems may mediate the formation of these biological fertilisers (Panpatte et al. 2016; Yadav et al. 2015). Consequently, these biological fertilisers can function effectively in the field as biocontrol and biofertiliser agents. Microbes utilised to produce microbial inoculants and nanoparticles could control a variety of diseases both individually and in conjunction with various other microorganisms as biofertilisers and biocontrol agents. Additionally, a few microorganisms have been extensively used in the production of microbial inoculants and nanoparticles (Mani and Mondal 2016; Belal and ElRamady 2016; Thilakarathna and Raizada 2017; Shalaby et al. 2016) (Fig. 15.1).

In developing nations, nanotechnology is a novel strategy that attracts interest in the agriculture industry. Nanoparticles serve as stimuli that enable plants under adverse conditions to activate a variety of defence mechanisms. Because nanoparticles are light, highly soluble, and have a large surface area, they are superior fertilisers to regular salts. Nanoparticles and biofertilisers are combined to create nanobiofertiliser. This method involves encasing biofertilisers in an appropriate nanomaterial. This lessens the negative impacts of external stresses and releases the nutrients into the soil in a controlled way. They boost nutrient availability and uptake, reduce the need for large amounts of chemical fertilisers, are economical, and environmentally friendly (Pudake et al. 2019; Eliaspour et al. 2020).

Nanotechnology, as a rapidly emerging and promising field, has introduced various approaches to boost agricultural production and productivity (El-Shetehy et al. 2021). According to recent research, nanoscale systems with unique characteristics are more readily and efficiently assimilated by plant roots and leaves, which benefits crops and their niche and can be used to create 'smart' and 'sustainable' agricultural systems (Shao et al. 2022; Hong et al. 2021; Li et al. 2016).

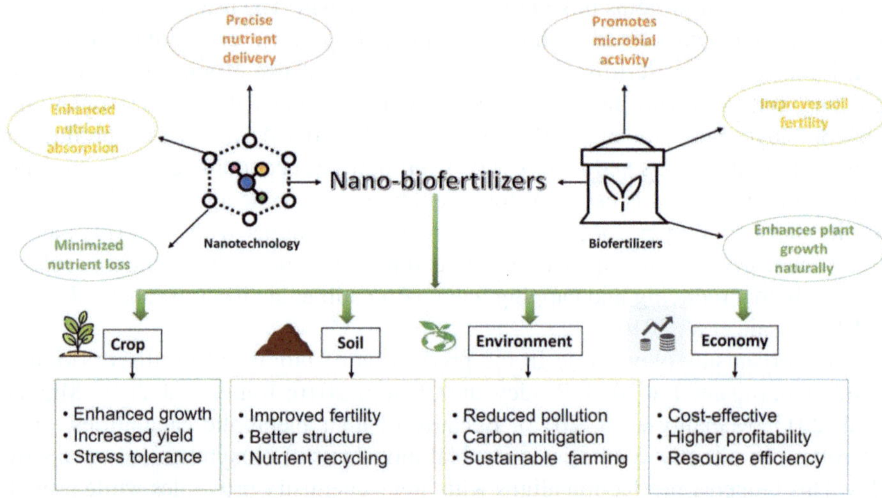

Fig. 15.1 Benefits of nanobiofertilisers (Ali et al. 2021)

Recently, the integration of biofertilisers with nanoparticles to create nanobiofertilisers has emerged as a promising strategy to enhance nutrient use efficiency (NUE) and improve stress resilience in crops. This innovation represents a convergence of nanotechnology and biotechnology in agriculture. Nanobiofertilisers are typically composed of nanoscale materials that are closely associated with beneficial microbes. Their dual purpose is to improve nutrient uptake by plants while minimising environmental impact of conventional farming practices. The advanced formulations may involve plant growth-promoting substances encapsulated within nanoparticles or bacterial cells adhering to the surface of the nanoparticles. This synergistic approach not only addresses the limitations of traditionally used chemical fertilisers but also offers eco-friendly and cost-effective alternatives. Despite the development of various nanobiofertiliser formulations, little is known about their interactions with plants, endophytes, soil bacteria, and soil constituents, particularly at the molecular level (Danish and Hussain 2019).

Further research may be necessary to fully understand the interactions between microorganisms (such as biofertilisers), nanoparticles (such as nanofertilisers), and plants. The beneficial effects of biofertilisers (or agriculturally beneficial microorganisms) in lowering the toxicity of nanoparticles (Boddupalli et al. 2017); plants responses to nanofertilisers or nanoparticles (Siddiqi and Husen 2017b), the connection between biofertilisers and plants (Mahanty et al. 2017; Tomer et al. 2016) and the role that microorganisms and plants play in the biosynthesis of biofertiliser and nanofertilisers (Siddiqi and Husen 2017a) are some examples of these interactions. Further research is required to better understand how bacterial biofertilisers interact with nanoparticles, which are also beneficial micronutrients for plants and microorganisms.

Considering the objective of sustainable agriculture, nanobiofertilisers hold enormous potential for enhancing crop productivity, biological fortification, and resilience to biotic and abiotic stresses because they are needed in considerably smaller amounts than other fertilisers. Therefore, the use of nanobiofertilisers maximises nutrient management while reducing environmental damage.

The purpose of this chapter is to explore how the integration of nanotechnology and biofertilisers can offer a sustainable alternative to conventional fertilisers. It discusses the fundamental concepts underlying the development of nanobiofertiliser, such as the use of microorganisms, nanoparticle synthesis and their effects on plant growth and nutrient availability. The chapter further explains recent advancements in nanoenabling biofertiliser formulations, emphasising their potential to improve crop yields, improve soil fertility, and support environmentally friendly agricultural practices. Special attention is given to their applicability within the African agricultural context. Overall, this chapter contributes to the broader discussion of sustainable agricultural intensification by providing the scientific basis and practical applications of nanobiofertilisers.

15.2 Constituents of Nanobiofertiliser

Nanoparticles and biofertiliser are the two primary components of nanobiofertiliser.

15.2.1 Nanoparticles

These are minuscule particles that typically have a size between 1 and 100 nm. When creating nanobiofertiliser, zinc, silicon, iron, copper, and silver nanoparticles are most often utilised. The chemical makeup, size, shape, and the nanoparticles aggregation phase determine their uptake and buildup in plants. The size of nanoparticles has an inverse relationship with their translocation. Because the receptors differ from plant to plant, they are also associated with the species of the plant. It is possible for a plant to act as both an accumulator and an excluder of various kinds of nanoparticles. Various nanoparticles enhance the growth of plants by a variety of processes (Baazaoui et al. 2021; Fatima et al. 2020). The nanoparticles penetrate the plant roots via plasmodesmata, carrier proteins, or endocytosis when they are added to the rhizosphere. They just diffuse through the stomata and permeate the vascular bundles when administered as a foliar mist spray. They also use apoplastic and symplastic channels to move throughout the plant. The kind of vegetation and the type of nanoparticles determine the migration pathway (Anjum et al. 2019; Pérez-de-Luque 2017).

15.2.1.1 Silicon Nanoparticles (SiNPs)

Silicon is one of the most abundant elements in the Earth's crust and plays a vital role in supporting plant health. It mostly exists in the soil as monosilicic acid [$Si(OH)_4$], which is the form that plants can most easily take up (Thakral et al. 2024). This absorption happens on its own and is affected by the amount of silicon in the soil solution and how fast water is lost through transpiration (Handreck and Jones 1967). Some plants, like *Trifolium incarnatum*, take up silicon, but most plants do not. Silicon makes plants better able to handle a variety of biotic and abiotic stresses, including drought, salinity, and heavy metal toxicity (Ma 2004; Mir et al. 2022). Once it gets into plants, it accumulates in their tissues, strengthens cell walls, controls oxidative stress, and affects hormone activity (Prisa 2023; Basu and Kumar 2021). However, modern farming methods, such as continuous cropping and heavy irrigation, have caused silicon reserves in soils to run out, especially in tropical areas, through leaching, erosion, and harvesting (Meena et al. 2014; Haynes 2014). Conventional silicon fertilisers, often industrial byproducts, tend to have limited availability to plants depending on soil characteristics like pH and organic matter content (Babu et al. 2016). Silicon nanoparticles are a promising new development in nanotechnology in this context. SiNPs are made using methods like agro-waste valorisation or green synthesis (Su et al. 2016; Barani et al. 2024). Because they are so small and have a large surface area, they are better at absorbing and interacting with plants (Yan et al. 2024; Bhat et al. 2021). These particles can also function as carriers for fertilisers, and agrochemicals, improving delivery efficiency (Asgher et al. 2023). They stick to toxic metals in polluted soils, which makes it harder for plants to take them up and boosts their internal antioxidant defences (Emamverdian et al. 2018; Singh et al. 2024). When applied judiciously, SiNPs offer a sustainable approach to crop enhancement and ecological restoration by bridging the strengths of nanotechnology and biofertilisers.

15.2.1.2 Zinc Nanoparticles

Zinc nanoparticles (ZnNPs) have gained significant attention in agriculture due to their ability to enhance crop growth, improve nutrient use efficiency, and support biofortification (Ahmed et al. 2024). These nanoparticles can be synthesised using chemicals or environmentally friendly methods, plant extracts-based synthesis being the most sustainable approach (Sabir et al. 2014; Parmar 2024). Zinc nanoparticles, in combination with biochar, improve soil health, enhance nutrient uptake, and increase plant resilience to abiotic stress (Ahmed et al. 2024). They are important in addressing zinc deficiency by making the micronutrient more bioavailable in soil and crops (Parmar 2024). Zinc nanoparticles positively influence seed germination, seedling vigour, chlorophyll content, and stress tolerance, especially when there is salinity or drought (Timilsina and Chen 2021; Zhou et al. 2023). Their antibacterial properties also contribute to postharvest food safety (Zhou et al. 2023).

However, appropriate dosages must be maintained as excessive concentrations may exhibit phytotoxicity (Timilsina and Chen 2021).

Field studies demonstrate the practical adaptability of zinc nanoparticles in agriculture. Using biosynthesised zinc nanoparticles on amaranth leaves at a level of 10 ppm helped the plant grow better and use nutrients more effectively. In maize, ZnNPs increased yield and zinc uptake, with foliar application being the most effective (Umar et al. 2020). A combination of zinc nanoparticles and zinc-solubilising bacteria significantly improved wheat growth and zinc accumulation without disrupting the balance of microbes (Khan et al. 2024). Their nanoscale size and high surface area allow for reduced chemical input and greater efficiency. Additionally, zinc nanoparticles mitigate salinity stress by enhancing antioxidant activity, modulating plants defence mechanisms (Seleiman et al. 2023; Francis et al. 2024). While beneficial for sustainable agriculture, concerns remain regarding their environmental impacts, warranting further research (Ranjan et al. 2025; Francis et al. 2024).

15.2.1.3 Copper Nanoparticles

Copper nanoparticles (CuNPs) are emerging as a valuable tool in sustainable agriculture because of their nanoscale properties, strong antimicrobial activity, and potential to enhance crop productivity (Bakshi and Kumar 2022). Typically ranging from 5–280 nm, these particles are biosynthesised using plant extracts or microorganisms, offering an environmentally friendly alternative to conventional chemical synthesis methods (Kasana et al. 2016; Rautela et al. 2025). Their small size facilitates efficient uptake by plants, enabling systemic action and effective mitigation of both biotic and abiotic stresses (Rautela et al. 2025).

Copper nanoparticles have demonstrated strong antibacterial and antifungal activity against a wide range of phytopathogens. When applied at concentrations below 100 ppm, they have been reported to improve seed germination, plant vigour, and nutrient assimilation (Kasana et al. 2016; Bhuvaneshwari et al. 2022). In agricultural practice, copper nanoparticles are used not only as nanoparticles and nanofertilisers but also in advanced nanoformulations that enable the slow and targeted release of agrochemicals (Bakshi and Kumar 2022; Rai et al. 2018). Their versatility extends beyond crop protection: they function as nanosensors for pollutant detection and are incorporated into antimicrobial food packaging, demonstrating broad applicability across the agri-food system (Bakshi and Kumar 2022; Rai et al. 2018).

Recent innovations have focused on multifunctional copper-based nanomaterials that act as plant stimulants, regulators, and protectants—helping crops become more resilient to stress while improving yield potential (Su et al. 2024). However, like other metal-based nanomaterials, overuse or poor management of CuNPs can lead to phytotoxicity effects and copper buildup in the soil, with potential implications for soil health and food safety (Tortella et al. 2024). To fully harness their benefits, future research must address long-term environmental interactions, impacts on soil microbial communities and dosage precision (Tortella et al. 2024; Su et al. 2024).

Overall, when applied judiciously within a well-regulated framework, copper nanoparticles hold considerable promise for advancing sustainable agriculture.

15.2.1.4 Iron Nanoparticles

Iron nanoparticles (FeNPs) have emerged as a promising tool for enhancing agricultural productivity due to their ability to support seed germination, improve vegetative growth, and bolster plant resilience under stress conditions (Hafiz et al. 2024). By addressing iron deficiency—a common micronutrient limitation in many crops—iron nanoparticles contribute to more robust plant development, especially in nutrient-poor or stressful environments (Hafiz et al. 2024; Yuan et al. 2018). Research has shown that at low concentrations, iron nanoparticles can positively influence plant morphology by modifying leaf structure, increasing the number of chloroplasts, and enhancing the organisation of vascular bundles (Yuan et al. 2018). These structural improvements are accompanied by an upregulation in photosynthetic capacity and enzymatic activity, both of which are critical for healthy plant function (Satdev 2020).

Studies comparing crops like rice and peanuts show that iron nanoparticles work better than regular iron in fixing iron deficiency, helping plants grow, and reducing stress symptoms. Their nanoscale size allows for efficient uptake through the roots and facilitates transport to aerial parts of the plant in a bioavailable form (Yuan et al. 2018). However, using too many iron nanoparticles needs to be controlled because high amounts can clump together in plant cell walls, which can block nutrient movement and cause harm. To mitigate these risks and enhance the stability of FeNP suspensions, surface functionalisation with biocompatible polymers, such as chitosan and casein, has been explored. These coatings help reduce nanoparticle agglomeration and improve delivery efficiency (Hafiz et al. 2024). With these advancements, iron nanoparticles are gaining recognition as a controlled-release, targeted iron fertiliser alternative that holds significant potential for sustainable agriculture.

15.2.1.5 Silver Nanoparticles

Silver nanoparticles (AgNPs), which are usually between 1 and 100 nm in size (Rai et al. 2009). Recent studies have reported that AgNps can occasionally reach sizes up to 1000 nm, depending on synthesis conditions (Islam et al. 2024; Marathe et al. 2023). While physical and chemical methods are commonly used to synthesise silver nanoparticles, biological methods are becoming more popular because they are better for the environment (Islam et al. 2024). AgNPs are well known for their broad-spectrum antimicrobial activity against bacteria, fungi, viruses, and inflammation in addition to agriculture use, they are widely studied for biomedical applications due to their anti-inflammatory and antimicrobial properties (Prabhu and Poulose 2012; Roy 2017; Islam et al. 2024).

Silver nanoparticles have been shown to enhance seed germination, promote plant growth, and improve photosynthetic efficiency in various crop species (Khan et al. 2023; Kamal 2025). They serve as nanopesticides and nanofertilisers, offering controlled nutrient release and effective disease suppression (Alfosea-Simón et al. 2025). AgNPs have demonstrated efficacy in both in vitro and field conditions, where they contribute to improved plant stress tolerance. Beyond pant systems, AgNPs are also applied in animal husbandry to reduce the number of harmful microbes and help mitigate the development of antibiotic resistance (Perfilyeva et al. 2019). Plant responses to AgNPs vary significantly depending on their size, shape, concentration, and exposure duration. At lower concentrations, AgNPs have been reported to stimulate root and shoot growth, enhance chlorophyll synthesis and upregulate antioxidant enzyme (Sami et al. 2020; Xalxo et al. 2021). However, higher doses may inhibit photosynthesis, interfere with nutrient uptake, and disrupt hormone signalling (Syu et al. 2014; Noori et al. 2024).

Recent studies show that integrating silver nanoparticles with beneficial microbes like Piriformospora indica or plant growth-promoting bacteria (PGPB) can improve the performance and stress tolerance of rice and wheat (Solanki et al. 2023; Karunakaran et al. 2024; Tripathi et al. 2024). However, addition of silicon in combination with AgNPs has been reported to exacerbate their phytotoxic effects in certain cases (Vishwakarma et al. 2019). To ensure environmental safety, the adoption of green synthesis and long-term accumulation is safe to use; it is important to use green synthesis methods that use algae, microbes, or agricultural waste is strongly recommended (Jangid et al. 2025). Further research is necessary to assess the impact of AgNPs on soil microbial communities, ecological safety and their long-term accumulation in the environment (Francis et al. 2024).

15.2.2 Biofertiliser

Beneficial microbes like fungal mycorrhizae, plant growth-promoting rhizobacteria (PGPR), blue-green algae (BGA), or processed plant material are found in biofertiliser (Vafa et al. 2021). The concept of biofertiliser originated with the ancient Greeks and Romans, who described several agricultural techniques to increase plant yield. Microbes were first introduced and promoted as biofertilisers in the late nineteenth century (Malusà et al. 2016). Currently, the agroecosystem is being contaminated by the widespread use of chemical fertilisers and these fertilisers are some of the primary causes of air, soil, and water contamination. There is an urgent need to develop a low-cost, nontoxic and environmentally friendly alternative in order to achieve high agricultural output without any negative consequences on the environment (Sagar et al. 2022). The impacts of the biofertilisers are long-lasting, sustainable, economical, and environmentally beneficial. They enhance nutrient availability in the plant rhizosphere, which promotes their growth (Enrico et al. 2020).

In addition to performing phosphate solubilisation (Ahmed et al. 2021a; Sharma et al. 2013) and potassium solubilisation (Baba et al. 2021), they also release chemicals that stimulate plant growth and fix atmospheric nitrogen in the soil (Kusale et al. 2021a, b). A siderophore released by certain biofertilisers sequesters iron (Nithyapriya et al. 2021; Kusale et al. 2021a, b), rendering it inaccessible to deleterious organisms and preventing their proliferation. They generate a variety of phytohormones that assist in the development of plants, including gibberellins, cytokinins, abscisic acid, and indole-3-acetic acid (Hamid et al. 2021). However, the drawbacks of biofertilisers include their poor quality, poor performance in a changing environment, low shelf life, and high dose requirements to cover a wide area (Kalam et al. 2020; Basu et al. 2021). Numerous problems have been resolved by the use of nanobiofertiliser and agriculture relies on nanotechnology to improve crop growth and yield while safeguarding the natural ecosystem. They preserve the purity of fertiliser and improve its gradual release. They are a sustainable method to reduce the price of chemical synthetic fertiliser and have enduring outcomes on the field (Du et al. 2018) (Fig. 15.2).

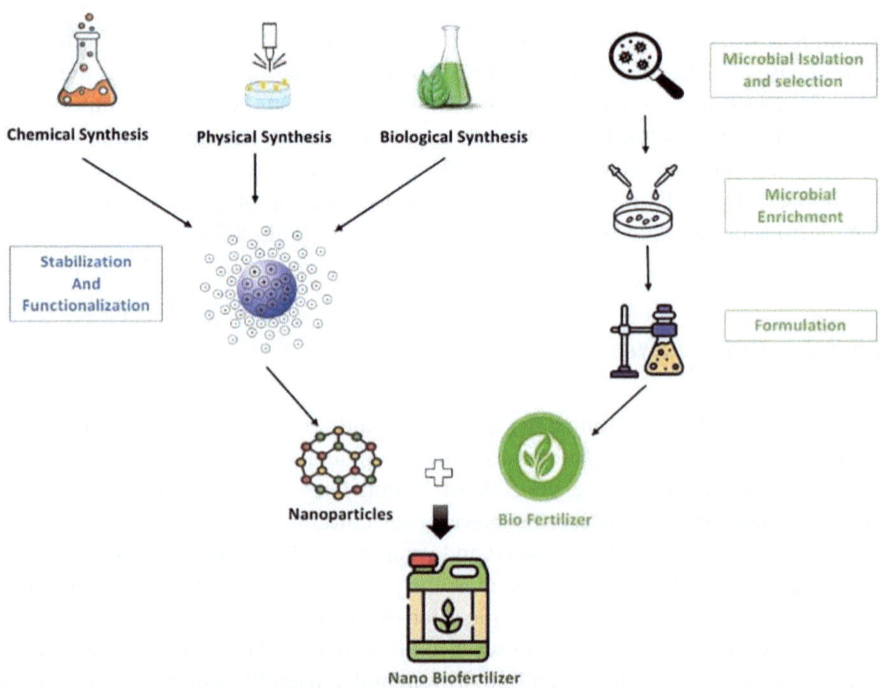

Fig. 15.2 Process of synthesis of nanobiofertiliser (Mahapatra et al. 2022)

15.3 Development of Nano-Based Fertilisers

The selection process for the manufacture of nanofertilisers is based on the desired attributes (primarily size and physicochemical parameters) of the necessary nanomaterials. The following describes the three different methods used to produce nanofertilisers (Khan et al. 2021).

15.3.1 Top–Down Approach

The initial substance is broken down into nanoparticles using physical techniques, including crushing, grinding, etching, milling, and other lithographic techniques in the top–down process of creating nanofertilisers. The top–down method can be used to create large quantities of nanofertilisers from any precursor material; however, it is not appropriate for designer customised nanoparticles because it is a physical technique of creating nanoscale particles. Because of this, only a certain type of nanofertiliser can be made, and this method frequently makes it impossible to manufacture a desired diversity in the shape of nanomaterials/nanofertilisers (Khan et al. 2021; Zulfiqar et al. 2019; Raliya et al. 2017).

15.3.2 Bottom–Up Approach

The bottom–up method forms nanoparticles from atomic and molecular levels using a variety of chemical and physical techniques. To create nanoparticles from gaseous and liquid precursors, physical techniques such as chemical vapour deposition, electrolysis, flame pyrolysis, etc., are employed. Nanofertilisers are made chemically using sol–gel, hydrothermal, microemulsion, and polyol synthesis techniques. The capacity to generate mass-scale nanoparticles with controlled physicochemical characteristics makes the bottom-up approach the most promising technique for creating nanofertilisers (Khan et al. 2021; Zulfiqar et al. 2019). This method has the advantages of being less harmful and more environmentally friendly than the other two.

15.3.3 Biological/Green Synthesis Approach

This method uses a variety of angiosperms, bacteria, algae, and fungi to produce nanofertilisers for plant growth and development. It has the advantage of being less harmful and more environmentally friendly than the other two. The primary drawback of the recovered nanofertiliser from these biological sources is the sluggish rate of nanoparticle formation, which is then stabilised by encapsulation, nanofilms, synthetic carriers, nanotubes, polymer coatings, dispersion in an emulsion, etc. (Mahapatra et al. 2022; Zulfiqar et al. 2019).

15.4 Nanobiofertiliser Formulation for Sustainable Production

Nanobiofertilisers are engineered by integrating the nanoparticles with biofertiliser to combine the synergistic benefits of both technologies. Through the gradual release of nutrients, it can enhance the effectiveness of plant nutrient utilisation. It can enhance the properties of soil by bringing about long-lasting changes to the chemical, physical, and biological attributes of the soil (Table 15.1).

Nanoparticles are encapsulated with biofertiliser as part of the formulation of nanobiofertiliser. It shields the strains of bacteria from mechanical stress, and the slow release of nutrients makes this product even more effective. The process of encapsulation involves allowing cells of biofertiliser to reside within the nanomaterial capsule. It uses a biodegradable, nontoxic substance, like starch and calcium

Table 15.1 Characteristics and advantages of nanofertiliser formulations

Category	Description	Advantages
Nanoscale fertilizer	Fertilizers reformulated into nanoscale size through mechanical or chemical methods.	Requires reduced quantity
		Extended shelf life
		Multifunctional: Acts as pesticides and heavy metal scavengers
Nanoscale additives	Addition of nanoparticles (e.g., fertilizers, micronutrients, or supplements) to bulk macroscale fertilizers (>100 nm). These enhance bulk fertilizer activity.	Improved water retention
		Pathogen control in plants and soil
		Example: Carbon nanotubes (CNT) enhance seed germination and biomass in tomatoes by increasing water intake and moisture content.
Nanoscale coatings	Use of nano-thin films or nanoporous materials (e.g., zeolites, clay, polymer coatings) for controlled nutrient release.	Controlled nutrient release
		Enhanced mechanical robustness and thermal resilience
		Example: Nanoclays serve as fillers for nanocomposite formations

alginate. Starch accelerates the growth of bacterial strains (Vafa et al. 2021; Du et al. 2018).

The process of creating nanobiofertilisers consists of three crucial steps:

(a) Growing the culture of biofertiliser.
(b) Encasing it in nanoparticles.
(c) Assessing its efficiency, purity, quality, and shelf life (Panichikkal et al. 2021a, b).

Additionally, nanobiofertiliser can be made as a microcapsule. PGPR suspension is mixed in a 2:1 ratio with a solution of 1.5% sodium alginate, 3% starch, and 4% bentonite to create its synthesis. After coating the mixture with the crosslinking calcium chloride solution, sterile distilled water is used to rinse the microcapsules (Saberi-Rise and Moradi-Pour 2020). Nanobiofertiliser has also been made using a mixture of salicylic acid and nanoparticles. This approach involves mixing the biofertiliser with salicylic acid (1.5 mM), sodium alginate (2%) and ZnONPs (1 μg/mL). After coating the solution with a 3% calcium chloride solution, 1-mm beads are created, allowed to air dry and then incubated at 4 °C (Panichikkal et al. 2021a, b). An efficient nanobiofertiliser that improves soil fertility can be created by combining the nanoparticles with organic waste, such as cow dung, flowers and residue from the kitchen. After being cleaned with water to get rid of contaminants, the organic waste can be cut into tiny pieces and put through pyrolysis or decomposition. Nanobiofertiliser can be created by combining nanoparticles with this partially degraded or pyrolysed waste (Singh et al. 2019) (Table 15.2).

15.5 Mode of Nanobiofertiliser Application

Efforts to enhance plant growth-promoting fertilisers are driven by the inefficient uptake of traditional fertilisers by crops and the harmful impact of their overuse on soil health (Kalia et al. 2019). Conventional methods often lead to significant nutrient loss without benefiting plants. Nanomaterials, with their higher efficiency, offer a promising alternative but require cautious use to avoid environmental and health risks. A notable advancement is the development of nanobiofertilisers, which combine nanomaterials with PGPR i.e. plant growth promoting rhizobacteria, for targeted and sustained nutrient delivery (Gouda et al. 2018). However, large-scale production and application face challenges due to limited understanding of the relationships between nanoparticles, plant systems, and biofertiliser microflora (Kalia and Kaur 2019). Nanofertilisers can be applied either separately or as integrated nanobiofertilisers, with individual nanoparticles showing significant benefits such as enhanced seed germination, enzyme activity, carbon sequestration, photosynthesis, and nitrogen fixation (Gouda et al. 2018; Khodakovskaya et al. 2013; Lahiani et al. 2013). Various studies have highlighted that multi-walled carbon nanotubes and copper nanoparticles (CuNPs) have improved the growth of crops like tomatoes (Bhanvase et al. 2017). Furthermore, nanoparticles can indirectly promote

Table 15.2 Types of biofertiliser: their microrganisms, functions, and agronomic benefits (Malusà et al. 2016; Basu et al. 2021)

Biofertilizer type	Key microorganism	Functions	Additional benefits
Rhizobium biofertilizer	Rhizobium	Fixes atmospheric nitrogen in symbiotic relationship with legume plants.	Enhances nitrogen availability for legumes.
Azotobacter biofertilizer	Azotobacter	Fixes atmospheric nitrogen independently (free-living).	Solubilizes phosphates, enhances plant hormone production, suppresses plant pathogens, and aids in pesticide degradation.
Azospirillum biofertilizer	Azospirillum	Fixes atmospheric nitrogen and forms a symbiotic relationship with non-legume plants.	Produces antifungal and antimicrobial substances, enhances phytohormone synthesis, and improves systemic resistance in plants.
Phosphorus-solubilizing bacteria biofertilizer	PSB	Solubilizes inorganic and organic phosphorus, making it available to plants.	Chelates cations, lowers soil pH, and mineralizes organic phosphate through phosphatase enzyme production.
Mycorrhizae biofertilizer	Arbuscular Mycorrhizal fungi (AMF)	Enhances nutrient absorption by forming a symbiotic relationship with plant roots.	Improves tolerance to salinity and drought stress, confers disease resistance, and alleviates heavy metal toxicity.

microbial growth by boosting cell viability under stress conditions, encouraging the secretion of stress-alleviating enzymes, increasing nodule formation and protecting inoculants from dehydration. For instance, zinc oxide nanoparticles (ZnO-NPs) improved PGPR siderophore production, while CuNPs enhanced IAA production in a dose-dependent manner (Haris and Ahmad 2017). Two popular modes of application of nanofertilisers are foliar application and soil application.

15.5.1 Foliar Application

Applying nanofertilisers foliarly entails sprinkling them on the leaves, where they immediately get absorbed by the stomata on the leaf surface. Various macro and micronutrients from plants have been transformed into nanoforms and utilised as nanofertilisers. Research on foliar application shows that the cuticle prevents nutrients from being absorbed through the leaf surface, but nanofertilisers increase the effectiveness of nutrient absorption, removing the barrier effect (Hong et al. 2021). The stomata and vascular bundles that move throughout the plant via symplastic and apoplastic routes are thought to be the mechanisms underlying the absorption of nanofertiliser foliar spray (Anjum et al. 2019).

15.5.2 *Soil Application*

Plant growth and soil quality can both be enhanced by the application of nanofertilisers (Mahapatra et al. 2022). Applying nanofertilisers to the soil can be done via spraying or direct application. Following application, crop plants in the rhizosphere absorb nanofertilisers via plasmodesmata, endocytosis, or carrier protein assistance, following which, the nutrients move via apoplastic and symplastic pathways (Akhtar et al. 2022).

15.6 Mode of Action of Nanobiofertiliser

One promising way to address the present issues with environmental and nutrient safety is through nanobiofertilisers. The apoplastic pathway is used by nanobiofertilisers to enter the rhizosphere and reach the vascular tissue, which includes the root's core and central region. Meanwhile, the symplastic pathway is used to get beyond the Casparian strip barrier in this pathway (Sharma et al. 2023). In contrast, phyllospheric application of nanobiofertilisers to the leaves makes them adhere to the phloem translocation pathway. The silicon nanocomposites help plants become more capable of photosynthetic processes by increasing the size of the chloroplasts, the amount of grana, and the amount of chlorophyll (Sharma et al. 2023). Both abiotic and biotic stress have an adverse effect on plant growth by raising the production of highly reactive molecules derived from oxygen, i.e. Reactive Oxygen Species (ROS) and lowering chlorophyll concentration, which damages DNA (Abdelaal et al. 2021). The ability of nanobiofertilisers to replicate the actions of antioxidant enzymes, such as nanoenzymes, can help eliminate oxidative stress. Moreover, the harmful heavy metals would interact with these particles, reducing them, depending on the surface charge, area and small size (Mittal et al. 2020).

The advantages of both biofertilisers and nanofertilisers are combined to create nanobiofertilisers, which offer multiple uses and better results. Because a higher concentration might result in phytotoxicity, the concentration of the nanomaterial utilised is a crucial consideration. Some PGPRs can serve as a remedy to lessen this effect. The extracellular polymeric substance (EPS) produced by Azotobacter salinestris, for example, efficiently traps nanoparticles and forms a metal–EPS complex, guaranteeing their steady presence in the rhizosphere (Ahmed et al. 2021b). Consequently, these EPSs aid in mitigating the adverse effects linked to high concentrations of certain nanoparticles. Enhancing plant growth-promoting qualities, increasing bioavailability and improving nutrient usage efficiency can all be done to further maximise the advantages of nanobiofertilisers. Nanobiofertilisers can additionally enhance the production of PGP compounds and secondary metabolites, safeguard microbial inoculants from dehydration, boost cell viability, and prolong shelf life.

15.7 Role of Nanobiofertilisers in Improving Soil Fertility

Both the environment and human wellness are negatively impacted by the unregulated application of chemical fertilisers, which also pollute water resources, induce eutrophication, have toxic effects on soil health and leaching of nutrients. For agriculture to be productive, economically viable and sustainable, soil health management must be improved. Proper monitoring of soil nutrients is essential to satisfy crop needs without adversely affecting the ecosystem (Mahapatra et al. 2022; Janmohammadi et al. 2016).

Nanobiofertiliser improves and promotes the fertility of the soil through a number of methods, such as:

(a) Replenishing nutrients through fixation of nitrogen and solubilisation of phosphorus.
(b) Enhancing the nutrient absorption capacity of soil.
(c) Synthesising siderophores, which chelate iron and trap heavy metals to prevent plants from accessing them.
(d) Developing solutes and phytohormones that are compatible.
(e) Conserving the moisture content of soil.
(f) Fostering biotic and abiotic stress resistance or tolerance in plants.

Furthermore, they improve plants' efficiency of utilising nutrients (Shafiei-Masouleh 2022; Sharma et al. 2021).

15.8 Response of Plants to Nanobiofertiliser

The application of nanobiofertiliser has been shown to boost plant growth, development and tolerance to stress. Coating nanoparticles with biofertilisers improves the efficacy of biofertilisers and results in the slow and sustained release of nanoparticles in the rhizosphere of the plant, maintaining fertiliser dissolution and lowering the possibility of leaching (Ali et al. 2021). They further enhance the crop quality by boosting the synthesis of secondary metabolites, such as enzymatic antioxidants (peroxidase, superoxide dismutase and catalase) and nonenzymatic antioxidants (flavonoids and phenols), which extend the shelf life of vegetables and fruits and have a number of health benefits (Paschalidis et al. 2021). When applied to plants, biofertiliser and nanoparticles work in concert to produce stronger results. Nanobiofertilisers cause several plant systems to become active, which improves plant development and yield. Additionally, they lessen or inhibit the formation of pathogenic organisms in the rhizosphere of plants and lessen the negative impacts of harmful substances. In addition to replenishing vital nutrients in the soil, nanobiofertilisers facilitate bioremediation. They lessen the detrimental effects that ROS have on the plant system, promote the production of osmolytes, antioxidants and stress-associated proteins, and preserve the structure and functionality of cells

(Khundi et al. 2025). Additionally, they sustain membrane transporter activity and boost hormone synthesis (Shcherbakova et al. 2017; Vedamurthy et al. 2021). Nanobiofertilisers activate the antioxidant mechanisms of the plants, safeguarding the cell membrane and organelles from the damaging consequences of stress. Additionally, they produce fewer stress hormones (abscisic acid) along with additional growth hormones, such as indole acetic acid and cytokinin. These modifications improve the plant's ability to withstand stress and raise the likelihood that it will establish as a crop in adverse and unfavourable environmental conditions (Bibi et al. 2022).

15.9 Potential Application of Nanobiofertiliser in Africa

Nanobiofertilisers present a good opportunity to address critical agricultural challenges in Africa, including poor soil fertility, low input efficiency, and climate-related stress (Irewale et al. 2024). Dominated by smallholder systems, African agriculture often suffers from poor fertility and limited access to conventional fertilisers (Thierfelder et al. 2015). The integration of nanotechnology with biofertilisers enables controlled nutrient release, enhanced uptake, and improved microbial activity—making it a sustainable solution for such constraints (Rai et al. 2023).

Sub-Saharan soils are typically acidic, low in organic matter, and deficient in micronutrients like zinc and iron (Kihara et al. 2020). According to Guardiola-Márquez et al. 2023, nanoformulations of these nutrients, particularly zinc oxide and iron oxide, combined with plant growth-promoting rhizobacteria (PGPR), have demonstrated improved root development and nutrient efficiency—especially valuable for resource-poor farmers. Nanobiofertilisers also enhance plant tolerance to abiotic stress such as salinity and drought (Sharma et al. 2023). For instance, nanosilicon and nanocopper formulations boost crop resilience in arid zones where traditional inputs are less effective when encapsulated with mycorrhizae or *Bacillus* spp., (Shalaby et al. 2022). Africa's microbial diversity provides a strong foundation for region-specific formulations. Leveraging native strains in nanobiofertiliser development aligns with agroecological and regenerative agriculture principles (Dimkpa et al. 2023).

To enable broader adoption, policy support, and investment in research infrastructure are essential. Institutions like the Forum for Agricultural Research in Africa (FARA) (Mkomwa et al. 2022) and initiatives such as the African Union's Agenda 2063 offer pathways for integrating nanobiofertiliser innovations into national agricultural strategies. Pilot trials, public–private partnerships, and clear regulatory frameworks will be key to translating research into practice (Jalango et al. 2024).

15.10 Limitations and Future Prospects for Nanobiofertiliser

It is projected that the utilisation of nanobiofertiliser will rise as agriculture advances. Nonetheless, the formulation needs to be friendly to the environment, plants, and mankind. Despite the fact that several articles describe the creation of nanobiofertiliser, no thorough investigation has been conducted into how it functions to enhance the development of plants (Patel et al. 2023). Biofertilisers and nanoparticles are the key constituents of nanobiofertiliser. Nanoparticles are currently being evaluated and are not yet widely used (Zhou et al. 2020; Bhari et al. 2021). The creation and use of nanobiofertiliser have received less attention up to this point. Nanoparticles and biofertiliser must be correctly combined to create nanobiofertiliser. The recommendations for future use are as follows:

1. To create nanobiofertiliser that is superior, inexpensive, durable, and easy to use, sophisticated tools and procedures must be used.
2. To evaluate the influence of nanobiofertilisers on human health and well-being, a meticulous examination should be conducted.
3. It is essential to develop a multifunctional nanobiofertiliser that works well for a variety of crops.
4. It is necessary to perform an economic analysis of nanobiofertiliser. It is important to test the nanobiofertiliser's suitability for different soil types and environmental conditions.
5. Field tests should be conducted, and industries should work together to accomplish large-scale production.
6. Farmers must be made aware of the drawbacks of synthetic fertilisers in addition to how nanobiofertiliser can lower costs while producing long-lasting results.

15.11 Conclusion

The use of synthetic fertilisers is compromising the health of human beings and disrupting the ecological balance. Plant growth is enhanced by nanobiofertilisers, which also boost the plants' nutritional value, shelf life, yield and defence against stresses, both abiotic and biotic. By triggering many processes, the application of nanobiofertilisers improves crop development and yield while preserving the amount of soil nutrients. Enhancing both the volume and quality of products is mostly dependent on nanobiofertilisers. Because nanobiofertilisers travel and function more efficiently, they produce good outcomes at extremely low concentrations and have a slight possibility of bioaccumulating in the soil. Nanobiofertilisers is an emerging component of climate-smart agricultural systems and contribute extensively to soil fertility. In comparison to a single dose of biofertiliser or nanoparticles, they stimulate soil fertility and depict a unique combination with favourably desirable interactions. Although nanoparticles and biofertilisers are subjects of extensive research, nanobiofertiliser formulation and application have received less attention. When compared to their individual treatments, their combination promises to produce outstanding results. Available research on the molecular genesis and mechanism of action of nanobiofertiliser in plants is not sufficient. To regulate dispersion and release at the intended location, the association between biofertiliser and nanoparticles should be optimised to create nanobiofertiliser. The development of a more effective nanobiofertiliser is necessary, and its practical application should receive further exploration. Its effectiveness is linked to the integration of creative strategies and extensive testing. Understanding how nanobiofertiliser works for agricultural development is essential.

References

Abdelaal K, AlKahtani M, Attia K, Hafez Y, Király L, Künstler A (2021) The role of plant growth-promoting bacteria in alleviating the adverse effects of drought on plants. Biology 10(6):520

Ahmed S, Roy Choudhury A, Roy SK, Choi J, Sayyed RZ, Sa T (2021a) Biomolecular painstaking utilization and assimilation of phosphorus under indigent stage in agricultural crops. In: Antioxidants in plant-microbe interaction, pp 565–588

Ahmed B, Syed A, Rizvi A, Shahid M, Bahkali AH, Khan MS, Musarrat J (2021b) Impact of metal-oxide nanoparticles on growth, physiology and yield of tomato (*Solanum lycopersicum* L.) modulated by Azotobacter salinestris strain ASM. Environ Pollut 269:116218

Ahmed N, Deng L, Narejo M-u-N, Baloch I, Deng L, Chachar S, Li Y et al (2024) Bridging agro-science and human nutrition: zinc nanoparticles and biochar as catalysts for enhanced crop productivity and biofortification. Front Plant Sci 15:1435086. https://doi.org/10.3389/fpls.2024.1435086

Akhtar N, Ilyas N, Meraj TA, Pour-Aboughadareh A, Sayyed RZ, Mashwani ZUR, Poczai P (2022) Improvement of plant responses by nanobiofertilizer: a step towards sustainable agriculture. Nanomaterials 12(6):965

Alfosea-Simón FJ, Burgos L, Alburquerque N (2025) Silver nanoparticles help plants grow, alleviate stresses, and fight against pathogens. Plants 14(3):428. https://doi.org/10.3390/plants14030428

Ali SS, Darwesh OM, Kornaros M, Al-Tohamy R, Manni A, El-Shanshoury AERR, Metwally MA, Elsamahy T, Sun J (2021) Nanobiofertilizers: synthesis, advantages, and applications. In: Biofertilizers. Woodhead Publishing, pp 359–370

Anjum S, Anjum I, Hano C, Kousar S (2019) Advances in nanomaterials as novel elicitors of pharmacologically active plant specialized metabolites: current status and future outlooks. RSC Adv 9(69):40404–40423

Asgher M, Rehaman A, Nazar Ul Islam S, Khan NA (2023) Multifaceted roles of silicon nano particles in heavy metals-stressed plants. Environ Pollut 341:122886

Baazaoui N, Sghaier-Hammami B, Hammami SB, Khefacha R, Chaari S, Elleuch L, Messaoud M, Abdelly C (2021) A handbook guide to better use of nanoparticles in plants. Commun Soil Sci Plant Anal 52(4):287–321

Baba ZA, Hamid B, Sheikh TA, Alotaibi SH, El Enshasy HA, Ansari MJ, Zuan ATK, Sayyed RZ (2021) Psychrotolerant Mesorhizobium sp. isolated from temperate and cold desert regions solubilizes potassium and produces multiple plant growth promoting metabolites. Molecules 26(19):5758

Babu T, Tubana BS, Datnoff LE, Yzenas JJ, Maiti K (2016) Release and sorption pattern of Monosilicic acid from silicon fertilizers in different soils of Louisiana: a laboratory incubation study. Commun Soil Sci Plant Anal 47:1559–1577

Bagherzade G, Tavakoli MM, Namaei MH (2017) Green synthesis of silver nanoparticles using aqueous extract of saffron (*Crocus sativus* L.) wastages and its antibacterial activity against six bacteria. Asian Pac J Trop Biomed 7(3):227–233

Bakshi M, Kumar A (2022) Applications of copper nanoparticles in plant protection and pollution sensing: toward promoting sustainable agriculture. In: Copper nanostructures: next-generation of agrochemicals for sustainable agroecosystems. Elsevier

Barani S, Sebastian SP, Dhevagi P, Prasanthrajan M, Suganthi A (2024) Synthesis of silica nanoparticles (SiNPs) from agro-wastes for removal of heavy metals from an aqueous medium—a mini review. Green Chem Lett Rev 17(1). https://doi.org/10.1080/17518253.2024.2422416

Basu S, Kumar G (2021) Exploring the significant contribution of silicon in regulation of cellular redox homeostasis for conferring stress tolerance in plants. Plant Physiol Biochem 166:393–404

Basu A, Prasad P, Das SN, Kalam S, Sayyed RZ, Reddy MS, El Enshasy H (2021) Plant growth promoting rhizobacteria (PGPR) as green bioinoculants: recent developments, constraints, and prospects. Sustainability 13(3):1140

Belal ES, El-Ramady H (2016) Nanoparticles in Water, Soils and Agriculture. In: Ranjan S, Dasgupta N, Lichtfouse E (eds) Nanoscience in Food and Agriculture 2. Sustainable Agriculture Reviews, vol 21. Springer, Cham. https://doi.org/10.1007/978-3-319-39306-3_10

Bhanvase BA, Shende TP, Sonawane SH (2017) A review on graphene–TiO2 and doped graphene–TiO2 nanocomposite photocatalyst for water and wastewater treatment. Environ Technol Rev 6(1):1–14

Bhari R, Kaur M, Sarup Singh R (2021) Chicken feather waste hydrolysate as a superior biofertilizer in agroindustry. Curr Microbiol 78(6):2212–2230

Bhat JA, Rajora N, Raturi G, Sharma S, Dhiman P, Sanand S, Shivaraj SM, Sonah H, Deshmukh R (2021) Silicon nanoparticles (SiNPs) in sustainable agriculture: major emphasis on the practicality, efficacy and concerns. Nanoscale Adv 3:4019–4028. https://doi.org/10.1039/D1NA00233C

Bhuvaneshwari V, Ramasamy NK, Kumar S, Kalaivani S, Vaidehi D, Kumar DK (2022) Antimicrobial activity of copper nanomaterials: current status and future perspectives. In: Copper nanostructures: next-generation of agrochemicals for sustainable agroecosystems. Elsevier

Bibi F, Ilyas N, Arshad M, Khalid A, Saeed M, Ansar S, Batley J (2022) Formulation and efficacy testing of bio-organic fertilizer produced through solid-state fermentation of agro-waste by Burkholderia cenocepacia. Chemosphere 291:132762

Bisht N, Chauhan PS (2020) Excessive and disproportionate use of chemicals cause soil contamination and nutritional stress. In: Soil contamination. IntechOpen

Boddupalli A, Tiwari R, Sharma A, Singh S, Prasanna R, Nain L (2017) Elucidating the interactions and phytotoxicity of zinc oxide nanoparticles with agriculturally beneficial bacteria and selected crop plants. Folia Microbiol 62:253–262

Chaudhary S, Dheri GS, Brar BS (2017) Long-term effects of NPK fertilizers and organic manures on carbon stabilization and management index under rice-wheat cropping system. Soil Tillage Res 166:59–66

Danish M, Hussain T (2019) Nanobiofertilizers in crop production. In: Nanotechnology for agriculture: crop production & protection. Springer, pp 107–118

Davarpanah S, Tehranifar A, Davarynejad G, Abadía J, Khorasani R (2016) Effects of foliar applications of zinc and boron nano-fertilizers on pomegranate (*Punica granatum* cv. Ardestani) fruit yield and quality. Sci Hortic 210:57–64

Dimkpa C, Adzawla W, Pandey R, Atakora WK, Kouame AK, Jemo M, Bindraban PS (2023) Fertilizers for food and nutrition security in sub-Saharan Africa: an overview of soil health implications. Front Soil Sci 3:1123931

Drechsel P, Heffer P, Magen H, Mikkelsen R, Wichelns D (2015) Managing water and fertilizer for sustainable agricultural intensification

Du C, Abdullah JJ, Greetham D, Fu D, Yu M, Ren L, Li S, Lu D (2018) Valorization of food waste into biofertiliser and its field application. J Clean Prod 187:273–284
Eliaspour S, Seyed Sharifi R, Shirkhani A, Farzaneh S (2020) Effects of biofertilizers and iron nano-oxide on maize yield and physiological properties under optimal irrigation and drought stress conditions. Food Sci Nutr 8(11):5985–5998
Elkhatib EA, Mahdy AM, Salama KA (2015) Green synthesis of nanoparticles by milling residues of water treatment. Environ Chem Lett 13:333–339
El-Shetehy M, Moradi A, Maceroni M, Reinhardt D, Petri-Fink A, Rothen-Rutishauser B, Mauch F, Schwab F (2021) Silica nanoparticles enhance disease resistance in Arabidopsis plants. Nat Nanotechnol 16(3):344–353
Emamverdian A, Ding Y, Xie Y, Sangari S (2018) Silicon mechanisms to ameliorate heavy metal stress in plants. Biomed Res Int 2018(1):8492898
Enrico JM, Piccinetti CF, Barraco MR, Agosti MB, Eclesia RP, Salvagiotti F (2020) Biological nitrogen fixation in field pea and vetch: response to inoculation and residual effect on maize in the Pampean region. Eur J Agron 115:126016
Fatima A, Singh S, Prasad SM (2020) Interaction between copperoxide nanoparticles and plants: uptake, accumulation and phytotoxicity. In: Sustainable agriculture reviews 41: nanotechnology for plant growth and development. Springer, pp 143–161
Francis DV, Abdalla AK, Mahakham W, Sarmah AK, Ahmed ZF (2024) Interaction of plants and metal nanoparticles: exploring its molecular mechanisms for sustainable agriculture and crop improvement. Environ Int 190:108859
Gouda S, Kerry RG, Das G, Paramithiotis S, Shin HS, Patra JK (2018) Revitalization of plant growth promoting rhizobacteria for sustainable development in agriculture. Microbiol Res 206:131–140
Guardiola-Márquez CE, López-Mena ER, Segura-Jiménez ME, Gutierrez-Marmolejo I, Flores-Matzumiya MA, Mora-Godínez S et al (2023) Development and evaluation of zinc and iron nanoparticles functionalized with plant growth-promoting rhizobacteria (PGPR) and microalgae for their application as bio-nanofertilizers. Plants 12(20):3657
Hafiz NA, Raja Abdul Malek RNEH, Osman MS, So'aib MS, Rosdi H, Abu Bakar NF, Burhan NS, Abdul Wahab MS, Ismail AK, Lisak G, Liang L (2024) Enhancing agricultural productivity through iron oxide nanoparticle priming: opportunities, challenges, and surface modification strategies. J Adv Res Micro Nano Eng 22(1):56–67. https://doi.org/10.37934/armne.22.1.5667
Hamid B, Zaman M, Farooq S, Fatima S, Sayyed RZ, Baba ZA, Sheikh TA, Reddy MS, El Enshasy H, Gafur A, Suriani NL (2021) Bacterial plant biostimulants: a sustainable way towards improving growth, productivity, and health of crops. Sustainability 13(5):2856
Handreck KA, Jones LH (1967) Uptake of Monosilicic acid by Trifolium Incarnatum (L.). Aust J Biol Sci 20:483–486
Haris Z, Ahmad I (2017) Impact of metal oxide nanoparticles on beneficial soil microorganisms and their secondary metabolites. Int J Life-Sci Sci Res 3(April):1020–1030
Haynes RJ (2014) A contemporary overview of silicon availability in agricultural soils. J Plant Nutr Soil Sci 177:831–844
Hong J, Wang C, Wagner DC, Gardea-Torresdey JL, He F, Rico CM (2021) Foliar application of nanoparticles: mechanisms of absorption, transfer, and multiple impacts. Environ Sci Nano 8(5):1196–1210
Hussain I, Singh NB, Singh A, Singh H, Singh SC (2016) Green synthesis of nanoparticles and its potential application. Biotechnol Lett 38:545–560
Irewale AT, Dimkpa CO, Agunbiade FO, Oyetunde OA, Elemike EE, Oguzie EE (2024) Unlocking sustainable agricultural development in Africa via bio-Nanofertilizer application-challenges, opportunities and prospects. Sci Afr 25:e02276
Islam MM, Kumar A, Kumar A, Raikwar S (2024) Addressing toxicity concerns: state-of-the-art synthesis methods and emerging multifaceted applications of silver nanoparticles. Curr Nanomed 14. https://doi.org/10.2174/0124681873309773240515043837

Jalango DA, Tesfaye LA, Yaregal Y, Ndetu V, Ogutu E (2024) Strategic opportunities for scaling climate smart agriculture investments: the policy landscape in Eastern and Southern Africa
Jangid H, Joshi HC, Dutta J, Ahmad A, Alshammari MB, Hossain K, Pant G, Kumar G (2025) Advancing food safety with biogenic silver nanoparticles: addressing antimicrobial resistance, sustainability, and commercial viability. Food Chem X 26:102298. https://doi.org/10.1016/j.fochx.2025.102298
Janmohammadi M, Navid A, Segherloo AE, Sabaghnia N (2016) Impact of nano-chelated micronutrients and biological fertilizers on growth performance and grain yield of maize under deficit irrigation condition. Biologija 62(2)
Kalam S, Basu A, Ahmad I, Sayyed RZ, El-Enshasy HA, Dailin DJ, Suriani NL (2020) Recent understanding of soil acidobacteria and their ecological significance: a critical review. Front Microbiol 11:580024
Kalia A, Kaur H (2019) Nano-biofertilizers: harnessing dual benefits of nano-nutrient and biofertilizers for enhanced nutrient use efficiency and sustainable productivity. In: Nanoscience for sustainable agriculture. Springer, pp 51–73
Kalia A, Sharma SP, Kaur H (2019) Nanoscale fertilizers: harnessing boons for enhanced nutrient use efficiency and crop productivity. In: Nanobiotechnology applications in plant protection, vol 2. Springer, pp 191–208
Kamal KP (2025) Silver nanoparticles: synthesis, characterization, and emerging applications in agriculture and biomedicine for enhancing crop production and human health. J Agric Sustain Environ 4(1):1–37. https://doi.org/10.56556/jase.v4i1.1140
Karunakaran A, Fathima Y, Singh P, Beniwal R, Singh J, Ramakrishna W (2024) Next-generation biofertilizers: nanoparticle-coated plant growth-promoting bacteria biofertilizers for enhancing nutrient uptake and wheat growth. Agriculture 14(4):517
Kasana RC, Panwar NR, Kaul RK, Kumar P (2016) Copper nanoparticles in agriculture: biological synthesis and antimicrobial activity. Springer
Khan ST, Adil SF, Shaik MR, Alkhathlan HZ, Khan M, Khan M (2021) Engineered nanomaterials in soil: their impact on soil microbiome and plant health. Plants 11(1):109
Khan S, Zahoor M, Khan RS, Ikram M, Islam NU (2023) The impact of silver nanoparticles on the growth of plants: the agriculture applications. Heliyon 9(6)
Khan MA, Sarfraz W, Ditta A (2024) Applications of nano-based fertilizers, pesticides, and biosensors in sustainable agriculture and food security. In Molecular Impacts of Nanoparticles on Plants and Algae (pp. 277–303). Academic Press
Khodakovskaya MV, Kim BS, Kim JN, Alimohammadi M, Dervishi E, Mustafa T, Cernigla CE (2013) Carbon nanotubes as plant growth regulators: effects on tomato growth, reproductive system, and soil microbial community. Small 9(1):115–123
Khundi Q, Jiang Y, Sun Y, Rui Y (2025) Nanofertilizers for sustainable African agriculture: a global review of agronomic efficiency and environmental sustainability. Nanomaterials 15(5):390
Kihara J, Bolo P, Kinyua M, Rurinda J, Piikki K (2020) Micronutrient deficiencies in African soils and the human nutritional nexus: opportunities with staple crops. Environ Geochem Health 42(9):3015–3033
Kourgialas NN, Karatzas GP, Koubouris GC (2017) A GIS policy approach for assessing the effect of fertilizers on the quality of drinking and irrigation water and wellhead protection zones (Crete, Greece). J Environ Manag 189:150–159
Kusale SP, Attar YC, Sayyed RZ, El Enshasy H, Hanapi SZ, Ilyas N, Elgorban AM, Bahkali AH, Marraiki N (2021a) Inoculation of Klebsiella variicola alleviated salt stress and improved growth and nutrients in wheat and maize. Agronomy 11(5):927
Kusale SP, Attar YC, Sayyed RZ, Malek RA, Ilyas N, Suriani NL, Khan N, El Enshasy HA (2021b) Production of plant beneficial and antioxidants metabolites by Klebsiella variicola under salinity stress. Molecules 26(7):1894
Lahiani MH, Dervishi E, Chen J, Nima Z, Gaume A, Biris AS, Khodakovskaya MV (2013) Impact of carbon nanotube exposure to seeds of valuable crops. ACS Appl Mater Interfaces 5(16):7965–7973

Li H, Ye X, Guo X, Geng Z, Wang G (2016) Effects of surface ligands on the uptake and transport of gold nanoparticles in rice and tomato. J Hazard Mater 314:188–196
Liu R, Lal R (2015) Potentials of engineered nanoparticles as fertilizers for increasing agronomic productions. Sci Total Environ 514:131–139
Ma JF (2004) Role of silicon in enhancing the resistance of plants to biotic and abiotic stresses. Soil Sci Plant Nutr 50(1):11–18. https://doi.org/10.3389/fpls.2022.819658
Mahanty T, Bhattacharjee S, Goswami M, Bhattacharyya P, Das B, Ghosh A, Tribedi P (2017) Biofertilizers: a potential approach for sustainable agriculture development. Environ Sci Pollut Res 24:3315–3335
Mahapatra DM, Satapathy KC, Panda B (2022) Biofertilizers and nanofertilizers for sustainable agriculture: Phycoprospects and challenges. Sci Total Environ 803:149990
Malusà E, Pinzari F, Canfora L (2016) Efficacy of biofertilizers: challenges to improve crop production. In: Microbial inoculants in sustainable agricultural productivity: Vol. 2: functional applications. Springer, pp 17–40
Satdev, Mandal N (2020) A review on effect of copper and iron nanoparticle on agricultural crop. Int J Incl Develop 6(1):31–36
Mani PK, Mondal S (2016) Agri-nanotechniques for plant availability of nutrients. In: Plant nanotechnology: principles and practices. Springer, pp 263–303
Marathe VS, Nikum AP, Marathe GS, Patil SG (2023) A review on silver nanoparticles. Res J Pharmacol Pharmacodyn 15(2):87–95
Meena VD, Dotaniya ML, Coumar V et al (2014) A case for silicon fertilization to improve crop yields in tropical soils. Proc Natl Acad Sci, India, Sect B Biol Sci 84:505–518. https://doi.org/10.1007/s40011-013-0270-y
Meena MD, Joshi PK, Jat HS, Chinchmalatpure AR, Narjary B, Sheoran P, Sharma DK (2016) Changes in biological and chemical properties of saline soil amended with municipal solid waste compost and chemical fertilizers in a mustard–pearl millet cropping system. Catena 140:1–8
Mikhak A, Sohrabi A, Kassaee MZ, Feizian M (2017) Synthetic nanozeolite/nanohydroxyapatite as a phosphorus fertilizer for German chamomile (*Matricariachamomilla* L.). Ind Crop Prod 95:444–452
Mir RA, Bhat BA, Yousuf H, Islam ST, Raza A, Rizvi MA, Charagh S, Albaqami M, Sofi PA, Zargar SM (2022) Multidimensional role of silicon to activate resilient plant growth and to mitigate abiotic stress. Front Plant Sci 13:819658. https://doi.org/10.3389/fpls.2022.819658
Mittal D, Kaur G, Singh P, Yadav K, Ali SA (2020) Nanoparticle-based sustainable agriculture and food science: recent advances and future outlook. Front Nanotechnol 2:579954
Mkomwa S, Kassam A, Bwalya M, Shula RK (2022) The Malabo declaration and agenda 2063: making climate smart agriculture real with conservation agriculture in Africa. In: Conservation agriculture in Africa: climate smart agricultural development. CABI, GB, pp 1–16
Nithyapriya S, Lalitha S, Sayyed RZ, Reddy MS, Dailin DJ, El Enshasy HA, Luh Suriani N, Herlambang S (2021) Production, purification, and characterization of bacillibactin siderophore of Bacillus subtilis and its application for improvement in plant growth and oil content in sesame. Sustainability 13(10):5394
Noori A, Hasanuzzaman M, Roychowdhury R, Sarraf M, Afzal S, Das S, Rastogi A (2024) Silver nanoparticles in plant health: physiological response to phytotoxicity and oxidative stress. Plant Physiol Biochem 209:108538
Pandey V, Chandra K (2016) Agriculturally important microorganisms as biofertilizers: commercialization and regulatory requirements in Asia. In: Agriculturally important microorganisms: commercialization and regulatory requirements in Asia. Springer, pp 133–145
Panichikkal J, Prathap G, Nair RA, Krishnankutty RE (2021a) Evaluation of plant probiotic performance of pseudomonas sp. encapsulated in alginate supplemented with salicylic acid and zinc oxide nanoparticles. Int J Biol Macromol 166:138–143
Panichikkal J, Puthiyattil N, Raveendran A, Nair RA, Krishnankutty RE (2021b) Application of encapsulated Bacillus licheniformis supplemented with chitosan nanoparticles and rice starch

for the control of Sclerotium rolfsii in *Capsicum annuum* (L.) seedlings. Curr Microbiol 78:911–919

Panpatte DG, Jhala YK, Shelat HN, Vyas RV (2016) Nanoparticles: the next generation technology for sustainable agriculture. In: Microbial inoculants in sustainable agricultural productivity: Vol. 2: functional applications. Springer, pp 289–300

Parmar K (2024) Green synthesis of zinc oxide nanoparticles and its application as a green fertilizer in agriculture. Holistic Appr Environ 14(3):109–113

Paschalidis K, Fanourakis D, Tsaniklidis G, Tzanakakis VA, Bilias F, Samara E, Kalogiannakis K, Debouba FJ, Ipsilantis I, Tsoktouridis G, Matsi T (2021) Pilot cultivation of the vulnerable cretan endemic *Verbascum arcturus* L.(scrophulariaceae): Effect of fertilization on growth and quality features. Sustainability 13(24):14030

Patel C, Singh J, Karunakaran A, Ramakrishna W (2023) Evolution of nano-biofertilizer as a green technology for agriculture. Agriculture 13(10):1865

Patil HJ, Solanki MK (2016) Microbial inoculant: modern era of fertilizers and pesticides. In: Microbial inoculants in sustainable agricultural productivity: vol. 1: research perspectives. Springer, pp 319–343

Pérez-de-Luque A (2017) Interaction of nanomaterials with plants: what do we need for real applications in agriculture? Front Environ Sci 5:12

Perfilyeva AI, Graskova IA, Nozhkina OA, Zabanova NS, Sukhov BG, Shkil NN, Nefedova EV (2019) Current aspects of the use of chemically synthesized compounds of silver nanoparticles in animal husbandry and agrochemistry. Sci Rev Biol Sci 9–10:85–93. https://doi.org/10.21517/1992-7223-2019-9-10-85-93

Prabhu S, Poulose EK (2012) Silver nanoparticles: mechanism of antimicrobial action, synthesis, medical applications, and toxicity effects. Int Nano Lett 2(1). https://doi.org/10.1186/2228-5326-2-32

Prisa D (2023) Increasing plant resistance with silicon applications. World J Adv Res Rev 17(3):602–608. https://doi.org/10.30574/wjarr.2023.17.3.0413

Pudake RN, Chauhan N, Kole C (2019) Nanoscience for sustainable agriculture, vol 711. Springer International Publishing, Cham

Rai M, Yadav A, Gade AK (2009) Silver nanoparticles as a new generation of antimicrobials. Biotechnol Adv 27(1):76–83

Rai M, Ingle AP, Pandit R, Paralikar PV, Shende SS, Gupta IR, Biswas JK, da Silva SS (2018) Copper and copper nanoparticles: role in management of insect-pests and pathogenic microbes. Nanotechnol Rev 7:303–315

Rai PK, Rai A, Sharma NK, Singh T, Kumar Y (2023) Limitations of biofertilizers and their revitalization through nanotechnology. J Clean Prod 418:138194

Raliya R, Saharan V, Dimkpa C, Biswas P (2017) Nanofertilizer for precision and sustainable agriculture: current state and future perspectives. J Agric Food Chem 66(26):6487–6503

Ranjan A, Rajput VD, Shende SS, Saxena P, Prazdnova EV, Sushkova S, Arora J, Chauhan A, Jindal T, Zargar SM, Minkina TM (2025) Eco-friendly fabrication of Zn-based nanoparticles: implications in agricultural advancement and elucidation of toxicity aspects. J Sci Food Agric 105:4130

Rautela G, Rizvi R, Ansari S (2025) Role of biosynthesized copper nanoparticles in agriculture and other fields. J Nanosci Technol 10:990–994

Roy A (2017) Synthesis of silver nanoparticles from medicinal plants and its BiologicalApplications: a review. Res Rev Biosci 12:1–11

Saberi-Rise R, Moradi-Pour M (2020) The effect of Bacillus subtilis Vru1 encapsulated in alginate–bentonite coating enriched with titanium nanoparticles against Rhizoctonia solani on bean. Int J Biol Macromol 152:1089–1097

Sabir S, Arshad M, Chaudhari SK (2014) Zinc oxide nanoparticles for revolutionizing agriculture: synthesis and applications. Sci World J 2014:925494. https://doi.org/10.1155/2014/925494. 8 pages

Sagar A, Rathore P, Ramteke PW, Ramakrishna W, Reddy MS, Pecoraro L (2021) Plant growth promoting rhizobacteria, arbuscular mycorrhizal fungi and their synergistic interactions to counteract the negative effects of saline soil on agriculture: key macromolecules and mechanisms. Microorganisms 9(7):1491

Sagar A, Rai S, Ilyas N, Sayyed RZ, Al-Turki AI, El Enshasy HA, Simarmata T (2022) Halotolerant rhizobacteria for salinity-stress mitigation: diversity, mechanisms and molecular approaches. Sustainability 14(1):490

Sami F, Siddiqui H, Hayat S (2020) Impact of silver nanoparticles on plant physiology: a critical review. In: Hayat S (ed) Sustainable agriculture reviews, vol 41. Springer, pp 111–127. https://doi.org/10.1007/978-3-030-33996-8_6

Seleiman MF, Ahmad A, Battaglia ML, Bilal HS, Alhammad BA, Khan NU (2023) Zinc oxide nanoparticles: a unique saline stress mitigator with the potential to increase future crop production. S Afr J Bot 159:208–218

Shafiei-Masouleh SS (2022) Use of magnetic nanochitosan as biofertilizer to reduce production period in three cyclamen cultivars. J Soil Sci Plant Nutr 22(1):281–293

Shalaby TA, Bayoumi Y, Abdalla N, Taha H, Alshaal T, Shehata S, Amer M, Domokos-Szabolcsy E, El-Ramady H (2016) Nanoparticles, soils, plants and sustainable agriculture. In: Nanoscience in food and agriculture, vol 1. Springer, pp 283–312

Shalaby TA, Bayoumi Y, Eid Y, Elbasiouny H, Elbehiry F, Prokisch J, El-Ramady H, Ling W (2022) Can nanofertilizers mitigate multiple environmental stresses for higher crop productivity? Sustainability 14(6):3480

Shao C, Zhao H, Wang P (2022) Recent development in functional nanomaterials for sustainable and smart agricultural chemical technologies. Nano Converg 9(1):11

Sharma SB, Sayyed RZ, Trivedi MH, Gobi TA (2013) Phosphate solubilizing microbes: sustainable approach for managing phosphorus deficiency in agricultural soils. Springerplus 2:1–14

Sharma A, Patel S, Menghani E (2021) Synthesis, application and prospects of nano-biofertilizers: a reappraisal. J Phytol Res 34(1)

Sharma B, Tiwari S, Kumawat KC, Cardinale M (2023) Nano-biofertilizers as bio-emerging strategies for sustainable agriculture development: potentiality and their limitations. Sci Total Environ 860:160476

Shcherbakova EN, Shcherbakov AV, Andronov EE, Gonchar LN, Kalenskaya SM, Chebotar VK (2017) Combined pre-seed treatment with microbial inoculants and Mo nanoparticles changes composition of root exudates and rhizosphere microbiome structure of chickpea (Cicer arietinum L.) plants. Symbiosis 73:57–69

Siddiqi KS, Husen A (2017a) Recent advances in plant-mediated engineered gold nanoparticles and their application in biological system. J Trace Elem Med Biol 40:10–23

Siddiqi KS, Husen A (2017b) Plant response to engineered metal oxide nanoparticles. Nanoscale Res Lett 12:1–18

Singh P, Ghosh D, Manyapu V, Yadav M, Majumder S (2019) Synergistic impact of iron (III) oxide nanoparticles and organic waste on growth and development of *Solanum lycopersicum* plants: new paradigm in nanobiofertilizer. Plant Arch 19(1):339–344

Singh A, Shrivastava R, Verma S, Agnihotri RK (2024) A review on overcoming plant stress by silicon nanoparticles. J Sci Innov Nat Earth 5:21–25

Solanki S, Gupta S, Kapoor R, Varma A (2023) Chemically synthesized AgNPs and Piriformospora indica synergistically augment nutritional quality in black rice. J Fungi 9(6):611. https://doi.org/10.3390/jof9060611

Su Y, Ji X, He Y (2016) Water-dispersible fluorescent silicon nanoparticles and their optical applications. Adv Mater 28:10567

Su C, Chen A, Liang W, Xie W, Xu X, Zhan X, Zhang W, Peng C (2024) Copper-based nanomaterials: opportunities for sustainable agriculture. Sci Total Environ 926:171948

Subbarao CV, Kartheek G, Sirisha D (2013) Slow release of potash fertilizer through polymer coating. Int J Appl Sci Eng 11(1):25–30

Syu Y, Hung J, Chen J, Chuang H (2014) Impacts of size and shape of silver nanoparticles on Arabidopsis plant growth and gene expression. Plant Physiol Biochem 83:57–64

Thakral V, Raturi G, Sudhakaran S, Mandlik R, Sharma Y, Shivaraj SM, Tripathi DK, Sonah H, Deshmukh R (2024) Silicon, a quasi-essential element: availability in soil, fertilizer regime, optimum dosage, and uptake in plants. Plant Physiol Biochem 208:108459. https://doi.org/10.1016/j.plaphy.2021.108459

Thierfelder C, Rusinamhodzi L, Ngwira AR, Mupangwa W, Nyagumbo I, Kassie GT, Cairns JE (2015) Conservation agriculture in Southern Africa: advances in knowledge. Renew Agric Food Syst 30(4):328–348

Thilakarathna MS, Raizada MN (2017) A meta-analysis of the effectiveness of diverse rhizobia inoculants on soybean traits under field conditions. Soil Biol Biochem 105:177–196

Timilsina A, Chen H (2021) The emerging applications of zinc-based nanoparticles in plant growth promotion. In: Nanotechnology in plant growth promotion and protection. Wiley

Tomer S, Suyal DC, Goel R (2016) Biofertilizers: a timely approach for sustainable agriculture. In: Plant-microbe interaction: an approach to sustainable agriculture. Springer, pp 375–395

Tortella GR, Rubilar O, Fincheira P, Parada J, Caixeta de Oliveira H, Benavides-Mendoza A, Leiva S, Fernández-Baldo MA, Seabra AB (2024) Copper nanoparticles as a potential emerging pollutant: divergent effects in the agriculture, risk-benefit balance and integrated strategies for its use. Emerg Contam 10:100352

Tripathi S, Sharma S, Rai P, Mahra S, Tripathi DK, Sharma S (2024) Synergy of plant growth promoting rhizobacteria and silicon in regulation of AgNPs induced stress of rice seedlings. Plant Physiol Biochem 213:108720. https://doi.org/10.1016/j.plaphy.2024.108720

Umar W, Hameed MK, Aziz T, Maqsood MA, Bilal HM, Rasheed N (2020) Synthesis, characterization and application of ZnO nanoparticles for improved growth and Zn biofortification in maize. Arch Agron Soil Sci 67(9):1164–1176. https://doi.org/10.1080/03650340.2020.1782893

Vafa ZN, Sohrabi Y, Sayyed RZ, Luh Suriani N, Datta R (2021) Effects of the combinations of rhizobacteria, mycorrhizae, and seaweed, and supplementary irrigation on growth and yield in wheat cultivars. Plants 10(4):811

Vedamurthy AB, Bhattacharya S, Das A, Shruthi SD (2021) Exploring nanomaterials with rhizobacteria in current agricultural scenario. In: Advances in nano-fertilizers and nano-pesticides in agriculture. Woodhead Publishing, pp 487–503

Vishwakarma K, Singh VP, Prasad SM, Chauhan DK, Tripathi DK, Sharma S (2019) Silicon and plant growth promoting rhizobacteria differentially regulate AgNP-induced toxicity in Brassica juncea: implication of nitric oxide. J Hazard Mater 390:121806

Xalxo R, Yadu B, Chandrakar V, Kumar M, Chandra J, Keshavkant S (2021) Silver nanoparticles and their morpho-physiological responses on plants. In: Plant responses to nanomaterials: recent interventions, and physiological and biochemical responses. Springer, pp 183–216

Yadav A, Kon K, Kratosova G, Duran N, Ingle AP, Rai M (2015) Fungi as an efficient mycosystem for the synthesis of metal nanoparticles: progress and key aspects of research. Biotechnol Lett 37:2099–2120

Yan G, Huang Q, Zhao S, Xu Y, He Y, Nikolic M, Nikolic N, Liang Y, Zhu Z (2024) Silicon nanoparticles in sustainable agriculture: synthesis, absorption, and plant stress alleviation. Front Plant Sci 15:1393458. https://doi.org/10.3389/fpls.2024.1393458

Yuan J, Chen Y, Li H et al (2018) New insights into the cellular responses to iron nanoparticles in Capsicum annuum. Sci Rep 8:3228. https://doi.org/10.1038/s41598-017-18055-w

Zhou P, Adeel M, Shakoor N, Guo M, Hao Y, Azeem I, Li M, Liu M, Rui Y (2020) Application of nanoparticles alleviates heavy metals stress and promotes plant growth: an overview. Nano 11(1):26

Zhou X-Q, Hayat Z, Zhang D-D, Li M-Y, Hu S, Wu Q, Cao Y-F, Yuan Y (2023) Zinc oxide nanoparticles: synthesis, characterization, modification, and applications in food and agriculture. PRO 11(4):1193. https://doi.org/10.3390/pr11041193

Zulfiqar F, Navarro M, Ashraf M, Akram NA, Munné-Bosch S (2019) Nanofertilizer use for sustainable agriculture: advantages and limitations. Plant Sci 289:110270

Chapter 16
Biosynthesis of Nanoparticles for Enhancing Plant Growth, Development, and Yield

Thobo Motlhalamme, Amani Gabriel Kaningini, and Mashapa Elvis Malobane

Abstract The extensive usage of conventional chemical fertilisers in agriculture to enhance crop yields has subsequently led to many unintended effects, such as water and air pollution and soil quality degradation. For this reason, innovative technologies such as nanotechnology have emerged to help produce quality and sufficient food sustainably for the ever-growing world population. Nanotechnology, as the science of the production of nanoparticles with a size less than 100 nm has been adopted among other solutions. Nanoparticles can play a vital role in enhancing plant nutrition use efficiency, protection development, and yield. Nanoparticles are applied in small quantities due to their sizes, thus reducing the risk of leaching and unintended release in the environment. The synthesis of nanoparticles is traditionally done through chemical and physical methods that involve toxic chemicals, which can also lead to detrimental effects on the environment in the long run. Thus, biosynthesis using living systems such as plant extracts, fungi, and algae has equipped nanoparticles with their eco-friendliness and cytocompatibility, which are key for their usage in the field of agriculture as nanofertilisers for targeted delivery of several minerals that are hard to supply using conventional methods. Although their mechanism of action is not yet well understood, their benefits have been highlighted in several studies that have reported beneficial effects of several nanoparticles such as calcium carbonate, magnesium oxide, copper oxide, iron oxide, and zinc oxide in increasing tomato, chickpea, groundnut and soybean growth, yield, and fruit quality. Therefore, this chapter focuses on the use of green synthesised nanoparticles for plant fertilisation.

Keywords Nanofertilisers · Crop nutrition · Green synthesis · Environmentally friendly

---

T. Motlhalamme (✉) · A. G. Kaningini · M. E. Malobane
Department of Agriculture and Animal Health, College of Agriculture and Environmental Sciences, University of South Africa, Florida, South Africa

A. D. Nciizah et al. (eds.), *Biofertilizers for Sustainable Agriculture in Africa*,
Sustainability Sciences in Asia and Africa, https://doi.org/10.1007/978-981-95-6009-7_16

16.1 Introduction

Chemical fertilisers play an essential role in enhancing crop productivity and soil fertility. However, their intensified use reduces soil fertility, alters soil pH, decreases organic matter and leads to soil, water, and air pollution, which negatively affects the environment (Pahalvi et al. 2021; Savci 2012). This poses a threat to agricultural sustainability as the demand for food and feed continues to grow (Zulfiqar et al. 2024). In addition, the production of chemical fertilisers requires high energy, which has the potential to increase their greenhouse gas emission footprint (Eyhorn et al. 2019; Janmohammadi et al. 2018). Thus, the development of novel technologies, such as nanotechnology, can help achieve sustainable food production (Zulfiqar et al. 2024).

Nanotechnology refers to the study of particles with dimensions between 1 and 100 nm. Nanoparticles have gained a reputation in several fields of study due to their various applications (Hulla et al. 2015; Misra et al. 2010), including cancer therapy (Misra et al. 2010), environmental studies (Ali Mansoori et al. 2008), cosmetics (Mu and Sprando 2010), food and agriculture industry (He et al. 2019), and dentistry (Moraes et al. 2021). In agriculture, nanoparticles can be applied as fertilisers to promote plant growth and yield. Many authors have reported the positive impact on crop yield in agriculture on many crops, including tomato, chickpea, groundnuts, and cowpea (Motlhalamme et al. 2023; Nelwamondo et al. 2023). Additionally, because of their small size and high surface area, which increases their reactivity with microbial cells, some nanoparticles have antimicrobial properties that make them good candidates for pesticide formulations (Behl et al. 2024; Singh et al. 2022).

Nanoparticle synthesis follows two methods: the top–down methods, which refer to the processes that involve breaking down bulk materials into their monomers and the bottom–up synthesis, which involves the processes in which atoms react with other substrates to produce nanoparticles (Jamkhande et al. 2019). However, for many decades, these methods have used high vacuum, pressure, temperature and toxic chemicals (Altammar, 2023). Therefore, modern studies aim to synthesise nanoparticles at ambient temperature, a neutral pH, and reasonable expenses, without using hazardous chemicals that can adversely harm the environment (Chopra et al. 2022; Parveen et al. 2016). Thus, making nanoparticle synthesis using living entities a suitable alternative for the application of nanoparticles in agriculture (Bahrulolum et al. 2021).

Green synthesis of nanoparticles has become an extremely interesting subject, extending their application in many fields (Hussain et al. 2016; Ramesh et al. 2014). Nanoparticles produced using green methods are much safer for human usage and the environment, owing to the use of less hazardous chemicals (Kumar and Yadav 2009). Several living organisms, including bacteria, fungi, algae, and plants have been exploited to date for the synthesis of nanoparticles (Pandit et al. 2022).

However, green synthesis using plant extracts gives good results, offering particles with diverse structures, high stability, and suitable sizes (Jeevanandam et al. 2022). Plants such as *Allium sativum*, *Aloe vera*, *Carica papaya*, *Hyphaene thebaica*, *Moringa oleifera*, *Athrixia phylicoides* have been used in the synthesis process due to their secondary metabolites like flavonoids, proteins, alkaloids, phenols, polyphenols, sugars, saponins, tannin, and terpenoids that act as capping or reducing agents of nanoparticles (Kaningini et al. 2022, 2023b; Motlhalamme et al. 2023; Nelwamondo et al. 2023). Thus, this chapter aims to highlight the use of green synthesised nanoparticles in the agricultural field for the enhancement of plant growth and protection.

16.2 Green Synthesis of Nanoparticles

Green synthesis of nanoparticles is a safe, environmentally friendly, and less costly method with a reduced risk of failure when compared to chemical and physical approaches (Banjara et al. 2024; Rasool et al. 2024). Plants, algae, fungi, and bacteria are used in this method, making it a very ecologically friendly choice (Kiarashi et al. 2024). Plants contain phytochemicals such as alkaloids, phenolic acids, polyphenols, proteins, sugars, and terpenoids crucial for nanoparticles synthesis due to their reducing and capping properties (Ettadili et al. 2022).

Plants are increasingly being used in nanoparticle synthesis due to their ability to promote nanoparticle formation and boost the efficacy of the synthesis, as researchers try to establish scalable procedures for monodispersed and stabilised nanoparticles (Iravani 2019). The standard method of producing metallic nanoparticles using plants uses a reducing agent obtained from dehydrated plant material and a metal salt as a precursor (Parveen et al. 2016). Biosynthesised nanoparticles have remarkable photocatalytic activity, chemical stability, and oxidation potential, owing to their varied nature. They are therefore better than physical and chemical approaches. Additionally, the strong antibacterial and antimicrobial qualities of green-produced NPs allow for their application in a variety of sectors (Bahri et al. 2021). Thus, green synthesis has lately received substantial interest because of its ecologically beneficial characteristics (Durán and Seabra 2012). Several plants have already been employed in nanoparticle synthesis. As shown in Table 16.1, various parts of the plant can be used for the preparation of plant extract, however, leaves and fruits are the commonly used parts. The most commonly synthesised nanoparticles include zinc oxide, copper oxide, iron oxide, titanium oxide, gold nanoparticles, and silver nanoparticles.

Table 16.1 Summary of plant species used as reducing agents in nanoparticle synthesis

Plant species	Part of the plant used	Nanoparticle synthesized	Reference
Pinus brutia	Leaves	Zinc oxide	Mahdi Ismail et al. (2023)
Camelia sinensis (Iranian green tea)	Leaves	Zinc oxide	Gonbadi et al. (2023)
Persea americana	Seeds	Gold nanoparticles	Ngungeni et al. (2023)
Thymus vulgaris	Leaves	Zinc oxide	Haspulat Taymaz et al. (2023)
Solanum lycopersicum	Fruit juice	Zinc oxide	Elbrolesy et al. (2023)
Watermelon	Seeds shells	Copper	Ekinci et al. (2023)
Ephedra alata (alenda)	Aerial parts	Copper oxide	Atri et al. (2023)
Spinacia oleracea	Leaves	Copper oxide	Thandapani et al. (2023)
Murraya koenigii	Leaves	Calcium oxide	Bano and Pillai (2020)
Persea americana	Fruit peel	Cadmium sulfide	Turki et al. (2022)
Dalbergia sissoo (Indian rosewood)	Leaves	Magnesium oxide	Khan et al. (2020)
Psidium guajava	Leaves	Titanium oxide	Santhoshkumar et al. (2014)
Lawsonia inermis (Henna) *and Gardenia jasminoides* (Cape jasmine)	Leaves	Iron nanoparticle	Naseem and Farrukh (2015)
Trigonella foenum-graecum (fenugreek)	Leaves	Iron oxide	Vergheese and Vishal (2018)
Moringa oleifera	Leaves	Iron oxide	Katata-Seru et al. (2018)
Origanum majorana L.	Leaves	Cerium oxide	Aseyd Nezhad et al. (2020)
Acorus calamus (sweet flag)	Rhizomes	Cerium oxide	Altaf et al. (2021)
Camellia sinensis	Leaves	Magnesium	Kumar et al. (2022)
Rhododendron arboreum	Leaves	Magnesium	Singh et al. (2019)
Capsicum annum L.	Fruits(proteins)	Silver nanoparticles	Li et al. (2007)
Jatropha curcas	Green stem (crude latex)	Silver nanoparticles	Bar et al. (2009)
Euphorbia prostrata	Leaves	Silver and TiO_2	Zahir et al. (2015)
Banana (*Musa* spp.)	Stem, leaf and peel	Cadmium sulfide	Zhou et al. (2014)
Cymbopogon citratus	Leaves	Gold	Murugan et al. (2015)
Catharanthus roseus	Leaves	Palladium	Kalaiselvi et al. (2015)
Citrus sinensis	Orange peel	Zinc oxide	Thi et al. (2020)

(continued)

Table 16.1 (continued)

Plant species	Part of the plant used	Nanoparticle synthesized	Reference
Panax ginseng	Leaves	Silver and gold	Singh et al. (2016)
Hibiscus sabdariffa	Flowers	Europium (III) oxide	Kweyama (2018)
Psidium guajava	Leaves	Titanium oxide and Eu-doped TiO_2	Wahid et al. (2022)
Ananas comosus	Leaves	Nickel oxide	Olajire and Mohammed (2020)
Citrus sinensis	Orange peel	Silver and platinum nanoparticles	Castro et al. (2015)
Solanum trilobatum	Leaves	Nickel oxide	Ezhilarasi et al. (2020)
Citrus sinensis	Orange peel	Platinum nanoparticles	Karim et al. (2019)
Persea americana	Fruit peel	Silver and gold nanoparticles; silver-gold alloy	Adebayo et al. (2019)

16.3 Effect of Nanoparticles on Crop Development and Yield

16.3.1 Growth Stimulation Attributes

Nanoparticles can be used as nanofertilisers or nanopesticides. They mainly come in contact with plants through leaves, seed priming, and roots following soil drenching, causing several morphological, physiological, and cellular changes (Ali et al. 2021; Balusamy et al. 2023). These changes can be beneficial to plants as they promote growth and development. The effects of nanoparticles on crops depend on their size, composition, surface coating, as well as the applied concentration (Ali et al. 2021). The use of nanofertilisers reduces nutrient losses and while increasing nutrient use efficiency due to their size, allowing easy transport and distribution between cells through the plasmodesmata channels (Iqbal 2019; Solanki et al. 2015). Their ability to promote plant germination, growth, development, and yield has been reported in the literature. They can stimulate photosynthesis through their beneficial effects on chlorophyll and carotenoid biosynthesis (Landa 2021). Most common nanofertilisers include metal and metal oxides of different nutrient formulations, which promote different plant mechanisms. Copper, cerium, silica, zinc, titanium, iron, and silver, among others, have been applied to plants such as wheat, peanut, lavender, soybean, tomato, barley, etc. (Arora et al. 2022; Hyder et al. 2023; Servin et al. 2015).

Johns (2018), reported that silicon nanoparticles increased the germination rate, root length, and root weight of soybean, maize, wheat, and lupine at concentrations lower than 500 mg/L. Also, Motlhalamme et al. (2025) reported increased total flavonoids, lycopene, and vitamin C contents in tomatoes after application of $CaCO_3$ nanoparticles. In a separate study, the application of $CaCO_3$ nanoparticles led to an improved gas exchange in tomato plants (Motlhalamme et al. 2023). Similarly, the growth parameters of *Cicer arietinum* were increased by iron oxide nanoparticles Kaningini et al. (2023).

16.3.2 Nanoparticles Phytotoxicity

Nanoparticles present various beneficial effects on plant growth and yield. However, they also present certain levels of phytotoxicity which depend on their composition, size, concentration, and plant species (Jogaiah et al. 2021). Research shows that lower concentrations of nanoparticles are associated with reduced phytotoxicity, highlighting the importance of optimising nanoparticle dosage to mitigate adverse effects on plant systems (Rizwan et al. 2021). Also, depending on their chemical composition and due to their small size and large surface area, certain nanoparticles are readily taken up by plant cells and can be toxic to plants (Khan et al. 2021). Metal-based nanoparticles tend to be more toxic to plants compared to their carbon-based and polymeric counterparts (Peng et al. 2024). Also, different plant species have varied reactions towards nanoparticle toxicity. While various detoxification mechanisms that plants employ to counteract nanoparticles-induced toxicity have been reported, these mechanisms are not common to all plants (Khan et al. 2021). These mechanisms include the production of enzymatic and nonenzymatic antioxidants that allow plants to mitigate oxidative stress caused by the production of high levels of reactive oxygen species (Singh et al. 2024).

16.4 Nanoparticles and Plant Water Use Efficiency

Climate change remains one of the main threats to food security in Africa and the world at large (Nhemachena et al. 2020; Pickson and Boateng 2022). The effect of climate change on food security is severe, mostly in arid and semi-arid regions of the world (Hirwa et al. 2022; Mirzabaev et al. 2023). In recent years, climate change has led to an increase in food insecurity in many countries in Africa (Masipa 2017; Oyelami et al. 2023). One of the key impacts of climate change on food production is by negatively impacting plants water use efficiency (Chaudhry and Sidhu 2022; Farooq et al. 2022). This is due to the increase in carbon dioxide (CO_2) concentrations and temperatures, decline in precipitation, and variations in humidity (Hatfield and Dold 2019; Kabir et al. 2023). Climate change has been reported to decrease the water use efficiency (WUE) of many crops, thus resulting in a decline in yields (Liu et al. 2020; Mbewu et al. 2024). WUE in crops is defined as biomass or grain production per amount of water used (Hatfield and Dold 2019; Cooley et al. 2022; Hoover et al. 2023). In order to increase crop yields under climate change, there's a need to increase WUE (Hatfield and Dold 2019; Harder et al. 2023). Cropping systems and management have the potential to enhance crop WUE and lead to an increase in crop yields (Hatfield et al. 2011; Alharbi et al. 2024; Luo et al. 2024).

The application of nanoparticles (NPs) as part of sustainable practices has been reported to enhance the WUE of many crops (Ghiyasi et al. 2023; Wahab et al. 2023; Raza et al. 2024). Several studies have been conducted within the African continent to demonstrate the potential benefits of NPs on WUE under different conditions. For example, a study by Rakgotho et al. (2022) in South Africa found that the seed priming with green-synthesised ZnO NPs increased sorghum WUE under salt stress conditions by increasing sorghum shoot lengths and fresh weights compared with the control. The same authors found that the shoot lengths and fresh weights increased by 51% and 60% compared to the control, respectively. Another study by Semida et al. (2021) in Egypt also found that the foliar application of ZnO NPs increased eggplant fruit yield under salt conditions by up to 22.6% compared to the control without NPs. An increase in WUE was also reported by Ndou et al. (2023) after the application of hematite (αFe_2O_3) NPs on sorghum under drought stress in South Africa. The foliar application of bio-synthesised calcium carbonate (CaCO3) NPs was reported to increase tomato WUE in South Africa (Motlhalamme et al. 2023). However, the authors urged that the extent of the impact of NPs is also influenced by tomato cultivars (Motlhalamme et al. 2023). Water use efficiency of chickpea was found to be high under Fe_2O_3 NPs compared with control without NPs under optimum water conditions in South Africa. On the contrary, other authors in South Africa found that treatments with no ZnO NPs had higher chickpea WUE compared to treatments with ZnO NPs under optimum water conditions (Kaningini et al. 2024). This suggests that the impact of NPs on crop WUE is specific and can differ across different NPs, even on the same crop and conditions. Thus, general conclusions on WUE cannot be made across different NPs. Generally, the application of NPs enhances crop WUE by improving crop antioxidant defence mechanisms, photosynthetic rate, stomatal conductance, root shoots, and stress-tolerant (Singh et al. 2021a, b; El-Saadony et al. 2022; Zulfiqar et al. 2024).

16.5 Antimicrobial, Cytotoxicity, and Antioxidant Properties of Nanoparticles

Microorganisms are microscopic living organisms that are visible with the help of aided microscopic devices. Most of these microorganisms are unicellular, such as bacteria, but a few are multicellular, such as algae and fungi (Rajak et al. 2020). Though there are beneficial effects linked to them, some present several harmful properties detrimental to animals, humans, and plants. They can cause several plant diseases which lead to reduced nutrient uptake and utilisation, ultimately causing yield losses and plant death if untreated (Nazarov et al. 2020). Several pesticides have been used to counter their effects on plants; however, due to the generation of resistant strains, the use of nanoparticles has been a promising future in this field (Pathak et al. 2022; Ruddaraju et al. 2020). The antimicrobial activity of nanoparticles such as TiO_2, ZnO, and CuO, to mention a few, has been reported in several studies (Janaki et al. 2015; Mageshwari and Sathyamoorthy 2013; Mathew et al. 2018; Pişkin et al. 2013). In controlling diseases, they have different modes of action such as generating reactive oxygen species, releasing ions that are toxic to microorganisms, size exclusion, depth filtration, and releasing antibiotics or antimicrobial peptides. This leads to the destruction of the cell membrane, nucleic acid and enzyme damage, which causes cell death (Ball et al. 2019).

Nanotechnology has been a breakthrough tool for various fields such as pharmaceuticals for antimicrobial, anticancer drug delivery, and tumour detection, as well as agriculture for plant protection and nutrition (Mousa and Bharali 2011; Zhang et al. 2010). The use of nanoparticles is effective in the treatment of infectious diseases in response to antibiotic failure due to microbial resistance (Albadi et al. 2013; Usman et al. 2013). This has led to their usage for drug delivery and the treatment of several diseases (Wilczewska et al. 2012). Potential candidates comprise carbon materials, metal and metal oxide nanoparticles, and magnetic nanoparticles with a size of less than 100 nm in diameter (Cole et al. 2011). For antimicrobial purposes, metal-based nanomaterials such as copper, silver, gold, titanium, and zinc have been investigated and have demonstrated a broad spectrum of activity against various bacterial strains (Usman et al. 2013; Zhang et al. 2010). Despite their dependency on the synthesis method, nanoparticles such as ZnO, Ag, TiO_2, CuO (CuO_2), etc., have demonstrated strong antimicrobial activity, thus making them perfect candidates for the mitigation of microbial multidrug resistance (Wang et al. 2017). Hence, the use of green technology through plants and other natural and living systems in the synthesis of nanoparticles has been proposed as a way to stop the use of toxic chemicals in the production process. This increases their suitability for human consumption in case they remain in the cells of treated plants (Ali et al. 2020; Shukla et al. 2016).

16.6 Conclusion

Green synthesis of nanoparticles has the potential to be used as nanofertilisers. This is because they can enhance plant growth and yield. In addition, the use of NPs can be used to enhance growth yields under drought conditions as they enhance plant water use efficiency. However, the impact of nanoparticles on plant growth, development and yield is nanoparticle-specific. Further research should be undertaken to compare green synthesised nanoparticles with commonly used fertlisers.

References

Adebayo AE, Oke AM, Lateef A, Oyatokun AA, Abisoye OD, Adiji IP, Fagbenro DO, Amusan TV, Badmus JA, Asafa TB (2019) Biosynthesis of silver, gold and silver–gold alloy nanoparticles using Persea americana fruit peel aqueous extract for their biomedical properties. Nanotechnol Environ Eng 4:1–15

Albadi J, Mansournezhad A, Abbaszadeh H (2013) CuO-CeO2 nanocomposite: a highly efficient recyclable catalyst for the green synthesis of 1, 8-Dioxooctahydroxanthenes in water. J Chin Chem Soc 60(10):1193–1196

Alharbi S, Felemban A, Abdelrahim A, Al-Dakhil M (2024) Agricultural and technology-based strategies to improve water-use efficiency in arid and semiarid areas. Water 16(13):1842

Ali Mansoori G, Bastami TR, Ahmadpour A, Eshaghi Z (2008) Environmental application of nanotechnology. In: Annual review of nano research. World Scientific Publishing, pp 439–493

Ali MA, Ahmed T, Wu W, Hossain A, Hafeez R, Islam Masum MM, Wang Y, An Q, Sun G, Li B (2020) Advancements in plant and microbe-based synthesis of metallic nanoparticles and their antimicrobial activity against plant pathogens. Nanomaterials 10(6):1146

Ali S, Mehmood A, Khan N (2021) Uptake, translocation, and consequences of nanomaterials on plant growth and stress adaptation. J Nanomater 2021(1):6677616

Altaf M, Manoharadas S, Zeyad MT (2021) Green synthesis of cerium oxide nanoparticles using Acorus calamus extract and their antibiofilm activity against bacterial pathogens. Microsc Res Tech 84(8):1638–1648

Altammar KA (2023) A review on nanoparticles: characteristics, synthesis, applications, and challenges. Frontiers in microbiology, 14, p.1155622

Arora S, Murmu G, Mukherjee K, Saha S, Maity D (2022) A comprehensive overview of nanotechnology in sustainable agriculture. J Biotechnol 355:21–41

Aseyd Nezhad S, Es-haghi A, Tabrizi MH (2020) Green synthesis of cerium oxide nanoparticle using Origanum majorana L. leaf extract, its characterization and biological activities. Appl Organomet Chem 34(2):e5314

Atri A, Echabaane M, Bouzidi A, Harabi I, Soucase BM, Chaâbane RB (2023) Green synthesis of copper oxide nanoparticles using ephedra Alata plant extract and a study of their antifungal, antibacterial activity and photocatalytic performance under sunlight. Heliyon 9(2)

Bahri SS, Harun Z, Hubadillah SK, Salleh WNW, Rosman N, Kamaruddin NH, Azhar FH, Sazali N, Ahmad RAR, Basri H (2021) Review on recent advance biosynthesis of TiO2 nanoparticles from plant-mediated materials: characterization, mechanism and application. IOP Conf Ser Mater Sci Eng 1142(1):012005

Bahrulolum H, Nooraei S, Javanshir N, Tarrahimofrad H, Mirbagheri VS, Easton AJ, Ahmadian G (2021) Green synthesis of metal nanoparticles using microorganisms and their application in the agrifood sector. Journal of Nanobiotechnology, 19(1), p.86

Ball AS, Patil S, Soni S (2019) Introduction into nanotechnology and microbiology. In: Methods in microbiology, vol 46. Elsevier, pp 1–18

Balusamy SR, Joshi AS, Perumalsamy H, Mijakovic I, Singh P (2023) Advancing sustainable agriculture: a critical review of smart and eco-friendly nanomaterial applications. J Nanobiotechnol 21(1):372

Banjara RA, Kumar A, Aneshwari R, Satnami ML, Sinha SK (2024) A comparative analysis of chemical vs green synthesis of nanoparticles and their various applications. Environ Nanotechnol Monit Manag 22:100988

Bano S, Pillai S (2020) Green synthesis of calcium oxide nanoparticles at different calcination temperatures. World J Sci Technol Sustain Develop 17(3):283–295

Bar H, Bhui DK, Sahoo GP, Sarkar P, De SP, Misra A (2009) Green synthesis of silver nanoparticles using latex of Jatropha curcas. Colloids Surf A Physicochem Eng Asp 339(1–3):134–139

Behl K, Jaiswal P, Pabbi S (2024) Recent advances in microbial and nano-formulations for effective delivery and agriculture sustainability. Biocatal Agric Biotechnol 58:103180

Castro L, Blázquez ML, González F, Muñoz JÁ, Ballester A (2015) Biosynthesis of silver and platinum nanoparticles using orange peel extract: characterisation and applications. IET Nanobiotechnol 9(5):252–258

Chaudhry S, Sidhu GPS (2022) Climate change regulated abiotic stress mechanisms in plants: a comprehensive review. Plant Cell Rep 41(1):1–31

Chopra H, Bibi S, Singh I, Hasan MM, Khan MS, Yousafi Q, Baig AA, Rahman MM, Islam F, Emran TB (2022) Green metallic nanoparticles: biosynthesis to applications. Front Bioeng Biotechnol 10:548

Cole AJ, David AE, Wang J, Galbán CJ, Hill HL, Yang VC (2011) Polyethylene glycol modified, cross-linked starch-coated iron oxide nanoparticles for enhanced magnetic tumor targeting. Biomaterials 32(8):2183–2193

Cooley SS, Fisher JB, Goldsmith GR (2022) Convergence in water use efficiency within plant functional types across contrasting climates. Nat Plants 8(4):341–345

Durán N, Seabra AB (2012) Metallic oxide nanoparticles: state of the art in biogenic syntheses and their mechanisms. Appl Microbiol Biotechnol 95:275–288

Ekinci A, Kutluay S, Şahin Ö, Baytar O (2023) Green synthesis of copper oxide and manganese oxide nanoparticles from watermelon seed shell extract for enhanced photocatalytic reduction of methylene blue. Int J Phytoremediation 25(6):789–798

Elbrolesy A, Abdou Y, Elhussiny FA, Morsy R (2023) Novel green synthesis of UV-sunscreen ZnO nanoparticles using Solanum Lycopersicum fruit extract and evaluation of their antibacterial and anticancer activity. J Inorg Organomet Polym Mater:1–10

El-Saadony MT, Saad AM, Soliman SM, Salem HM, Desoky ESM, Babalghith AO, El-Tahan AM, Ibrahim OM, Ebrahim AA, Abd El-Mageed TA, Elrys AS (2022) Role of nanoparticles in enhancing crop tolerance to abiotic stress: a comprehensive review. Front Plant Sci 13:946717

Ettadili FE, Aghris S, Laghrib F, Farahi A, Saqrane S, Bakasse M, Lahrich S, El Mhammedi MA (2022) Recent advances in the nanoparticles synthesis using plant extract: applications and future recommendations. J Mol Struct 1248:131538

Eyhorn F, Muller A, Reganold JP, Frison E, Herren HR, Luttikholt L, Mueller A, Sanders J, Scialabba NEH, Seufert V, Smith P (2019) Sustainability in global agriculture driven by organic farming. Nature sustainability, 2(4), pp.253–255

Ezhilarasi AA, Vijaya JJ, Kaviyarasu K, Zhang X, Kennedy LJ (2020) Green synthesis of nickel oxide nanoparticles using Solanum trilobatum extract for cytotoxicity, antibacterial and photocatalytic studies. Surf Interfaces 20:100553. https://doi.org/10.1016/j.surfin.2020.100553

Farooq MS, Uzair M, Raza A, Habib M, Xu Y, Yousuf M, Yang SH, Ramzan Khan M (2022) Uncovering the research gaps to alleviate the negative impacts of climate change on food security: a review. Front Plant Sci 13:927535

Ghiyasi M, Rezaee Danesh Y, Amirnia R, Najafi S, Mulet JM, Porcel R (2023) Foliar applications of ZnO and its nanoparticles increase safflower (Carthamus tinctorius L.) growth and yield under water stress. Agronomy 13(1):192

Gonbadi M, Sabbaghi S, Rasouli J, Rasouli K, Saboori R, Narimani M (2023) Green synthesis of ZnO nanoparticles for spent caustic recovery: adsorbent characterization and process optimization using I-optimal method. Inorg Chem Commun 158:111460. https://doi.org/10.1016/j.inoche.2023.111460

Harder P, Helgason WD, Johnson B, Pomeroy JW (2023) Crop water use efficiency from eddy covariance methods in cold water-limited regions. Agric For Meteorol 341:109657

Haspulat Taymaz B, Demir M, Kamış H, Orhan H, Aydoğan Z, Akıllı A (2023) Facile and green synthesis of ZnO nanoparticles for effective photocatalytic degradation of organic dyes and real textile wastewater. Int J Phytoremediation 25(10):1306–1317

Hatfield JL, Dold C (2019) Water-use efficiency: advances and challenges in a changing climate. Front Plant Sci 10:103

Hatfield JL, Boote KJ, Kimball BA, Ziska LH, Izaurralde RC, Ort D, Thomson AM, Wolfe D (2011) Climate impacts on agriculture: implications for crop production. Agron J 103(2):351–370

He X, Deng H, Hwang H (2019) The current application of nanotechnology in food and agriculture. J Food Drug Anal 27(1):1–21

Hirwa H, Li F, Qiao Y, Measho S, Muhirwa F, Tian C, Leng P, Ingabire R, Itangishaka AC, Chen G, Turyasingura B (2022) Climate change–drylands–food security nexus in Africa: from the perspective of technical advances, challenges, and opportunities. Front Environ Sci 10:851249

Hoover DL, Abendroth LJ, Browning DM, Saha A, Snyder K, Wagle P, Witthaus L, Baffaut C, Biederman JA, Bosch DD, Bracho R (2023) Indicators of water use efficiency across diverse agroecosystems and spatiotemporal scales. Sci Total Environ 864:160992

Hulla JE, Sahu SC, Hayes AW (2015) Nanotechnology: history and future. Hum Exp Toxicol 34(12):1318–1321. https://doi.org/10.1177/0960327115603588

Hussain I, Singh NB, Singh A, Singh H, Singh SC (2016) Green synthesis of nanoparticles and its potential application. Biotechnol Lett 38:545–560

Hyder S, Ul-Nisa M, Shahid H, Gohar F, Gondal AS, Riaz N, Younas A, de Los Santos-Villalobos S, Montoya-Martínez AC, Sehar A (2023) Recent trends and perspectives in the application of metal and metal oxide nanomaterials for sustainable agriculture. Plant Physiol Biochem:107960

Iravani S (2019) Bio-based synthesis of magnetic nanoparticles and their applications. In Magnetic nanostructures: environmental and agricultural applications (pp. 13-31). Cham: Springer International Publishing

Iqbal MA (2019) Nano-fertilizers for sustainable crop production under changing climate: a global perspective. Sustain Crop Product 8:1–13

Jamkhande PG, Ghule NW, Bamer AH, Kalaskar MG (2019) Metal nanoparticles synthesis: an overview on methods of preparation, advantages and disadvantages, and applications. J Drug Delivery Sci Technol 53:101174. https://doi.org/10.1016/j.jddst.2019.101174

Janaki AC, Sailatha E, Gunasekaran S (2015) Synthesis, characteristics and antimicrobial activity of ZnO nanoparticles. Spectrochim Acta A Mol Biomol Spectrosc 144:17–22

Janmohammadi M, Abdoli H, Sabaghnia N, Esmailpour M, Aghaei A (2018) The effect of iron, zinc and organic fertilizer on yield of chickpea (Cicer artietinum L.) in Mediterranean climate. Acta Universitatis Agriculturae Et Silviculturae Mendelianae Brunensis, 66(1)

Jeevanandam J, Kiew SF, Boakye-Ansah S, Lau SY, Barhoum A, Danquah MK, Rodrigues J (2022) Green approaches for the synthesis of metal and metal oxide nanoparticles using microbial and plant extracts. Nanoscale 14(7):2534–2571

Jogaiah S, Paidi MK, Venugopal K, Geetha N, Mujtaba M, Udikeri SS, Govarthanan M (2021) Phytotoxicological effects of engineered nanoparticles: an emerging nanotoxicology. Sci Total Environ 801:149809

Johns DA (2018) Effect of silicon dioxide nanoparticles on seed germination and growth of four different plant species. Deakin University

Kabir M, Habiba UE, Khan W, Shah A, Rahim S, Patricio R, Ali L, Shafiq M (2023) Climate change due to increasing concentration of carbon dioxide and its impacts on environment in 21st century; a mini review. J King Saud Univ-Sci 35(5):102693

Kalaiselvi A, Roopan SM, Madhumitha G, Ramalingam C, Elango G (2015) Synthesis and characterization of palladium nanoparticles using Catharanthus roseus leaf extract and its application in the photo-catalytic degradation. Spectrochim Acta A Mol Biomol Spectrosc 135:116–119

Kaningini AG, Nelwamondo AM, Azizi S, Maaza M, Mohale KC (2022) Metal nanoparticles in agriculture: a review of possible use. Coatings 12(10):1586

Kaningini AG, Motlhalamme T, Cloete KJ, More GK, Mohale KC, Maaza M (2023a) Fe2O3 nanoparticles from Athrixia phylicoides DC and their effect on Cicer arietinum L. growth performance. Biocatal Agric Biotechnol:102948. https://doi.org/10.1016/j.bcab.2023.102948

Kaningini AG, Motlhalamme T, More GK, Mohale KC, Maaza M (2023b) Antimicrobial, antioxidant, and cytotoxic properties of biosynthesized copper oxide nanoparticles (CuO-NPs) using Athrixia phylicoides DC. Heliyon 9(4):e15265

Kaningini AG, Motlhalamme T, More GK, Azizi S, Mohale KC, Maaza M (2024) Evaluation of the antimicrobial activity and chickpea growth stimulation effects of green synthesized zinc oxide nanoparticles. Plant Stress 12:100442

Karim NA, Rubinsin NJ, Burukan MAA, Kamarudin SK (2019) Sustainable route of synthesis platinum nanoparticles using orange peel extract. Int J Green Energy 16(15):1518–1526

Katata-Seru L, Moremedi T, Aremu OS, Bahadur I (2018) Green synthesis of iron nanoparticles using Moringa oleifera extracts and their applications: removal of nitrate from water and antibacterial activity against Escherichia coli. J Mol Liq 256:296–304

Khan MI, Akhtar MN, Ashraf N, Najeeb J, Munir H, Awan TI, Tahir MB, Kabli MR (2020) Green synthesis of magnesium oxide nanoparticles using Dalbergia sissoo extract for photocatalytic activity and antibacterial efficacy. Appl Nanosci 10:2351–2364

Khan M, Khan MSA, Borah KK, Goswami Y, Hakeem KR, Chakrabartty I (2021) The potential exposure and hazards of metal-based nanoparticles on plants and environment, with special emphasis on ZnO NPs, TiO2 NPs, and AgNPs: a review. Environ Adv 6:100128

Kiarashi M, Mahamed P, Ghotbi N, Tadayonfard A, Nasiri K, Kazemi P, Badkoobeh A, Yasamineh S, Joudaki A (2024) Spotlight on therapeutic efficiency of green synthesis metals and their oxide nanoparticles in periodontitis. J Nanobiotechnol 22(1):21

Kumar V, Yadav SK (2009) Plant-mediated synthesis of silver and gold nanoparticles and their applications. J Chem Technol Biotechnol Int Res Proc Environ Clean Technol 84(2):151–157

Kumar SA, Jarvin M, Inbanathan SSR, Umar A, Lalla NP, Dzade NY, Algadi H, Rahman QI, Baskoutas S (2022) Facile green synthesis of magnesium oxide nanoparticles using tea (Camellia sinensis) extract for efficient photocatalytic degradation of methylene blue dye. Environ Technol Innovation 28:102746

Kweyama Z (2018) Green synthesis of europium (III) oxide nanoparticles using Hibiscus sabdariffa flower extract. University of Zululand

Landa P (2021) Positive effects of metallic nanoparticles on plants: overview of involved mechanisms. Plant Physiol Biochem 161:12–24

Li S, Shen Y, Xie A, Yu X, Qiu L, Zhang L, Zhang Q (2007) Green synthesis of silver nanoparticles using Capsicum annuum L. extract. Green Chem 9(8):852–858

Liu Y, Zhang J, Qin Y (2020) How global warming alters future maize yield and water use efficiency in China. Technol Forecast Soc Chang 160:120229

Luo Z, Zhang S, Zhao Z, Minasny B, Chang J, Huang J, Li B, Shi Z, Wang E, Wang M, Wu Y (2024) Soil-smart cropping for climate-smart production. Geoderma 451:117061

Mageshwari K, Sathyamoorthy R (2013) Flower-shaped CuO nanostructures: synthesis, characterization and antimicrobial activity. J Mater Sci Technol 29(10):909–914

Mahdi Ismail SM, Ahmed SM, Abdulrahman AF, AlMessiere MA (2023) Characterization of green synthesized of ZnO nanoparticles by using pinus brutia leaves extracts. J Mol Struct 1280:135063. https://doi.org/10.1016/j.molstruc.2023.135063

Masipa T (2017) The impact of climate change on food security in South Africa: current realities and challenges ahead. Jàmbá J Disas Risk Stud 9(1):1–7

Mathew S, Ganguly P, Rhatigan S, Kumaravel V, Byrne C, Hinder SJ, Bartlett J, Nolan M, Pillai SC (2018) Cu-doped TiO2: visible light assisted photocatalytic antimicrobial activity. Appl Sci 8(11):2067

Mbewu AD, Elephant DE, Motsi H, Nyambo P (2024) Climate change effects on water footprint of crop production: a meta-analysis. Environ Challenges 17:101033

Mirzabaev A, Kerr RB, Hasegawa T, Pradhan P, Wreford A, von der Pahlen MCT, Gurney-Smith H (2023) Severe climate change risks to food security and nutrition. Clim Risk Manag 39:100473

Misra R, Acharya S, Sahoo SK (2010) Cancer nanotechnology: application of nanotechnology in cancer therapy. Drug Discov Today 15(19–20):842–850

Moraes G, Zambom C, Siqueira WL (2021) Nanoparticles in dentistry: a comprehensive review. Pharmaceuticals 14(8):752

Motlhalamme T, Mohamed H, Kaningini AG, More GK, Thema FT, Mohale KC, Maaza M (2023) Bio-synthesized calcium carbonate (CaCO3) nanoparticles: their anti-fungal properties and application as nanofertilizer on Lycopersicon esculentum growth and gas exchange measurements. Plant Nano Biol 6:100050

Motlhalamme T, Kaningini AG, Maropeng R, Thema FT, Mohale KC, Maaza M (2025) Tomato lycopene, total flavonoids, vitamin C content and mineral composition as affected by biosynthesized calcium carbonate nanoparticles foliar application. J Food Compos Anal 139:107173

Mousa SA, Bharali DJ (2011) Nanotechnology-based detection and targeted therapy in cancer: nano-bio paradigms and applications. Cancer 3(3):2888–2903

Mu L, Sprando RL (2010) Application of nanotechnology in cosmetics. Pharm Res 27:1746–1749

Murugan K, Benelli G, Panneerselvam C, Subramaniam J, Jeyalalitha T, Dinesh D, Nicoletti M, Hwang J-S, Suresh U, Madhiyazhagan P (2015) Cymbopogon citratus-synthesized gold nanoparticles boost the predation efficiency of copepod Mesocyclops aspericornis against malaria and dengue mosquitoes. Exp Parasitol 153:129–138

Naseem T, Farrukh MA (2015) Antibacterial activity of green synthesis of iron nanoparticles using Lawsonia inermis and Gardenia jasminoides leaves extract. J Chem 2015

Nazarov PA, Baleev DN, Ivanova MI, Sokolova LM, Karakozova MV (2020) Infectious plant diseases: etiology, current status, problems and prospects in plant protection. Acta Nat 12(3):46

Ndou N, Rakgotho T, Nkuna M, Doumbia IZ, Mulaudzi T, Ajayi RF (2023) Green synthesis of iron oxide (hematite) nanoparticles and their influence on Sorghum bicolor growth under drought stress. Plants 12(7):1425

Nelwamondo AM, Kaningini AG, Ngmenzuma TY, Maseko ST, Maaza M, Mohale KC (2023) Biosynthesis of magnesium oxide and calcium carbonate nanoparticles using Moringa oleifera extract and their effectiveness on the growth, yield and photosynthetic performance of groundnut (Arachis hypogaea L.) genotypes. Heliyon 9

Ngungeni Y, Aboyewa JA, Moabelo KL, Sibuyi NRS, Meyer S, Onani MO, Meyer M, Madiehe AM (2023) Anticancer, antioxidant, and catalytic activities of green synthesized gold nanoparticles using avocado seed aqueous extract. ACS Omega 8(29):26088–26101

Nhemachena C, Nhamo L, Matchaya G, Nhemachena CR, Muchara B, Karuaihe ST, Mpandeli S (2020) Climate change impacts on water and agriculture sectors in Southern Africa: threats and opportunities for sustainable development. Water 12(10):2673

Olajire AA, Mohammed AA (2020) Green synthesis of nickel oxide nanoparticles and studies of their photocatalytic activity in degradation of polyethylene films. Adv Powder Technol 31(1):211–218. https://doi.org/10.1016/j.apt.2019.10.012

Pahalvi HN, Rafiya L, Rashid S, Nisar B, Kamili AN (2021) Chemical Fertilizers and Their Impact on Soil Health. In: Dar GH, Bhat RA, Mehmood MA, Hakeem KR. (eds) Microbiota and Biofertilizers, Vol 2. Springer, Cham. https://doi.org/10.1007/978-3-030-61010-4_1

Oyelami LO, Edewor SE, Folorunso JO, Abasilim UD (2023) Climate change, institutional quality and food security: sub-Saharan African experiences. Scientific African 20:e01727

Pandit C, Roy A, Ghotekar S, Khusro A, Islam MN, Emran TB, Lam SE, Khandaker MU, Bradley DA (2022) Biological agents for synthesis of nanoparticles and their applications. J King Saud Univ Sci 34(3):101869

Parveen K, Banse V, Ledwani L (2016) Green synthesis of nanoparticles: their advantages and disadvantages. AIP Conf Proc 1724. https://doi.org/10.1063/1.4945168

Pathak VM, Verma VK, Rawat BS, Kaur B, Babu N, Sharma A, Dewali S, Yadav M, Kumari R, Singh S (2022) Current status of pesticide effects on environment, human health and it's eco-friendly management as bioremediation: a comprehensive review. Front Microbiol:2833

Peng W, Li P, Zia M, Bhuiyan SA, Liu Y, Chen D, Tudi M, Gao Y, Yan X, Jia Y (2024) A comparative review on carbon nanomaterials and their impact on plant growth with the lens of green chemistry principles. Glob Chall 8(11):2400008

Pickson RB, Boateng E (2022) Climate change: a friend or foe to food security in Africa? Environ Dev Sustain 24:1–26

Pişkin S, Palantöken A, Yılmaz MS (2013) Antimicrobial activity of synthesized TiO2 nanoparticles. In: International conference on emerging trends in engineering and technology (ICETET'2013), pp 7–8

Rajak BL, Kumar R, Gogoi M, Patra S (2020) Antimicrobial activity of nanomaterials. Nanosci Med 1:147–185

Rakgotho T, Ndou N, Mulaudzi T, Iwuoha E, Mayedwa N, Ajayi RF (2022) Green-synthesized zinc oxide nanoparticles mitigate salt stress in Sorghum bicolor. Agriculture 12(5):597

Ramesh P, Rajendran A, Meenakshisundaram M (2014) Green syntheis of zinc oxide nanoparticles using flower extract cassia auriculata. J Nanosci Nanotechnol 2(1):41–45

Rasool A, Sri S, Zulfajri M, Krismastuti FSH (2024) Nature inspired nanomaterials, advancements in green synthesis for biological sustainability. Inorg Chem Commun 169:112954

Raza MAS, Amin J, Valipour M, Iqbal R, Aslam MU, Zulfiqar B, Muhammad F, Ibrahim MA, Al-Ghamdi AA, Elshikh MS, Iqbal J (2024) Cu-nanoparticles enhance the sustainable growth and yield of drought-subjected wheat through physiological progress. Sci Rep 14(1):14254

Rizwan M, Ali S, ur Rehman MZ, Riaz M, Adrees M, Hussain A, Zahir ZA, Rinklebe J (2021) Effects of nanoparticles on trace element uptake and toxicity in plants: a review. Ecotoxicol Environ Saf 221:112437

Ruddaraju LK, Pammi SVN, sankar Guntuku G, Padavala VS, Kolapalli VRM (2020) A review on anti-bacterials to combat resistance: from ancient era of plants and metals to present and future perspectives of green nano technological combinations. Asian J Pharm Sci 15(1):42–59

Santhoshkumar T, Rahuman AA, Jayaseelan C, Rajakumar G, Marimuthu S, Kirthi AV, Velayutham K, Thomas J, Venkatesan J, Kim S-K (2014) Green synthesis of titanium dioxide nanoparticles using Psidium guajava extract and its antibacterial and antioxidant properties. Asian Pac J Trop Med 7(12):968–976

Savci S (2012) An agricultural pollutant: chemical fertilizer. International Journal of Environmental Science and Development, 3(1):73

Semida WM, Abdelkhalik A, Mohamed GF, Abd El-Mageed TA, Abd El-Mageed SA, Rady MM, Ali EF (2021) Foliar application of zinc oxide nanoparticles promotes drought stress tolerance in eggplant (Solanum melongena L.). Plants 10(2):421

Servin A, Elmer W, Mukherjee A, De la Torre-Roche R, Hamdi H, White JC, Bindraban P, Dimkpa C (2015) A review of the use of engineered nanomaterials to suppress plant disease and enhance crop yield. J Nanopart Res 17:1–21

Shukla PK, Misra P, Kole C (2016) Uptake, translocation, accumulation, transformation, and generational transmission of nanoparticles in plants. In: Plant nanotechnology: principles and practices. Springer, pp 183–218

Singh P, Kim YJ, Yang DC (2016) A strategic approach for rapid synthesis of gold and silver nanoparticles by Panax ginseng leaves. Artif Cells Nanomed Biotechnol 44(8):1949–1957

Singh A, Joshi NC, Ramola M (2019) Magnesium oxide nanoparticles (MgONps): green synthesis, characterizations and antimicrobial activity. Res J Pharm Technol 12(10):4644–4646

Singh P, Arif Y, Siddiqui H, Sami F, Zaidi R, Azam A, Alam P, Hayat S (2021a) Nanoparticles enhances the salinity toxicity tolerance in Linum usitatissimum L. by modulating the antioxidative enzymes, photosynthetic efficiency, redox status and cellular damage. Ecotoxicol Environ Saf 213:112020

Singh RP, Handa R, Manchanda G (2021b) Nanoparticles in sustainable agriculture: an emerging opportunity. J Control Release 329:1234–1248

Singh G, Ramadass K, Sooriyakumar P, Hettithanthri O, Vithange M, Bolan N, Tavakkoli E, Van Zwieten L, Vinu A (2022) Nanoporous materials for pesticide formulation and delivery in the agricultural sector. J Control Release 343:187–206

Singh R, Choudhary P, Kumar S, Daima HK (2024) Mechanistic approaches for crosstalk between nanomaterials and plants: plant immunomodulation, defense mechanisms, stress resilience, toxicity, and perspectives. Environ Sci Nano 11:2324

Solanki P, Bhargava A, Chhipa H, Jain N, Panwar J (2015) Nano-fertilizers and their smart delivery system. In: Nanotechnologies in food and agriculture. Springer, pp 81–101

Thandapani G, Arthi K, Pazhanisamy P, John JJ, Vinothini C, Rekha V, Santhanalakshmi K, Sekar V (2023) Green synthesis of copper oxide nanoparticles using Spinacia oleracea leaf extract and evaluation of biological applications: antioxidant, antibacterial, larvicidal and biosafety assay. Mater Today Commun 34:105248

Thi TUD, Nguyen TT, Thi YD, Thi KHT, Phan BT, Pham KN (2020) Green synthesis of ZnO nanoparticles using orange fruit peel extract for antibacterial activities. RSC Adv 10(40):23899–23907

Turki ZT, Al Hindawi AM, Shiltagh NM (2022) Green synthesis of CdS nanoparticles using avocado Peel extract. NanoWorld J 8(3):73–78

Usman MS, el Zowalaty ME, Shameli K, Zainuddin N, Salama M, Ibrahim NA (2013) Synthesis, characterization, and antimicrobial properties of copper nanoparticles. Int J Nanomed:4467–4479

Vergheese M, Vishal SK (2018) Green synthesis of magnesium oxide nanoparticles using Trigonella foenum-graecum leaf extract and its antibacterial activity. J Pharmacogn Phytochem 7(3):1193–1200

Wahab A, Batool F, Muhammad M, Zaman W, Mikhlef RM, Naeem M (2023) Current knowledge, research progress, and future prospects of phyto-synthesized nanoparticles interactions with food crops under induced drought stress. Sustainability 15(20):14792

Wahid SM, Malini TP, Puthilibai G, Chitradevi S (2022) Green synthesis of europium doped titanium dioxide nanoparticle and its photocatalytic application. AIP Conf Proc 2464(1)

Wang L, Hu C, Shao L (2017) The antimicrobial activity of nanoparticles: present situation and prospects for the future. Int J Nanomedicine 12:1227

Wilczewska AZ, Niemirowicz K, Markiewicz KH, Car H (2012) Nanoparticles as drug delivery systems. Pharmacol Rep 64(5):1020–1037. https://doi.org/10.1016/S1734-1140(12)70901-5

Zahir AA, Chauhan IS, Bagavan A, Kamaraj C, Elango G, Shankar J, Arjaria N, Roopan SM, Rahuman AA, Singh N (2015) Green synthesis of silver and titanium dioxide nanoparticles using Euphorbia prostrata extract shows shift from apoptosis to G0/G1 arrest followed by necrotic cell death in Leishmania donovani. Antimicrob Agents Chemother 59(8):4782–4799

Zhang L, Pornpattananangkul D, Hu C-M, Huang C-M (2010) Development of nanoparticles for antimicrobial drug delivery. Curr Med Chem 17(6):585–594

Zhou GJ, Li SH, Zhang YC, Fu YZ (2014) Biosynthesis of CdS nanoparticles in banana peel extract. J Nanosci Nanotechnol 14(6):4437–4442

Zulfiqar B, Raza MAS, Akhtar M, Zhang N, Hussain M, Ahmad J, Abdel-Maksoud MA, Ebaid H, Iqbal R, Aslam MU, El-Tayeb MA (2024) Combined application of biochar and silicon nanoparticles enhance soil and wheat productivity under drought: insights into physiological and antioxidant defense mechanisms. Curr Plant Biol 40:100424

Chapter 17
Socio-Economic and Policy Obstacles to Biofertiliser Adoption in African Agriculture: A Thematic Chapter

Ebere Lovelyn Udeh

Abstract In African agriculture, biofertilisers offer vital answers to the problems of diminishing soil fertility, food poverty, and environmental degradation. They are a potential and environmentally benign substitute for chemical fertilisers. However, there is still a lack of broad adoption throughout the continent in spite of their demonstrated agronomic and ecological advantages. This chapter looks at the social and policy challenges that stop farmers from using biofertiliser, using research, reports, and examples from different African countries. Low farmer awareness, cultural opposition, poor extension services, restricted market access, and insufficient policy assistance are some of the main challenges. Enabling variables like government–NGO partnerships, community-led demonstration projects, and favourable subsidy regimes are also identified in the review. To enhance comprehension of farmer decision-making and institutional dynamics, three conceptual frameworks are used: the Sustainable Livelihoods Framework, the Theory of Planned Behaviour, and the Diffusion of Innovations. Ethiopian, Nigerian, and Kenyan case studies emphasise the significance of multi-stakeholder cooperation and context-specific interventions. To enable inclusive and sustainable agricultural transformation in Africa, the analysis ends with policy suggestions and identifies research gaps, such as the need for cross-country policy assessments, gender-sensitive evaluations, and longitudinal studies.

E. L. Udeh (✉)
Department of Agriculture and Animal Health, University of South Africa, Florida, South Africa
e-mail: udehel@unisa.ac.za

A. D. Nciizah et al. (eds.), *Biofertilizers for Sustainable Agriculture in Africa*, Sustainability Sciences in Asia and Africa, https://doi.org/10.1007/978-981-95-6009-7_17

17.1 Introduction

Africa confronts a pivotal moment in its agricultural advancement, exacerbated by rising population growth, land degradation, and food insecurity. The continent's agricultural output is suboptimal due to many constraints, such as restricted access to advanced technologies, deteriorated soil quality, and the impacts of climate change (Makinde and Ayoola 2019; FAO 2020). With the population expanding and arable land diminishing, the necessity for sustainable agriculture practices is becoming more pressing.

Biofertilisers offer a viable answer to these issues and include advantageous microorganisms such as nitrogen-fixing bacteria, phosphate-solubilising microbes, and mycorrhizal fungi. These biological agents augment nutrient cycling, enhance soil fertility, and increase crop yields without the detrimental environmental effects linked to artificial fertilisers (Khan et al. 2015; Ayoola 2022). Their capacity to enhance long-term soil health and resilience closely corresponds with worldwide demands for sustainable agricultural intensification, especially in relation to the UN Sustainable Development Goals (SDGs).

Notwithstanding their shown advantages and long-term economic efficiency, the adoption of biofertilisers in African agricultural systems remains remarkably constrained. A range of socio-economic, institutional, and policy-related obstacles persist in hindering their extensive adoption. Factors encompass inadequate awareness and knowledge among farmers, cultural prejudices, deficient infrastructure, ineffective extension services, and insufficient regulatory frameworks (Abebaw et al. 2010; Atuahene et al. 2020; Mfilinge et al. 2014). Moreover, government subsidies typically favour synthetic fertilisers, thereby disincentivizing the adoption of bio-based alternatives (Baah-Boateng 2016).

The shift to more sustainable agricultural inputs, such as biofertilisers, necessitates technical innovation, legislative reforms, stakeholder collaboration, and socio-economic incentives. Because of this, international organisations such as FAO, IFDC, and UNEP have encouraged countries to include biofertilisers in their agricultural policies, especially in relation to climate-smart agriculture and managing soil fertility effectively.

This chapter consolidates existing material to analyse the socio-economic and policy factors affecting biofertiliser use in Africa. This chapter aims to guide future research and policy actions by concentrating on identified difficulties, facilitating variables, and case-specific insights. It seeks to offer a comprehensive grasp of the limitations and possibilities in this domain, thereby fostering a more sustainable and productive agricultural future for the continent.

Africa is at a crucial juncture in its agricultural revolution. The continent faces escalating pressure to guarantee food security and maintain environmental health, as its population is expected to quadruple by 2050. Agricultural land degradation, diminished soil fertility, unpredictable rainfall patterns resulting from climate change, and significant reliance on chemical fertilisers have collectively created an urgent demand for sustainable agricultural solutions. Most African farmers, especially smallholders who are essential to the continent's food supply, function under resource-limited conditions with restricted access to inputs, new technologies, and extension services.

Biofertilisers are natural products that contain helpful microorganisms like nitrogen-fixing bacteria, phosphate-solubilising bacteria, and mycorrhizal fungi, offering an eco-friendly alternative to chemical fertilisers. These bacteria enhance nutrient availability, promote plant growth, and improve soil structure, thereby supporting long-term agricultural sustainability. In contrast to synthetic fertilisers, which frequently deteriorate soil quality and contaminate water resources, biofertilisers are ecologically sustainable and economically advantageous in the long term.

Nonetheless, despite their recognised benefits and conformity with global sustainable development goals (SDGs), the adoption rate of biofertilisers in African agricultural systems remains remarkably low. Multiple interconnected obstacles impede their extensive adoption. This includes socio-economic limitations such as poverty, insufficient agricultural education, and inadequate market access, as well as policy-related challenges like deficient regulatory frameworks, minimal investment in agricultural research and development, and a lack of specific governmental incentives.

This chapter aims to critically analyse the socio-economic and policy factors influencing biofertiliser uptake in Africa. This thematic chapter brings together information from different research papers, policy documents, reports from development agencies, and case studies, unlike studies that rely on collecting new data. It aims to deliver a thorough comprehension of the fundamental obstacles and facilitators of biofertiliser use while pinpointing avenues for subsequent study and policy action.

This chapter seeks to enhance discussions on agricultural innovation and sustainability in Africa by examining these issues from a multidimensional perspective. It aims to provide pragmatic insights for policymakers, development practitioners, researchers, and farmers collaboratively striving for a more sustainable and productive agricultural sector throughout the continent.

17.2 Scope and Structure of the Review

This chapter synthesises a wide range of scholarly sources, including peer-reviewed journal articles, institutional publications, and documented case studies to present a comprehensive understanding of the socio-economic and policy factors influencing the adoption of biofertilisers in African agriculture. Sources were drawn from leading agricultural databases and international development repositories, covering contributions from across Sub-Saharan and North Africa. Reports from organisations such as the Food and Agriculture Organisation (FAO 2020), the International Fertiliser Development Centre (IFDC 2021), and the United Nations Environment Programme (UNEP) were instrumental in capturing policy frameworks and donor-driven intervention models.

The literature reviewed spans multiple disciplines—including agricultural science, rural sociology, policy studies, and development economics—reflecting the multifaceted nature of biofertiliser adoption. Particular attention is given to contextual case studies from Ethiopia, Nigeria, Ghana, Kenya, and South Africa, where research and pilot interventions provide meaningful insights into real-world applications and limitations (Makinde and Ayoola 2019; Atuahene et al. 2020; Abebaw et al. 2010).

This chapter is organised thematically to ensure clarity and depth of analysis. The core themes include the following: (1) socio-economic barriers to adoption such as poverty, education, and market access; (2) institutional and policy dynamics including regulatory frameworks, extension services, and government subsidies; (3) roles of NGOs and international agencies in promoting adoption; (4) insights from successful intervention models; and (5) the application of theoretical frameworks such as the Diffusion of Innovations Theory, Sustainable Livelihoods Framework, and the Theory of Planned Behaviour in understanding farmer decision-making (Rogers 2003; Ajzen 1991; DFID 2001).

Using search terms like 'biofertilizer adoption Africa', 'agricultural policy', 'extension services', and 'sustainable farming', sources were retrieved from top agricultural databases and international development archives, such as Scopus, Web of Science, and Google Scholar. The inclusion criteria prioritised papers with empirical or policy relevance that were published between 2000 and 2023.

In addition to identifying barriers and opportunities, this chapter highlights existing knowledge gaps and future research directions, with a focus on empirical validation, cross-country comparative studies, and longitudinal impact assessments. This structure aims to provide both an analytical framework and a practical roadmap for policymakers, researchers, and stakeholders engaged in sustainable agricultural development.

17.3 Thematic Analysis

17.3.1 Socio-Economic Barriers to Adoption

The implementation of biofertilisers in Africa is markedly hindered by socio-economic issues, including low household incomes, restricted access to finance, insufficient education, and deficient agricultural infrastructure. A multitude of smallholder farmers function under subsistence conditions and are unable to bear the initial expenses of alternative inputs, despite the potential for long-term savings, and improvements in soil health (Makinde and Ayoola 2019). The cost-effectiveness of biofertilisers is exacerbated by inadequate market frameworks and the lack of financial options designed for small-scale farmers (Mogues et al. 2009).

The transmission of knowledge regarding the utilisation of biofertilisers is notably limited in terms of education and awareness. Agriculturalists frequently possess insufficient understanding of application methodologies, advantages, and compatibility with various crops and soil types (Khan et al. 2015). The knowledge gap is worsened by inadequate extension networks, which are either poorly funded or restricted in their outreach, particularly in isolated rural regions (Mfilinge et al. 2014). These socio-economic limitations collectively impede farmers from embracing environmentally sustainable solutions, such as biofertilisers.

17.3.2 Cultural and Social Norms

Cultural attitudes and traditional customs significantly influence farmers' decisions concerning the adoption of agricultural advances. In numerous African communities, traditional agricultural practices are firmly established, and there is significant dependence on chemical fertilisers due to their familiarity and perceived efficacy (Atuahene et al. 2020). Scepticism regarding novel agricultural inputs, particularly those devoid of robust marketing or local endorsements, frequently hinders the uptake of biofertilisers.

Moreover, agricultural societies are generally shaped by the perspectives of peers and specialised knowledge networks. The resistance of important influencers or early adopters within the community might markedly diminish the probability of widespread acceptance. In contrast, community-based demonstration projects and participatory extension approaches have been beneficial in overcoming reluctance and fostering trust in biofertilisers.

17.3.3 Policy and Regulatory Gaps

The lack of policy consistency and particular regulatory frameworks for biofertilisers presents significant obstacles to their commercialisation and widespread use. In many African countries, farming rules have usually supported synthetic fertilisers through financial aid and government buying programs, which makes it hard to encourage more sustainable options.

In regions with biofertiliser legislation, enforcement is frequently inadequate, and methods to ensure product quality are often absent, leading to market scepticism. Countries like Ethiopia and South Africa have commenced the implementation of methods for integrated soil fertility management that include biofertilisers within comprehensive agricultural development plans (FAO 2020). These policy examples illustrate the capacity for public–private collaborations and institutional dedication to facilitate adoption.

17.3.4 Institutional and Infrastructural Constraints

Institutional deficiencies and infrastructural shortcomings continue to pose substantial obstacles to the distribution and efficient utilisation of biofertilisers. Agricultural research organisations in several African nations are inadequately funded and deficient in capacity for product development, quality assessment, and dissemination of biofertiliser breakthroughs (Ayoola 2022). Moreover, extension systems frequently have a lack of personnel and insufficient training in the use of bio-based technology.

Poor road networks, inadequate supply chains, and insufficient storage and transportation facilities impede access to biofertilisers, especially in rural regions. In the absence of dependable distribution systems and consistent supply, farmers are unlikely to utilise biofertilisers sustainably.

17.3.5 Role of International Organisations and NGOs

International organisations and nongovernmental organisations (NGOs) have been instrumental in facilitating the adoption of biofertilisers by providing research funds, enhancing capacity, and engaging in advocacy. Organisations, including FAO, UNEP, IFDC, and AGRA, have launched pilot projects, financed farmer education initiatives, and advocated for the incorporation of sustainable inputs into national policy frameworks (IFDC 2021; FAO 2020).

These groups frequently provide technical skills and financial resources that local governments may be deficient in. Collaborative initiatives between FAO and local ministries have resulted in experimental schemes in East Africa designed to showcase the effectiveness of microbial biofertilisers. Such initiatives often promote multi-stakeholder forums that unite researchers, policymakers, farmers, and private sector participants to establish more inclusive and evidence-based adoption routes.

17.4 Conceptual and Theoretical Frameworks

Understanding how biofertilisers are used in African farming requires knowledge of existing ideas and theories. These frameworks offer a perspective for analysing the motivations, constraints, and behavioural inclinations of farmers. This section examines three pertinent theoretical frameworks that support the socio-economic and policy analysis outlined in this review.

Diffusion of Innovations Theory

The diffusion of innovations theory, formulated by Rogers (2003), elucidates the mechanisms, motivations, and velocity of the dissemination of novel ideas and technology within a social system. The theory delineates five principal criteria that affect the acceptance of innovations: relative advantage, compatibility, complexity, trialability, and observability. Farmers are more likely to use biofertilisers if they see clear benefits compared to chemical options, like being cheaper and better for the environment, if the technology fits with their current farming methods, and if they can test it out on a small scale before fully adopting it. Demonstration plots and farmer field schools have been shown to be useful in improving trialability and observability.

Sustainable Livelihoods Framework
The Sustainable Livelihoods Framework (SLF), established by DFID in 2001, highlights the interaction among five types of capital: human, financial, natural, physical, and social. This model is particularly useful for understanding how a lack of resources or support in different areas affects a household's ability to use new farming methods, like biofertilisers. A farmer may have natural capital (land and water) but may lack human capital (knowledge), financial capital (credit), or social capital (networks for information sharing) necessary to execute a biofertiliser-based approach (Makinde and Ayoola 2019). The SLF emphasises the necessity for comprehensive interventions that enhance capacities across several aspects of farmer livelihoods.

Theory of Planned Behaviour (TPB)
The theory of planned conduct, introduced by Ajzen (1991), asserts that individual actions are influenced by behavioural intentions, which are determined by attitudes toward the conduct, subjective norms, and perceived behavioural control. This hypothesis elucidates why certain farmers may be reluctant to use biofertilisers, even when acknowledging their advantages, within the realm of agricultural innovation. Adverse past experiences, apprehension of failure, or societal pressure from conservative community members may diminish the motivation to adopt. On the other hand, when farmers believe they can successfully use biofertilisers and get support from positive community attitudes and expert advice, they are more likely to start using them (Khan et al. 2015).

These frameworks combined offer a sophisticated comprehension of the diverse factors influencing farmer decision-making processes. They assert that adoption encompasses not just the provision of new inputs but also the necessity of addressing cognitive, social, and institutional aspects of behavioural change. Using these models for biofertiliser adoption methods can help create more suitable and effective solutions.

In addition to functioning separately, these frameworks have significant points of intersection. For example, the Diffusion of Innovations emphasises how peer pressure and visibility cause such habits to spread, whereas the Theory of Planned Behaviour describes how personal beliefs and social norms impact willingness to use biofertilisers. Together, these theories offer a thorough perspective: diffusion of innovations describes how adoption spreads among societies, whereas TPB describes why people choose to use biofertilisers. The sustainable livelihoods framework enhances both by illustrating how resource availability either facilitates or limits these behavioural changes.

17.5 Case Study Insights

Evidence from case studies in Kenya, Nigeria, and Ethiopia shows that tailored strategies based on local conditions significantly influence how well biofertilisers are adopted. These case studies offer practical insights into the socio-economic and institutional factors that either promote or hinder adoption among smallholder farmers.

In Kenya, integrated initiatives for managing soil fertility endorsed by NGOs and agricultural research entities have demonstrated that demonstration plots, coupled with participatory training modules, enhance awareness and confidence among farmers. Community-based groups facilitated farmer field schools that promoted experiential learning, thereby increasing confidence in the utilisation and advantages of biofertilisers. These programs proved most efficacious when they engaged local agricultural leaders and synchronised with seasonal planting schedules (FAO 2020).

In Nigeria, insights from programs conducted by entities like the National Biofertiliser Development Centre (NBDC) demonstrate that government-supported initiatives aimed at smallholder cooperatives can enhance adoption rates. Farmers were more inclined to experiment with and maintain the use of biofertilisers when they were provided with packages that included seeds, training, and microfinance. The absence of continuous policy support and funding discontinuity frequently undermines these initiatives, highlighting the necessity of institutional stability (Makinde and Ayoola 2019).

Ethiopia exemplifies a scenario where community-driven solutions have yielded substantial results. Localised extension services and subsidies for bio-based inputs were implemented as components of comprehensive initiatives to promote climate-smart agriculture. Research indicated that peer learning and community-based training enhanced the perceived worth of biofertilisers, especially in regions facing soil deterioration. Abebaw et al. (2010) underscored that the existence of local champions—esteemed farmers who embrace and advocate for new technologies—was essential in shaping community views and decision-making.

These national examples underscore many overarching success elements. The engagement of reputable local institutions and leaders mitigates scepticism and fosters behavioural transformation. Secondly, when biofertilisers are incorporated into comprehensive agricultural support packages, farmers regard them as less hazardous and more financially feasible. Ultimately, continuity in program design, ensuring that interventions are not isolated but integrated into enduring agricultural policy frameworks, promotes long-term adoption.

These case studies highlight the importance of customising interventions to specific local settings while integrating them into wider institutional and policy frameworks. They emphasise that methods focused on farmers, which build empowerment and trust, are crucial for successfully spreading biofertiliser technology in African agriculture (Table 17.1).

Table 17.1 Comparative case studies on agricultural extension strategies in Kenya, Nigeria, and Ethiopia

Country	Strategy used	Key outcomes	Barriers noted
Kenya	Plots of demonstration and interactive instruction	Enhanced consciousness and acceptance by peers	Issues with scalability and seasonal alignment
Nigeria	Cooperatives with government support and packaged inputs	Enhanced uptake through seed-training-finance packages	Lack of institutional backing and irregular funding
Ethiopia	Subsidies and local extension services	Strong community-based education and increased value perception	Limitations in infrastructure and access

17.6 Knowledge Gaps and Research Needs

Despite heightened focus on the potential of biofertilisers to enhance soil fertility and agricultural sustainability, substantial knowledge deficiencies persist in both empirical research and policy development. A significant gap in data exists about the long-term effects of biofertiliser application in various agro-ecological zones in Africa. The majority of existing research concentrates on short-term effects or pilot interventions, resulting in ambiguity about the durability of benefits over several planting seasons and climatic conditions (Makinde and Ayoola 2019).

Moreover, although anecdotal evidence and small-scale initiatives demonstrate favourable results, there is a paucity of extensive empirical research that quantifies adoption rates, productivity improvements, and cost–benefit evaluations across many agricultural contexts. Research is often fragmented and specialised, limiting its generalisability throughout the continent (FAO 2020; IFDC 2021). There is an urgent need for cross-country comparative research and longitudinal assessments to comprehend the broader dynamics of biofertiliser uptake.

A significant deficiency exists in the assessment of policy tools and governance frameworks. Limited research systematically investigates the impact of various subsidy regimes, regulatory standards, and extension systems on adoption behaviour. The policy landscape is frequently marked by inconsistent execution and redundant mandates among ministries and development organisations, hence complicating program results (Baah-Boateng 2016).

Furthermore, although multi-stakeholder alliances comprising NGOs, government entities, research institutions, and the commercial sector are frequently advocated, their true efficacy remains inadequately examined. Inquiries remain regarding coordination, accountability, and enduring financial viability. Comprehensive policy frameworks that synchronise agricultural innovation with rural development and environmental objectives are infrequently executed in their entirety (Abebaw et al. 2010; Mfilinge et al. 2014).

There is a deficiency of gender-disaggregated data and analysis. Women are integral to agricultural work in Africa, but their unique obstacles and contributions in the adoption of new technology, such as biofertilisers, remain little recorded.

Gender-sensitive research may provide significant insights for the development of inclusive and equitable intervention methods.

Finally, increased scrutiny is required to examine the function of information and communication technologies (ICTs) in augmenting farmer knowledge, overseeing field implementation, and facilitating feedback mechanisms. The use of mobile platforms, digital extension services, and remote sensing to improve the use of biofertilisers in Africa has not been fully taken advantage of.

Rectifying these deficiencies necessitates collaborative endeavours among academic institutions, development organisations, and national governments. Establishing a robust evidence foundation would enhance policy formulation and execution while promoting increased trust and engagement among farmers.

17.7 Conclusion and Policy Implications

Biofertilisers are a vital means of restoring soil fertility, reducing reliance on chemicals, and achieving food security despite increasing ecological and demographic challenges, placing them at the forefront of Africa's sustainable agricultural revolution.

Biofertilisers provide a revolutionary potential for the agricultural future of Africa. Their capacity to improve soil health, augment crop yield, and conform to sustainable agricultural techniques renders them a key asset in combating food insecurity, land degradation, and excessive reliance on chemicals. Nonetheless, actualising this potential necessitates confronting a complex array of socio-economic, institutional, and policy-related obstacles.

This chapter illustrates that insufficient awareness, inadequate farmer education, and restricted market access continue to be significant impediments. Furthermore, in the absence of robust extension networks and financing access, even knowledgeable and motivated farmers may find it challenging to incorporate biofertilisers into their practices. Governments and development partners must prioritise inclusive policy measures that reduce entry barriers for smallholders.

The primary policy action involves the reorganisation of agricultural subsidies. Existing frameworks unfairly favour chemical fertilisers, unintentionally deterring investment in sustainable alternatives. Allocating a segment of these subsidies to biofertilisers could encourage uptake and enhance long-term soil health. Moreover, governmental investments ought to focus on establishing regulatory frameworks to guarantee product quality, eradicate counterfeit inputs, and enhance consumer confidence.

The enhancement of agricultural extension systems is equally crucial. Training and equipping extension personnel with the necessary tools to advocate for and facilitate biofertiliser application can address knowledge deficiencies and empower farmers. Integrating biofertilisers into the curricula of agricultural colleges and technical institutes can cultivate a new generation of informed practitioners.

Collaborative strategies, including governments, NGOs, research institutions, and the commercial sector are crucial. These collaborations ought to facilitate local biofertiliser production, enhance capacity building, establish monitoring and evaluation systems, and incorporate digital technologies for farmer engagement. Moreover, inclusive methods must account for gender dynamics and the specific issues encountered by women farmers, who constitute a substantial segment of the agricultural labour force yet frequently experience limited access to resources and decision-making authority.

Ultimately, policy formulation and execution must be grounded in empirical evidence. Additional longitudinal studies and multinational assessments are required to evaluate the economic, environmental, and social effects of biofertiliser usage. These insights will be essential in formulating strategies that are both contextually relevant and scalable.

In conclusion, realising the complete potential of biofertilisers in Africa necessitates a transformative approach to agricultural development—one that prioritises sustainability, inclusion, and innovation. Through concerted efforts and smart investments, biofertilisers can serve as a fundamental element in Africa's pursuit of resilient and self-sufficient food systems.

References

Abebaw D, Fentie Y, Kassa B (2010) The impact of a food security program on household food consumption in Northwestern Ethiopia: a matching estimator approach. Food Policy 35(4):286–293. https://doi.org/10.1016/j.foodpol.2010.01.002

Ajzen I (1991) The theory of planned behavior. Organ Behav Hum Decis Process 50(2):179–211

Atuahene F, Donkoh SA, Ankamah-Yeboah I (2020) Determinants of biofertilizer adoption among smallholder farmers in Ghana. Agric Res Technol 24(3):556268. https://doi.org/10.19080/ARTOAJ.2020.24.556268

Ayoola JB (2022) Agricultural input subsidy policies and sustainable soil fertility management in sub-Saharan Africa. Int J Agric Pol 14(2):67–74

Baah-Boateng W (2016) The youth unemployment challenge in Africa: what are the drivers? Econ Labour Relat Rev 27(4):413–431. https://doi.org/10.1177/1035304616674292

Food and Agriculture Organization of the United Nations (2020) The State of Food Security and Nutrition in the World 2020: transforming food systems for affordable healthy diets. Rome: FAO. https://doi.org/10.4060/ca9692en

DFID (2001) Sustainable livelihoods guidance sheets. Department for International Development, London. Available at: https://www.livelihoodscentre.org.uk/resources/guidance-sheets/

International Fertilizer Development Center (IFDC) (2021) Biofertilizer use in sub-Saharan Africa: scaling up through public-private partnerships. IFDC report. The Ghana Fertilizer Platform Study (No. 3, p. 3). IFDC FERARI Research Report

Khan MS, Zaidi A, Wani PA, Oves M (2015) Role of plant growth promoting rhizobacteria in the remediation of metal contaminated soils. Environ Chem Lett 13(1):1–19. https://doi.org/10.1007/s10311-014-0458-0

Makinde EA, Ayoola JB (2019) Adoption of biofertilizer technology among smallholder farmers in Nigeria: constraints and strategies for improvement. Afr J Agric Res 14(20):871–878. https://doi.org/10.5897/AJAR2019.13943

Mfilinge P, Muzuka ANN, Shilla DA (2014) Influence of land use patterns on the quality of groundwater in Dar es Salaam, Tanzania. J Water Resource Prot 6(7):617–626. https://doi.org/10.4236/jwarp.2014.67060

Mogues T, Benin S, Cudjoe G (2009) Agricultural public spending in Ghana. IFPRI discussion paper 00987. International Food Policy Research Institute (IFPRI), Washington. Available at: https://doi.org/10.2499/0896296200

Rogers EM (2003) Diffusion of innovations, 5th edn. Free Press

Chapter 18
Biostimulants and Biopesticides: Multifunctional Biological Solutions for African Agriculture

Makomborero Nyoni, Norman Muzhinji, Kahsay Tadesse Mawcha, Reward Muzerengwa, and Tatenda Justice Gunda

Abstract Global agriculture faces increasing demands for sustainable food production, leading to a greater focus on innovative solutions that enhance crop performance while minimising impact on the environment. In Africa, excessive chemical pesticide use has caused severe environmental and health impacts, including biodiversity loss, soil degradation, water pollution, and risks to human health and nontarget species. Forthwith, consumers have advanced their awareness of food quality, health, ethics, and environmental impact. This stint has prompted the agricultural industry to move towards sustainability, with an aim to reduce environmental footprints and promote high-value crops. One promising approach involves the use of biostimulants and biopesticides that are substances or microorganisms derived from natural resources like microbials and botanicals. These emerging tools in sustain-

M. Nyoni (✉)
Innov Bio_M Ltd, 131 Rhodesville, Greendale, Harare, Zimbabwe

Innov Bio_M Ltd, 131 Rhodesville, Harare, Zimbabwe

N. Muzhinji
University of the Free State, Department of Plant Sciences, Division of Plant Pathology, Bloemfontein, South Africa

K. T. Mawcha
Biopesticides Group, International Centre for Genetic Engineering and Biotechnology, Cape Town, South Africa

Department of Plant and Soil Sciences, University of Pretoria, Pretoria, South Africa

R. Muzerengwa
Institute of Medicine, Department of Microbiology and Virology, Peoples' Friendship of Russia University, Moscow, Russia

African Genome Centre (AGC), University Mohammed VI Polytechnic (UM6P), Ben Guerir, Morocco

T. J. Gunda
Golden Helix Life Sciences Ltd, UNN Hse, Community 25 Peace Be Tema, Greater Accra, Ghana

A. D. Nciizah et al. (eds.), *Biofertilizers for Sustainable Agriculture in Africa*, Sustainability Sciences in Asia and Africa, https://doi.org/10.1007/978-981-95-6009-7_18

able agriculture aim to boost crop productivity and resilience. When applied to plants or soils, biostimulants promote plant growth, and stress tolerance, while biopesticides control specific pests and diseases. Growing interest in biostimulants and biopesticides in the last decade highlights their high potential as an essential component of sustainable agriculture, especially when used alongside traditional pesticides. However, despite the promising potential of biostimulants and biopesticides, several significant challenges continue to hinder their widespread adoption across Africa. Key among these is the lack of harmonised regional regulatory policies, limited awareness and education regarding their use and benefits, inadequate infrastructure to support their distribution and application, and the absence of standalone registration guidelines. Additionally, continuous improvement in product efficacy remains essential to build trust and encourage broader uptake among farmers and stakeholders. Promotion of local microbial strains in biostimulants and biopesticides development and manufacturing is highly advisable. This chapter explores the multifunctionality of biostimulants and biopesticides in the context of enhancing their use in Africa. However, it will primarily focus on microbials and botanicals (phytopesticides), as the later have greater potential for development and application in Africa, given the availability of resources and indigenous knowledge systems.

Keywords Chemical pesticides · Biostimulants · Biopesticides · Microbials · Botanicals · Sustainable agriculture

18.1 Introduction

Since the introduction of the Green Revolution in the 1960s, pest control has primarily relied on extensive chemical pesticide application (Sharma et al. 2019). According to the FAOSTAT database, pesticide consumption in Africa is relatively low, constituting less than 5% of the global total. However, the region experiences disproportionate misuse, leading to severe environmental and health implications. This dependence on chemical pesticides has contributed to biodiversity loss, soil degradation, water pollution, and adverse health effects on humans and nontarget species (Matyjaszczyk 2018). Furthermore, pesticide residues on export crops create trade barriers (Bailey et al. 2010). Currently, consumers have become increasingly aware of the quality, health, ethics, and environmental impact of food. This shift has prompted the agricultural industry to move towards sustainability, with a focus on reducing environmental footprints and promoting high-value crops (Rodrigues et al. 2025).

The growing demand for sustainable agricultural methods has spurred the adoption of innovative solutions such as biostimulants and biopesticides. Biostimulants, including humic acids, amino acids, and microbial inoculants, enhance plant physiological processes, improve growth, yield, and stress resilience (Calvo et al. 2014). Likewise, biopesticides derived from microbes, plants, and minerals incite targeted pest and disease management with reduced ecological impact. Research indicates that the integration of biostimulants and biopesticides can significantly boost crop yields through improved water use efficiency, enhanced nutrient uptake, and decreased losses from pests and diseases. These strategies align with the principles of organic agriculture, with emphasis on natural input usage to promote crop growth and soil health while minimising dependence on synthetic chemicals (Jindo et al. 2022).

However, despite the promising potential of biostimulants and biopesticides, significant challenges hinder their widespread adoption across Africa. There is a critical need for supportive legislation, education on the usage and benefits of biopesticides, and infrastructure development to facilitate their distribution and use (FAO 2022). This chapter examines the potential of biostimulants and biopesticides in addressing Africa's pest management challenges, highlighting their advantages, applications, and success stories. The primary focus will be on microbials and botanicals (phytopesticides), the latter of which has more potential for development and application in Africa, given the availability of resources and indigenous knowledge systems.

18.2 Definition and Mechanisms of Biostimulants and Biopesticides

The 'plant biostimulant' lacks a universally accepted definition, with various scholars defining it differently. Yakhin et al. (2017) described biostimulant as 'a diverse range of compounds and microorganisms that, when applied to plants, seeds, or growing substrates, improve vigour, yields, quality, and stress tolerance without providing a direct nutrient function or acting as pesticides'. Similarly, Du Jardin (2015) defined biostimulants as 'substances or microorganisms applied to plants to enhance nutrient efficiency, stress resilience, and overall crop quality. When used in small amounts, these substances enhance plant growth and development by stimulating natural processes like nutrient uptake, hormone regulation, and tolerance to abiotic stresses'.

18.2.1 What Are Biopesticides?

Biopesticides are products that are mainly made from microorganisms, biochemicals, botanicals, or nanoparticles that help manage agricultural pests and enhance crop production (Kidd et al. 2017; Samada and Tambunan 2020; Adeleke et al. 2022). The US Environmental Protection Agency (EPA) defines biopesticides as 'derived from natural materials such as animals, plants, bacteria, and certain minerals'. Similarly, the European Environment Agency defines biopesticides as 'pesticides in which the active ingredient is a virus, fungus, bacteria, or a natural product derived from a plant source'. According to Jhala et al. (2020), 'biopesticide' encompasses living materials, including animals, plants, and microorganisms. An ideal biopesticide includes the following: (i) high selectivity for target species with low toxicity to nontarget organisms, (ii) high efficacy at low application rates, and (iii) low environmental persistence or biodegradability to minimise the risk of bioconcentration and biomagnification within the food chain (Dhangar and Choudhury 2021).

18.2.2 Mode of Action of Biopesticides

Biopesticides employ various mechanisms to control pests. Table 18.1 illustrates the primary modes of action for biopesticide products. Biopesticides disrupt pest physiology through mechanisms such as gut disruption, growth regulation, and metabolic interference. They can denature proteins, cause metabolic disorders and paralysis, activate toxic pathways, demonstrate multisite inhibitory actions, and release neuromuscular toxins and bioactive compounds (Fenibo et al. 2021).

18.2.3 Microbial Biopesticides

Microbial biopesticides are biological formulations containing microorganisms such as viruses, fungi, bacteria, and protozoa as active ingredients (Kiewnick 2007; Adeleke et al. 2022). These microbes utilise toxic metabolites to eliminate and suppress pest populations. These microbial pesticides are applied in various agricultural settings to target specific pests while minimising environmental impact. Species such as *Bacillus thuringiensis* (Bt) and *Bacillus sphaericus* are widely used due to their production of delta-endotoxins, such as parasporal crystals, that disrupt insect digestive systems upon ingestion (Roh et al. 2007; Xiao and Wu 2019). Bt has been used as a commercial biopesticide, valued for its specificity and safety (Ujvary 2010). Other bacteria, including *Pseudomonas, Burkholderia*, and *Serratia*, target pests through similar mechanisms (Kachhawa 2017). Fungi such as *Trichoderma* sp. inhibit soil-borne pathogens causing root rot in crops like chickpeas and groundnuts (Samada and Tambunan 2020), while *Beauveria bassiana* and *Metarhizium brunneum* control a broad range of pests, including thrips, beetles, and aphids, in ornamental and food crops (Dara 2017; Arthurs and Dara 2019).

These examples underscore the versatility and specificity of microbial biopesticides, positioning them as important tools for sustainable pest management.

Table 18.1 The mode of action of biopesticides and their descriptions

Class of biopesticides	Descriptions and examples	Mode of action	References
Microbial biopesticides	Bacterial biopesticides (e.g. *Bacillus thuringiensis*, *Bacillus subtilis*).	Produce insecticidal toxins (e.g. Cry and Vip proteins) that disrupt the midgut epithelial cells of insects, leading to their death. Compete with pathogens via antibiosis, nutrient competition, or induced systemic resistance (ISR) in plants. Secreted lipopeptides (e.g. surfactin and iturin) that lyse fungal membranes.	Mnif and Ghribi (2015), Osman et al. (2015), Apurva et al. (2019), Kumar et al. (2021a, b, c), Rocha et al. (2023), and Beltrán Pineda and Castellanos-Rozo (2025)
	Fungal biopesticides (e.g. *Beauveria bassiana, Metarhizium anisopliae*).	Spores adhere to insect cuticle, germinate, and penetrate using enzymes (chitinases, proteases). Proliferate in the hemocoel and kill hosts by nutrient depletion and toxin release. Some species (e.g. *Trichoderma*) also act against plant pathogens via mycoparasitism or ISR.	Meyling and Eilenberg (2007), Oliveira et al. (2014), Song et al. (2019), and Bapfubusa et al. (2023)
	Viral biopesticides (e.g. Nucleopolyhedroviruses—NPVs).	Infect specific insect larvae via ingestion. Replicate within host cells, causing tissue degradation and larval death. Highly species-specific and environmentally safe.	Nawaz et al. (2019) and Kalawate (2014)

(continued)

Table 18.1 (continued)

Class of biopesticides	Descriptions and examples	Mode of action	References
Botanical biopesticides	Extracted from plants or plant secondary metabolites (e.g. neem, pyrethrin, essential oils).	Disrupts insect hormonal balance (e.g. azadirachtin acts as an insect growth regulator). Block feeding, oviposition, or mating via repellence or antifeedant effects. Act as contact poisons (e.g. pyrethrins target insect nervous systems). Some (e.g. allelochemicals) suppress pathogen germination or growth.	Pino et al. (2013), Regnault-Roger and Philogène (2008), Ramirez (2013), and Lengai et al. (2020)
Biochemical biopesticides	Naturally occurring substances that do not target pests directly but interfere with their behaviour. Examples *include pheromones, kairomones, and insect growth regulators (such as* juvenile hormone mimics).	Disrupt mating communication by confusing or trapping individuals. Inhibit moulting, metamorphosis, or egg development. Attract beneficial predators or parasitoids.	Ryan (2002), Ghoneim and Bakr (2018), and Poornakala et al. (2023)
RNA-based biopesticides	These are based on RNA interference (RNAi) mechanisms that target specific pest or pathogen genes.	Double-stranded RNA (dsRNA) is ingested or sprayed on target organisms. Triggers the RNAi pathway, leading to the silencing of essential genes (e.g. toxin synthesis, development, and virulence). Highly specific with minimal off-target effects.	de Andrade and Hunter (2016), Svoboda (2020), and Zhu and Palli (2020)

(continued)

Table 18.1 (continued)

Class of biopesticides	Descriptions and examples	Mode of action	References
Endophytic and symbiotic biopesticides	Live microbes that colonise plant tissues or the rhizosphere to confer protection.	Outcompete pathogens through niche exclusion and antimicrobial secretion. Enhance host immunity via induced systemic resistance (ISR). Some release volatile organic compounds (VOCs) that repel pests or inhibit pathogens.	Andrés et al. (2017) and Aravinthraju et al. (2024)
GMO-based biopesticides	Utilising genetic engineering to confer pest resistance to crops, the most prominent and widely adopted examples are: Bt Corn, Bt Cotton, Bt Soybean, Bt Potato.	Bt toxins are highly specific to certain insect groups due to the requirement for specific gut pH and the presence of particular receptors. This makes them relatively safe for nontarget organisms. For the toxins to be effective, the insect must ingest the plant material containing the Bt proteins.	Yanni et al. (2010), Ravensberg and Ravensberg (2011), Rank and Koch (2021), and Luo et al. (2024)

18.2.4 Phytopesticides

Phytopesticides comprise a variety of phytochemical compounds with diverse mechanisms of action. Notable examples of commercial botanical pesticides include neem (*Azadirachta indica*), sabadilla (*Schoenocaulon officinale*), pyrethrum (*Tanacetum cinerariifolium*), tobacco (*Nicotiana tabacum*), and ryania (*Ryania speciosa*) (Mawcha et al. 2025). Additionally, essential oils and extracts from various plant parts have been effectively utilised to combat plant diseases (Ayilara et al. 2023). Plant extracts and oils are derived from chemical substances in plants. These biochemicals differ from pheromones in their composition, targeted pests, and mechanisms of action (Byrappa et al. 2012).

18.2.5 Molecular Mechanisms of the Application of Biopesticides

Although the actions of commonly used biopesticides are documented, the entomopathogenic fungus *Beauveria bassiana* has emerged as a prominent model for investigating the molecular mechanisms of biopesticides due to its broad efficacy against arthropod pests. Understanding its pathogenicity at the molecular level is essential for optimising its use and developing durable pest control solutions. The pathogenic process of *B. bassiana* starts with its attachment to the host pest, progressing with cuticle penetration and subsequent colonisation of the pest's hemocoel (Wojda et al. 2009).

Bacterial and fungal biopesticides, such as *Bacillus thuringiensis* and *Beauveria bassiana*, target pests through diverse molecular mechanisms, including disruption of protein synthesis, membrane integrity, and metabolic processes. These mechanisms underpin their efficacy as sustainable alternatives to synthetic pesticides (Svidritskiy et al. 2013; Parker et al. 2019). Specific biopesticides, such as blasticidin produced by *Streptomyces* species, bind to the 50S subunit of prokaryotic ribosomes, the cellular machinery responsible for protein production. These binding blocks peptide transfer, preventing the formation of proteins essential for pest survival (Svidritskiy et al. 2013; Parker et al. 2019). Additionally, these biopesticides compromise plasma membrane permeability, leading to cell death through the leakage of electrolytes and amino acids, as demonstrated by natamycin, derived from *Streptomyces* species. They also block trehalase to prevent glucose formation (validamycin) (Gwinn 2018).

The hydrophobicity of *B. bassiana*'s aerial conidia, coated with hydrophobins, facilitates its interaction with insects' cuticles (Holder and Keyhani 2005). This hydrophobic property is influenced by several genes expressed by *B. bassiana* that play a role in lipid homeostasis. Transcriptomic analyses have shown an upregulation of gene expressions for hydrophobins and *Metarhizium* adhesion-like proteins 1 and 2 (MAD1 and MAD2), which are critical for adherence to insect cuticles (Wang and St. Leger 2007). The transport and storage of lipids in conidia, as well as the maintenance of lipid homeostasis, depend on the overexpression of mammalian-like perilipin 1 (MPL1) genes (Chen et al. 2019). The MPL1 gene is crucial; its deletion results in a decrease in turgor pressure within the appressoria, negatively affecting the adhesiveness of *B. bassiana* (Wang and Leger 2007).

Once *B. bassiana* successfully adheres to the host insect, its conidia germinate and develop appressoria, facilitating penetration into the cuticle. The penetration efficiency increases when the appressorium's structure allows enzymatic digestion and mechanical pressure to work synergistically (Singh et al. 2017). As penetration occurs, the hyphae of *B. bassiana* grow within the insect's exoskeleton and produce secondary hyphae inside the cuticle. When exposed to the hyperosmotic environment of the hemocoel, these hyphae convert into blastospores, which are motile, more hydrophilic, and better equipped to evade the insect's immune responses (Ortiz-Urquiza and Keyhani 2016).

18.3 Biopesticides as an Alternative to Chemical Pesticides in Africa

Biopesticides provide an eco-friendly alternative to conventional chemical pesticides, as they are less toxic and break down more rapidly. Several biopesticides have been commercialised (Tables 18.2 and 18.3) in Africa and used in Integrated Pest Management (IPM) programs for various crops. For example, the adoption of biopesticides in Kenya, such as the fungal bioherbicide Kichawi Kill, has demonstrated success in controlling the parasitic plant witchweed, significantly improving maize yields (Reuters 2024). Additionally, biopesticides provide economic advantages for smallholder farmers by reducing their dependency on costly chemical inputs. A study by the FAO in West Africa highlighted that over 2000 farmers in the Sahel region have successfully transitioned to biopesticides, improving both crop health and productivity (FAO 2024). However, despite the promising potential of biopesticides, challenges in scaling up their use across Africa remain the biggest hurdle. There is a critical need for supportive legislation, proper education on the use and benefits of biopesticides, and infrastructure development to facilitate their distribution and application (FAO 2022).

18.3.1 Current Status of Biopesticides in Africa

The biopesticides market in Africa is expanding, driven by the wide adoption of sustainable agricultural practices. In 2024, the market was valued at USD 160.85 million and is projected to reach USD 269.16 million by 2029, reflecting a compound annual growth rate (CAGR) of 10.84% during the forecast period. However, the adoption of biopesticides across the continent faces several challenges, particularly the lack of harmonised regulatory frameworks, which can impede the registration and distribution of biopesticide products. To address this, various initiatives are underway to streamline and harmonise regulations and promote innovation in the development of biopesticides. For instance, a regional workshop titled ‘Advancing Regulatory Harmonisation and Biopesticide Innovation in Africa’ was held to share knowledge and experiences related to the regulatory landscape for biopesticides and biocontrol agents (ICGEB 2024).

18.4 Synergistic Benefits: How Biostimulants and Biopesticides Work Together?

Recently, the integration of biostimulants and biopesticides represents an innovative and promising strategy in sustainable agriculture, leading to improved plant health and vigour that make crops more resilient to both abiotic stresses, such as drought and salinity, and biotic stresses, such as pest and disease attacks. For instance, the

Table 18.2 Some commercial microbial-based biopesticides commercialised in Africa

Product name	Species	Target pest	Country	Status	References
Green Muscle®	*Metarhizium anisopliae*	Desert locust (*Schistocerca gregaria*)	Across the Sahel region.	Widely used	Rada and Atreby (2017)
Novacrid	*Metarhizium acridum*	Locusts and grasshoppers.	Africa-wide	Under development	Luke (2023)
VectoBac	*Bacillus thuringiensis israelensis*	Mosquito larvae	Africa-wide	Commercialised	Diédhiou et al. (2017)
BioPower	*Beauveria bassiana*	Aphids, thrips, whiteflies, weevils	South Africa	Commercialised	Maina et al. (2018)
Kichawi Kill	*Fusarium oxysporum*	Witchweed (*Striga* spp.).	Kenya	Commercialised	Baker et al. (2024)
MycoPesticide 1	*Metarhizium anisopliae*	Termites	Ghana	Under Development	Akutse et al. (2020)
NemOut	*Paecilomyces lilacinus*	Plant-parasitic nematodes.	Kenya, Uganda	Commercialised	Wann et al. (2016)
NemaPower	*Heterorhabditis bacteriophora*	Plant-parasitic nematodes.	Kenya	Under Development	Singh et al. (2022)
AflaSafe	Nontoxigenic strains of *Aspergillus flavus*	Aflatoxin-producing fungi in maize and groundnuts.	Nigeria, Kenya, Senegal, and others	Commercialised	Hoffmann et al. (2020) and Ola et al. (2022)
Various Bt Products	*Bacillus thuringiensis*	Lepidopteran larvae	Kenya, Ghana, South Africa	Registered	Akotsen-Mensah et al. (2017). Botha et al. (2019), and Lalah et al. (2022)
Bio-Sect	*Beauveria bassiana*	Whiteflies, aphids	Nigeria	Registered	Tantawy et al. (2018)
Eco-T™	*Trichoderma* spp.	Soil-borne pathogens (e.g. *Fusarium*, *Pythium*)	South Africa	Registered	Kubheka et al. (2018)
Trichoderma Max	*Trichoderma harzianum*	Soil-borne pathogens (e.g. Fusarium)	South Africa	Commercialised	Gerber (2010)
RhizoVital®	*Bacillus amyloliquefaciens*	Root pathogens (*Rhizoctonia*, *Fusarium*)	South Africa	Commercialised	Gerber (2010)

Table 18.3 Some commercial biochemical biopesticides commercialised in Africa

Plant	Active compounds	Target pests/ diseases	Country	References
Neem (*Azadirachta indica*)	Azadirachtin and Nimbin	Aphids, whiteflies, and fungal pathogens.	Kenya, Nigeria, Ghana	Isman (2006) and Mugisha-Kamatenesi et al. (2008)
Pyrethrum (*Chrysanthemum cinerariifolium*)	Pyrethrins	Mosquitoes, aphids, and beetles.	Rwanda, Kenya, Tanzania	Dayan et al. (2009)
Garlic (*Allium sativum*)	Allicin	Nematodes and fungal diseases	South Africa, Egypt	Rehman et al. (2019)
Tephrosia (*Tephrosia vogelii*)	Rotenoids	Aphids, weevils, and caterpillars.	Malawi, Uganda, Tanzania	Neuwinger (2004)
Chili pepper (*Capsicum* spp.)	Capsaicin	Aphids and termites	Ghana, Malawi, Kenya	Sharma and Sharma (2021)
Castor bean (*Ricinus communis*)	Ricinoleic acid	Stored grain pests	Ethiopia, Sudan	Saxena et al. (1992) and Devappa et al. (2010)
Siam weed (*Chromolaena odorata*)	Alkaloids and flavonoids	Stored grain pests and soil pests	Nigeria, Ghana	Iwu (1993)
African marigold (*Tagetes* spp.)	Terpenoids and thiophenes	Nematodes and whiteflies	Kenya, South Africa	Ploeg (2002)
Violet tree (*Securidaca longipedunculata*)	Methyl salicylate	Stored grain pests	Various African Countries	Jayasekara et al.
Tobacco (*Nicotiana tabacum*)	Nicotine	Aphids and caterpillars	Various African Countries	Isman (1997)
Mexican Marigold (*Tagetes minuta*)	α-Terthienyl and Ocimenone	Nematodes and insect pests	South Africa, Zimbabwe	Maradufu (1982)
African Pepper (*Piper guineense*)	Piperine and Chavicine	Stored product insects and mosquitoes	Nigeria, Ghana	Ivbijaro and Agbaje (1986)
African Mahogany (*Khaya senegalensis*)	Limonoids	Termites and beetles	Senegal, Burkina Faso	Bekele (2002)
Drumstick Tree (*Moringa oleifera*)	Pterygospermin	Aphids and fungal pathogens	Kenya, Malawi	Fuglie (2001)
Sodom's Apple (*Calotropis gigantea*)	Calotropin and Uscharin	Aphids and grasshoppers	Ethiopia, Kenya	Kabaru and Gichia (2001)

(continued)

Table 18.3 (continued)

Plant	Active compounds	Target pests/ diseases	Country	References
Wild Soursop (*Annona senegalensis*)	Annonacin and Squamocin	Weevils and beetles	Burkina Faso, Nigeria	Seck et al. (1993)
African Locust Bean (*Parkia biglobosa*)	Tannins and Saponins	Storage pests and termites	Ghana, Nigeria	Boeke et al. (2004)
Bush Tea (*Lippia javanica*)	Linalool and Camphor	Mosquitoes and ticks	Zimbabwe, South Africa	Mawela et al. (2013)
Sour Plum (*Ximenia americana*)	Ximenynic acid	Stored grain pests	Botswana, Namibia	Lale and Abdulrahman (1999)
African Wormwood (*Artemisia afra*)	Artemisinin and Thujone	Mosquitoes and moths	South Africa, Lesotho	Kamatou et al. (2013)
Speargrass (*Imperata cylindrica*)	Cylindrene and Isoarborinol	Termites and soil pests	Nigeria, Cameroon	Ukeh et al. (2009)

co-application of microbial biostimulants such as mycorrhizal fungi with biopesticides containing *Bacillus thuringiensis* (Bt) has been shown to improve plant growth and pest resistance simultaneously (Calvo et al. 2014). Additionally, combining microbial species that do not naturally share the same ecological niche has gained attention (Arif et al. 2020). In agriculture, microbial consortia are valued for their ecological functions, including plant growth promotion (Muthuraja and Muthukumar 2022; Kaur et al. 2022), soil bioremediation (Siripattanakul-Ratpukdi et al. 2014), and pest and disease suppression (Nawaz et al. 2020; Nafady et al. 2022; Muthuraja and Muthukumar 2022; Soth et al. 2022; Upamanya et al. 2020). Microbial consortia offer distinct advantages, such as overcoming biological limitations of single organisms and improving stability across diverse ecological conditions (Trivedi et al. 2020; Tabacchioni et al. 2021). Combining microorganisms with complementary modes of action, nutritional requirements, and growth characteristics fosters synergistic interactions. However, challenges such as microbial incompatibility, competition with native organisms, and habitat adaptation must be addressed for successful implementation (Arif et al. 2020; Xu et al. 2011).

18.4.1 Mechanisms of Action of Biostimulants and Biopesticides

Biostimulants and biopesticides operate through distinct but complementary mechanisms, which, when combined, can amplify their effectiveness. Biostimulants function by promoting root development and microbial symbiosis, enhancing nutrient uptake and assimilation, and alleviating abiotic stress through improved hormonal balance and antioxidant activity (du Jardin 2015). On the other hand, biopesticides act through target-specific inhibition of pests and pathogens, inducing systemic resistance in plants, and disrupting pest life cycles without harming beneficial organisms (Reddy 2017). The co-application of biostimulants and biopesticides can yield synergistic interactions. For instance, biostimulants boost plant vigour, making them less susceptible to pests and diseases, while biopesticides enhance the plant's systemic resistance. Additionally, these products foster beneficial soil microbiota, creating an environment unfavourable for pathogens (Rouphael and Colla 2020).

18.5 Economic and Environmental Benefits of Biostimulants and Biopesticides in African Agriculture

Global population growth and the ensuing economic activity have led to significant environmental issues, including air, water, and soil pollution (Roy et al. 2022). The use of chemical fertilisers to increase agricultural output has increased due to population growth, resulting in a loss of soil physico-chemical and microbiological equilibrium. The need for environmentally friendly agricultural products that enhance the physical, chemical, and microbiological properties of the soil and subsequently provide plants with essential nutrients has increased due to these issues (Roy et al. 2022). Biostimulants do not supply necessary nutrients such as nitrogen, phosphorus, or potassium; they are not regarded as fertilisers. Rather, they act by enhancing several plant functions, such as the intake of nutrients, resilience to stress, root development, and the general health of the plant. Biostimulants can be produced using a variety of ingredients, such as amino acids, seaweed, microbial cultures, plant extracts, humic and fulvic acids, and others. Numerous methods, such as foliar sprays, seed treatments, and soil additives, can be employed to make use of them. The primary goal of biostimulants is to increase a plant's ability to respond to environmental changes, endure stress, and function to the best of its ability. This will eventually lead to increased crop yields and higher-quality final products. By increasing crop yields and reducing input costs, these products can boost food security and farmer livelihoods while promoting sustainable agricultural practices that protect the environment. As public awareness and adoption of the new technologies grow, they hold the potential to transform agricultural systems across the African continent and beyond.

18.6 Challenges and Barriers to the Adoption of Biostimulants and Biopesticides in Africa

Current challenges include climate change, emerging insect pests and diseases, diminished agricultural productivity, and ecological transition processes. These issues have opened significant economic and practical avenues for integrating biostimulants and biopesticides (Marrone 2023). For instance, Bt-based biopesticides have effectively controlled *Tuta absoluta*, a significant pest affecting tomato crops and leading to considerable yield losses in Africa and worldwide. Research in greenhouses and field settings indicates that these biopesticides outperform synthetic insecticides, such as Abamectin, Biolep, and Indoxacarb, in managing *T. absoluta* larvae (Abolghasemi et al. 2023). Furthermore, implementing integrated pest management (IPM) in African nations has been crucial in promoting biostimulants and biopesticides across the continent. IPM advocates for diverse pest control methodologies while discouraging the overuse of chemical pesticides to support sustainable pest management practices (Vlaiculescu and Varrone 2022).

18.6.1 Lack of Awareness and Knowledge Among Farmers

Mounaimi et al. (2024) extensively reviewed the benefits of biostimulants and biopesticides and highlighted how they can help reduce agriculture's environmental impact by minimising the use of chemical fertilisers and pesticides. However, several challenges to their adoption in Africa include limited accessibility, insufficient awareness and knowledge, and a lack of research specifically tailored to African agriculture. Additionally, the lack of appropriate formulations and production difficulties hinder the scalability and adoption of biopesticides. The need for biopesticides to fit into existing value chains further complicates their integration into agricultural practices (Dahabieh et al. 2018). Biopesticides and biostimulants have the potential to play a significant role in sustainable agriculture in Africa. However, more research is needed to address the challenges associated with their adoption. Kirksey (2021) found that many growers do not understand plant biostimulants and how they work. There is significant ambiguity among growers regarding the definition and efficacy of plant biostimulants. Growers rely on various information sources, including university-based research and extension, industry representatives, and other growers. However, they place the most trust in sources perceived as unbiased or with established expert credentials (Kirksey 2021).

18.6.2 Regulatory Challenges and the Slow Pace of Approval for New Products

The regulatory landscape for biopesticides presents several challenges. The registration and authorisation process for new biopesticides is often slow and tedious, partly because these products have a delayed mode of action. In many countries, stringent regulations limit the availability of these products. In Africa, the slow pace of approval for new biopesticide and biostimulant products is attributed to several regulatory challenges, including inconsistent regulatory frameworks, limited expertise, high data demands, bureaucratic challenges, and lack of awareness. This results in numerous challenges such as delays and inconsistencies in the approval process, prolonged review timelines, scepticism, and resistance. Additionally, the absence of standardised preparation methods, guidelines for application, and challenges in determining the appropriate dosage of active ingredients complicate the use of biopesticides (Chandler et al. 2011). Concerns regarding their speed of action, short persistence, and overall cost also hinder adoption (Ndakidemi et al. 2021).

18.7 Regulatory Frameworks for Biostimulants and Biopesticides in Africa

The regulatory frameworks for biostimulants and biopesticides in Africa are often fragmented across different countries. There is a significant need for harmonisation and the development of robust guidelines to ensure the quality and safe use of these products. This situation is particularly evident in the lack of dedicated regulatory pathways for biostimulants, which contrasts with the more established pesticide registration systems. Typically, biopesticides rely on existing legislation with adaptations. South Africa has a relatively stronger regulatory environment for pesticides, including biopesticides, due to its Fertiliser, Farm Feeds, Agricultural Remedies, and Stock Remedies Act (Act No. 36 of 1947) and associated regulations. The International Centre for Genetic Engineering and Biotechnology (ICGEB), in collaboration with other regional organisations (African Union, CABI, CropLife, and STDF), has been working on harmonising biopesticides in Africa. As part of this effort, ICGEB has organised a workshop on 'Advancing Regulatory Harmonisation and Biopesticide Innovation in Africa' that brought together regulatory professionals, researchers, industry experts, and policymakers to discuss various aspects of regional and continental harmonisation of registration guidelines for biopesticides in Africa. Mounaimi et al. (2024) highlighted several key challenges related to regulatory elements, including a lack of harmonised regulations, limited technical capacity, insufficient data requirements, bureaucratic hurdles, and a lack of awareness and understanding. By addressing these regulatory challenges and capitalising on the opportunities, Africa can create a favourable environment for developing, approving, and adopting biopesticides and biostimulants.

A general overview of successful applications of biostimulants and biopesticides in different African agroecologies is provided in Table 18.4. Each example underlines the potential of these technologies to address critical problems in crop production, such as pest management, yield improvement, and soil health recovery. The diversity of interventions from microbial inoculants, such as *Azospirillum* and *Rhizobia*, to natural extracts, such as neem and seaweed, illustrates the versatility of biostimulants and biopesticides in diverse agroecological areas and cropping systems. Notable results include the significant reduction in infestations by pests and diseases, such as in the use of neem extract in Kenya and *Trichoderma*-based biopesticides in Ghana, which demonstrate that biopesticides can be used to either replace or supplement synthetic chemical use. The yield increases, as seen in a 35% increase in chickpea yield in Ethiopia due to rhizobium biofertilisation, are examples of the potential for biostimulants to enhance agricultural productivity in a sustainable manner (Ochieng et al. 2024).

Table 18.5 outlines a comprehensive strategic plan designed to increase the use of biostimulants and biopesticides in Africa, in line with the continent's ambitious

Table 18.4 Case studies of biostimulants and biopesticides in Africa

Country	Type of biostimulant/biopesticide	Target crop(s)	Outcome/impact	Lessons learnt	References
Kenya	Biopesticide (Neem extract, *Bacillus* spp.)	Tomato and cabbage	Reduced pest damage by 60%, leading to a 30% increase in tomato yield. Improved soil microbial balance with *Bacillus* spp. applications.	Neem extracts are cost-effective but require consistent application. *Bacillus* spp. enhances disease resistance and soil health.	Fenibo et al. (2023) and Ochieng et al. (2024)
South Africa	Mycorrhizal-based biostimulant	Maize and wheat	20% yield increase due to better nutrient uptake. Enhanced soil structure through increased mycorrhizal networks. Improved drought resilience in crops.	Farmer education is critical for understanding application techniques and timing.	Agbodjato et al. (2022) and Khetsha et al. (2024)
Nigeria	*Azospirillum*-based biofertilisers	Rice and maize	Increased yield by 25% through improved nitrogen fixation. Reduced dependency on synthetic fertilisers, enhancing long-term soil fertility.	Effective under smallholder farming systems, collaboration with extension officers is crucial for adoption.	MallÃ et al. (2020)
Uganda	Biostimulant (Seaweed extract)	Banana and coffee	15–20% yield increase due to enhanced drought resilience. Improved root development and soil organic matter retention.	Seaweed extract is highly effective but requires regional production facilities to reduce costs.	Fertahi et al. (2023)

Ethiopia	*Rhizobia*-based biofertiliser	Legumes (chickpea)	35% yield improvement through increased nitrogen fixation. Restored degraded soils by enriching microbial biodiversity.	Biofertilisers support sustainable agriculture but face storage and transport challenges in remote areas.	Fertahi et al. (2023)
Rwanda	Pyrethrum-based biopesticide	Tobacco and vegetables	Controlled aphid infestations by 80%, protecting crops and reducing pesticide residues. Maintained soil ecosystem balance by reducing synthetic chemical inputs.	Developing local production facilities ensures affordability and accessibility for farmers.	Hategekimana and Erler (2022)
Tanzania	*Bacillus thuringiensis* (Bt)-based bioinsecticide	Cotton	Reduced bollworm infestations by 85%, decreasing yield losses and pesticide costs. No adverse impacts on nontarget soil organisms.	Bt products require proper farmer training for application to avoid resistance development.	Moshi and Matoju (2017)
Morocco	Humic acid-based biostimulant	Olive and citrus	Enhanced fruit size and quality while reducing water usage by 15%. Improved soil moisture retention and nutrient availability in arid regions.	Humic acids are effective in arid climates but require integration with other soil health strategies for maximum benefit.	Akensous et al. (2022)

Table 18.5 Future outlook: scaling the use of biostimulants and biopesticides in Africa

Key area	Strategic action	Expected outcomes	Challenges	Examples	References
The future role of biostimulants and biopesticides	Promote adoption as key tools for sustainable agriculture and climate resilience.	Enhanced food security, reduced reliance on synthetic inputs, and improved environmental health.	Limited awareness among policymakers and farmers, as well as competing agricultural priorities.	Integration into sustainable agriculture programs in Kenya.	Mounaimi et al. (2024)
	Develop region-specific formulations for crops and conditions unique to African agroecosystems.	Increased efficacy and relevance of products; wider adoption across diverse farming systems.	High cost of research and development; limited technical capacity in some regions.	Rhizobia-based formulations for legumes in Ethiopia.	Negash et al. (2024)
Scaling up production and distribution	Establish decentralised production hubs to reduce costs and ensure timely availability.	Increased local access, reduced costs for farmers, and enhanced regional supply chains.	High capital requirements; lack of skilled labour.	Seaweed extract production facilities in Uganda.	Fertahi et al. (2023)
	Build infrastructure for cold storage and transport of sensitive biological products.	Minimised spoilage; extended shelf life of products; greater access for remote farmers.	High initial costs for infrastructure development; inconsistent power supply in rural areas.	Cold chain development for Bacillus-based biopesticides in Tanzania.	Moshi and Matoju (2017)
	Foster partnerships between local manufacturers and multinational corporations.	Accelerated technology transfer; improved access to advanced formulations.	Risk of dependency on foreign entities; competition with imported products.	Collaboration for Trichoderma production in Ghana.	coyne et al.

Leveraging international trade and investment	Facilitate trade agreements to ease import/export of biostimulants and biopesticides.	Expanded markets for local manufacturers; access to advanced formulations from global markets.	Complex trade regulations; risk of undercutting local industries.	Export of chitosan-based biostimulants from Egypt.	Khaled et al. (2023)
	Provide tax incentives and subsidies for foreign and local investments in biostimulant production.	Increased investment in R&D and production, leading to the creation of jobs in the sector.	Budgetary constraints for governments; potential misuse of subsidies.	Tax incentive schemes in South Africa.	Khetsha et al. (2024)
Education, extension services, and farmer training	Incorporate biostimulant and biopesticide training into existing agricultural extension services.	Increased awareness and correct usage among farmers, as well as integration into broader agricultural systems.	Inadequate funding for extension services, language, and literacy barriers among farmers.	Extension training for Bacillus-based biostimulants in Kenya.	Fenibo et al. (2023)
	Use digital platforms and apps to disseminate knowledge and provide support to farmers.	Wider reach for training programs; improved farmer decision-making through accessible information.	Limited Internet access in remote areas; low digital literacy among farmers.	Mobile app for biopesticide application in Uganda.	Lillian (2024)
	Develop demonstration farms and participatory research initiatives to build farmer confidence.	Faster adoption due to visible benefits; better understanding of local conditions and requirements.	Logistical challenges in setting up demonstration farms; difficulty in scaling participatory research efforts.	Community trials of neem-based biopesticides in Zimbabwe.	Kamunhukamwe et al. (2022)

(continued)

Table 18.5 (continued)

Key area	Strategic action	Expected outcomes	Challenges	Examples	References
Research and development	Invest in long-term field trials to assess efficacy under varying environmental conditions.	Stronger evidence base for adoption; enhanced farmer confidence in product reliability.	Long trial durations; funding limitations for extensive research.	Multi-season trials in rice in Nigeria.	MallÃ et al. (2020)
	Explore innovative delivery mechanisms such as seed coatings and drip irrigation for biostimulants.	Increased efficiency and ease of use; reduced wastage of products.	High cost of development; limited technical expertise in some regions.	Drip irrigation-based biostimulant delivery in Egypt.	Saber et al. (2024)

agricultural transformation agenda. Through an examination of key areas including production and supply chain improvement, international trade, policy frameworks, and farmer education, the table offers a multi-dimensional approach to overcoming challenges and exploiting opportunities for the adoption of biostimulants and biopesticides (Mounaimi et al. 2024).

18.8 Conclusions: The Path Forward for Biostimulants and Biopesticides in African Agriculture

Biostimulants and Biopesticides represent a viable alternative to chemical pesticides, offering solutions to the pressing challenges of pesticide resistance, environmental degradation, and health risks in Africa. By leveraging their benefits, African farmers can adopt sustainable pest management practices that enhance agricultural productivity and protect natural resources. However, the successful integration of biostimulants and biopesticides into African agriculture requires concerted efforts to improve accessibility, build farmer awareness, and strengthen regulatory frameworks. With these measures in place, biostimulants and biopesticides have the potential to transform pest management and contribute to the continent's agricultural sustainability (Glare and O'Callaghan 2019). For continuous improvement in the use of biostimulants and biopesticides, there is a high need for regional harmonisation of regulatory policies, the development of standalone registration guidelines for biostimulants and biopesticides, and continuous improvement in efficacy. Promoting local microbial strains in the development and manufacturing of biostimulants and biopesticides is highly advisable. Furthermore, more research is needed to combine modes of action for additive benefits, as well as the use of nanobiostimulants and biopesticides.

Author Contributions MN, NM, KTM, RM and TJG designed the review outline, wrote and reviewed the chapter. All authors listed have made a substantial, direct and intellectual contribution to the work and approved it for publication.

Conflict of Interest The authors declare that the research was conducted in the absence of any commercial or financial relationships that could be construed as a potential conflict of interest.

References

Abolghasemi S, Ahmadi K, Takalozadeh HM (2023) Efficacy of Bacillus thuringiensis compared with some chemical insecticides in controlling tomato leafminer, Tuta absoluta (Meyrick). JAST 25. https://doi.org/10.22034/jast.25.4.16

Adeleke BS, Ayilara MS, Akinola SA, Babalola OO (2022) Biocontrol mechanisms of endophytic fungi. Egypt J Biol Pest Control 32(1):46

Agbodjato NA, Assogba SA, Babalola OO et al (2022) Formulation of biostimulants based on arbuscular mycorrhizal fungi for maize growth and yield. Front Agron 4:894489

Akensous F-Z, Anli M, Boutasknit A, Ben-Laouane R, Ait-Rahou Y, Ahmed HB, Nasri N, Hafidi M, Meddich A (2022) Boosting date palm (Phoenix dactylifera L.) growth under drought stress: effects of innovative biostimulants. Gesunde Pflanzen 74(4):961–982

Akotsen-Mensah C et al (2017) Pest management knowledge and practices of mango farmers in Southeastern Ghana. J Integr Pest Manag 8(1):13

Akutse KS et al (2020) Biopesticide research and product development in Africa for sustainable agriculture and food security–experiences from the International Centre of Insect Physiology and Ecology (icipe). Front Sustain Food Syst 4:563016

Andrés MF, Diaz CE, Giménez C, Cabrera R, González-Coloma A (2017) Endophytic fungi as novel sources of biopesticides: the Macaronesian Laurel forest, a case study. Phytochem Rev 16:1009–1022

Apurva K, Singh C, Kumar V, Jha VB (2019) Bacterial biopesticides and their use in agricultural production. In: Biofertilizers and biopesticides in sustainable agriculture. Apple Academic Press, pp 23–42

Aravinthraju K, Shanthi M, Murugan M, Srinivasan R, Maxwell LA, Boopathi NM, Anandham R (2024) Endophytic entomopathogenic fungi: their role in enhancing plant resistance, managing insect pests, and synergy with management routines. J Fungi 10(12):865

Arif I, Batool M, Schenk PM (2020) Plant microbiome engineering: expected benefits for improving sustainable agriculture. Front Microbiol 11:341

Arthurs S, Dara SK (2019) Microbial biopesticides for invertebrate pests and their markets in the United States. J Invertebr Pathol 165(2019):13–21

Ayilara MS, Adeleke BS, Akinola SA, Fayose CA, Adeyemi UT, Gbadegesin LA, Omole RK, Johnson RM, Uthman QO, Babalola OO (2023) Biopesticides as a promising alternative to synthetic pesticides: a case for microbial pesticides, phytopesticides, and nanobiopesticides. Front Microbiol 14:104090

Bailey A, Chandler D, Grant WP, Greaves J, Prince G, Tatchell M (2010) Biopesticides: pest management and regulation. CAB, Wallingford

Baker CS, Sands DC, Nzioki HS (2024) The Toothpick Project: commercialization of a virulence-selected fungal bioherbicide for Striga hermonthica (witchweed) biocontrol in Kenya. Pest Manag Sci 80(1):65–71

Bapfubusa Niyibizi IA, Hanna R, Kekeunou S, Membang G, Fiaboe KK, Mahot HC, Fomumbod Abang A, Kumar PL, Fotso Kuate A (2023) Potential of Cameroon-indigenous isolates of the entomopathogenic fungi Beauveria bassiana and Metarhizium anisopliae as microbial control agents of the flea beetle Nisotra uniformis. Biocontrol Sci Technol 33(3):226–40

Bekele T (2002) Studies on seasonal dynamics of ticks of Ogaden cattle and individual variation in resistance to ticks in eastern Ethiopia. J Vet Med Ser B 49:285–288

Beltrán Pineda ME, Castellanos-Rozo J (2025) Bacterial insecticides beyond Bacillus thuringiensis. Phytopathology Res 7(1):19

Boeke SJ, Baumgart IR, Van Loon JJA, et al (2004) Toxicity and repellence of African plants traditionally used for the protection of stored cowpea against Callosobruchus maculatus. J Stored Prod Res 40:423–438

Botha AS et al (2019) Efficacy of Bt maize for control of Spodoptera frugiperda (Lepidoptera: Noctuidae) in South Africa. J Econ Entomol 112(3):1260–1266

Bronzwaer S, de Coen W, Heuer O, Marnane I, Vidal A (2024) The framework for action of the Cross-agency One Health Task Force. One Health. 19:100925

Byrappa AM, Kumar NG, Divya M (2012) Impact of biopesticides application on pod borer complex in organically grown field bean ecosystem. J Biopest 5(2):148

Calvo P, Nelson L, Kloepper JW (2014) Agricultural uses of plant biostimulants. Plant Soil 383(1):3–41

Chandler D, Bailey AS, Tatchell GM, Davidson G, Greaves J, Grant WP (2011) The development, regulation and use of biopesticides for integrated pest management. Philos Trans R Soc B 366:1987–1998. https://doi.org/10.1098/rstb.2010.0390

Chaudhary P & Choudhary V (2022) Biopesticides: an alternative tool for sustainable agriculture. International Journal of Research in Analytical and Applied Sciences, 5(8), 1–8

Chen Y, Wang L, Dai F, Tao M, Li X, Tan Z (2019) Biostimulants application for bacterial metabolic activity promotion and sodium dodecyl sulfate degradation under copper stress. Chemosphere 226:736–43

Dahabieh MS, Broring S, Maine E (2018) Overcoming barriers to innovation in food and agricultural biotechnology. Trends Food Sci Technol 79:204–213. https://doi.org/10.1016/j.tifs.2018.07.004

Dara SK (2017) Microbial control of arthropod pests in small fruits and vegetables in temperate climate. In: Microbial control of insect and mite pests. Academic Press, pp 209–221

Dayan FE, Cantrell CL, Duke SO (2009) Natural products in crop protection. Bioorg Med Chem 17:4022–4034

de Andrade EC, Hunter WB (2016) RNA interference–natural gene-based technology for highly specific pest control (HiSPeC). In: RNA interference. IntechOpen

Devappa RK, Makkar HPS, Becker K (2010) Biodegradation of Jatropha curcas phorbol esters in soil. J Sci Food Agric 90:2090–2097

Dhangar NV, Choudhury D (2021) Aspects of biopesticides: a review. Pharma Innov J 10:962–966

Diédhiou SM et al (2017) Effectiveness of three biological larvicides and of an insect growth regulator against Anopheles arabiensis in Senegal. Bull Soc Pathol Exot 110:102–115

Du Jardin P (2015) Plant biostimulants: definition, concept, main categories and regulation. Sci Total Environ 468–469:1–7

FAO (2022) Pesticide use in Africa: challenges and opportunities. FAO, Rome

FAO (2024) Sustainable pest management practices in the Sahel. FAO, Rome

Fenibo EO, Ijoma GN, Matambo T (2021) Biopesticides in sustainable agriculture: A critical sustainable development driver governed by green chemistry principles. Front Sustain Food Syst 5:619058

Fenibo EO, Christian R, Matambo TS (2023) Biopesticide commercialization in African countries: successful case studies. In: Development and commercialization of biopesticides. Elsevier, pp 297–328

Fertahi S, Elalami D, Tayibi S, Taarji N, Lyamlouli K, Bargaz A, Oukarroum A, Zeroual Y, El Bouhssini M, Barakat A (2023) The current status and challenges of biomass biorefineries in Africa: a critical review and future perspectives for bioeconomy development. Sci Total Environ 870:162001

Fuglie KO, Kascak CA (2001) Adoption and diffusion of natural-resource-conserving agricultural technology. Appl Econ Perspect Policy 23:386–403

Gerber J (2010) Yield response of Fusarium infected maize seed treated with biological control agent formulations. University of South Africa Pretoria

Ghoneim K, Bakr RFA (2018) Physiological activities of anti-Juvenile hormone agents against insects and their role for devising fourth generation insecticides: a comprehensive review. Egypt Acad J Biol Sci A Entomol 11(3):45–138

Glare TR, O'Callaghan M (2019) Microbial biopesticides for control of invertebrates: Progress from New Zealand. J Invertebr Pathol 165:82–88

Gwinn KD (2018) Bioactive natural products in plant disease control. Stud Nat Prod Chem 56:229–246

Hategekimana A, Erler F (2022) Laboratory and field evaluation of a pyrethrum-based product (Agrothrin®) for the protection of stored beans from Acanthoscelides obtectus (Say) (Coleoptera: Chrysomelidae) in Rwanda. Int J Trop Insect Sci 42(1):471–478

Hoffmann V et al (2020) Technologies and strategies for aflatoxin control in Kenya: a synthesis of emerging evidence. International Food Policy Research Institute

Holder DJ, Keyhani NO (2005) Adhesion of the entomopathogenic fungus Beauveria (Cordyceps) bassiana to substrata. Appl Environ Microbiol 71(9):5260–6

ICGEB (2024) Advancing regulatory harmonisation and biopesticide innovation in Africa: report of a regional workshop. Retrieved from https://www.icgeb.org/wp-content/uploads/2024/10/

Advancing-Regulatory-Harmonisation-and-Biopesticide-Innovation-in-Africa_Report-of-a-Regional-Workshop.pdf
Isman MB (1997) Neem and other botanical insecticides: barriers to commercialization. Phytoparasit 25:339–344
Isman MB (2006) Pesticides based on plant essential oils: from traditional practice to modern development. Phytochem Rev 5(1):143–152
Ivbijaro MF, Agbaje M (1986) Insecticidal activities of Piper guineense Schum and Thonn, and Capsicum species on the cowpea bruchid, Callosobruchus maculatus F. Int J Trop Insect Sci 7:521–524
Iwu MM (1993) African medicinal plants. CRC Press Maryland Jamkhande, PG, Ajgunde, BR, Jadge, DR (2017) Annona cherimola Mill(Custard apple) A Rev its plant profile, Nutr values, Tradit claims ethnomedicinal Prop Orient Pharm Exp Med 17:189–201
Jindo K, Goron TL, Pizarro-Tobías P, Sánchez-Monedero MÁ, Audette Y, Deolu-Ajayi AO, van der Werf A, Goitom Teklu M, Shenker M, Pombo Sudré C, Busato JG (2022) Application of biostimulant products and biological control agents in sustainable viticulture: A review. Front Plant Sci 13:932311
Jhala J, Baloda AS, Rajput VS (2020) Role of bio-pesticides in recent trends of insect pest management: a review. J Pharmacogn Phytochem 9(1):2237–2240
Kachhawa D (2017) Microorganisms as a biopesticides. J Entomol Zool Stud 5(3):468–473
Kabaru JM, Gichia L (2001) Insecticidal activity of extracts derived from different parts of the mangrove tree Rhizophora mucronata (Rhizophoraceae) Lam. against three arthropods. African J Sci Technol 2(2)
Kalawate AS (2014) Microbial viral insecticides. In: Basic and applied aspects of biopesticides. Springer India, pp 47–68
Kamatou GPP, Vermaak I, Viljoen AM, Lawrence BM (2013) Menthol: A simple monoterpene with remarkable biological properties. Phytochem 96:15–25
Kamunhukamwe T, Nzuma JK, Maodzeka A, Gandawa CG, Matongera N, Madzingaidzo L, Muturiki L (2022) Efficacy of neem bio-pesticide and synthetic insecticides against control of fall armyworm (Spodoptera frugiperda) in Maize. JEZS 10(4):109–114
Kaur H, Ohri P, Bhardwaj R (2022) The interplay of microbial consortia in sustainable agriculture. Microb Ecol 84(1):59–71
Khaled AM, Abdel-Razik AB, Tonbol K (2023) Enhancing heat stress resilience of tomato during flowering stage using biostimulants: nano Chitosan and Spirulina platensis extract. J Agric Chem Biotechnol 14(7):79–87
Khetsha Z, Van Der Watt E, Masowa M, et al (2024) Phytohormone-Based Biostimulants as an Alternative Mitigating Strategy for Horticultural Plants Grown Under Adverse Multi-Stress Conditions: Common South African Stress Factors. Caraka Tani J Sustain Agric 39:167–193
Kidd PS, Álvarez-López V, Becerra-Castro C, Cabello-Conejo M, Prieto-Fernández Á (2017) Potential role of plant-associated bacteria in plant metal uptake and implications in phytotechnologies. Adv Bot Res 83:87–126. https://doi.org/10.1016/bs.abr.2016.12.004
Kiewnick S (2007) Practicalities of developing and registering microbial biologicalcontrol agents. CABI Rev 2007:11
Kirksey JA (2021) Understanding grower perceptions of plant biostimulants in Georgia blueberry production. Master's thesis, University of Georgia, 2021
Koutroubas SD, Antoniadis V, Damalas CA, Fotiadis S (2020) Sunflower growth and yield response to sewage sludge application under contrasting water availability conditions. Ind Crops Prod 154:112670
Kubheka BP, Laing MD, Yobo KS (2018) An integrated approach to managing Rhizoctonia solani on maize in marginal soils: field studies. Arch Phytopathol Plant Protect 51(7–8):373–386
Kumar P et al (2021a) Bacillus thuringiensis as microbial biopesticide: uses and application for sustainable agriculture. Egypt J Biol Pest Control 31(1):95

Kumar P, Kamle M, Borah R, Mahato DK, Sharma B (2021b) Bacillus thuringiensis as microbial biopesticide: uses and application for sustainable agriculture. Egypt J Biol Pest Control 31(1):95
Kumar S, Kaushik N, Goud GA (2021c) Microbial biopesticides: status and advancement for sustainable agriculture. J Environ Biol 42(6):1515–1524
Lalah JO et al (2022) Pesticides: chemistry, manufacturing, regulation, usage and impacts on population in Kenya. Intech Open
Lale NES, Abdulrahman HT (1999) Evaluation of neem (Azadirachta indica A. Juss) seed oil obtained by different methods and neem powder for the management of Callosobruchus maculatus (F.)(Coleoptera: Bruchidae) in stored cowpea. J Stored Prod Res 35:135–143
Lengai GMW, Muthomi JW, Mbega ER (2020) Phytochemical activity and role of botanical pesticides in pest management for sustainable agricultural crop production. Sci Afr 7:e00239
Lillian NR (2024) Organic farming techniques: enhancing sustainability and resilience in East Africa. Res Invent J Biol Appl Sci 3(2):1
Luke B (2023) Improving uptake of biopesticides for Desert Locust control. Plant Health Cases 2023:phcs20230020
Luo X, Satyabrata N, Zhang Y, Zhou X, Yang C, Pan H (2024) Risk assessment of RNAi-based biopesticides. New Crops 1:100019
Maina UM et al (2018) A review on the use of entomopathogenic fungi in the management of insect pests of field crops. J Entomol Zool Stud 6(1):27–32
MallÃ I, KassoguÃ A, Babana AH, de Oliveira Paiva CA, Murriel II (2020) A Malian native Azospirillum sp. Az6-based biofertilizer improves growth and yield of both rice (Oryza sativa L.) and maize (Zea mays L.). Afr J Microbiol Res 14(7):286–293
Maradufu A (1982) Furanosesquiterpenoids of Commiphora erythraea and C. myrrh. Phytochemistry 21:677–680
Marrone PG (2023) Status of the biopesticide market and prospects for new bioherbicides. Pest Manag Sci 80:81–86. https://doi.org/10.1002/ps.7403
Matyjaszczyk E (2018) Chemical crop protection—where now? Pest Manag Sci 74(3):249–254
Mawcha TK, Malinga L, Muir D, Ge J, Ndolo D (2025) Recent advances in biopesticide research and development with a focus on microbials. F1000Res 13:1071. https://doi.org/10.12688/f1000research154392.2
Mawela K V, Kfir R, Krüger K (2013) Effect of temperature and host species on parasitism, development time and sex ratio of the egg parasitoid Trichogrammatoidea lutea Girault (Hymenoptera: Trichogrammatidae). Biol Control 64:211–216
Meyling NV, Eilenberg J (2007) Ecology of the entomopathogenic fungi Beauveria bassiana and Metarhizium anisopliae in temperate agroecosystems: potential for conservation biological control. Biol Control 43(2):145–155
Mnif I, Ghribi D (2015) Potential of bacterial derived biopesticides in pest management. Crop Prot 77:52–64
Moshi AP, Matoju I (2017) The status of research on and application of biopesticides in Tanzania. Rev Crop Prot 92:16–28
Mounaimi S, Hamim A, El Boukhari MEM, et al (2024) A comprehensive review of integrating biostimulants and biopesticides for organic berry farming: exploring challenges and opportunities for Africa. Front Sustain Food Syst 8:1452823
Mugisha-Kamatenesi M, Deng AL, Ogendo JO, et al (2008) Indigenous knowledge of field insect pests and their management around Lake Victoria basin in Uganda. African J Environ Sci Technol 2:342–348
Muthuraja R, Muthukumar T (2022) Co-inoculation of halotolerant potassium solubilizing Bacillus licheniformis and Aspergillus violaceofuscus improves tomato growth and potassium uptake in different soil types under salinity. Chemosphere 294:133718
Nafady NA, Abdelhameed RE, Hussein MH (2022) Biopesticides and their role in integrated pest management. J Plant Prot Res 62(1):1–15

Nawaz A, Ali H, Sufyan M, Gogi MD, Arif MJ, Ranjha MH, Arshid M et al (2019) Comparative bio-efficacy of nuclear polyhedrosis virus (NPV) and Spinosad against American bollwormm, Helicoverpa armigera (Hubner). Rev Bras Entomol 63(4):277–282
Nawaz A, Shahbaz M, Asadullah, et al (2020) Potential of salt tolerant PGPR in growth and yield augmentation of wheat (Triticum aestivum L.) under saline conditions. Front Microbiol 11:2019
Ndakidemi BJ, Mbega ER, Ndakidemi PA, Stevenson PC, Belmain SR, Arnold SEJ, Woolley VC (2021) Natural pest regulation and its compatibility with other crop protection practices in smallholder bean farming systems. Biology 10:805. https://doi.org/10.3390/biology10080805
Negash S, Elema R, Tolosa A (2024) Evaluation of Rhizobium Bio-fertilizer inoculation on yield and yield components of Faba bean (Vicia faba L.) at Jarso District, East Hararghe, and Oromia, Ethiopia. Int J Trend Sci Res Dev 8:398–406. Regional review workshop on completed research activities, 151
Neuwinger HD (2004) Plants used for poison fishing in tropical Africa. Toxicon 44:417–430
Ochieng TA, Akutse KS, Ajene IJ, Kilalo DC, Muiru M, Khamis FM (2024) Interactions between Bacillus thuringiensis and selected plant extracts for sustainable management of Phthorimaea absoluta. Sci Rep 14(1):9299
Ola OT et al (2022) Aflatoxin biocontrol effectiveness in the real world—private sector-led efforts to manage aflatoxins in Nigeria through biocontrol-centered strategies. Front Microbiol 13:977789
Oliveira DGP, Alves LFA, Sosa-Gómez DR (2014) Advances and perspectives of the use of the entomopathogenic fungi Beauveria bassiana and Metarhizium anisopliae for the control of arthropod pests in poultry production. Braz J Poult Sci 16:01–12
Ortiz-Urquiza A, Keyhani NO (2016) Molecular genetics of Beauveria bassiana infection of insects. Adv Genet 94:165–249
Osman GE, El-Ghareeb D, Already R, Assaeedi AS, Organji SR, Abulreesh HH, Althubiani AS (2015) Bioinsecticide Bacillus thuringiensis a comprehensive review. Egypt J Biol Pest Control 25(1):271–288
Parker KM, Borrero VB, Van Leeuwen DM, Lever MA, Mateescu B, Sander M (2019) Environmental fate of RNA interference pesticides: adsorption and degradation of double-stranded RNA molecules in agricultural soils. Environ Sci Technol 53(6):3027–3036
Ploeg AT (2002) Effects of selected marigold varieties on root-knot nematodes and tomato and melon yields. Plant Dis 86:505–508
Pino O, Sánchez Y, Rojas MM (2013) Plant secondary metabolites as an alternative in pest management. I: background, research approaches and trends. Rev Prot Veg 28(2):81–94
Poornakala T, Muniasamy S, Rajagopal T, Archunan G, Ponmanickam P (2023) Insect pheromones: chemistry and applications. Curr Adv Biosci 78:149–155
Rada WM, Atreby EL (2017) Efficiency of three natural products for controling Desert Locust, Schistocerca Gregaria (Forsk.)(Orthoptera: Acrididae). J Environ Sci 38(2):93–104
Ramirez A (2013) Pyrethrum secondary metabolism: biosynthesis, localization and ecology of defence compounds. Wageningen University and Research
Rank AP, Koch A (2021) Lab-to-field transition of RNA spray applications–how far are we? Front Plant Sci 12:755203
Ravensberg WJ, Ravensberg WJ (2011) General introduction and outline. In: A roadmap to the successful development and commercialization of microbial pest control products for control of arthropods. Springer Science & Business Media, pp 1–21
Reddy PP (2017) Biointensive integrated pest management in horticultural ecosystems. Springer. https://doi.org/10.1007/978-81-322-2783-9
Regnault-Roger C, Philogène BJR (2008) Past and current prospects for the use of botanicals and plant allelochemicals in integrated pest management. Pharm Biol 46(1–2):41–52
Rehman S, Shahzad B, Bajwa AA, et al (2019) Utilizing the allelopathic potential of Brassica species for sustainable crop production: a review. J Plant Growth Regul 38:343–356

Reuters (2024) Bioherbicide helps lift Kenya's Witchweed curse for farmers. Retrieved from https://www.reuters.com/world/africa/bioherbicide-helps-lift-kenyas-witchweed-curse-farmers-2024-10-30/. Accessed 24 October 2024
Rocha GT, Queiroz PRM, Grynberg P, Togawa RC, Ferreira ADC d L, Do Nascimento IN, Gomes ACMM, Monnerat R (2023) Biocontrol potential of bacteria belonging to the Bacillus subtilis group against pests and diseases of agricultural interest through genome exploration. Antonie Van Leeuwenhoek 116(7):599–614
Roh J-Y, Choi J-Y, Li M-S, Jin B-R, Je Y-H (2007) Bacillus thuringiensis as a specific, safe, and effective tool for insect pest control. J Microbiol Biotechnol 17(4):547–559
Rodrigues MC, Brito Brito ML, da Silva VD, Mendonça DP, Ribeiro MJ, da Silva GB, de Castro GL (2025) Combined application of Bacillus subtilis and seaweed extracts in açaí palm seedlings subjected to water deficit and recovery. J Appl Phycol 1–3
Ryan MF (2002) Plant chemicals. In: Insect chemoreception: fundamental and applied. Springer Netherlands, Dordrecht, pp 27–72
Rouphael Y, Colla G (2020) Biostimulants in agriculture. Front. Plant Sci. 11:40
Roy PS, Ramachandran RM, Paul O, et al (2022) Anthropogenic land use and land cover changes—A review on its environmental consequences and climate change. J Indian Soc Remote Sens 50:1615–1640
Saber EA, Elbagory M, Abdel-Kader NI, Ahmed ME, Abd El-Rahman LA, Khalifa TH, Omara AE-D (2024) Sustainable practices for arid climates: evaluating combined mulches with biostimulant in combating soil salinity and cowpea cultivation. Horticulturae 10(11):1213
Samada LH, Tambunan US (2020) Biopesticides as promising alternatives to chemical pesticides: A review of their current and future status. Online J Biol Sci 20(2):66–76
Saxena RC, Dixit OP, Harshan V (1992) Insecticidal action of Lantana camara against Callosobruchus chinensis (Coleoptera: Bruchidae). J Stored Prod Res 28:279–281
Seck D, Lognay G, Haubruge E, et al (1993) Biological activity of the shrub Boscia senegalensis (Pers.) Lam. ex Poir.(Capparaceae) on stored grain insects. J Chem Ecol 19:377–389
Sharma A, Kumar V, Shahzad B, Tanveer M, Sidhu GP, Handa N, Kohli SK, Yadav P, Bali AS, Parihar RD, Dar OI (2019) Worldwide pesticide usage and its impacts on ecosystem. SN Appl Sci 1(11):1446
Sharma R, Sharma P (2021) Fungal entomopathogens: a systematic review. Egypt J Biol Pest Control 31:1–13
Singh MD (2017) Nano-fertilizers is a new way to increase nutrients use efficiency in crop production. Int J Agric Sci ISSN 9:975–3710
Singh AK et al (2022) Entomopathogenic nematodes: a sustainable option for insect pest management. In: Biopesticides. Elsevier, pp 73–92
Siripattanakul-Ratpukdi S, Fürhacker M (2014) Issues of silver nanoparticles in engineered environmental treatment systems. Water, Air, Soil Pollut 225:1939
Song MH, Jeong Seon Y, Kim S, Lee SJ, Kim JC, Nai Y-S, Shin TY, Kim JS (2019) Downstream processing of Beauveria bassiana and Metarhizium anisopliae-based fungal biopesticides against Riptortus pedestris: solid culture and delivery of conidia. Biocontrol Sci Tech 29(6):514–532
Soth S, Glare TR, Hampton JG, et al (2022) Biological control of diamondback moth—increased efficacy with mixtures of Beauveria fungi. Microorg 10:646
Svidritskiy E, Ling C, Ermolenko DN, Korostelev AA (2013) Blasticidin S inhibits translation by trapping deformed tRNA on the ribosome. Proc Natl Acad Sci 110(30):12283–12288
Svoboda P (2020) Key mechanistic principles and considerations concerning RNA interference. Front Plant Sci 11:1237
Tabacchioni S, Passato S, Ambrosino P, et al (2021) Identification of beneficial microbial consortia and bioactive compounds with potential as plant biostimulants for a sustainable agriculture. Microorg 9:426
Tantawy MM et al (2018) An attempt to design a chemical control program to suppression piercing sucking pests infesting cucumber plants at Sohag region. Minia J Agric Res Dev 38(2):349–362

Trivedi P, Leach JE, Tringe SG, et al (2020) Plant–microbiome interactions: from community assembly to plant health. Nat Rev Microbiol 18:607–621

Ujváry I (2010) Pest control agents from natural products. InHayes' handbook of pesticide toxicology 19–229

Ukeh DA, Birkett MA, Pickett JA, et al (2009) Repellent activity of alligator pepper, Aframomum melegueta, and ginger, Zingiber officinale, against the maize weevil, Sitophilus zeamais. Phytochemistry 70:751–758

Upamanya GK, Bhattacharyya A, Dutta P (2020) Consortia of entomo-pathogenic fungi and bio-control agents improve the agro-ecological conditions for brinjal cultivation of Assam. 3 Biotech 10:450

Vlaiculescu A, Varrone C (2022) Sustainable and eco-friendly alternatives to reduce the use of pesticides. In: Pesticides in the natural environment, pp 329–364. https://doi.org/10.1016/B978-0-323-90489-6.00014-8

Wang C, St Leger RJ (2007) A scorpion neurotoxin increases the potency of a fungal insecticide. Nat Biotechnol 25:1455–1456

Wann SB et al (2016) Isolation, characterization of nematode-controlling bacteria and fungi from nature. In: Microbial inoculants in sustainable agricultural productivity, vol 1: Research perspectives. Springer, pp 271–295

Wojda I, Kowalski P, Jakubowicz T (2009) Humoral immune response of Galleria mellonella larvae after infection by Beauveria bassiana under optimal and heat-shock conditions. J Insect Physiol 55(6):525–31

Xiao Y, Wu K (2019) Recent progress on the interaction between insects and Bacillus thuringiensis crops. Philos Trans R Soc Lond B Biol Sci 374(1767):20180316

Xu R, Xiao S, Yuan J, Zhao A (2011) Adsorption of methyl violet from aqueous solutions by the biochars derived from crop residues. Bioresour Technol 102:10293–10298

Yakhin OI et al (2017) Biostimulants in plant science: a review. Front Plant Sci 8:220

Yanni SF, Whalen JK, Ma BL (2010) Crop residue chemistry, decomposition rates, and CO 2 evolution in Bt and non-Bt corn agroecosystems in North America: a review. Nutr Cycl Agroecosyst 87:277–293

Zhu KY, Palli SR (2020) Mechanisms, applications, and challenges of insect RNA interference. Annu Rev Entomol 65(1):293–311

Chapter 19
Biofertilisers and Aquaculture in Africa: Enhancing Sustainable Fish Farming-Crop and Animal Production Synergies

Takudzwa C. Madzivanzira, Munetsi Zvavahera, Makomborero Nyoni, Norman Muzhinji, and Tatenda J. Gunda

Abstract The integration of aquaculture-crop and animal production promotes sustainable resource use by enhancing soil fertility, improving crop yields, recycling nutrients, and reducing environmental waste. In this synergised system, aquaculture by-products can be repurposed as biofertilisers for crop and animal production, and vice versa. This chapter provides an in-depth analysis of the current state of aquaculture-crop and animal production synergies in Africa. Aquaculture-crop and animal production synergies have significant benefits across the continent, including improved water quality, enhanced nutrient cycling increased yields. Aquaculture systems in Africa generate substantial amounts of solid and liquid waste, with solid biofertilisers derived from fish waste, algae, unused feed, and pond sediments, while liquid biofertilisers are produced from fish effluents and fermented fish waste. Crop and animal production provides biofertilisers such as compost and livestock manure, which enhance natural primary production and nutrient recycling

---

T. C. Madzivanzira (✉)
Freshwater Research Group, Department of Biological Sciences and Ecology,
University of Zimbabwe, Harare, Zimbabwe

South African Institute for Aquatic Biodiversity (SAIAB),
Makhanda, Eastern Cape, South Africa

M. Zvavahera
Centre for Global Change, Walter Sisulu University, Mthatha, Eastern Cape, South Africa

M. Nyoni
Innov BioM, Rhodhesville, Harare, Zimbabwe

N. Muzhinji
Department of Plant Sciences, Division of Plant Pathology, University of the Free State,
Bloemfontein, South Africa

T. J. Gunda
Golden Helix Life Sciences Ltd, Unn HSE Community 25, Peace Be Tema, Greater Accra,
Ghana, India

A. D. Nciizah et al. (eds.), *Biofertilizers for Sustainable Agriculture in Africa*,
Sustainability Sciences in Asia and Africa, https://doi.org/10.1007/978-981-95-6009-7_19

in ponds. However, the commercial production of biofertilisers from aquaculture in Africa has not been fully realised due to a wide range of challenges, such as a lack of capital and government support. Given Africa's rapid population growth, high food demands, and increasing environmental pressures, aquaculture-crop and animal production synergies are vital for sustainable resource management practices that ensure food security and environmental protection.

19.1 Introduction

The African continent possesses abundant natural resources and vast expanses of land suitable for agricultural production. This wealth of resources presents significant opportunities for economic growth and food security across the African region. The continent relies heavily on food production systems that include crop and animal production (Bjornlund et al. 2020) and aquaculture (El-Sayed and Fitzsimmons 2023) as primary sources of food security, livelihoods, and economic stability. The crop and animal production sector employs a significant population, particularly in the rural areas dominated by smallholder farming. Aquaculture, on the other hand, is gaining importance as a complementary strategy to meet the increasing demand for protein while addressing declining wild fish stocks (Adeleke et al. 2020; Menezes ct al. 2024). All the agriculture sectors, however, face challenges related to sustainability, productivity and resource management, necessitating innovative approaches to ensure long-term food security.

The concept of synergising biofertilisers and agriculture represents a promising avenue for sustainable development in Africa, addressing critical issues of food security and aquatic ecosystem health (Kahle and Palni 2023). Biofertilisers, as defined in the previous chapters, are organic substances that contain beneficial microbes, which enhance soil fertility and plant health by fixing atmospheric nitrogen (N), solubilising phosphates, and stimulating the growth of plants (Raimi et al. 2021; Kahle and Palni 2023). They present an environmentally friendly alternative to inorganic fertilisers, which have been shown to degrade soil quality and pollute water resources, leading to biodiversity loss when misapplied or overused (Win et al. 2018; Hasimuna et al. 2023; Zhao et al. 2024). Common biofertilisers include N-fixing bacteria such as *Rhizobium*, phosphate-solubilising bacteria like *Pseudomonas*, and mycorrhizal fungi that facilitate nutrient absorption in plants (Soumare et al. 2020; Zhao et al. 2024). The application of biofertilisers has been increasingly recognised for improving crop yields and reducing the environmental footprint of conventional farming practices (Zhao et al. 2024).

Aquaculture, defined as the cultivation of aquatic animals and plants in controlled environments, is emerging as a vital component of Africa's food security strategy (El-Sayed and Fitzsimmons 2023). With the increasing demand for protein-rich diets as well as the decline of wild fish stocks, aquaculture provides an efficient means of producing fish, shellfish, and other aquatic products (Heimann and Delzeit 2024). The industry, however, faces challenges related to sustainability, water quality management, and nutrient cycling. Integrating aquaculture with biofertilisers can create a closed-loop system where nutrient-rich aquaculture effluents are recycled as fertilisers, reducing waste and enhancing productivity across both sectors.

As the continent grapples with challenges posed by rapid population growth, climate change, and resource depletion, innovative strategies that enhance productivity while maintaining ecological stability have become imperative. Biofertilisers and agriculture, when implemented synergistically, offer a holistic approach to sustainable food production and environmental stewardship. This chapter explores the potential of biofertilisers and aquaculture in Africa, emphasising their roles in sustainable aquaculture-crop and animal production while highlighting examples where these practices have been successfully implemented. This chapter further examines the opportunities, benefits and challenges associated with their adoption, advocating for integrated, ecologically sound approaches that balance productivity, resource conservation, and socio-economic well-being in line with achieving Africa's sustainable development goals.

19.2 Overview of Aquaculture in Africa: Current Trends and Challenges

In Africa, aquaculture significantly enhances food security by providing essential protein and micronutrients while creating income opportunities and employment, particularly in riparian communities (Musinguzi et al. 2019; Menezes et al. 2024). African governments are increasingly recognising the importance of aquaculture for food security and economic development (Adeleke et al. 2020). Although global fish consumption has been increasing in recent years, partly due to improved preference for fish as a source of high-quality protein, the per capita consumption of fish in Africa still remains low (Musinguzi et al. 2019; Menezes et al. 2024). Aquaculture production in Africa has steadily increased in recent years, yet progress remains slow despite Africa's vast untapped potential for growth in the sector (Hinrichsen et al. 2022; Menezes et al. 2024). The region experienced a remarkable surge from 1995 to 2018, tin aquaculture production, expanding from 110,200 to 2,196,000 tons with a compound annual growth rate of 15.55% (Adeleke et al. 2020). This expansion has been primarily driven by the rise of private sector-led small and medium enterprises and increased foreign direct investment (Adeleke et al. 2020). Reasons for the slow development of aquaculture in Africa, despite the availability of vast natural resources and a growing demand for food, include a lack of infrastructure and development capital, limited technological know-how, inadequate information and poor governance, among others (Hinrichsen et al. 2022).

19.2.1 Aquaculture Production Trends Across Africa

Aquaculture production in Africa varies between countries, with Egypt standing out as the dominant producer, contributing over 1.5 million metric tons. This makes Egypt far and away the largest producer on the continent, reflecting its well-established infrastructure for aquaculture. Nigeria, with a production of 291,233 metric tons, ranks second, while substantially lower than Egypt, still represents a notable contribution to Africa's aquaculture output. Other countries in the top five include Uganda, Ghana, and Zambia, with aquaculture production of 103,737; 76,630 and 24,300 metric tons, respectively. Tunisia and Kenya, with production levels of 21,756 and 15,124 metric tons respectively, also feature in the top 10, and Malawi (9014 metric tons) and Madagascar (7421 metric tons) follow closely behind, while South Africa (6181 metric tons) rounds out the list.
These nations play a crucial role in the continent's overall production, driven by both domestic consumption and export potential. With respect to the global share of aquaculture production, a similar trend to that observed in aquaculture production within Africa is evident. Amongst African countries, Egypt leads, contributing 1.9% to the global aquaculture industry. In comparison, Nigeria, the second-largest African producer, contributes a smaller 0.35% to global aquaculture. Other African countries show even smaller global contributions: Uganda (0.13%), Ghana (0.09%), Zambia (0.03%), Tunisia (0.03%), Kenya, Malawi, Madagascar, and South Africa (ranging from 0.02% to 0.01%). Although these countries are part of Africa's top ten aquaculture producers, their global aquaculture contributions are very minimal.

The economic impact of aquaculture in Africa is substantial. Employment plays a significant role in evaluating the socio-economic impact of aquaculture (Menezes et al. 2024). Aquaculture jobs in Africa involve various roles such as fish farming, processing, distribution, and value chain activities that support livelihoods and food security. The sector employs about 6.2 million people in Africa (FAO 2022) from both capture fisheries and aquaculture. Women play a prominent role, particularly in postharvest processing and marketing activities (Adeleke et al. 2020). Research by Chan et al. indicates that if current trends in aquaculture expansion continue, the African fish sector will support 22 million jobs by 2050.

19.2.2 Common Aquaculture Species in Africa

Aquaculture is perceived as the primary solution for addressing the growing fish supply gap in Africa (Musinguzi et al. 2019). Freshwater aquaculture dominates the aquaculture sector in Africa, with tilapia and catfish native to the continent being the most commonly cultivated species, while mariculture remains a minor but promising sector (Adeleke et al. 2020). Eleven tilapia species are currently farmed in Africa with the Nile tilapia (*Oreochromis niloticus* Linnaeus 1758) being by far, the most widely cultured tilapia species in the continent (Adeleke et al. 2020; FAO

2021), due to its rapid growth rate, tolerance to varying environmental conditions, high reproduction, ability to thrive in diverse aquaculture systems, and nutritional profile (El-Sayed and Fitzsimmons 2023). *Oreochromis niloticus* is naturally distributed in the Nilo-Sudanian region, the Ethiopian Rift Valley, the western Rift Lakes (Albert, Edward, George, Kivu, and Tanganyika) and Lake Turkana in the eastern Rift Valley. *Oreochromis niloticus* is also naturally established in Central and Western Africa (Benue, Chad, Gambia, Niger, Senegal, and Volta river basins) (McAndrew 2000). The species has been widely introduced across the African continent for aquaculture purposes, with feral populations establishing in various freshwater systems (Ellender et al. 2014).

The African sharptooth catfish (*Clarias gariepinus* Burchell 1822) is another widely cultivated fish species in African aquaculture, playing a significant role in the continent's food security and economic development. *Clarias gariepinus* is native to most African regions and small parts of Asia (Syria, Israel, and Southern Turkey) (Weyl et al. 2016) where it inhabits a broad range of habitats such as rivers, lakes, impoundments, swamps, floodplains, and even some estuaries (Weyl et al. 2016). Known for its hardiness, *C. gariepinus* can thrive in various aquaculture systems, making it suitable for both small-scale and commercial fish farming (Oké and Goosen 2019). *Clarias gariepinus* are omnivorous and can consume a wide range of feed types, such as macrophytes, plankton, invertebrates, fish, amphibians, detritus, molluscs, birds, and sediment (Tesfahun and Alebachew 2023). This diet plasticity makes *C. gariepinus* cultivation more cost-effective, especially in regions where access to high-protein feeds is limited. Nutritionally, *C. gariepinus* is valued for its high protein content, which supports muscle growth and overall health, while its low-fat content makes it a heart-healthy protein choice (Mahboob et al. 2019).

In addition to *O. niloticus*, *C. gariepinus*, other species like the Common carp (*Cyprinus carpio* Linnaeus 1758) and rainbow trout (*Oncorhynchus mykiss* Walbaum 1792) are also cultivated in African aquaculture. Both species are non-native to the African continent: *Cyprinus carpio* is native to Asia and Eastern Europe; and *O. mykiss* is native to the cold waters of the Pacific Ocean in Asia and North America. African aquaculture also includes molluscs and crustaceans. Prawns and crayfish are popular crustaceans, both high in protein, low in fat, and rich in minerals. Prawns, with their high market value, are often farmed in ponds or tanks. Molluscs such as oysters, abalone, and mussels are commonly farmed in coastal areas. Abalone valued for their high protein and mineral content, have a high market value, while mussels, equally nutritious, are more affordable and widely consumed.

The diversification of species in African aquaculture reflects a growing industry that seeks to meet rising fish demand while promoting sustainable practices. Expanding the cultivation of multiple species helps reduce overfishing pressure on wild stocks and enhances food security across the continent. The next subsection focuses on aquaculture production systems, examining the best suitable, to integrate with biofertilisers for enhanced productivity and sustainability.

19.2.3 Aquaculture Production Systems

In Africa, aquaculture production systems are diverse, reflecting the continent's varied economic capacities, environmental conditions, and cultural practices. Aquaculture predominantly takes place in inland waters, which accounts for 82% of the production volume (FAO 2022). Ocean-based aquaculture is mainly practiced in the Black and Mediterranean and Seas accounting for 70% of marine production (Menezes et al. 2024).

Aquaculture production systems in Africa include pond culture, cage culture, recirculating aquaculture systems (RAS) with water filtration and reuse, raceway systems, integrated aquaculture, water channels, pen and enclosure systems, and hatchery-based systems for producing fingerlings (Hinrichsen et al. 2022). Pond culture involves the cultivation of fish in artificial earthen ponds and is widely used as a cost-effective system for smallholder fish farmers (Verdegem et al. 2023). Cage culture utilises enclosures in lakes and reservoirs, enabling high production but requiring careful management to prevent pollution. The RASs are advanced setups where water is filtered and reused, allowing for intensive fish production in controlled environments. Raceway systems use continuously flowing water channels, ideal for fish species that require high levels of dissolved oxygen (DO), and lotic waters. Integrated aquaculture combines fish farming with crop and/or livestock production, maximising resource use and promoting nutrient recycling. Pen and enclosure systems confine fish within netted areas of natural water bodies, providing a cost-effective alternative for large-scale production. Hatchery-based systems focus on producing quality fingerlings for stocking other aquaculture systems, ensuring a steady supply of fish for grow-out operations (Verdegem et al. 2023).

In North Africa, the use of extensive aquaculture systems involves the stocking of juvenile fish in reservoirs, a practice seen in Algeria, Egypt, Libya, and Morocco (El-Sayed 2017). Semi-intensive aquaculture, mainly conducted in earthen ponds with supplementary feed, dominates in Egypt, contributing to approximately 80% of the country's fish production (Hinrichsen et al. 2022). Southern Africa employs a range of production methods that include raceways for trout, ponds for extensive and semi-intensive fish farming, cages for grow-out, longlines for mussel and oyster culture, baskets for oyster shallow flow-through systems, and RASs (Adeleke et al. 2020; FAO 2021). The RAS technology has been widely adopted in abalone farming in South Africa (Satia 2017). Abalone farming in the region utilises land-based systems with advanced hatchery technologies and specialised tanks (Satia 2017). Cage fish farming is also prominent in Southern Africa, with a good example from Lake Kariba (Ellender et al. 2014). East African aquaculture mainly involves extensive pond-based systems due to simpler technical requirements and lower infrastructure costs compared to intensive systems (Hinrichsen et al. 2022). Some aquaculture farms, however, have adopted the cage-based intensive production methods, which offer profitability and lower capital investment per unit of fish produced (Satia 2017). Uganda leads East Africa in inland cage farms, followed by Kenya, Tanzania, and Rwanda (Musinguzi et al. 2019). West African aquaculture features dam ponds in countries like Cameroon, Côte d'Ivoire, and Guinea while integrated agri-aquaculture systems are reported in Benin and Nigeria (Satia 2017; Hinrichsen et al. 2022). For the purpose of this chapter, emphasis will be placed on integrated aquaculture systems and pond culture, as these production methods closely align with the use of biofertilisers in sustainable agriculture and aquaculture.

The African continent is not immune to several challenges that affect aquaculture around the world. The continent faces even more pressing challenges that hinder the full aquaculture potential despite the continent's vast water resources and favourable climatic conditions (Satia 2017). One significant barrier is limited access to quality inputs, including feed, equipment, infrastructure and fingerlings, which affects productivity and profitability (Soliman 2017). Many small-scale aquaculture farmers rely on poorly bred fingerlings and low-quality leading to slow growth rates and high mortality (Emam et al. 2024). Additionally, insufficient training and technical knowledge among aquaculture farmers limit the adoption of best practices in aquaculture, such as proper pond management, water quality monitoring, and disease control (Gwazani et al. 2022). Financial constraints further compound the challenges, as many aquaculture farmers lack access to affordable credit and investment opportunities to expand or modernise their operations (Mangeni and Mhlanga 2019). Poor infrastructure, including inadequate refrigeration facilities and poor road networks, restricts efficient distribution and market access, causing serious postharvest losses and reduced profitability (Abelti and Teka 2024). Environmental challenges also play a critical role, as some aquaculture practices have led to habitat degradation, water pollution (Ojewole et al. 2024), and the spread of invasive species (Ellender et al. 2014; Madzivanzira et al. 2020, 2021), while climate change has introduced unpredictable weather patterns affecting water temperatures and availability (Magadza et al. 2020). Policy and regulatory frameworks in many African countries remain underdeveloped or poorly enforced, leading to challenges in monitoring compliance and supporting sustainable growth (Ellender et al. 2014). Additionally, competition with capture fisheries and imported seafood often makes it difficult for locally farmed fish to compete in price and quality (Zajicek et al. 2021). Social and geopolitical challenges, such as conflicts over water use, gender disparities, and limited community involvement in decision-making processes, further complicate aquaculture development. By overcoming these barriers, aquaculture in Africa can contribute significantly to food security, economic development, and poverty alleviation.

19.3 The Intersection of Aquaculture-Crop and Animal Production in Africa

Integrated aquaculture-crop and animal production systems (hereafter referred to as integrated systems) involve the combination of various components that interact and are complementary to one another to fulfil human needs while maintaining harmony with ecological systems (Zajdband 2011). This approach can be applied in multiple ways: fish cultured alongside crops, livestock, or a mix of both within the same production system (Singh et al. 2021). Alternatively, crops, livestock, and fish, may be cultivated separately on the same farm, with by-products from each section serving as inputs for the others. This method enhances food quality, generates affordable nutrition, creates employment, and boosts household income (Hasimuna et al. 2023).

A key advantage of the integrated system is the synergistic relationship between its components, where the collective benefits exceed those of the individual parts. Materials that would typically be discarded as waste in one system are instead repurposed as resources for another, increasing resource efficiency and overall productivity (Hasimuna et al. 2023). Farm assets such as land, water, and organic waste can be optimised for improved production outcomes (Maulu et al. 2019). Furthermore, this integrated system provides farmers with a reliable source of income from the diverse elements within their production system (Zajdband 2011).

19.3.1 Integrated Production Systems

(a) Aquaculture-Crop System

The aquaculture-crop system involves simultaneously cultivating fish and crops, with both components benefiting from shared resources and waste recycling (Zajdband 2011). In this system, fishponds or tanks are integrated into crop fields, allowing nutrient-rich water from the fish culture to be used as irrigation and biofertiliser for crops (Mariscal-Lagarda et al. 2012). The effluents from the fish tanks, rich in nutrients such as N and phosphorus (P); and organic matter, improve soil fertility, and promote crop growth (Melaku and Natarajan 2019). This system minimises the need for expensive inorganic fertilisers, reduces water usage, and enhances overall farm productivity (Hasimuna et al. 2023).

Common crops grown in such systems include rice, vegetables, and fruits, with rice–fish farming being one of the most prominent examples, especially in Asian countries (Hasimuna et al. 2023). *Oreochromis niloticus* has been the predominant fish species in these systems, with a few reports of *C. gariepinus* (Melaku and Natarajan 2019). The rice–fish farming integration has been found to greatly increase the productivity of both rice and fish farming systems (Melaku and Natarajan 2019; Hasimuna et al. 2023). In this system, rice filters potentially hazardous compounds such as ammonia, nitrates, and phosphates, which are important nutrients for the growth of rice (Lemma et al. 2014). These nutrients come from the fish waste and decomposing organic matter (Melaku and Natarajan 2019). By filtering the hazardous nutrients, rice fields provide a favourable habitat for fish and other important invertebrates to flourish. Rice–fish farming significantly reduces pesticide use, hence preserving biodiversity (Hasimuna et al. 2023).

(b) Aquaculture-Livestock System

In the aquaculture-livestock system, fish are cultured alongside livestock such as cattle, pigs, or poultry, creating a mutually beneficial relationship (Melaku and Natarajan 2019). The main link between fish and livestock production is the usage of nutrients, especially the application of livestock manure for fish production. This application of livestock manure is aimed mainly at supplying nutrients like N and P, which are the precursors of the primary production in fish ponds (Melaku and Natarajan 2019). Livestock manure is also directly fed to detritus-feeding fishes. This system can accommodate a variety of livestock species, including pigs, ducks, chickens, and ruminants such as cattle, goats, and sheep (Hasimuna et al. 2023). Aquaculture-livestock system is predominantly utilised in many Asian nations, but have also been documented in several African nations, including Ethiopia, Kenya, and Malawi (Melaku and Natarajan 2019).

(c) Aquaculture-Crop and Livestock System

The aquaculture-crop–livestock system represents a comprehensive approach where all three production components: crops, livestock and fish are integrated into a single farming operation (Nhan et al. 2007). This system maximises resource efficiency by utilising by-products and waste from one component to support the others, creating a sustainable and closed-loop agricultural model (Ndjadi et al. 2023). In such systems, livestock waste can be composted or directly applied as organic fertiliser for both fish ponds and crops (Nhan et al. 2007). The nutrient-rich water from fish ponds can, in turn, be used for irrigating crops, while crop residues may serve as supplementary feed for livestock. For instance, rice bran and vegetable waste can be fed to livestock or used as fish feed, creating a continuous cycle of nutrient exchange (Nhan et al. 2007).

The holistic integration of the three components enhances agricultural sustainability by improving soil fertility, reducing chemical input dependency, and lowering production costs (Ndjadi et al. 2023). It also contributes to diversified farm income, increased food security, and reduced environmental impact through effective waste recycling (Nhan et al. 2007). The successful implementation of the integration, however, requires proper design and management to ensure nutrient balance, avoid overloading ponds with organic matter, and maintain optimal water quality (Melaku and Natarajan 2019).

19.3.2 Integrated Production Systems Across Africa

Various integrated systems have been implemented across Africa with notable success, focusing on fish species such as *O. niloticus*, *C. carpio* redbreast tilapia (*Coptodon rendalli*) and *C. gariepinus* combined with livestock species like cattle, goats, chickens, and ducks, well summarised in Hasimuna et al. (2023). Prinsloo and Schoonbee (1987) demonstrated that a combination of *C. carpio*, Peking ducks, and the vegetables tomato, spinach, and lettuce resulted in ducks gaining an average of 2.6 kg in 55 days. Fish production surpassed 19.5 tons/ha/year. Ten batches of Peking ducks were stocked at a density of 2500 ducks per hectare of water over 6 months, yielding an average of 32,184 tons/ha/year with a feed conversion ratio (FCR) of 3.05.

Kang'ombe et al. (2006) investigated a system integrating *C. rendalli* with cattle, pig and chicken manure. After 84 days, *C. rendalli* treated with chicken manure were significantly larger and produced higher yearly yields than those treated with cattle or pig manure or no manure. Dey et al. (2010) reported a significant increase in chlorophyll and zooplankton in waters treated with chicken manure, further improving fish growth.

Mohammed et al. (2015) examined the integration of *O. niloticus* with White Leghorn egg-laying chickens. They observed an increased net present value of the project, which facilitated funding from the local national bank. Endebu et al. (2016) explored a combination of *O. niloticus*, *C. carpio*, and *C. gariepinus* with red layers chickens, goats' manure, and Rhode Island chickens. Similar to Dey et al. (2010), poultry manure significantly enhanced primary productivity by increasing zooplankton abundance. Fish grown in poultry-manured ponds exhibited significantly better growth performance compared to non-manured ponds.

Getu et al. (2017) evaluated *O. niloticus* integrated with White Leghorn chicken, and vegetable crops such as Vikima cabbage, red bombey onion, and the ROMA VFN tomato variety. The net profit of the integrated farming system was significantly improved per ha. Daba et al. (2017) further investigated *O. niloticus*, *C. carpio*, and *C. gariepinus* integrated with pullets of Lohman brown chicken and the onion variety *Allium cepa*. The total profit from this integrated system increased compared to the single farming system. Egg, fish, and vegetable productivity rose dramatically, with 8 tons of fish per ha per year, 233 eggs per hen per year, and 10,800 kg of onions per ha per year.

Waktola et al. (2016) analysed an integrated system of *O. niloticus* with cattle, poultry, and tomato and Bombay red onion crops. The water's physicochemical properties were maintained within the optimal range for *O. niloticus* growth. Nitrate and phosphate concentrations were higher in the integrated systems, and organic matter content was superior to nonintegrated systems. The integrated system produced higher quantities of tomatoes and onions, while the expense and income analysis revealed this method to be more profitable than conventional farming using chemical fertilisers.

Lemma (2017) studied *O. niloticus* integrated with Rhode Island Red chicken and a diverse vegetable selection including Bombay red onion, Allium cepa, *Brassica oleracea* cabbage and Malkashola tomato. Cabbage production increased from 9800 kg/ha to 22,100 kg/ha, while Bombay red onion yields increased from 16,500 to 37,100 kg/ha under the integrated system. Tomato yields also improved, with the integrated system producing 45,800 kg/ha compared to 17,100 kg/ha in nonintegrated plots.

Ugwumba (2010) demonstrated that *C. gariepinus* integrated with maize, yam, and cassava resulted in a net return on investment 0.5 times greater for crop–fish integrated systems compared to crop and fish systems alone. Miller et al. (2006)

observed a 10% increase in rice yield and over 50% revenue gains from fish sales in *O. niloticus* integrated with rice systems. Lemma et al. (2014, 2015) studied *O. niloticus* and rice systems, reporting increased paddy grain and plant biomass production in integrated plots (112.3 kg) compared to control plots (95.65 kg). Nutrient recycling contributed to higher fish yields with larger fish size and better nutrient retention. Rasowo and Auma (2006) concluded that *O. niloticus* integrated with rice led to improved fish production and recovery rates, while total rice yields were slightly lower compared to nonintegrated systems.

The integrated system combining fish, livestock, and crop production offers significant benefits for productivity and sustainability. Common fish species that include *O. niloticus*, *C. carpio*, and *C. gariepinus* integrated with poultry such as White Leghorn, Rhode Island Red, and Peking ducks, along with crops including rice, maize, tomatoes, onions, spinach, and yams are common examples across the African continent. Studies consistently demonstrate increased yields in livestock, crop, and fish products, along with enhanced water quality and nutrient cycling. Manure from livestock serves as a valuable nutrient source, promoting primary productivity in ponds, which supports fish growth while reducing nutrient depletion rates. The diversity of crops tested indicates that integrated systems can be adapted to various ecological conditions with positive economic returns. These systems not only improve profitability but also contribute to more efficient use of resources, making them a promising strategy for sustainable agriculture and food security in diverse regions.

19.4 Aquaculture as a Source of Biofertilisers

While the rapid expansion of aquaculture on the African continent helps to meet rising food demands in the region, it also produces voluminous amounts of waste and wastewater rich in dissolved nutrients and organic matter (Zhang et al. 2023). Approximately 50% of the fish produced from aquaculture is lost as waste (Sahu et al. 2014), and the waste presents a major challenge of disposal (Sahu et al. 2014), and in most cases contributes to pollution of aquatic ecosystems. This nutrient load leads to environmental pollution in nearby water bodies, posing risks to both aquatic and human health (Ndiaye et al. 2019). In China, Li et al. (2020) estimated that aquaculture waste effluents release approximately 1 million tons of N and 173,000 tons of P into water bodies annually. This is of concern as similar effluents could also be released into aquatic systems in Africa, which otherwise could be utilised as biofertilisers.

Fish farming serves as a valuable source of biofertilisers through recycling of organic waste generated during fish farming operations (Jaies et al. 2024). Aquaculture generates vast amounts of fishery waste during culturing, catching, and processing of fish (Jaies et al. 2024), which can potentially make biofertilisers. Aquaculture provides several biofertiliser sources that can be classified into solid and liquid biofertilisers. Each type of biofertiliser offers distinct benefits and is suitable for different agricultural applications. Solid biofertilisers are ideal for long-term soil health improvement, while liquid biofertilisers provide immediate nutrient availability and stress resistance. Solid and liquid biofertilisers are discussed in the following subsections with examples from African countries.

19.4.1 Solid Biofertilisers

Solid biofertilisers are generally produced from waste using faecal matter from piggery, dairy, and poultry animals (Arshad 2023). Unfinished food matter and wasted food from animal husbandry contain high value of nutrients that can be used as biofertilisers (Sahu et al. 2014). Soiled bedding matter used for poultry and rabbits, for example, can be a good source of fertiliser for production. (Arshad 2023). In aquaculture, solid biofertilisers are primarily derived from unused feed, fish faeces, algae, dead fish, and enriched sediment (Sahu et al. 2014; Muendo et al. 2014; Nabti et al. 2016; Win et al. 2018; Jaies et al. 2024). These materials are rich in organic matter and essential plant nutrients, including NPK and trace minerals. Fish faeces from *O. niloticus* contain $\approx 18\%$ dry matter, which is slightly higher than that typically in pig manure, which contains $\approx 7\%$ dry matter (Ndiaye et al. 2019). Solid biofertilisers are either applied directly or prepared using compost heaps where microbial decomposition occurs, leading to the mineralisation of nutrients (Muendo et al. 2014). Beneficial microbial strains such as *Bacillus subtilis* and *Trichoderma* sp. Among others are commonly present in solid biofertilisers. These microbes improve the soil structure, enhance water retention, and promote microbial activity (Agwa et al. 2017).

In Senegal, Ndiaye et al. (2019) demonstrated that a biofertiliser from *O. niloticus* produced similar effects on the growth parameters of onion and tomato plants compared to the unfertilised control and NPK treatments in the first 30 days. Some growth parameters, however, changed during the second measurement period (90 days) where plant height and the distance between nodes became similar to the NPK treatment and higher compared with the unfertilised treatment (Ndiaye et al. 2019).

The use ofalgae-based biofertilisers has become central to green agriculture worldwide (Win et al. 2018). Algae-based biofertilisers have also been shown to yield significant benefits in agriculture. The use of algae-based biofertilisers comes from the fact that algae are good nitrogen fixers. Besides N fixation, algae-based biofertilisers increase soil fertility and give the plant-growth-promoting rhizobacteria effect to crops (Win et al. 2018). In Nigeria, Agwa et al. (2017) investigated the biofertiliser potential of the green algal species *Chlorella vulgaris* by comparing its effects with NPK commercial fertiliser, and poultry manure, on the microbiological and physicochemical properties of the rhizosphere soil of okra *Hibiscus esculentus*. Species of *Aspergillus*, *Bacillus*, *Citrobacter*, *Fusarium*, *Micrococcus*, *Penicillium*, *Pseudomonas*, and *Staphylococcus*, were found in rhizosphere soil inoculated with *C. vulgaris*. These species collectively improve soil fertility by solubilising phosphate and cycling N (Agwa et al. 2017). The germination time of *H. esculentus* seeds inoculated with *C. vulgaris* was shortened by 3 days. NPK application before planting inhibited the germination of *H. esculentus,* which could be due to the high concentrations of chemical fertilisers elicited on the tender seeds. *Hibiscus esculentus* treated with *C. vulgaris* showed an increased mean pod yield per plant of 14, becoming ready for harvest 8 weeks after seed germination. In contrast, pods from NPK treatments were ready for harvest only after the 9–10th week following seed germination. The highest crude protein, crude lipid, moisture, and chlorophyll contents were associated with *H. esculentus* cultivated in soil amended with *C. vulgaris* inoculant (Agwa et al. 2017).

Sediment rich in NPK and organic matter accumulates in aquaculture fish ponds (Muendo et al. 2014), affecting the water quality and ichthyofaunal health. To maintain a pond environment conducive to fish growth, enriched sediment needs to be periodically replaced. Disposal of pond sediments to the environment is a waste of valuable nutrients (Muendo et al. 2014). In Asia, pond sediments have been used to fertilise crops and higher yields have been realised (Kong et al. 2023). Despite fish farming expanding in Africa, utilisation of pond sediment is not fully explored. Experiments in Africa have been conducted by Ihejirika et al. (2012) and Muendo et al. (2014) in Nigeria and Egypt, respectively, which demonstrated that sediment from aquaculture ponds is rich in NPK and organic matter beneficial for plant growth.

From the marine environment, seaweed, a macroalga biotic component, emerges as a valuable resource, with its applications as a biofertiliser (Nabti et al. 2016). Seaweeds improve soil conditions and crop performance by supplying NPK, and various micronutrients like Ca, Magnesium (Mg), and Iron (Fe) (Raghunandan et al. 2019). They are rich in organic compounds such as amino acids, vitamins and fibre, contributing to overall plant health. Additionally, seaweeds have high mineral content and offer diverse polysaccharides beneficial for plant growth (Raghunandan et al. 2019). The African continent possesses vast offshore, coastal and land resources well-suited for seaweed cultivation, particularly in pollution-free regions (Msuya et al. 2022). Over 2000 seaweed species have been recorded in Africa, with only 13 countries currently involved in seaweed farming and harvesting for domestic utilisation and export. However, compared with Asia, Africa lags behind in production and utilisation of seaweeds as biofertilisers (Msuya et al. 2022). In South Africa, research efforts are underway to optimise seaweed cultivation techniques to manufacture biofertilisers to meet agricultural demands.

19.4.2 Liquid Biofertilisers

This includes urine and wastewater from animal husbandry. Wash water is another commonly used ingredient to produce biofertilisers (Arshad 2023). In aquaculture, liquid biofertilisers are derived from fish effluents and fermented fish waste (Maquén-Perleche et al. 2023). The fermentation process, typically carried out in fermentation chambers, breaks down organic matter, releasing soluble NP, amino acids, and microbial cultures beneficial to plants (Maquén-Perleche et al. 2023). Microbial strains such as *Bacillus* sp. and *Azotobacter* sp. are commonly involved in liquid biofertilisers, facilitating N fixation and nutrient solubilisation (Mponda and Kim 2023). Liquid biofertilisers are often used as foliar sprays or through fertigation, providing rapid nutrient delivery and boosting plant metabolism (Maquén-Perleche et al. 2023). They have also been shown to improve crop resilience to environmental stress (Otaiku 2022).

In Tanzania, more than 70% of fish waste and 43.0% of wastewater is generated during the industrial processing of Nile perch (*Lates niloticus* Linnaeus 1758) (Mponda and Kim 2023). In Northeastern and Central African countries with high *L. niloticus* consumption, only a small portion of *L. niloticus* waste is repurposed for animal feed production. The remaining waste is discarded, whereas wastewater is chemically treated or drained into aquatic systems, which poses a threat to aquatic life owing to environmental pollution. Mponda and Kim (2023) further investigated the effectiveness of processing *L. niloticus* waste to produce a high-quality liquid biofertiliser, achieving complete reuse of the waste. Barley crops were used with three replicated treatments of: (1) commercial fertiliser; (2) biofertiliser; and (3) control. Barley crops treated with biofertiliser had significantly higher leaf and stem growth than the other treatments. The biofertiliser-fed barley leaves exhibited a higher production of photosynthetic pigments and significantly higher antioxidant activity than the leaves in the commercial fertiliser and control groups (Mponda and Kim 2023).

The most commonly used biofertiliser from aquaculture in Africa is wastewater effluent as described in Sect. 19.3.1. Aquaculture wastewater contains undecomposed feed, faeces, organic matter, and other elements such as N, P, and Fe (Al-Wabel et al. 2024). Aquaculture wastewater can cause serious negative environmental impacts due to nutrient enrichment. Aquaculture wastewater contains high levels of nitrates and phosphates, which cause eutrophication (Enwereuzoh et al. 2021). This process, marked by excessive algal proliferation, depletes oxygen in water, kills aquatic life, promotes the proliferation of harmful algal blooms, and releases greenhouse gases. Eutrophication is costly, with estimated expenses of $2.4 billion annually in US freshwater systems (Kashem et al. 2023). The widespread use of antibiotics in aquaculture also leads to the emergence of antibiotic-resistant organisms, which can degrade water quality and pose health risks. The accumulation of antibiotics in water can cause resistant bacteria to thrive, leading to infections in humans and animals, contributing to an alarming rise in drug-resistant diseases and fatalities (Kashem et al. 2023).

The application of nutrient-enriched wastewater may result in improved soil health and nutrient availability to the plants (Al-Wabel et al. 2024). In Egypt, El-Kady and Suloma (2013) determined the effects of effluent from *O. niloticus* and *C. gariepinus* ponds as irrigation and biofertiliser sources on African mahogany (*Khaya senegalensis*) trees. Findings revealed that there were no significant differences in the growth of *K. senegalensis* between those treated with NPK fertiliser and those irrigated with portable water. Still in Egypt, Kimera et al. (2021) conducted a field experiment to determine the effect of *O. niloticus* pond effluent as an irrigation water source on the growth, yield, and composition of essential oils of sweet marjoram (*Origanum majorana*). Kimera et al. (2021) revealed that irrigating *O. majorana* with fish effluents and supplementing with 50% inorganic fertilisers results in higher maximum oil content and crop herbage yield.

In Uganda, Alupo et al. (2016) investigated the effects of applying aquaculture effluent from *C. gariepinus* hatchery and ponds on nutrient uptake and grain yield of the common bean (*Phaseolus vulgaris*). Other treatments were rain and lake water. Results showed that bean plants irrigated with pond and hatchery effluent had a significantly higher concentration of N in the roots and shoots compared to lake and rainwater. Further, bean plants that received pond and hatchery effluent had a significantly higher bean yield (Alupo et al. 2016).

Under liquid biofertilisers is the aquaponics concept, which is a sustainable food production that applies principles of circular economy to minimise input and waste (Obirikorang et al. 2021). Aquaponics also falls under the aquaculture-crop integrated system described in Sect. 19.3.1(a). The aquaponics system combines two primarily productive systems: RAS and hydroponic cultivation. The RAS involves the farming of fish in a tank, while hydroponic cultivation involves the cultivation of crops particularly vegetables, in a medium other than soil (Vasdravanidis et al. 2022). Effluent is generated by the fish and microbes in the fish tank, which is used as a nutrient source for the plants. The effluent water is pumped over the plants, which clean the water by filtering metabolic byproducts and nutrients that are toxic to the fish. The cleaned water is then returned to the fish tank (Stegelmeier et al. 2022).

The benefits of aquaponics are substantial (Love et al. 2015). The natural recycling of nutrients reduces the need for organic fertilisers for the plants, thereby lowering nutrient runoff and minimising water pollution (Vasdravanidis et al. 2022). The aquaponics system uses significantly less water compared to conventional farming (Stegelmeier et al. 2022), making the system ideal for regions facing water scarcity, such as Northern and Southern Africa. Additionally, aquaponics enhances food production efficiency by enabling year-round fish and crop harvests in controlled environments, reducing dependency on seasonal changes. However, vegetables are harvested at higher frequencies (1–3 months) compared to fish, which are usually harvested at periods >6 months (Love et al. 2015). As a green agricultural practice, aquaponics produces chemical-free food, reduces the carbon footprint by localising food production, and eliminates the need for herbicides and pesticides (Stegelmeier et al. 2022). This sustainable approach not only conserves natural systems but also promotes food security and resilience against climate change, making it a promising solution for the future of agriculture (Vasdravanidis et al. 2022).

A number of African countries have utilised the aquaponics system with significant benefits as summarised in Obirikorang et al. (2021). Commonly cultivated fish species in aquaponics across the continent include *O. niloticus* and *C. gariepinus*, with crops ranging from chives, lettuce, maize, pumpkin, spinach, and tomatoes. Fish biomass production varies widely, with figures such as 27.9 kg/year in Nigeria for a small-scale system up to 7.5 tons/year in Egypt for a commercial setup. Crop yields also varied, with examples like 3 kg/year for spinach in Nigeria and 81 kg/

month for tomatoes in Côte d'Ivoire (Obirikorang et al. 2021). Aquaponics is also rapidly evolving in other countries in Southern Africa, for example, South Africa (Mchunu et al. 2018) and Namibia (Obirikorang et al. 2021). Other countries such as Zimbabwe have shown interest in utilising aquaponics owing to the benefits associated with adopting the system (Obirikorang et al. 2021).

19.5 Agriculture as a Source of Biofertilisers for Aquaculture

Agriculture also serves as a vital source of biofertilisers for fish farming thereby promoting sustainability through nutrient recycling and resource optimisation. Biofertilisers derived from agriculture include materials such as compost, manure, crop residues, and green manure. These organic inputs contain essential nutrients like NPK, which are critical for the growth of the primary producers, macrophytes, and phytoplankton. Manure from livestock is a commonly used agricultural biofertiliser in aquaculture ponds (Nyberg et al. 2024). This was discussed in Sect. 19.3.1(b) under the aquaculture–livestock integrated system. When livestock waste is properly treated and applied, it gradually releases nutrients thereby promoting phytoplankton growth while preventing nutrient overload that could, otherwise lead to eutrophication. Similarly, green manure and composted crop residues serve as effective organic fertilisers, releasing nutrients slowly and supporting microbial activity in water systems.

In Malawi, Kang'ombe et al. (2006) evaluated the effect of using different types of organic manure on plankton abundance, and growth and survival of redbreast tilapia (*Coptodon rendalli*) juveniles in ponds. Replicates of fish were placed in tanks treated with chicken, cattle, and pig manure, with the fourth treatment being the control. After 90 days, *C. rendalli* individuals from the chicken manure treatment were significantly larger than those from other treatments. Survival rates were, however, not significantly different across the treatments. The chicken manure treatment had significantly higher amounts of chlorophyll-*a* and plankton densities. The overall results obtained by Kang'ombe et al. (2006) suggest that the use of chicken manure produces better results than other manures from domestic animals. In a separate study in Malawi, Minich et al. (2018) evaluated the microbial effects of manure-based fertilisation methods in aquaculture ponds and examined the influence on performance of the tilapia *Oreochromis shiranus* and *C. gariepinus* grown in polyculture. Minich et al. (2018) demonstrated that chicken manure significantly influenced the microbial composition of the water column and increased fish production.

Fish farmers in Egypt highly depend on the commercial feeds, although they often fertilise their ponds with chicken manure to enhance natural food in ponds, thereby reducing feed costs. Chicken manure is abundant, cheap, readily available and rich in N. Previous studies in Egypt (summarised in Nyberg et al. 2024) documented an increase in fish growth as a result of chicken manure fertilisation. Using chicken manure as a biofertiliser in aquaculture, however, poses significant risks, including the spread of antibiotic-resistant pathogens and water quality deterioration (Nyberg et al. 2024). Unregulated antibiotic use in Egypt's poultry sector leads to antibiotic residues and resistant bacteria, such as *Salmonella* and *E. coli*, in chicken manure. These residues can accumulate in fish tissues and pond sediments, even in farms with no history of antibiotic use. Application of chicken manure has also been linked to ammonia toxicity and pathogen outbreaks, causing fish mortalities. The practice raises biosecurity concerns, cross-sectoral management challenges, and potential human health risks due to antibiotic resistance gene proliferation (Nyberg et al. 2024).

In Kenya, Wanja et al. (2020), most fish farmers fertilise their ponds with livestock manure from cattle, goats, pigs and poultry. Fish farms that fertilised their ponds with livestock manure were 50% more likely to experience fish mortality compared to those using inorganic fertilisers (Wanja et al. 2020). In contrast to successful cases in the African region where livestock manure has been effectively used, the widespread use of livestock manure among fish farmers has been linked to increased risks of transmitting disease-causing agents, particularly bacteria capable of infecting fish (Wanja et al. 2020). Additionally, decomposing manure can deplete DO levels, further contributing to unfavourable pond conditions (Zeng et al. 2022) and highlighting varying outcomes across different regions.

In the previous section (Sect. 19.4), the discussion was on aquaculture as a source of biofertilisers for agriculture. Conversely, this section provided a discussion on agriculture as a source of biofertilisers for aquaculture. Agricultural biofertilisers, such as livestock manure, can enhance fish farming by promoting natural food production and reducing feed costs. While studies from Egypt and Malawi have demonstrated positive effects on fish growth and pond productivity, the use of untreated manure also presents significant risks. Proper manure management is essential to balance productivity and sustainability in aquaculture.

19.6 Biofertilisers Production Processes

The previous sections primarily focused on biofertilisers derived directly from aquaculture and agricultural sources. However, advancements in food production systems have led to the industrial processing of biofertilisers, where they are now manufactured and commercially distributed as standardised products for broader agricultural use. Commercial biofertiliser production involves a series of carefully coordinated steps aimed at ensuring the safety and quality of the final product. According to Raimi et al. (2021), there are fundamental steps that must be considered before the final product of a biofertiliser. The biofertiliser production process involves the isolation and identification of beneficial microbial strains, functionally characterising them for plant growth traits. Pure microbial strains are selected and tested through in vitro and greenhouse experiments, optimising formulation materials. The microbes are then propagated under controlled conditions, prototyping various formulations, and conducting extensive field trials for standardisation and commercialisation (Raimi et al. 2021).

While information on biofertilisers manufactured directly from aquaculture products is scarce, several biofertiliser products have been commercialised, for example, Kinyara Sugar Limited from Uganda. The company has commercialised the three biofertiliser products: Kinyspirillum, Kinyphos and Kinyazoto, suitable for a wide range of agricultural crops.

Biofertilisers can also be produced at a small-scale, making it accessible for smallholder farmers who seek sustainable and affordable alternatives to inorganic fertilisers (Mataba et al. 2024). The process involves composting and obtaining the solid biofertilisers using basic materials such as livestock dung, fish waste, and molasses as nutrient sources.

The demand for biofertilisers, particularly microbe-based, has risen significantly in Europe, North America and Asia (Raimi et al. 2021). The global biofertiliser market was worth over US$2 billion in 2019 and is predicted to reach up to US$3.8 billion by 2025 (Raimi et al. 2021). North America has the largest market share of biofertilisers, followed by Europe and Asia. These countries have strong regulations in place that encourage the biofertiliser industry (Raimi et al. 2021). Species that include: *Azotobacter*, *Azospirillum*, *Pseudomonas*, *Klebsiella*, *Bacillus*, and *Rhizobium* dominate the market in terms of microbial strains (Raimi et al. 2021). While comprehensive data on the demand for biofertilisers in Africa is lacking, their potential in sustainable agriculture is increasingly being recognised across Africa, as countries like Uganda, South Africa, Kenya, and Nigeria have already begun to invest in research and local production.

19.7 Challenges and Barriers to the Adoption of Biofertilisers in Aquaculture and Crop and Animal Production

There are several challenges hindering the widespread adoption of biofertilisers produced from aquaculture waste in Africa. These include technical, regulatory, and cultural dimensions and are linked to the real African scenarios of a complex interplay of socio-economic, environmental, and institutional factors.

The issue of cost is a major impediment for manufacturing biofertilisers particularly in terms of capital needed for biofertiliser-processing infrastructure. This has been documented in Nigeria, Uganda, Malawi, South Africa, Tanzania, and Kenya (Freyer et al. 2024). Capital costs for equipment and microbial inoculants are extremely high making production of biofertilisers expensive (Fadiji et al. 2024). The formulation of microbial inoculants requires specialised facilities and equipment (Fadiji et al. 2024). In both Uganda and Kenya, governments have yet to be sufficiently sensitised about the critical role of biofertilisers in sustainable agriculture and environmental protection. This lack of awareness has resulted in commercialisation and large-scale adoption of biofertiliser-based technologies, receiving limited or no funding thereby hindering the full realisation of their ecological and economic potential. Decentralised production centres, such as those implemented in Malawi, are an investment both into the lack of infrastructure as well as the challenge to distribute the product (Freyer et al. 2024). A similar model is seen through financial incentives, for instance, microloans granted to cassava farmers in Nigeria (Freyer et al. 2024). Creating tax breaks for biofertiliser adoption and having strong regulatory frameworks will enhance the drive for sustainability while inducing private sector participation.

A significant bottleneck is posed by the regulatory issues, such as a lack of harmonised standards for the production and application of biofertilisers. Concerns over safety issues, such as soil nutrient imbalances and contamination because of over application of inadequately processed biofertilisers have been raised. Application of biofertilisers derived from aquaculture can contaminate crops with heavy metals, *Salmonella* sp., and antibiotic residues used in fish farming, exposing humans to serious health risks. Waste material used as biofertiliser should be properly treated to minimise contamination risks. Harmonisation of biofertiliser standards across African countries, as advocated by South Africa's organic farming sector, could ensure consistent quality and minimise unfortunate health risks.

Cultural resistance to the adoption of biofertilisers is extremely high. This is often rooted in the general traditional and religious beliefs of communities, limited awareness of the benefits of processed biofertilisers, and scepticism regarding their effectiveness compared to traditional fertilisers (manure and inorganic fertilisers) (Odhiambo et al. 2023). Participatory approaches overcome such resistance, relying more on farmer-led demonstrations and success stories as tools in winning the hearts of individuals. For instance, a set of Ethiopian initiatives demonstrates that integration of biofertilisers into traditional systems has been possible by including locals (Zelleke et al. 2019).

The shelf life of biofertilisers is shorter than that of inorganic fertilisers (Fadiji et al. 2024). The shelf-life of solid biofertilisers is around 6 months, and that of liquid biofertilisers is approximately 2 years (Fadiji et al. 2024). The viability of microorganisms can also be affected by factors such as contamination, temperature, exposure to light, and transportation (Raimi et al. 2021). The addition of preservatives to ensure product stability might then raise concerns about their potential environmental impacts. Moreover, some farmers use biofertilisers in conjunction with inorganic fertilisers (Hu et al. 2023). Microbial cells in biofertilisers are highly susceptible to inorganic nutrients, for example, urea, ammonium nitrate, ammonium phosphate, and potassium chloride, among others (Hu et al. 2023). This often leads to negative perceptions and doubts towards biofertilisers.

19.8 Conclusion

Aquaculture-derived biofertilisers fall under the general bracket of organic fertilisers and represent a promising alternative to inorganic fertilisers, offering numerous benefits for soil health, crop yield, and environmental sustainability. Africa has the resources and knowledge to lead the world in the adoption of biofertilisers derived from aquaculture. By enriching the soil with essential nutrients and improving soil structure, microbial activity, and water retention, these aquaculture-derived biofertilisers support healthier plant growth and higher yields. Case studies presented in this chapter underscore the practical benefits and potential of integrating aquaculture waste into agricultural practices, paving the way for more sustainable and resilient farming systems.

Despite the obvious promises of aquaculture-derived biofertilisers, their full utilisation in Africa has not been realised. The challenges that hinder their adoption in several African countries are multifaceted, involving capital constraints, technical issues, and, most notably, a lack of government support. These challenges have left many African countries lagging behind in harnessing the full benefits of biofertilisers derived from aquaculture, despite the region's substantial aquatic and agricultural resources. Currently, vast amounts of aquaculture waste may be lost, which otherwise could be utilised to produce biofertilisers.

Given the pressing need to meet food security demands, environmental protection, and foster economic growth, African countries ought to utilise aquaculture waste to produce biofertilisers in line with conservation goals and sustainable agriculture. To unlock the full potential of aquaculture-derived biofertilisers, African governments, agricultural scientists, aquaculturists, conservationists, farmers, local communities, and various stakeholders must collaborate on capacity building, financial support and public awareness initiatives. Indeed, the examples presented in this chapter have shown that governments have supportive policies that promote sustainable aquaculture waste management. Incentives for biofertiliser production and use should be offered. Agricultural, aquaculture and environmental scientists need to focus on cost-effective technologies for aquaculture-derived biofertiliser production and widely document their agronomic and environmental benefits through scientific studies. Farmers should be educated on the ecological and economic benefits of aquaculture-derived biofertilisers through demonstration projects and extension services. By fostering cross-sector collaboration, Africa can reduce waste, improve soil fertility, boost crop productivity, and enhance food security through the sustainable adoption of biofertilisers. This holistic approach will not only drive environmental protection and sustainable agricultural practices, but also contribute significantly to the achievement of global environmental and development targets.

References

Abelti AL, Teka TA (2024) Intervening fish post-harvest losses to narrow the gap between demand and supply: a review on magnitude of fish post-harvest losses in some sub-Saharan African countries. Aquac Fish Fish 4(2):e168. https://doi.org/10.1002/aff2.168/

Adeleke B, Robertson-Andersson D, Moodley G, Taylor S (2020) Aquaculture in Africa: a comparative review of Egypt, Nigeria, and Uganda Vis-À-Vis South Africa. Rev Fish Sci Aquac 29(2):167–197. https://doi.org/10.1080/23308249.2020.1795615

Agwa OK, Ogugbue CJ, Williams EE (2017) Field evidence of Chlorella vulgaris potentials as a biofertilizer for *Hibiscus esculentus*. Int J Agric Res 12(3):181–189. https://doi.org/10.3923/ijar.2017.181.189

Alupo N, Majaliwa JGM, Nakileza B, Kizza CL, Bugenyi F, Iwadra M, Akol P, Majugu S (2016) Effect of application of aquaculture effluent on nutrient uptake and grain yield of the common bean. RUFORUM Working Document Series 14(3):221–227

Al-Wabel MI, Almutari MM, Ahmad M, Al-Swadi HA, Ahmad J, Al-Farraj ASF (2024) Impacts of aquaculture wastewater irrigation on soil health, nutrient availability, and date palm fruit quality. Sci Rep 14:18634. https://doi.org/10.1038/s41598-024-68774-0

Arshad M (2023) Chapter 1. Climate changes mitigation and sustainable bioenergy harvest through animal waste: sustainable environmental implications of animal waste in utilization of waste animal fat for sustainable biodiesel production. In: Ebook, pp 2–123. https://doi.org/10.1007/978-3-031-26224-1

Bjornlund V, Bjornlund H, Van Rooyen AF (2020) Why agricultural production in sub-Saharan Africa remains low compared to the rest of the world – a historical perspective. Int J Water Resour Dev 36(sup1):S20–S53. https://doi.org/10.1080/07900627.2020.1739512

Daba GM, Ishibashi N, Zendo T, Sonomoto K (2017) Functional analysis of the biosynthetic gene cluster required for immunity and secretion of a novel Lactococcus-specific bacteriocin, lactococcin Z. J Appl Microbiol 123(4):1124–1132

Dey MM, Paraguas FJ, Kambewa P, Pemsl DE (2010) The impact of integrated aquaculture–agriculture on small-scale farms in southern Malawi. Agric Econ 41:67–79. https://doi.org/10.1111/j.1574-0862.2009.00426.x

El-Kady AFY, Suloma A (2013) Towards wastewater-aquaculture-agriculture integration in arid and semi-arid regions: utilization of aquaculture effluents in the irrigation of Khaya and Mahogany seedlings. J Hortic Sci Ornam Plants 5(3):227–237. https://doi.org/10.5829/idosi.jhsop.2013.5.3.1130

Ellender BR, Woodford DJ, Weyl OLF, Cowx IG (2014) Managing conflicts arising from fisheries enhancements based on non-native fishes in southern Africa. J Fish Biol 85:1890–1906. https://doi.org/10.1111/jfb.12512

El-Sayed AFM (2017) Regional review on status and trends in aquaculture development in the Near East and North Africa – 2015. FAO Fisheries and Aquaculture Circular No. 1088, vol 6. Food and Agriculture Organization of the United Nations, Rome

El-Sayed AFM, Fitzsimmons K (2023) From Africa to the world—the journey of Nile tilapia. Rev Aquac 15(Suppl. 1):6–21. https://doi.org/10.1111/raq.12738

Emam W, Zantioti C, Prah J, Monckton V, Craig S, Lambert HS (2024) A review of Kenyan inland aquaculture with an eye to the status of animal welfare in the sector. Discov Agric 2:70. https://doi.org/10.1007/s44279-024-00092-5

Endebu M, Tugie D, Negisho T (2016) Fish growth performance in ponds integrated with poultry farm and fertilized with goat manure: a case in Ethiopian Rift Valley. Int J Fish Sci Aquac 3:40–45

Enwereuzoh UO, Harding KG, Low M (2021) Fish farm effluent as a nutrient source for algae biomass cultivation. S Afr J Sci 117(7–8):1–9. https://doi.org/10.17159/sajs.2021/8694

Fadiji AE, Xiong C, Egidi E, Singh BK (2024) Formulation challenges associated with microbial biofertilizers in sustainable agriculture and paths forward. J Sustain Agric Environ 3:e70006. https://doi.org/10.1002/sae2.70006

FAO (Food and Agriculture Organization) (2021) Global aquaculture production 1950–2019. Retrieved from: http://www.fao.org/fishery/statistics/global-aquaculture-production/query/en. Accessed 1 Feb 2025

FAO (Food and Agriculture Organization) (2022) The state of world fisheries and aquaculture. Towards blue transformation, Rome. Retrieved from: https://doi.org/10.4060/cc0461en. Accessed 1 Feb 2025

Freyer B, Ellssel P, Nyakanda F, Saussure S (2024) Exploring the off-farm production, marketing and use of organic and biofertilizers in Africa: a scoping study. Report to the European Commission. DeSIRA-LIFT European Union

Getu D, Amare F, Berhanu T, Kinfo H, Terefe T (2017) Evaluation of integrated fish farming with chicken and vegetables in Silte District of Southern Ethiopia. Adv Res J Multidiscip Discov 17:20–27

Gwazani R, Gandiwa E, Poshiwa X (2022) Rural small-scale aquaculture: an assessment of farmers' perceptions and technical efficiency in Masvingo, Zimbabwe. Cogent Food Agric 8(1):2139817. https://doi.org/10.1080/23311932.2022.2139817

Hasimuna OJ, Maulu S, Nawanzi K, Lundu B, Mphande J, Phiri CJ, Kikamba E, Siankwilimba E, Siavwapa S, Chibesa M (2023) Integrated agriculture-aquaculture as an alternative to improving small-scale fish production in Zambia. Front Sustain Food Syst 7:1161121. https://doi.org/10.3389/fsufs.2023.1161121

Heimann T, Delzeit R (2024) Land for fish: quantifying the connection between the aquaculture sector and agricultural markets. Ecol Econ 217:108090. https://doi.org/10.1016/j.ecolecon.2023.108090

Hinrichsen E, Walakira JK, Langi S, Ibrahim NA, Tarus V, Badmus O, Baumüller H (2022) Prospects for aquaculture development in Africa: a review of past performance to assess future potential. ZEF Working Paper Series No. 211. University of Bonn, Center for Development Research (ZEF), Bonn

Hu M, Hei R, Guo D, Luo J, Lu C, Xu W, Zhang Z, Xiao Q, Ma Y (2023) Shelf-life enhancement of bio-inoculants through synergistic effects of encapsulation technology and osmotic protectants. J Environ Chem Eng 11(5):110996. https://doi.org/10.1016/j.jece.2023.110996

Ihejirika CE, Onwudike SU, Nwaogu LA, Emereibeole EI, Ebe TE, Ejiogu CC (2012) Assessment of aquaculture sediment for agricultural fertilizer supplement and soil conditioner in Owerri Urban, Nigeria. J Res Agric 1(1):34–38

Jaies I, Qayoom I, Saba F, Khan S (2024) Fish wastes as a source of fertilizers and manures. In: Maqsood S, Naseer MN, Benjakul S, Zaidi AA (eds) Fish waste to valuable products, sustainable materials and technology. Springer Nature Singapore, pp 329–338. https://doi.org/10.1007/978-981-99-8593-7_15

Kahle P, Palni L (2023) Challenges and solutions in scaling up biofertilizer production. J Biodivers Biopros Dev 9(59):1–2. https://doi.org/10.37421/2376-0214.2023.9.59

Kang'ombe J, Brown JA, Halfyard LC (2006) Effect of using different types of organic animal manure on plankton abundance, and on growth and survival of Tilapia rendalli (Boulenger) in ponds. Aquac Res 37:1360–1371. https://doi.org/10.1111/j.1365-2109.2006.01569.x

Kashem AHM, Das P, Hawari AH, Mehariya S, Thaher MI, Khan S, Abduquadir M, Al-Jabri H (2023) Aquaculture from inland fish cultivation to wastewater treatment: a review. Rev Environ Sci Biotechnol 22:969–1008. https://doi.org/10.1007/s11157-023-09672-1

Kimera F, Sewilam H, Fouad WM, Suloma A (2021) Efficient utilization of aquaculture effluents to maximize plant growth, yield, and essential oils composition of Origanum majorana cultivation. Ann Agric Sci 66(1):1–7. https://doi.org/10.1016/j.aoas.2020.11.002

Kong W, Xu Q, Lyu H, Kong J, Wang X, Shen B, Bi Y (2023) Sediment and residual feed from aquaculture water bodies threaten aquatic environmental ecosystems: interactions among algae, heavy metals, and nutrients. J Environ Manag 326(Part B):116735. https://doi.org/10.1016/j.jenvman.2022.116735

Lemma AH (2017) Evaluation of integrated poultry-fish-horticulture production in Arsi zone, Ethiopia. Int J Fish Aquat Stud 5:562–565

Lemma D, PrabhaDevi L, Sreenivasa V, Tilahun A (2014) Studies on the ecology of the paddy and fish co-culture system at Dembi Gobu microwatershed at Bako, Ethiopia. Int J Fish Aquat Stud 1:49–53

Lemma D, PrabhaDevi L, Sreenivasa V, Abebe G (2015) Performance evaluation of paddy and fish integrated farming at Dambi-Gobu micro watershed at Bako, West Showa, Ethiopia. Adv J Agric Res 3:13–21

Li B, Su L, Zhang H, Deng H, Chen Q, Shi H (2020) Microplastics in fishes and their living environments surrounding a plastic production area. Sci Total Environ 727:138662. https://doi.org/10.1016/j.scitotenv.2020.138662

Love DC, Fry JP, Li X, Hill ES, Genello L, Semmens K, Thompson RE (2015) Commercial aquaponics production and profitability: findings from an international survey. Aquaculture 435:67–74. https://doi.org/10.1016/j.aquaculture.2014.09.023

Madzivanzira TC, South J, Wood LE, Nunes AL, Weyl OLF (2020) A review of freshwater crayfish introductions in continental Africa. Rev Fish Sci Aquac 29(2):218–241. https://doi.org/10.1080/23308249.2020.1802405

Madzivanzira TC, South J, Ellender BR, Chalmers R, Chisule G, Coppinger CR, Khaebeb HF, Jacobs FJ, Chomba M, Musando B, Mwale C, Nhiwatiwa T, Rennie CL, Richardson N, Weyl OLF (2021) Distribution and establishment of the alien Australian redclaw crayfish, *Cherax quadricarinatus*, in the Zambezi Basin. Aquat Conserv Mar Freshwat Ecosyst 31(11):3156–3168. https://doi.org/10.1002/aqc.3703

Magadza CHD, Madzivanzira TC, Chifamba PC (2020) Decline of zooplankton food resources of *Limnothrissa miodon* fishery in Lake Kariba: global warming induced ecosystem disruption by *Cylindrospermopsis raciborskii*. Lakes Reserv Sci Policy Manag 25:117–132. https://doi.org/10.1111/lre.12318

Mahboob S, Al-Ghanim KA, Al-Balawi HFA, Al-Misned F, Ahmed Z (2019) Study on assessment of proximate composition and meat quality of fresh and stored *Clarias gariepinus* and *Cyprinus carpio*. Braz J Biol 79(4):651–658. https://doi.org/10.1590/1519-6984.187647

Mangeni H, Mhlanga W (2019) The role of smallholder pond aquaculture in livelihoods diversification, income, and food security: a case of Kushinga Fish-Farmers, Masvingo, Zimbabwe. Int J Aquac 9(3):16–25. https://doi.org/10.5376/ija.2019.09.0003

Maquén-Perleche J, Aldana-Carbonel S, Muguerza LS, Sánchez-Purihuamán M, Caro-Castro J, Carreño-Farfán C (2023) Biofertilizer based on fish waste increases the yield of *Vigna unguiculata* L. Walp, *Zea mays* L., and the rhizospheric microbiota. Sci Agropec 14(4):529–538. https://doi.org/10.17268/sci.agropecu.2023.044

Mariscal-Lagarda M, Paez-Osuna F, Esquer-Mendez JL, Guerrero-Monroy I, Romo del Vivar A, Felix-Gastelum R (2012) Integrated culture of white shrimp (*Litopenaeus vannamei*) and tomato (*Lycopersicon esculentum* Mill) with low salinity groundwater: management and production. Aquaculture 366:76–84. https://doi.org/10.1016/j.aquaculture.2012.08.016

Mataba AL, Msuya N, Katima Z (2024) Production of biofertilizer from fruit and garden waste. Tanzania J Eng Technol 43(1):158–168. https://doi.org/10.52339/tjet.v43i1.990

Maulu S, Munganga BP, Hasimuna OJ, Haambiya LH, Seemani B (2019) A review of the science and technology developments in Zambia's aquaculture industry. J Aquac Res Dev 10:1–6. https://doi.org/10.4172/2155-9546.1000567

McAndrew BJ (2000) Evolution, phylogenetic relationships, and biogeography. In: Beveridge MCM, McAndrew BJ (eds) Tilapias: biology and exploitation. Kluwer Academic Publishers, Dordrecht, pp 1–32

Mchunu N, Lagerwall G, Senzanje A (2018) Aquaponics in South Africa: results of a national survey. Aquac Rep 12:12–19. https://doi.org/10.1016/j.aqrep.2018.08.001

Melaku S, Natarajan P (2019) Status of integrated aquaculture – agriculture systems in Africa. Int J Fish Aquat Stud 7(4):263–269

Menezes A, Ligeon C, Murekezi P, Jolly C (2024) Diagnosis of aquaculture employment governance in selected African countries. FAO Fisheries and Aquaculture Technical Paper, No. 715. FAO, Rome. https://doi.org/10.4060/cd0424en

Miller J, Atanda T, Asala G, Chen WH (2006) Integrated irrigation-aquaculture opportunities in Nigeria: the special programme for food security and rice-fish farming in Nigeria. In: Halwart M, van Dam AA (eds) Integrated irrigation and aquaculture in West Africa: concepts, practices and potential. FAO, Rome, pp 117–124

Minich JJ, Zhu Q, Xu ZZ, Amir A, Ngochera M, Simwaka M, Allen EE, Zidana H, Knight R (2018) Microbial effects of livestock manure fertilization on freshwater aquaculture ponds rearing tilapia (*Oreochromis shiranus*) and North African catfish (*Clarias gariepinus*). Microbiol Open:e00716. https://doi.org/10.1002/mbo3.716

Mohammed O, Dereje T, Erkie A (2015) Evaluation of integrated fish-rice farming in the Nile irrigation and drainage project areas, South Gonder, Ethiopia. Int J Fish Aquat Stud 3:5–8

Mponda SY, Kim JK (2023) Sustainable, quality biofertilizer produced from Nile perch (*Lates niloticus*) wastewater using three Bacillus species: demonstration in barley hydroponics. J Environ Chem Eng 11(3):110037. https://doi.org/10.1016/j.jece.2023.110037

Msuya FE, Bolton J, Pascal F, Narrain K, Nyonje B, Cottier-Cook EJ (2022) Seaweed farming in Africa: current status and future potential. J Appl Phycol 34:985–1005. https://doi.org/10.1007/s10811-021-02676-w

Muendo PN, Verdegem MCJ, Stoorvogel JJ, Milstein A, Gamal E-N, Duc PM, Verreth JAJ (2014) Sediment accumulation in fish ponds; its potential for agricultural use. Int J Fish Aquat Stud 1(5):228–241

Musinguzi L, Lugya J, Rwezawula P, Kamya A, Nuwahereza C, Halafo J, Kamondo S, Njaya F, Aura C, Shoko AP, Osinde R, Natugonza V, Ogutu-Ohwayo R (2019) The extent of cage aquaculture, adherence to best practices and reflections for sustainable aquaculture on African inland waters. J Great Lakes Res 45(6):1340–1347. https://doi.org/10.1016/j.jglr.2019.09.011

Nabti E, Jha B, Hartmann A (2016) Impact of seaweeds on agricultural crop production as biofertilizer. Int J Environ Sci Technol 14(5):1119–1134. https://doi.org/10.1007/s13762-016-1202-1

Ndiaye NA, Maiguizo-Diagne H, Diadhiou HD, Ndiayc WN, Diedhiou F, Cournac L, Gaye ML, Fall S, Brehmer P (2019) Methanogenic and fertilizing potential of aquaculture waste: towards freshwater farms energy self-sufficiency in the framework of blue growth. Rev Aquacult:1–10. https://doi.org/10.1111/raq.12390

Ndjadi SS, Matendo AF, Mwinja AC, Mondo JM, Civava RM, Mushagalusa GN (2023) Integrated crop-livestock-aquaculture farming systems: synergistic approaches for balancing agriculture's biophysical requirements and environmental supply. Afr J Rural Dev 6(4):167–190

Nhan DK, Phong LT, Verdegem MJC, Duong LT, Bosma RH, Little DC (2007) Integrated freshwater aquaculture, crop and livestock production in the Mekong delta, Vietnam: determinants and the role of the pond. Agric Syst 94(2):445–458. https://doi.org/10.1016/j.agsy.2006.11.017

Nyberg O, Novotny A, Sbaay AS, Nasr-Allah AM, Al-Kenawy DAR, Rossignoli CM, Henriksson PJG (2024) Poultry manure fertilization of Egyptian aquaculture ponds brings more cons than pros. Aquaculture 590:741040. https://doi.org/10.1016/j.aquaculture.2024.741040

Obirikorang KA, Sekey W, Gyampoh BA, Ashiagbor G, Asante W (2021) Front Sustain Food Syst 5. https://doi.org/10.3389/fsufs.2021.705549

Odhiambo KA, Ogola HJO, Onyango B, Tekere M, Ijoma GN (2023) Contribution of pollution gradient to the sediment microbiome and potential pathogens in urban streams draining into Lake Victoria (Kenya). Environ Sci Pollut Res 30(13):36450–36471

Ojewole AE, Ndimele PE, Oladele AH, Saba AO, Oladipupo IO, Ojewole CO, Ositimehin KM, Oluwasanmi AS, Kalejaye OS (2024) Aquaculture wastewater management in Nigeria's fisheries industry for sustainable aquaculture practices. Sci Afr 25:e02283. https://doi.org/10.1016/j.sciaf.2024.e02283

Oké V, Goosen NJ (2019) The effect of stocking density on profitability of African catfish (*Clarias gariepinus*) culture in extensive pond systems. Aquaculture 30:385–392. https://doi.org/10.1016/j.aquaculture.2019.04

Otaiku AA (2022) Biofertilizer impacts cassava (Manihot esculenta Crantz) cultivation crop yield and regenerative agriculture. Glob J Agric Res 10(1):1–90. https://doi.org/10.37745/gjar.2013

Prinsloo JF, Schoonbee H (1987) Investigations into the feasibility of a duck-fish-vegetable integrated agriculture-aquaculture system for developing areas in South Africa. Water SA 13:109–118

Raghunandan BL, Vyas RV, Patel HK, Jhala YK (2019) Perspectives of seaweed as organic fertilizer in agriculture. In: Panpatte DG, Jhala YK (eds) Soil fertility management for sustainable development soil fertility management for sustainable development. Springer Nature, Singapore, pp 267–289

Raimi A, Roopnarain A, Adeleke R (2021) Biofertilizer production in Africa: current status, factors impeding adoption and strategies for success. Sci Prog 11:e00694

Rasowo J, Auma EO (2006) On-farm trials with rice fish cultivation in the West Kano rice irrigation scheme, Kenya

Sahu BB, Barik NK, Mohapatra BC, Sahu BN, Sahu H, Sahoo P, Majhi D, Biswal NC, Mohanty PK, Jayasankar P (2014) Valorization of fish processing waste through natural fermentation with molasses for preparation of bio fertilizer and bio supplement. J Environ Sci Comput Sci Eng Technol 3(4):1849–1856

Satia PB (2017) Regional review on status and trends in aquaculture development in Sub-Saharan Africa – 2010. Food and Agriculture Organization of the United Nations, Rome

Singh S, Prabjeet S, Verma N, Kumar D (2021) Integrated fish farming – rationale and scope. College of Fisheries, G.B. Pant University of Agriculture and Technology, Pantnagar

Soliman NF (2017) Aquaculture in Egypt under changing climate. Alexandria Research Center for Adaptation to Climate Change (ARCA). Retrieved from: http://arca-eg.org/wp-content/uploads/2017/06/Working-Paper-4-Jan.2017.pdf. Accessed 2 Jan 2025

Soumare A, Diedhiou AG, Thuita M, Hafidi M, Ouhdouch Y, Gopalakrishnan S, Kouisni L (2020) Exploiting biological nitrogen fixation: a route towards sustainable agriculture. Plants 9(8):1011. https://doi.org/10.3390/plants9081011

Stegelmeier A, Rose D, Joris BR, Glick BR (2022) The use of PGPB to promote plant hydroponic growth. Plants 11(20):2783. https://doi.org/10.3390/plants11202783

Tesfahun A, Alebachew S (2023) Diet composition and feeding habits of the African sharptooth catfish *Clarias gariepinus* (Burchell, 1822) from Ribb Reservoir, South Gondar, Ethiopia. Cogent Food Agric 9(2):2284228. https://doi.org/10.1080/23311932.2023.2284228

Ugwumba COA (2010) Environmental sustainability and profitability of integrated fish cum crop farming in Anambra state, Nigeria. Agric J 5:229–233. https://doi.org/10.3923/aj.2010.229.233

Vasdravanidis C, Alvanou MV, Lattos A, Papadopoulos DK, Chatzigeorgiou I, Ravani M, Liantas G, Georgoulis I, Feidantsis K, Ntinas GK, Giantsis IA (2022) Aquaponics as a promising strategy to mitigate impacts of climate change on rainbow trout culture. Animals 12(19):2523. https://doi.org/10.3390/ani12192523

Verdegem M, Buschmann AH, Latt UW, Dalsgaard AJT, Lovatelli A (2023) The contribution of aquaculture systems to global aquaculture production. J World Aquacult Soc 54(2):206–250. https://doi.org/10.1111/jwas.12963

Waktola BA, Devi LP, Sreenivasa V, Lakew AA (2016) Study on the profitability of fish and horticrop integrated farming at Nono District, West Shoa Zone, Ethiopia. Greener J Agric Sci 6:41–48. https://doi.org/10.15580/GJAS.2016.2.112415163

Wanja DW, Mbuthia PG, Waruiru RM, Mwadime JM, Bebora LC, Nyaga PN, Ngowi HA (2020) Fish husbandry practices and water quality in Central Kenya: potential risk factors for fish mortality and infectious diseases. Vet Med Int 6839354:1–10. https://doi.org/10.1155/2020/6839354

Weyl OLF, Daga VS, Ellender BR, Vitule JRS (2016) A review of *Clarias gariepinus* invasions in Brazil and South Africa. J Fish Biol 89(1):386–402. https://doi.org/10.1111/jfb.12958

Win TT, Barone GD, Secundo F, Fu P (2018) Algal biofertilizers and plant growth stimulants for sustainable agriculture. Ind Biotechnol 14(4):203–211. https://doi.org/10.1089/ind.2018.0010

Zajdband AD (2011) Integrated agri-aquaculture systems. In: Lichtfouse E (ed) Genetics, biofuels and local farming systems, vol 7. Springer, Dordrecht, pp 87–127. https://doi.org/10.1007/978-94-007-1521-9_4

Zajicek P, Corbin J, Belle S, Rheault R (2021) Refuting marine aquaculture myths, unfounded criticisms, and assumptions. Rev Fish Sci Aquac 31(1):1–28. https://doi.org/10.1080/23308249.2021.1980767

Zelleke G, Agegnehu G, Abera D, Rashid S (2019) Fertilizer and soil fertility potential in Ethiopia. Gates Open Res 3(482):482. https://doi.org/10.21955/gatesopenres.1115635.1

Zeng J, Shen X, Yin H, Sun X, Dong H, Huang G (2022) Oxygen dynamics, organic matter degradation, and main gas emissions during pig manure composting: effect of intermittent aeration. Bioresour Technol 361:127697. https://doi.org/10.1016/j.biortech.2022.127697

Zhang J, Akyol Ç, Meers E (2023) Nutrient recovery and recycling from fishery waste and by-products. J Environ Manag 348:119266. https://doi.org/10.1016/j.jenvman.2023.119266

Zhao G, Zhu X, Zheng G, Meng G, Dong Z, Baek JH, Jeon CO, Yao Y, Xuan YH, Zhang J, Jia B (2024) Development of biofertilizers for sustainable agriculture over four decades (1980–2022). Geogr Sustain 5(1):19–28. https://doi.org/10.1016/j.geosus.2023.09.006

Chapter 20
Biofertilisers' Role in Enhancing Cropland Ecosystem Services and Mitigating Climate Change in Africa

Admire R. Dzvene, Leocadia Zhou, and Patrick Nyambo

Abstract African croplands face escalating ecological and agronomic challenges, including widespread soil degradation, nutrient depletion, and declining biodiversity, factors that collectively constrain agricultural productivity and threaten food security. Monoculture systems, often adopted in response to food demand pressures, have intensified these vulnerabilities by disrupting soil microbial networks and increasing dependency on synthetic inputs. In this chapter, we critically assess the capacity of biofertilisers to restore soil health, enhance nutrient cycling, and bolster ecosystem resilience. Drawing on recent advances in microbial ecology and agrobiotechnology, we synthesise evidence on the mechanisms by which nitrogen-fixing bacteria, phosphate-solubilising microbes, and arbuscular mycorrhizal fungi contribute to plant growth promotion, pathogen suppression, and abiotic stress tolerance. We highlight emerging insights into the role of biofertilisers in carbon sequestration, bioremediation, and the reconstitution of microbial diversity in degraded African soils. Despite their promise, adoption remains limited by socio-economic and institutional barriers. We argue that integrating biofertilisers into smallholder systems, alongside supportive policy and local production strategies, represents a scalable pathway towards climate-resilient and microbiome-informed agriculture across the continent.

A. R. Dzvene (✉)
Centre for Global Change, Faculty of Science and Agriculture, University of Fort Hare, Alice, South Africa

Department of Agronomy, Faculty of Science and Agriculture, University of Fort Hare, Alice, South Africa
e-mail: ADzvene@ufh.ac.za

L. Zhou
Centre for Global Change, Faculty of Science and Agriculture, University of Fort Hare, Alice, South Africa

P. Nyambo
Agricultural Research Council—Vegetables, Industrial and Medicinal Plants Institute, Pretoria, South Africa

A. D. Nciizah et al. (eds.), *Biofertilizers for Sustainable Agriculture in Africa*,
Sustainability Sciences in Asia and Africa, https://doi.org/10.1007/978-981-95-6009-7_20

Keywords Biofertilisers · Microbial diversity · Agroecosystem resilience · Soil health · Sustainable agriculture

20.1 Introduction

Cropland ecosystems in Africa are vital for ensuring food security and sustaining the livelihoods of millions (de Graaff et al. 2011; Meyfroidt 2018). These systems underpin agricultural productivity through the provision of critical ecosystem services, including biodiversity support and environmental regulation (Power 2010; Bommarco et al. 2018). Traditional agricultural practices, such as intercropping and cover cropping, have long been employed to enhance system resilience, promote soil health, and foster biodiversity (Mupambwa and Wakindiki 2012; Aloo et al. 2022; Dzvene et al. 2024; Sarkar et al. 2023). Increasingly, microbial communities influenced by these practices are recognised as key drivers of ecosystem function and resilience under changing environmental conditions.

Leguminous species are integral to sustainable agroecosystems due to their capacity to form mutualistic associations with rhizobia, facilitating biological nitrogen (N) fixation that significantly enhances soil N availability (Vanlauwe et al. 2019; Dzvene and Chiduza 2024). This process not only improves soil fertility but also reduces the need for synthetic N inputs, thereby lowering production costs and mitigating environmental externalities associated with fertiliser overuse. Beyond N enrichment, legumes contribute to broader agroecological functioning by fostering beneficial microbial communities (Schaedel et al. 2021). Their co-cultivation with other plant-growth-promoting microorganisms, such as *Streptomyces* spp., enhances the functional diversity of the soil microbiome, promoting plant health, suppressing soil-borne pathogens, and increasing crop resilience under stress conditions (Sylia et al. 2022; Omar et al. 2022). Moreover, the cultivation of climate-resilient indigenous crops like Bambara groundnut (*Vigna subterranea*), which thrive under low-input and marginal conditions, highlights the adaptive potential and ecological benefits of traditional cropping systems. These crops not only support food and nutrition security but also contribute to agrobiodiversity conservation and ecosystem stability (Sarkar et al. 2023).

However, the widespread shift towards intensified monoculture systems, often driven by the pursuit of short-term yields, has significantly compromised the ecological integrity of agricultural landscapes (Suarez and Gwozdz 2023). The excessive and prolonged use of synthetic fertilisers and pesticides, while initially boosting productivity, has had detrimental consequences for soil health, accelerating processes of soil degradation through nutrient imbalance, acidification, and loss of organic matter (Lal 2015). Simultaneously, the simplification of cropping systems has led to a marked decline in agrobiodiversity, reducing habitat heterogeneity, and disrupting trophic interactions essential for natural pest control (Groh et al. 2022; Liu et al. 2018). These ecological disruptions have collectively diminished the resilience of African farming systems to climate variability, weakening their adaptive capacity and impairing the delivery of critical ecosystem services such as nutrient cycling, pollination, and water regulation (Crews et al. 2018; Suarez and Gwozdz 2023; Enerijiofi et al. 2024).

In this context, sustainable agricultural approaches, including the application of biofertilisers, offer a promising pathway to restore soil functionality and ecological balance. Biofertilisers, which consist of living or dormant microbial inoculants, promote plant growth by enhancing nutrient availability, root development, and microbial activity (Dzvene and Chiduza 2024). They encompass plant-growth-promoting microbes (PGPMs) which not only improve nutrient cycling but also contribute to greenhouse gas (GHG) mitigation by reducing reliance on synthetic inputs (Leila and El-Hafid 2020; Raji et al. 2024). Moreover, integrating waste valorisation strategies, such as converting agricultural residues into biofuels, can augment agricultural sustainability through circular resource use and renewable energy generation (Afotey and Sarpong 2023).

Enhancing ecosystem services, such as carbon sequestration, nutrient cycling, and microbial diversity, through sustainable agricultural interventions is crucial for building climate-resilient food systems in Africa. However, knowledge gaps persist regarding the efficacy, scalability, and farmer adoption of biofertilisers across heterogeneous agroecological zones. Addressing these gaps is critical for aligning indigenous knowledge with science-based innovations. The objective was to review the potential of biofertilisers as a sustainable solution to address cropland soil degradation, enhance ecosystem services, and improve crop productivity in African croplands.

20.2 Production challenges in African Croplands

African agriculture is confronted by a nexus of challenges that undermine both food security and environmental sustainability. Extensive soil degradation has diminished the availability of essential nutrients such as N, phosphorus (P), potassium (K), and iron (Fe), constraining crop productivity (Rashid et al. 2016). These limitations are exacerbated by declining biodiversity, environmental pollution, and land degradation, which collectively intensify vulnerabilities across African food systems (Olanipekun et al. 2019; Mani et al. 2021). Food insecurity in sub-Saharan Africa (SSA) has markedly worsened since 2015, driven by economic instability, climate change, and regional conflicts, particularly in East Africa (Wudil et al. 2022; Abdullahi et al. 2024). Amid growing pressures to increase food output, farmers increasingly resort to simplified, less sustainable production strategies, notably monocropping (Giller 2020).

Monocropping, the continuous cultivation of a single crop species, has become prevalent in many regions (Loh et al. 2022; Hashakimana et al. 2023). Staples such as maize (*Zea mays*), rice (*Oryza sativa*), and wheat (*Triticum aestivum*) are extensively grown under monoculture systems. Although these systems may offer short-term yield gains, they severely compromise ecological integrity. Empirical evidence shows that monocultures reduce on-farm biodiversity, disrupt natural pest control, and erode long-term productivity (Selimovic et al. 2022; Suarez and Gwozdz 2023; Liu et al. 2023). Continuous cultivation of a single species depletes specific soil nutrients without replenishment (Aguégué et al. 2022), necessitating the use of synthetic fertilisers (Zhang et al. 2022), which further degrade soil quality and biological activity. Overreliance on chemical inputs also disrupts soil microbial diversity and structure, weakening nutrient cycling and soil fertility (Adetunji et al. 2023; Kapoor et al. 2024). These ecological imbalances heighten vulnerability to pests and diseases, impairing key ecosystem functions and threatening sustainable crop production (Bockstaller et al. 2024).

Economically, monocropping systems place increasing burdens on smallholder farmers, who must invest heavily in chemical inputs to maintain yields. As pest populations evolve resistance to pesticides, farmers are forced to escalate application rates or switch to more toxic alternatives (Olanrewaju et al. 2017). This intensifies input dependency, creates a feedback loop of diminishing returns, and escalates production costs, all while degrading environmental quality. Nontarget species, including beneficial insects and soil fauna, are frequently harmed by chemical exposure, leading to further destabilisation of agroecosystems (Selimovic et al. 2022; Sylia et al. 2021).

Climate change compounds these vulnerabilities (Dzvene et al., 2025). African agriculture is particularly sensitive to climatic variability (Connolly-Boutin and Smit 2016), with increasingly frequent droughts, floods, and erratic weather patterns disrupting traditional growing cycles and reducing crop yields (Kapoor et al. 2024). The adaptive capacity of farming systems is significantly constrained under monoculture regimes (Petersen-Rockney et al. 2021; Mukhovi and Jacobi 2022). In contrast, diversified cropping systems enhance agroecosystem resilience by supporting soil health, fostering beneficial biotic interactions, and buffering against climatic extremes (Benitez-Alfonso et al. 2023). The absence of such diversity in monoculture fields exacerbates susceptibility to environmental shocks.

Transitioning towards sustainable farming practices, including crop diversification, integrated pest management, and soil-enhancing strategies, is essential for restoring ecological balance and improving long-term productivity (Power 2010; Bommarco et al. 2018; Leila and El-Hafid 2020; Abioye et al. 2024). Nonetheless, structural barriers persist. Key among these are limited farmer education, financial constraints, and restricted access to sustainable agricultural inputs. Addressing these systemic challenges is imperative for enabling the widespread adoption of agroecological innovations and advancing climate-resilient food systems in Africa.

20.3 Biofertilisers as a Solution

Biofertilisers are biologically derived inputs composed of living microorganisms that (Table 20.1), when applied to seeds, plant surfaces, or soil, colonise the rhizosphere and promote plant growth by increasing the bioavailability of essential

Table 20.1 Biofertilisers role in carbon sequestration

Biofertiliser	Role in plant growth promotion	Reference
N fixing, *Azotobacter* spp., *Clostridium* spp.	Increases N concentration in soils and availability to plants, thereby promoting high biomass growth and adding to soil organic carbon input.	Yadav et al. (2013)
N fixing, *Rhizobium* spp., *Anabaena azollae*	Increases N concentration in soil and availability to plants, thereby promoting growth soil microbial community, thus improving carbon sequestration.	Ansari and Mahmood (2017)
P solubilisers, *Pseudomonas striata*, *Trichoderma* spp.	Increases the concentration of plant available P, thereby promoting high biomass growth and adding to soil organic carbon input	Singh et al. (2007)
P solubilisers, *Trichoderma* spp.	Solubilise insoluble forms of P, enhance fungal activity, and contribute to stable organic matter formation in soil	Liu et al. (2020)
Plant growth-promoting bacteria, Pseudomonas, Rhizobium spp., Azospirillum spp., Azotobacter spp.	Produces root growth-promoting hormones and increases nutrient assimilation, increasing crop biomass, and organic carbon addition to the soil	García-Fraile et al., 2015

Adapted from Just et al. (2024)

nutrients (Mahanty et al. 2017; Choudhary et al. 2022). These microbial inoculants commonly include N-fixing bacteria (e.g. *Rhizobium*, *Azotobacter*), P-solubilising bacteria, and arbuscular mycorrhizal fungi (AMF), which establish beneficial symbiotic associations with plant roots (Aguégué et al. 2022; Kapoor et al. 2024). Through these associations, biofertilisers improve nutrient cycling, enhance root development, and strengthen the plant's ability to absorb water and nutrients, particularly in nutrient-depleted or degraded soils. For instance, formulations containing *Rhizobium* enhance N acquisition in legumes, *Azotobacter* supports free-living N fixation, and mycorrhizal fungi expand root surface area, thereby facilitating P uptake (Sabannavar and Lakshman 2008). Beyond nutrient provision, biofertilisers contribute to soil health by promoting microbial diversity, enhancing soil structure, and suppressing soil-borne pathogens. Their strategic application is therefore essential not only for sustainable yield improvement but also for fostering resilient agroecosystems under conditions of environmental stress and reduced chemical input dependency (Nosheen et al. 2021).

The mechanisms through which biofertilisers enhance crop yields (Table 20.2), are complex and synergistic, encompassing a range of biochemical, physiological, and ecological processes (Rodŕiguez and Fraga 1999; Sturz et al. 2000). One of the primary pathways involves improving the bioavailability and uptake of key macronutrients, particularly N, P, and K, which are frequently deficient in many African soils (Yu et al. 2012; Bellenger et al. 2014; Shen et al. 2016). N-fixing bacteria such as *Rhizobium* and *Azotobacter* convert atmospheric N into ammonia, a plant-available form, thereby reducing reliance on synthetic N fertilisers (Adetunji et al.

Table 20.2 Mechanisms of biofertiliser action in enhancing crop yields

Mechanism	Description	Examples of microorganisms	References
N fixation	Conversion of atmospheric N_2 into ammonia, enhancing N availability to plants.	*Rhizobium*, *Azotobacter*, *Azospirillum*	Adetunji et al. (2023), Rodríguez and Fraga (1999), and Yu et al. (2012)
P solubilisation	Solubilisation of bound inorganic phosphates via organic acids and phosphatases.	*Bacillus*, *Pseudomonas*, *Aspergillus*	Bellenger et al. (2014) and Shen et al. (2016)
K mobilisation	Solubilising insoluble K-bearing minerals into bioavailable forms.	*Frateuria aurantia*, *Bacillus mucilaginosus*	Meena et al. (2016)
Mycorrhizal symbiosis	AMF enhance water and nutrient uptake via extraradical hyphae.	*Glomus*, *Acaulospora* spp.	Kothari et al. (1990) and Cheng et al. (2016)
Soil structure improvement	Excretion of glomalin and polysaccharides improves aggregation and porosity.	Mycorrhizal fungi	Demir (2020) and Malobane et al. (2021)
Enhanced root growth	Hormone production (e.g. auxins, cytokinins) stimulates root elongation and branching.	*Azospirillum*, *Bacillus subtilis*	Spaepen et al. (2007)
Biocontrol of pathogens	Competition, antibiosis, or induction of systemic resistance (ISR) suppresses diseases.	*Trichoderma*, *Pseudomonas fluorescens*	Sturz et al. (2000) and Vessey (2003)
Enhanced soil microbial activity	Increased microbial biomass, diversity, and enzyme activities catalyse nutrient cycling.	Mixed consortia	Enerijiofi et al. (2024) and Dzvene and Chiduza (2024)
Organic matter decomposition	Accelerates breakdown of crop residues and OM into humus, enriching soil fertility.	*Cellulomonas*, *Bacillus*, Actinomycetes	Saharan and Nehra (2011) and Adesemoye and Kloepper (2009)
Abiotic stress tolerance	Biofertilisers induce tolerance to drought, salinity, and heavy metals through osmolyte production and antioxidant enzymes.	*Bacillus*, *Enterobacter*, *Pseudomonas*	Sharaya et al. (2023) and Khan et al. (2021)

2023). Simultaneously, P-solubilising microbes release organic acids and enzymes that mobilise bound P, making it accessible to plant roots in otherwise P-limited soils.

Another critical mechanism involves the enhancement of root architecture and functional capacity through symbiosis with AMF. These fungi extend the effective surface area of plant roots via extraradical hyphae, significantly improving water and nutrient absorption, particularly under stress-prone or nutrient-depleted conditions (Kothari et al. 1990; Cheng et al. 2016). Furthermore, biofertilisers contribute to improved soil physical properties, including enhanced soil structure, porosity, and water-holding capacity, attributes essential for sustainable crop production in semi-arid and degraded environments (Saha et al. 2010; Demir 2020). For instance, mycorrhizal fungi secrete glomalin, a glycoprotein that functions as a soil aggregate stabiliser, thereby enhancing soil aggregation and reducing erosion (Malobane et al. 2021).

Emerging evidence also highlights the role of biofertilisers in stimulating microbial activity and diversity within the soil, catalysing nutrient cycling, and accelerating organic matter decomposition (Enerijiofi et al. 2024; Dzvene and Chiduza 2024). These microbial-driven processes not only replenish soil nutrients but also foster a biologically active and resilient soil ecosystem. This cumulative improvement in soil health supports sustained plant growth, increases productivity, and enhances agroecosystem resilience to environmental stressors, making biofertilisers a cornerstone of ecologically sustainable agriculture.

The integration of biofertilisers into traditional farming practices can significantly enhance food security in Africa while promoting environmental sustainability. However, successful adoption requires supportive policies, educational initiatives, and access to resources that empower farmers to implement biofertilisation strategies effectively. Bridging this knowledge gap is essential to fostering resilient farming systems that can adapt to both ecological challenges and changing climatic conditions. Biofertilisers represent a key component in the revitalisation of African agriculture, aligning the goals of productivity with sustainability. Their application, together with the promotion of indigenous agricultural practices, has the potential to create robust agroecosystems capable of sustaining food production into the future.

20.4 Impact on Cropland Ecosystem Services

20.4.1 Enhancement of Soil Micro-biodiversity

Soil microbial biodiversity underpins the functioning of agroecosystems (Fig. 20.1) (Bellenger et al. 2014; Shen et al. 2016; Enerijiofi et al. 2024), regulating essential processes such as nutrient cycling, soil structure formation, and plant productivity (Nizamani et al. 2024). This diversity encompasses a broad spectrum of microorganisms, including bacteria, archaea, fungi, and protists, that interact in complex, synergistic networks to sustain soil health. Biofertilisers have emerged as a sustainable intervention to enhance this microbiodiversity by introducing beneficial microbial consortia that stimulate biochemical processes, mobilise nutrients, and improve ecosystem resilience (Chaudhary et al. 2022). By altering microbial community composition, biofertilisers facilitate key ecosystem services such as N fixation, P solubilisation, organic matter mineralisation, and pathogen suppression, functions essential to long-term soil fertility and agricultural productivity (Altieri 1999; Nosheen et al. 2021).

The application of biofertilisers is particularly valuable in marginal croplands (Yadav et al. 2013), which are often degraded, nutrient-depleted, and exposed to

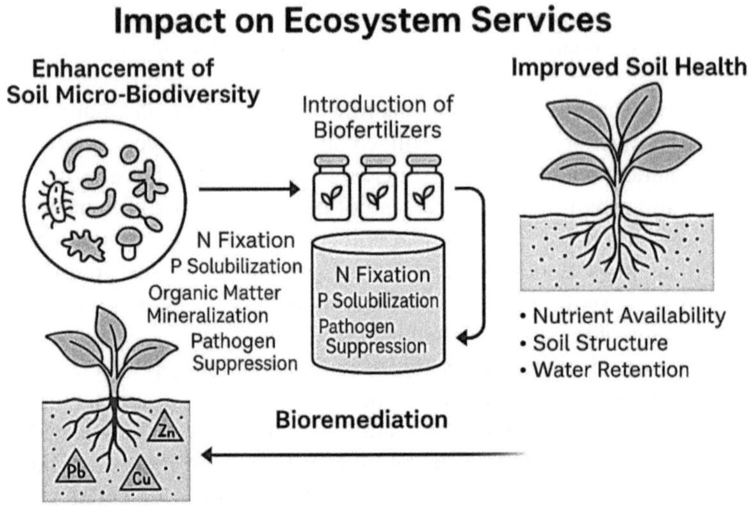

Fig. 20.1 Impact of biofertilisers on cropland ecosystem services

erratic moisture regimes (Rashid et al. 2016). In such fragile systems, microbial diversity enhances agroecosystem buffering capacity, mitigating the impacts of biotic stressors (e.g. pathogens) and abiotic stressors (e.g. drought, salinity) (Khan et al. 2021; Dzvene and Chiduza 2024). In SSA, where smallholder agriculture remains central to household food security, biofertiliser adoption offers an ecologically appropriate and cost-effective strategy for rehabilitating soil health. Recent studies by Kapoor et al. (2024) and Abioye et al. (2024) demonstrate the potential of biofertilisers to enhance microbial diversity and function, resulting in improved crop vigour, stress tolerance, and yield stability.

An expanding body of empirical evidence confirms the ability of biofertilisers to cultivate functionally diverse and resilient soil microbiomes. For example, inoculation with native *Rhizobium* strains significantly enhances biological N fixation, increasing nutrient availability and supporting plant performance under drought conditions (Aguégué et al. 2022; Basile and Lepek 2021). Similarly, AMF extend the root absorptive surface area and contribute to the development of soil aggregates and pore networks, improving aeration, infiltration, and water retention (Schlüter et al. 2020; Cheng et al. 2016). Adetunji et al. (2023) reported marked increases in microbial abundance and taxonomic richness following biofertiliser application, underscoring their role in restoring biological function in degraded soils.

These findings highlight the transformative potential of biofertilisers in revitalising soil microbial communities across African agroecosystems, thereby enhancing soil functionality, crop productivity, and food system resilience (Enerijiofi et al. 2024; Dzvene and Chiduza 2024). Yet, despite their agronomic and ecological benefits, the broader adoption of biofertilisers is constrained by several systemic barriers. These include limited farmer awareness of soil–microbe interactions, weak extension support, and inadequate access to high-quality, locally adapted microbial inoculants (Afotey and Sarpong 2023; Petersen-Rockney et al. 2021). Addressing these challenges will require integrated strategies spanning policy, research, and practice, including farmer training, investment in local biofertiliser production, and context-specific soil fertility management frameworks tailored to regional agroecological conditions.

20.4.2 Improved Soil Health

Enhancing soil health is fundamental to improving agricultural productivity and sustainability, particularly across the diverse ecological zones of Africa (Abdullahi et al. 2024). Soil health reflects the integrated physical, chemical, and biological properties of the soil, all of which directly influence plant growth, nutrient cycling, and ecosystem function (Canbolat et al. 2006; Berger et al. 2013; Demir 2020). Biofertilisers have emerged as an ecologically sound and cost-effective approach to restoring soil health by re-establishing microbial balance and promoting key biological functions that improve soil quality and fertility (Kapoor et al. 2024).

The chemical and biological dimensions of soil health are particularly responsive to biofertiliser interventions, which enhance nutrient cycling and bioavailability, two critical determinants of crop productivity (Dębska et al. 2016; Chaudhary et al. 2022; Kapoor et al. 2024; Abioye et al. 2024). In many African smallholder systems, soils are severely depleted, especially in N and P, limiting crop yields (Yu et al. 2012; Rashid et al. 2016; Olanipekun et al. 2019). These deficiencies are exacerbated by continuous nutrient export through harvesting, soil acidification from inappropriate fertiliser use, and land degradation driven by erosion and runoff (Giller 2020). Biofertilisers counter these challenges by stimulating beneficial microbial activity that decomposes organic matter and mobilises nutrients into plant-available forms (Rodríguez and Fraga 1999; Yu et al. 2012; Aguégué et al. 2022; Chaudhary et al. 2022; Abioye et al. 2024; Kapoor et al. 2024).

In addition to enhancing nutrient dynamics, biofertilisers contribute to improving soil physical properties, such as aeration, porosity, and water infiltration (Murugan and Chitraputhirapillai 2011; Mupambwa and Wakindiki 2012; Berger et al. 2013; Schlüter et al. 2020). These improvements are particularly relevant in African agricultural systems, where soil compaction, crusting, and erosion constrain root development and water-use efficiency. By enhancing soil structure, biofertilisers facilitate deeper root penetration, increased water retention, and more efficient nutrient acquisition, collectively supporting plant resilience and yield stability.

Despite their demonstrated benefits, the widespread adoption of biofertilisers in African agriculture remains limited. Key barriers include low levels of farmer awareness, insufficient access to high-quality biofertiliser products, and a lack of institutional support for knowledge dissemination. Overcoming these constraints will require targeted educational campaigns, capacity-building initiatives, and investments in local production and distribution networks. Integrating biofertilisers into African farming systems represents a strategic pathway towards building climate-resilient agriculture, improving productivity, and advancing sustainable land management practices that benefit both ecosystems and rural livelihoods.

20.4.3 Bioremediation

Heavy metals such as lead (Pb), zinc (Zn), Fe, manganese (Mn), and copper (Cu) are prevalent contaminants in mining and industrial waste tailings, posing a significant threat to soil quality and agricultural productivity (Okereafor et al. 2020; Haghighizadeh et al. 2024). These metals are particularly hazardous due to their nonbiodegradable nature, high water solubility, and long environmental half-life, leading to persistent accumulation in soils and crops (Selvi et al. 2019). In many African croplands, heavy metal toxicity has emerged as a critical constraint on food production, with detrimental effects on soil microbial communities and plant health (Yan et al. 2020; Okereafor et al. 2020; Seang-On et al. 2023; Haghighizadeh et al. 2024).

Biofertilisers offer a promising approach to in situ bioremediation through their role in phytoremediation, the use of specific plant–microbe interactions to extract, stabilise, or detoxify contaminants from soil and water (Yan et al. 2020). By restoring soil microbial activity and enhancing nutrient cycling, biofertilisers facilitate the degradation and immobilisation of pollutants while improving soil fertility and ecological function (Khoshru et al. 2024; Kapoor et al. 2024). This dual role positions biofertilisers as both remediation agents and soil health enhancers in contaminated agroecosystems.

Empirical studies have demonstrated that biofertiliser application can significantly improve crop growth and yield in heavy metal-contaminated soils. For instance, Du et al. (2022) reported that co-application of biofertilisers and organic amendments increased microbial diversity, particularly *Solicoccozyma* spp., and restored soil fertility in mine-affected areas. Similarly, Seang-On et al. (2023) found that biofertilisers reduced cadmium (Cd) toxicity in rice, resulting in enhanced growth and productivity. Magsayo et al. (2023) observed that native *Rhizobium* and AMF exhibited synergistic effects on plant development in nursery trials, highlighting their potential for bioremediation and ecological restoration.

Mycorrhizal fungi play a central role in this process by simultaneously promoting nutrient uptake and reducing the bioavailability of toxic metals (Cheng et al. 2016; Seang-On et al. 2023). Through mechanisms such as metal immobilisation, chelation, and root-mediated secretion of organic acids, these fungi enhance soil detoxification while improving plant resilience (Aguégué et al. 2022). The symbiotic relationship between plant roots and mycorrhizal networks contributes not only to contaminant mitigation but also to the re-establishment of soil biological function, thus supporting long-term agricultural sustainability in polluted environments.

20.5 Benefits of Carbon Sequestration for Climate Change Mitigation

Biofertilisers play a critical role in supporting ecosystem services by enhancing soil microbiodiversity, improving soil health, and increasing bioremediation capacity, all of which contribute to carbon sequestration and, consequently, climate change mitigation (Fig. 20.2). Central to this contribution is the ability of biofertilisers to stimulate microbial activity and diversity within the soil, processes that are essential for nutrient cycling, plant productivity, and the long-term accumulation and stabilisation of soil organic carbon (SOC) (Enerijiofi et al. 2024; Kapoor et al. 2024). This carbon-enhancing effect is mediated through a suite of interrelated mechanisms:

- (i) The promotion of soil aggregation and the physical protection of organic matter within microaggregates.
- (ii) Increased plant biomass and the subsequent return of organic residues to the soil.
- (iii) Stimulation of rhizosphere activity and root-derived carbon inputs that fuel microbial transformations.
- (iv) The restoration of biological function and carbon sink capacity in degraded or nutrient-depleted soils.
- (v) The reduction of greenhouse gas emissions, particularly through the substitution of synthetic nitrogen fertilisers.

Together, these mechanisms underscore the multifunctionality of biofertilisers as both soil health enhancers and climate mitigation tools.

The application of biofertilisers facilitates the formation of soil microaggregates, within which soil organic matter (SOM) is physically protected from microbial decomposition (Rashid et al. 2016; Enerijiofi et al. 2024; Kapoor et al. 2024). Microbial secretions such as glomalin and extracellular polysaccharides act as

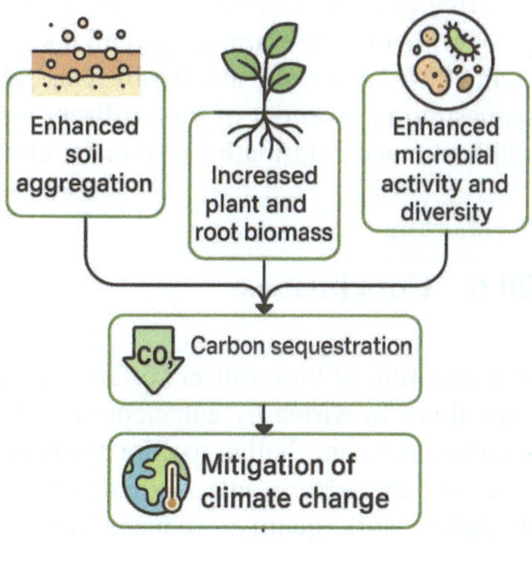

Fig. 20.2 Role of biofertilisers in carbon sequestration for climate change mitigation

natural binding agents, or 'biological glues', that stabilise soil particles and promote the development of robust soil structure (Wakindiki et al. 2019; Schlüter et al. 2020). Dębska et al. (2016) demonstrated that biofertilisers enhance soil carbon storage through a combination of biological and physicochemical processes. This aggregation not only safeguards SOM but also improves soil water retention, porosity, and aeration, fostering an environment conducive to microbial colonisation and long-term carbon stabilisation (Schlüter et al. 2020; Kapoor et al. 2024).

Biofertilisers containing N-fixing and P-solubilising microorganisms, including *Rhizobium*, *Azospirillum*, *Pseudomonas*, and *Bacillus* spp., enhance plant nutrient uptake and promote vigorous shoot and root development (Dutta et al. 2014; Liu et al. 2020). The resulting increase in biomass leads to a greater deposition of organic inputs into the soil system, both through aboveground litter and root turnover (Nyambo et al. 2020a). These inputs serve as primary carbon sources that, through microbial decomposition, are progressively incorporated into the soil organic carbon pool (Berger et al. 2013).

A key mechanism for soil carbon accumulation is rhizodeposition, the release of organic compounds from plant roots into the surrounding rhizosphere. Biofertilisers stimulate root growth and exudation, increasing the flow of carbon-rich substances such as sugars, amino acids, and organic acids into the soil (Lei et al. 2023). These exudates fuel microbial proliferation and drive the transformation of labile carbon into more stable humic substances, contributing to long-term carbon sequestration (Bindraban et al. 2015; Nyambo et al. 2020a).

In degraded and nutrient-depleted soils, which are widespread across African agroecosystems, biofertilisers help restore biological function and microbial diversity, thereby rebuilding the soil's capacity to act as a carbon sink (Dzvene and Chiduza 2024; Yadav et al. 2013). In such contexts, microbial inoculants improve the soil's buffering capacity against climate-induced stressors, while simultaneously enhancing carbon retention. This transformation positions degraded lands not as sources of emissions, but as active contributors to climate mitigation.

Moreover, biofertilisers reduce the need for synthetic nitrogen inputs, thereby lowering emissions of nitrous oxide (N_2O) and carbon dioxide (CO_2), greenhouse gases with a global warming potential significantly higher than carbon dioxide (Hu et al. 2024; Nyambo et al. 2020b). This substitution effect complements the carbon sequestration potential of biofertilisers, underscoring their dual role in enhancing soil health and contributing to broader climate change mitigation strategies.

20.6 Conclusions

The adoption of biofertilisers offers a transformative pathway towards sustainable agriculture in Africa by enhancing cropland ecosystem services, promoting food security, and contributing to climate change mitigation. They play a critical role in reversing the adverse effects of monocropping and synthetic inputs, which have degraded soil quality, reduced biodiversity, and increased environmental vulnerability. Additionally, they restore soil health through improved microbial diversity, nutrient cycling, and enhanced water-use efficiency. Biofertilisers also support bioremediation of heavy metal-contaminated soils, offering a low-cost, eco-friendly alternative for land restoration. To fully harness the benefits of biofertilisers, there is a pressing need for an integrated approach that combines indigenous knowledge, supportive policy frameworks, and sustained investment in research and extension services. Bridging knowledge gaps and increasing adoption will strengthen agroecosystem resilience and facilitate the transition to ecologically sound farming systems. Ultimately, biofertilisers present a viable, nature-based solution to promote sustainable agricultural productivity, environmental restoration, and long-term food security in Africa's diverse agricultural landscapes.

References

Abdullahi AM, Kalengyo RB, Warsame AA (2024) The unmet demand of food security in East Africa: review of the triple challenges of climate change, economic crises, and conflicts. Discov Sustain 5(1):1–15

Abioye OM, Olasehinde DA, Abadunmi T (2024) The role of biofertilizers in sustainable agriculture: an eco-friendly alternative to conventional chemical fertilizers. Appl Sci Eng Prog 17(1):6883–6883

Adesemoye AO, Kloepper JW (2009) Plant–microbes interactions in enhanced fertilizer-use efficiency. Applied microbiology and biotechnology, 85(1), pp.1–12

Adetunji CO, Oloke JK, Dwivedi N, Ummalyma SB, Dwivedi S, Hefft DI, Adetunji JB (2023) Next-generation algae, volume 1: Applications in agriculture, food and environment. John Wiley & Sons

Afotey B, Sarpong GT (2023) Estimation of biogas production potential and greenhouse gas emissions reduction for sustainable energy management using intelligent computing technique. Meas Sens 25:100650

Aguégué MR, Ahoyo Adjovi NR, Agbodjato NA, Noumavo PA, Assogba S, Salami H, Salako VK, Ramón R, Baba-Moussa F, Adjanohoun A, Glele Kakaï R (2022) Efficacy of native strains of arbuscular mycorrhizal fungi on maize productivity on ferralitic soil in Benin. Agric Res 11:627–641

Aloo BN, Tripathi V, Makumba BA, Mbega ER (2022) Plant growth-promoting rhizobacterial biofertilizers for crop production: the past, present, and future. Front Plant Sci 13:1002448

Altieri MA (1999) The ecological role of biodiversity in agroecosystems. In: Invertebrate biodiversity as bioindicators of sustainable landscapes. Elsevier, pp 19–31

Ansari RA, Mahmood I (2017) Optimization of organic and bio-organic fertilizers on soil properties and growth of pigeon pea. Sci Hortic 226:1–9

Basile LA, Lepek VC (2021) Legume–rhizobium dance: an agricultural tool that could be improved? Microb Biotechnol 14(5):1897–1917

Bellenger JP, Xu Y, Zhang X, Morel FM, Kraepiel AML (2014) Possible contribution of alternative nitrogenases to nitrogen fixation by asymbiotic N2-fixing bacteria in soils. Soil Biol Biochem 69:413–420

Benitez-Alfonso Y, Soanes BK, Zimba S, Sinanaj B, German L, Sharma V, Bohra A, Kolesnikova A, Dunn JA, Martin AC, u Rahman MK (2023) Enhancing climate change resilience in agricultural crops. Curr Biol 33(23):R1246–R1261

Berger LR, Stamford NP, Santos CERS, Freitas ADS, Franco LO, Stamford TCM (2013) Plant and soil characteristics affected by biofertilizers from rocks and organic matter inoculated with diazotrophic bacteria and fungi that produce chitosan. J Soil Sci Plant Nutr 13(3):592–603

Just BS, Marks EAN, Roquer-Beni L, Llenas L, Ponsà S, Vilaplana R (2024) Biofertilization increases soil organic carbon concentrations: results of a meta-analysis. Int J Agric Sustain 22(1):2361578. https://doi.org/10.1080/14735903.2024.2361578

Bindraban PS, Dimkpa C, Nagarajan L, Roy A, Rabbinge R (2015) Revisiting fertilisers and fertilisation strategies for improved nutrient uptake by plants. Biol Fertil Soils 51(8):897–911

Bockstaller C, Blatz A, Rapp O, Koller R, Slezack S, Schaub A (2024) Sustainable alternative to irrigated maize monoculture in a maize-dominated cropped area: lessons learned from a system experiment. Heliyon 10(10):e30400

Bommarco R, Vico G, Hallin S (2018) Exploiting ecosystem services in agriculture for increased food security. Glob Food Sec 17:57–63

Canbolat MY, Barik K, Çakmakçi R, Şahin F (2006) Effects of mineral and biofertilizers on barley growth on compacted soil. Acta Agric Scand Sect B Soil Plant Sci 56(4):324–332

Chaudhary P, Singh S, Chaudhary A, Sharma A, Kumar G (2022) Overview of biofertilizers in crop production and stress management for sustainable agriculture. Front Plant Sci 13:930340

Cheng L, Chen W, Adams TS, Wei X, Li L, McCormack ML, DeForest JL, Koide RT, Eissenstat DM (2016) Mycorrhizal fungi and roots are complementary in foraging within nutrient patches. Ecology 97(10):2815–2823

Connolly-Boutin L, Smit B (2016) Climate change, food security, and livelihoods in sub-Saharan Africa. Reg Environ Chang 16:385–399

Crews TE, Carton W, Olsson L (2018) Is the future of agriculture perennial? Imperatives and opportunities to reinvent agriculture by shifting from annual monocultures to perennial polycultures. Glob Sustain 1:e11

de Graaff J, Kessler A, Nibbering JW (2011) Agriculture and food security in selected countries in sub-Saharan Africa: diversity in trends and opportunities. Food Secur 3:195–213

Dębska B, Długosz J, Piotrowska-Długosz A et al (2016) The impact of a bio-fertilizer on the soil organic matter status and carbon sequestration—results from a field-scale study. J Soils Sediments 16:2335–2343. https://doi.org/10.1007/s11368-016-1430-5

Demir Z (2020) Effects of microbial bio-fertilizers on soil physicochemical properties under different soil water regimes in greenhouse grown eggplant (Solanum Melongena L.). Commun Soil Sci Plant Anal 51(14):1888–1903

Du TY, He HY, Zhang Q, Lu L, Mao WJ, Zhai MZ (2022) Positive effects of organic fertilizers and biofertilizers on soil microbial community composition and walnut yield. Appl Soil Ecol 175:104457

Dutta P, Biswas S, Kundu S (2014) Effect of organic manures and biofertilizers on production of organic litchi. J Eco-Friendly Agric 6(1):13–15

Dzvene AR, Chiduza C (2024) Application of biofertilizers for enhancing beneficial microbiomes in push–pull cropping systems: a review. Bacteria 3(4):271–286

Dzvene AR, Gura I, Tesfuhuney W, Walker S, Ceronio G (2024) Effect of intercropping maize and sunn hemp at different times and stand densities on soil properties and crop yield under in-field rainwater harvesting (IRWH) tillage in semi-arid South Africa. Plant Soil 505:363–379

Dzvene AR, Zhou L, Slayi M, Dirwai TL (2025) A scoping review on challenges and measures for climate change in arid and semi-arid agri-food systems. Discover Sustainability, 6(1), p.151

Enerijiofi KE, Musa SI, Igiebor FA, Lawani M, Chuka NE, Odozi EB, Ikhajiagbe B (2024) Nanotechnological approaches for enhanced microbial activities, sustainable agriculture, and the environment. In: Sustainable agriculture: nanotechnology and biotechnology for crop production and protection. Walter de Gruyter GmbH & Co KG, p 153

García-Fraile P, Menéndez E, Rivas R (2015) Role of bacterial biofertilizers in agriculture and forestry. Aims Bioengineering, 2(3), pp.183–205

Giller KE (2020) The food security conundrum of sub-Saharan Africa. Glob Food Sec 26:100431

Groh K, Vom Berg C, Schirmer K, Tlili A (2022) Anthropogenic chemicals as underestimated drivers of biodiversity loss: scientific and societal implications. Environ Sci Technol 56(2):707–710

Haghighizadeh A, Rajabi O, Nezarat A, Hajyani Z, Haghmohammadi M, Hedayatikhah S, Asl SD, Beni AA (2024) Comprehensive analysis of heavy metal soil contamination in mining environments: impacts, monitoring techniques, and remediation strategies. Arab J Chem 17(6):105777

Hashakimana L, Tessema T, Niyitanga F, Cyamweshi AR, Mukuralinda A (2023) Comparative analysis of monocropping and mixed cropping systems on selected soil properties, soil organic carbon stocks, and simulated maize yields in drought-hotspot regions of Rwanda. Heliyon 9(9):e19041

Hu M, Xue H, Wade AJ, Gao N, Qiu Z, Long Y, Shen W (2024) Biofertilizer supplements allow nitrogen fertilizer reduction, maintain yields, and reduce nitrogen losses to air and water in China paddy fields. Agriculture, Ecosystems & Environment, 362, p.108850

Kapoor D, Sharma P, Sharma MMM, Yadav S, Husen A (2024) Exploring soil microbiota and their role in plant growth, stress tolerance, disease control and nutrient immobilizer. Biocatal Agric Biotechnol 61:103358

Khan N, Ali S, Shahid MA, Mustafa A, Sayyed RZ, Curá JA (2021) Insights into the interactions among roots, rhizosphere, and rhizobacteria for improving plant growth and tolerance to abiotic stresses: a review. Cells 10(6):1551

Khoshru B, Fallah Nosratabad A, Mahjenabadi VAJ, Knežević M, Hinojosa AC, Fadiji AE, Enagbonma BJ, Qaderi S, Patel M, Baktash EM, Dawood MFAM (2024) Multidimensional role of Pseudomonas: from biofertilizers to bioremediation and soil ecology to sustainable agriculture. J Plant Nutr 48:1–27

Kothari SK, Marschner H, George E (1990) Effect of VA mycorrhizal fungi and rhizosphere microorganisms on root and shoot morphology, growth and water relations in maize. New Phytol 116(2):303–311

Lal R (2015) Restoring soil quality to mitigate soil degradation. Sustainability 7(5):5875–5895

Lei X, Shen Y, Zhao J, Huang J, Wang H, Yu Y, Xiao C (2023) Root exudates mediate the processes of soil organic carbon input and efflux. Plants 12(3):630. https://doi.org/10.3390/plants12030630

Leila B, El-Hafid N (2020) Biofertilizers and biopesticides: microbes for sustainable agriculture. In: Advances in plant microbiome and sustainable agriculture: diversity and biotechnological applications. Springer, pp 257–279

Liu CLC, Kuchma O, Krutovsky KV (2018) Mixed-species versus monocultures in plantation forestry: development, benefits, ecosystem services and perspectives for the future. Glob Ecol Conserv 15:e00419

Liu L, Gao Z, Yang Y, Gao Y, Mahmood M, Jiao H, Wang Z, Liu J (2023) Long-term high-P fertilizer input shifts soil P cycle genes and microorganism communities in dryland wheat production systems. Agric Ecosyst Environ 342:108226

Liu Q, Meng X, Li M, Raza W, Liu D, Shen Q (2020) The growth promotion of peppers (Capsicum annuum L.) by Trichoderma guizhouense NJAU4742-based biological organic fertilizer: possible role of i ncreasing nutrient availabilities. Microorganisms 8:1296

Loh SK, Asubonteng KO, Adanu SK (2022) Effects of monocropping on land cover transitions in the wet evergreen agro-ecological zone of Ghana. Land 11(7):1063

Magsayo B, Aggangan N, Bacosa H, Gilbero D, Guihawan J, Amparado RF Jr (2023) Growth response of Narra (Pterocarpus indicus willd.) seedlings to microbial biofertilizer applications in mined-out soils of Placer and Claver, Surigao del Norte, Philippines. Appl Ecol Environ Res 21(3):2277

Mahanty T, Bhattacharjee S, Goswami M, Bhattacharyya P, Das B, Ghosh A, Tribedi P (2017) Biofertilizers: a potential approach for sustainable agriculture development. Environ Sci Pollut Res 24:3315–3335

Malobane ME, Nciizah AD, Bam LC, Mudau FN, Wakindiki II (2021) Soil microstructure as affected by tillage, rotation and residue management in a sweet sorghum-based cropping system in soils with low organic carbon content in South Africa. Soil Tillage Res 209:104972

Mani S, Osborne CP, Cleaver F (2021) Land degradation in South Africa: justice and climate change in tension. People Nat 3(5):978–989

Meena VS, Maurya BR, Verma JP, Meena RS (2016) Potassium solubilizing microorganisms for sustainable agriculture, vol 331. Springer, New Delhi

Meyfroidt P (2018) Trade-offs between environment and livelihoods: bridging the global land use and food security discussions. Glob Food Sec 16:9–16

Mukhovi S, Jacobi J (2022) Can monocultures be resilient? Assessment of buffer capacity in two agroindustrial cropping systems in Africa and South America. Agric Food Secur 11(1):19

Mupambwa HA, Wakindiki IIC (2012) Winter cover crops effects on soil strength, infiltration and water retention in a sandy loam Oakleaf soil in Eastern Cape, South Africa. S Afr J Plant Soil 29(3–4):121–126

Murugan R, Chitraputhirapillai S (2011) Effects of combined application of biofertilizers with neem cake on soil fertility, grain yield and protein content of black gram (Vigna mungo (L.) Hepper). World J Agric Res 7(5):583–590

Nizamani MM, Hughes AC, Qureshi S, Zhang Q, Tarafder E, Das D, Acharya K, Wang Y, Zhang ZG (2024) Microbial biodiversity and plant functional trait interactions in multifunctional ecosystems. Appl Soil Ecol 201:105515

Nosheen S, Ajmal I, Song Y (2021) Microbes as biofertilizers, a potential approach for sustainable crop production. Sustainability 13(4):1868

Nyambo P, Chiduza C, Araya T (2020a) Carbon input and maize productivity as influenced by tillage, crop rotation, residue management and biochar in a semiarid region in South Africa. Agronomy 10(5):705

Nyambo P, Chiduza C, Tesfay A (2020b) Carbon dioxide fluxes and carbon stocks under conservation agricultural practices in South Africa. Agriculture (Basel) 10(9):374. https://doi.org/10.3390/agriculture10090374

Okereafor U, Makhatha M, Mekuto L, Uche-Okereafor N, Sebola T, Mavumengwana V (2020) Toxic metal implications on agricultural soils, plants, animals, aquatic life and human health. Int J Environ Res Public Health 17(7):2204

Olanipekun IO, Olasehinde-Williams GO, Alao RO (2019) Agriculture and environmental degradation in Africa: the role of income. Sci Total Environ 692:60–67

Olanrewaju OS, Glick BR, Babalola OO (2017) Mechanisms of action of plant growth promoting bacteria. World J Microbiol Biotechnol 33:1–16

Omar AF, Abdelmageed AH, Al-Turki A, Abdelhameid NM, Sayyed RZ, Rehan M (2022) Exploring the plant growth-promotion of four Streptomyces strains from rhizosphere soil to enhance cucumber growth and yield. Plants 11(23):3316

Petersen-Rockney M, Baur P, Guzman A, Bender SF, Calo A, Castillo F, De Master K, Dumont A, Esquivel K, Kremen C, LaChance J (2021) Narrow and brittle or broad and nimble? Comparing adaptive capacity in simplifying and diversifying farming systems. Front Sustain Food Syst 5:564900

Power AG (2010) Ecosystem services and agriculture: tradeoffs and synergies. Philos Trans R Soc Lond B Biol Sci 365(1554):2959–2971

Raji RO, Olaitan CO, Azeez Z, Zakari DA, Oyewole OA, Akinsola RO (2024) Marine greens as a source of biofertilizer. In: Marine greens. CRC Press, pp 66–74

Rashid MI, Mujawar LH, Shahzad T, Almeelbi T, Ismail IM, Oves M (2016) Bacteria and fungi can contribute to nutrients bioavailability and aggregate formation in degraded soils. Microbiol Res 183:26–41

Rodríguez H, Fraga R (1999) Phosphate solubilizing bacteria and their role in plant growth promotion. Biotechnol Adv 17(4–5):319–339

Sabannavar SJ, Lakshman HC (2008) Interactions between Azotobacter, Pseudomonas and arbuscular mycorrhizal fungi on two varieties of Sesamum indicum L. J Agron Crop Sci 194(6):470–478

Saha R, Mishra VK, Majumdar B, Laxminarayana K, Ghosh PK (2010) Effect of integrated nutrient management on soil physical properties and crop productivity under a maize (Zea mays)–

mustard (Brassica campestris) cropping sequence in acidic soils of Northeast India. Commun Soil Sci Plant Anal 41(18):2187–2200
Saharan BS, Nehra V (2011) Plant growth promoting rhizobacteria: a critical review. Life Sci Med Res, 21(1), p.30
Sarkar A, Fwanyanga FM, Horn LN, Welzel S, Diederichs M, Kerk LJ, Zimmermann M, Reinhold-Hurek B (2023) Towards inoculant development for Bambara groundnut (Vigna subterranean (L.) Verdc) pulse crop production in Namibia. Front Plant Sci 14:1270356
Schaedel M, Hidrobo G, Grossman J (2021) From microns to meters: exploring advances in legume microbiome diversity for agroecosystem benefits. Front Sustain Food Syst 5:668195
Schlüter S, Sammartino S, Koestel J (2020) Exploring the relationship between soil structure and soil functions via pore-scale imaging. Geoderma 370:114370
Seang-On L, Meeinkuirt W, Koedrith P (2023) Alleviation of cadmium toxicity in Thai Rice cultivar (PSL2) using biofertilizer containing indigenous cadmium-resistant microbial consortia. Plants 12(20):3651
Selimovic A, Tissier ML, Arnold W (2022) Maize monoculture causes niacin deficiency in free-living European brown hares and impairs local population development. Front Ecol Evol 10:1017691
Selvi A, Rajasekar A, Theerthagiri J, Ananthaselvam A, Sathishkumar K, Madhavan J, Rahman PK (2019) Integrated remediation processes toward heavy metal removal/recovery from various environments-a review. Frontiers in Environmental Science, 7, p.66
Sharaya R, Gill R, Kalwan G, Naeem M, Tuteja N, Gill SS (2023) Plant-microbe interaction mediated salinity stress tolerance for sustainable crop production. S Afr J Bot 161:454–471
Shen H, He X, Liu Y, Chen Y, Tang J, Guo T (2016) A complex inoculant of N2-fixing, P-and K-solubilizing bacteria from a purple soil improves the growth of kiwifruit (Actinidia chinensis) plantlets. Front Microbiol 7:841
Singh SR, Najar GR, Singh U (2007) Productivity and nutrient uptake of soybean (Glycine max) as influenced by bio- inoculants and farmyard manure under rainfed conditions. Indian J Agron 52(4):325–329
Spaepen S, Vanderleyden J, Remans R (2007) Indole-3-acetic acid in microbial and microorganism-plant signaling. FEMS Microbiol Rev 31(4):425–448
Sturz AV, Christie BR, Nowak J (2000) Bacterial endophytes: potential role in developing sustainable systems of crop production. Crit Rev Plant Sci 19(1):1–30
Suarez A, Gwozdz W (2023) On the relation between monocultures and ecosystem services in the global south: a review. Biol Conserv 278:109870
Sylia AB, Corrêa A, Cruz C, Yadav AN, Nabti E (2021) Plant growth promoting microbes as biofertilizers: promising solutions for sustainable agriculture under climate change associated abiotic stresses. Plant Sci Today 8(sp1):60–76
Sylia AB, Corrêa A, Cruz C, Yadav AN, Nabti E (2022) Plant growth promoting microbes as biofertilizers: Promising solutions for sustainable agriculture under climate change associated abiotic stresses. Plant Science Today, 8, pp.60–76
Vanlauwe B, Hungria M, Kanampiu F, Giller KE (2019) The role of legumes in the sustainable intensification of African smallholder agriculture: lessons learnt and challenges for the future. Agric Ecosyst Environ 284:106583
Vessey JK (2003) Plant growth promoting rhizobacteria as biofertilizers. Plant and soil, 255(2), pp.571–586
Wakindiki IIC, Malobane ME, Nciizah AD (2019) Integrating biofertilizers with conservation agri-culture can enhance its capacity to mitigate climate change: examples from southern Africa. In: Leal FW, Leal-Arcas R (eds) University initiatives in climate change mitigation and adaptation. Springer, Cham, pp 277–289
Wudil AH, Usman M, Rosak-Szyrocka J, Pilař L, Boye M (2022) Reversing years for global food security: a review of the food security situation in Sub-Saharan Africa (SSA). Int J Environ Res Public Health 19(22):14836

Yadav SK, Singh Y, Yadav MK, Babu S, Singh K (2013) Effect of organic nitrogen sources on yield, nutrient uptake and soil. Indian J Agric Sci 83(2):170–175
Yan A, Wang Y, Tan SN, Mohd Yusof ML, Ghosh S, Chen Z (2020) Phytoremediation: a promising approach for revegetation of heavy metal-polluted land. Front Plant Sci 11:359
Yu X, Liu X, Zhu TH, Liu GH, Mao C (2012) Co-inoculation with phosphate-solubilzing and nitrogen-fixing bacteria on solubilization of rock phosphate and their effect on growth promotion and nutrient uptake by walnut. Eur J Soil Biol 50:112–117
Zhang H, Zheng X, Wang X, Xiang W, Xiao M, Wei L, Zhang Y, Song K, Zhao Z, Lv W, Chen J (2022) Effect of fertilization regimes on continuous cropping growth constraints in watermelon is associated with abundance of key ecological clusters in the rhizosphere. Agric Ecosyst Environ 339:108135

MIX
Papier aus verantwortungsvollen Quellen
Paper from responsible sources
FSC® C105338

If you have any concerns about our products,
you can contact us on
ProductSafety@springernature.com

In case Publisher is established outside the EU,
the EU authorized representative is:
Springer Nature Customer Service Center GmbH
Europaplatz 3, 69115 Heidelberg, Germany

Printed by Libri Plureos GmbH
in Hamburg, Germany